Clean Air at Work

New Trends in Assessment and Measurement for the 1990s

Clean Air at Work

New Trends in Assessment and Measurement for the 1990s

The Proceedings of an International Symposium held in Luxembourg, 9–13 September 1991, organised by the Commission of the European Communities, Health and Safety Directorate and the Community Bureau of Reference

in collaboration with the

Royal Society of Chemistry, Analytical Division, Automatic Methods Group (UK)

and in cooperation with

Arbejdsmiljøinstituttet (København, DK)

Berufsgenossenschaftliches Institut für Arbeitssicherheit (St. Augustin, D)

Fondazione Clinica del Lavoro (Pavia, I)

Health and Safety Executive, Occupational Medicine and Hygiene Laboratories (London, UK)

INRS, Centre de Recherche (Vandoeuvre, F)

Instituto Nacional de Seguridad e Higiene en el Trabajo (Baracaldo, E)

WHO Regional Office for Europe (København, DK)

Edited by

R.H. Brown
Occupational Medicine and Hygiene Laboratories
Health and Safety Executive, London, UK

M. Curtis
Health and Safety Directorate
Commission of the European Communities, Luxembourg

K.J. Saunders
BP Research Centre
Sunbury-on-Thames, Middlesex, UK

S. Vandendriessche
Community Bureau of Reference (BCR)
Commission of the European Communities, Brussels

The Royal Society of Chemistry Special Publication No. 108

Published for the:

Commission of the European Communities
Directorate-General Telecommunications,
Information Industries and Innovation
Scientific and Technical Communication Unit, Luxembourg

by:

The Royal Society of Chemistry
Thomas Graham House
Science Park
Milton Road
Cambridge CB4 4WF, United Kingdom

EUR 14214

© ECSC, EEC, EAEC, Brussels and Luxembourg, 1992

Legal Notice

Neither the Commission of the European Communities nor any person acting on behalf of the Commission is responsible for the use which might be made of the following information.

ISBN: 0-85186-217-9

A catalogue record for this book is available from the British Library

Printed by Redwood Press Ltd, Melksham, Wiltshire

EDITORIAL NOTE

The papers in these proceedings have been the subject of minor editing in order to achieve consistency. The summary report gives an overview of the contributed papers and discussion, and contains the conclusions and recommendations for action agreed by the participants.

The editors would like to acknowledge the support given by the session chairmen and rapporteurs as well as the organising committee for the scientific planning of this symposium. Thanks are also due to S. Blair, L. Eisen, D. Nicolay and A. Poos (CEC) for their technical assistance.

ORGANISING COMMITTEE:

H. Blome (D), A. Braithwaite (UK), R.H. Brown (UK), B. Carton (F), D. Cottica (I), M. Curtis (CEC), J. Kristensson (S), K. Leichnitz (D), R. Lidgett (UK), M.J. Quintana (E), K.J. Saunders (UK), S. Vandendriessche (CEC), H. Walerius (CEC), P. Walkoff (DK).

CHAIRMEN:

A. Berlin (CEC), H. Blome (D), W. Coenen (D), J.G. Firth (UK), B. Griepink (CEC), R. Haigh (CEC), K.J. Saunders (UK), D.C.M. Squirrell (UK), P. Walkoff (DK).

RAPPORTEURS:

A. Braithwaite (UK), R.H. Brown (UK), E. Buringh (NL), J.C. Guichard (F), L.C. Kenny (UK), H. Knöppel (I), S. Luxon (UK), R. Narayanaswamy (UK), B. Striefler (D).

Contents

SESSION V DATA AND INFORMATION MANAGEMENT

SESSION VI INDOOR AIR

30

INTRODUCTION

This symposium was organised jointly by the Health and Safety Directorate (DGV) and the Community Bureau of Reference (DGXII) of the Commission of the European Communities, in conjunction with the UK Royal Society of Chemistry, Analytical Division, Automatic Methods Group and with the cooperation of the WHO and six national organisations involved in workplace health and safety.

The Commission of the European Communities in 1986 organised a symposium on diffusive sampling, which not only set out a new trend, but also catalysed cooperation in Europe in the field of workplace air monitoring. The importance of this cooperation has increased to such an extent that it was felt that there was a need for a new symposium with a much wider scope.

With this in mind, a team of speakers was brought together to outline the latest developments in regulations, in standards, in methods of assessment and measurement in both the industrial and non-industrial working environment, and in quality assurance for laboratory techniques.

It was the aim of the symposium in each of these areas to review the state of the art, predict future needs and, in particular, to promote European cooperation in the establishment of agreed performance targets, common methodology and consistent reporting and information management.

The symposium was attended by approximately 320 participants from all the Member States of the Community and a number of other countries such as Australia, Austria, Canada, Finland, Japan, Norway, Poland, South Africa, Sweden, Switzerland, and the United States of America.

The summary report is based on the reports of the Session Rapporteurs and gives an overview of the contributed papers and discussion, and contains the conclusions and recommendations for action agreed by the participants.

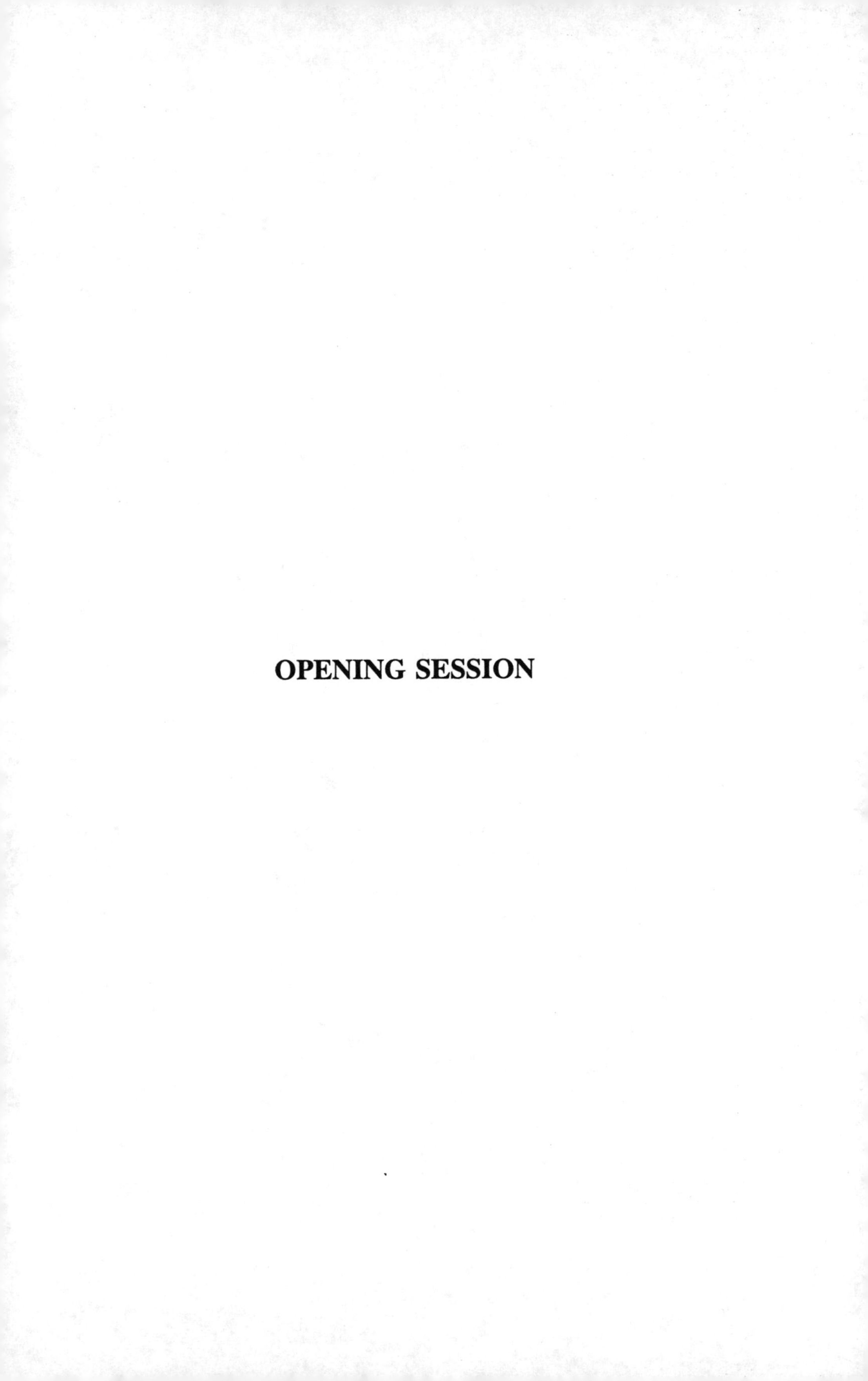

OPENING SESSION

WELCOMING ADDRESS AND OPENING OF THE CONFERENCE

W.J. HUNTER

Director
Commission of the European Communities
Health and Safety Directorate
Jean Monnet Building, L-2920 Luxembourg

Ladies and gentlemen, it is my pleasure, as Director of the Health and Safety Directorate of the Commission of the European Communities, to welcome you to this conference on 'Clean Air at Work' and to make the opening remarks.

First of all I should like to welcome you on behalf of Directorate General V and our Director Mr Degimbe. Also on behalf of the Commissioner responsible for health and safety affairs, Madame Papandreou, who unfortunately is unable to be present here today to open the Conference.

I must extend the welcome also on behalf of Directorate General XII, the Directorate General responsible for Science and Technology who, through their Community Bureau of Reference (BCR) have been responsible, with DG V and the other co-sponsors for organising this week's events.

Thanks go to the Organising Committee for their hard work over the past 18 months. You will see the list of co-sponsors in the conference literature. I would like to single out for special mention the Royal Society of Chemistry in the UK who have been especially involved in developing the programme for the conference and for administrative support and who will later on be arranging publication of the proceedings.

This conference comes at a time when the Community is placing renewed emphasis on the importance of health and safety as an integral part of its social policy. Plans of Action published by the Commission and endorsed by the Council have been prepared since the mid 1970s. But more recently the Treaty base on which the Community's actions are based has been strengthened in the area of health and safety. The Commission and in particular Mr Delors and our own Commissioner Mme Papandreou have stressed that health and safety must not be allowed to fall behind the economic aspects in the development of the internal market.

It is against this background that the Commission is pleased to support a conference that brings together experts in a rapidly developing field of health and safety such as 'clean air' and measurement technology. A number of you were present at an earlier conference on diffusive monitoring which took place here in Luxembourg in 1986. Since that time diffusive monitoring has become an established tool of the hygienist. In this week's sessions you will be taking a much broader look at the issues involved in measurement:

- the background in law,
- the developing standards,
- the important aspect of quality and reliability,
- future needs.

I cannot stress too strongly the need to keep alert to the changing technical, political and social needs and to be ready to meet the challenges they present.

The programme this week addresses the subject of Indoor Air reflecting the changing priorities and expectations of people at work, as the proportion of people who work in large, often air-conditioned, buildings increases. The focus is tending to move away from the traditional industries and increasingly to address 'well-being' whether in terms of comfort at work or the health of the population, of whom workers make up an important part. The interface between the workplace and the general environment has become very blurred and the demands on those responsible for measurement will be both changing and challenging as the nature and level of pollutants to be measured changes. Often you will be working near to limits of detection.

I am pleased to see that in your conference this week you will be devoting an important session to the quality and reliability of measurement. If we are to achieve our aims in health protection at work then it is vital that the time, the effort and the money put into measurement will yield results that can be relied on by employers, by employees and by the enforcing authorities alike. Unless we achieve this aim a vital part of our preventive approach will fail.

Reliable measurements of exposure are also an essential part of the data gathering which we, in the Community, need for setting our policies, our legislation and our limit values - and hence achieving the aims of the health and safety aspects of the Community Charter of the Fundamental Social Rights of Workers which was adopted by 11 Heads of State or Government in 1989.

Ladies and gentlemen, I am pleased to open your conference; to wish you a warm welcome on behalf of the Commission and all the sponsoring organisations; and to wish you an enjoyable and fruitful conference. I look forward with interest to the ideas and suggestions that will come out of this conference and in particular the session on Friday when you look to the needs for the future.

While you are here at the conference I hope that you will make the most of the opportunity to meet and to talk informally with colleagues who have a similar interest, or who share common problems. It is often these informal contacts that lead to new ideas and also avoid duplication and overlap. Please also support the Poster Session and the Exhibition - a great deal of effort has gone into making these a useful and integral part of the whole event.

I am now very pleased to hand over to Dr Alex Berlin who will chair the first Session on Legal Background and Standards.

Session I

LEGAL BACKGROUNDS AND STANDARDS

REGULATORY REQUIREMENTS FOR WORKPLACE AIR MONITORING IN EUROPE: THE CURRENT POSITION

A. BERLIN

Commission of The European Communities,
Health and Safety Directorate
DG V/E
Jean Monnet Building
L-2920 Luxembourg

SUMMARY

During the last 15 years the European Community has developed and adopted a number of Directives which impose a specific or an implied need for the measurement of exposure at the workplace. These requirements are outlined. In the future greater use is likely to be made of standards and legislation will need to take account of the rapid developments in measurement techniques, as well as the need for reliability in measurements. The importance of these aspects is discussed.

In this first session of the Conference we are setting out to provide an outline of the legislation on which Community legal requirements for measurement are based and to explain the content of the CEN standards that are under preparation by Technical Committee CEN/TC 137. Two of these standards are now available for public comment and their appearance marks an important development in the area of measurement of pollutants in workplace air within Member States. The EFTA countries are also members of CEN and will adopt these standards but will not, of course, be subject to Community law, although frequently they take note of it in their own legislation.

In order to trace the development of Community requirements relating to workplace air measurement we need to go back to the mid-1970's. The extension of Community activities into the health and safety at work arena (other than in the area of mining) really began in earnest with the setting up in 1974 of the tripartite Advisory Committee on Safety, Hygiene and Health Protection at Work. This was followed in 1978 by the first Directive involving a chemical agent and the duty to monitor exposure - Council Directive 78/610/EEC on vinyl chloride monomer. The Directive remains in force today.

Exposure limits

In the European Community; since the adoption in 1978 of the first directive concerned specifically with the control of a toxic substance, VCM, at work, it was considered that due to the diversity of approaches used by the 12 Member States to protect the health of their workers there was a need to have measurable criteria against which to assess the measures taken by Member States.

The need for these criteria was two-fold:

- ensure in the social context a minimal but high level of protection of the worker's health
- prevent the use of "health and safety" to gain competitive advantages.

Of course Member States may set higher standards if they wish, as long as such standards do not lead to distorsions in free movement of goods.

Among the different measures that could be used to protect the health of workers handling chemical agents, in the case of inhalation the establishment of Community-wide agreed occupational exposure limits or other guide value, was one way to have objective standards against which control of exposure could be assessed.

To my knowledge such a wide range of reference periods is unique in workplace legislation - it certainly is for the Community. Setting the various limits was done on the basis of an assumed relationship involving a normal distribution and variance as a function of measurement time. Such "precision" in a relationship could only be assumed for a unique chemical species used essentially in a single well-defined process. We can normally make no such assumptions.

The use of Exposure Limits to enhance the protection of the health of workers has seen a remarkable evolution in the Community since the adoption by the Council of Ministers in 1978 of the first legislative measure in this area.

The 1978 Directive on Vinyl Chloride Monomer introduces for the first time the concept of "technical long-term limit value".

This "technical long-term limit value" is defined as follows: the value which shall not be exceeded by the mean concentration, integrated with respect to time, of the chemical in the atmosphere of the working area, the reference period being the year.

Account was only to be taken of the concentrations measured during the periods in which the plant is in operation.

For guidance and practical purposes, a table of corresponding limit values for shorter periods was established (1 month, 1 week, 8 hours, 1 hour).

At the same time Alarm Threshold levels were established requiring essentially continuous monitoring in implying the immediate use of personal protective measures (1 hour, 20 min, 2 min).

In 1980 a framework legislation was adopted concerning the protection of workers from the risks related to exposure to chemical, physical and biological agents at work.

This text gives the following broad definition to "limit value": the exposure limit or biological indicator limit in the appropriate medium depending on the agent.

It identifies a short list of chemicals for which limit values will have to be laid down by directives at Community level, as well as the requirement for Member States to establish limit values and sampling procedures, measuring procedures and procedures for evaluating results.

On this basis in 1982 was adopted the first Directive for a specific compound: metallic lead and its ionic compounds.

Limit values were established in terms of both:

- concentrations in the air as time-weighted average over 40 hours per week.
- biological parameters in terms of PbB, ALAW, ZPP and ALAD.

The measurements shall be representative of worker exposure (spelled out for the first time).

Measurement strategies, technical specifications for the measurements and the methodology for evaluating results are laid down.

This is furthermore a graduated approach regarding the technical measures to be taken, and the health surveillance for 3 additional sets of exposure values.

In 1983 another application Directive was adopted concerning Asbestos.

These time limit values were defined in terms of fibers/cm^3 measured or calculated in relation to an 8 hour reference period.

This time it is clearly stated that the samples shall be taken within the individual worker's breathing zone: i.e. within a hemisphere of 300mm radius extending in front of the face and measured from the mid-point of a line joining the ears.

Detailed technical specifications for sampling and measuring are given, as results are very sensitive to these.

Further exposure values are given which establish the graduation of measures to be taken; one of these exposure values is in terms no. of fiber-days per cm^3 over a 3 month period (to cover intermittent work).

In 1988 a new Directive was adopted amending the 1980 Framework Directive.

It clarifies and defines in general the concept of limit values both technically and legally and it introduces the possibility to limit excursions for periods shorter than 8 hours. The limit value is stated as the eight-hour time-weighted (STLV) average concentration of exposure of a substance in gaseous, vaporous or suspended form in the air at the workplace.

Exposure means the presence of a chemical agent in the air within the breathing area of a worker: It is described in terms of concentration over a reference period.

Limit values are also defined in terms of legally binding or indicative.

Legally binding values are laid down by Council Directives.

Indicative limit values shall reflect expert evaluations based on scientific data. They are set by Commission Directives. Member States shall take them into account when establishing their own limit values.

In 1990 the Council adopted a Directive on carcinogens at work.

This Directive also covers limit values as follows:

Limit values shall be set on the basis of available information, including scientific and technical data, in respect of all those carcinogens for which this is possible, and, where necessary, other directly related provisions. These however have not yet been spelled out.

Finally in 1991 (29.5.91 published 5.7.91) the Commission adopted a first Directive on indicative limit values, based on the 1988 modification of the 1980 framework Directive.

This Directive gives indicative limit values for occupational exposure for 27 compounds and classes of compounds measured or calculated in relation to a reference period of eight hours.

For 17 of these compounds it is clearly stated that the existing scientific data on health effects appear to be particularly limited.

This is to be considered the first of a series of lists of indicative limit values promulgated by the Commission as reference for Member States.

Stages in the Procedure for Establishing "Indicative Limit Values"

The key element which drives this procedure is the requirement in the 1988 Council Directive (modification of the 1980 framework Directive) that Indicative limit values shall reflect expert evaluations based on scientific data.

As a consequence an approach based on the following 5 steps had to be established:

- collection of available data and preparation of review (Criteria Document)
- evaluation of Scientific Expert Group
- Opinion of Tripartite Advisory Committee
- Adoption by Committee on Adaptation to Technical Progress
- Commission Directive

Dr Haigh will no doubt discuss the implementation of these steps by the Commission when he looks forward to the future during your Friday session on future needs.

Another important part of 1980 Framework Directive 80/1107/EEC is the Annex which contains details of the content and requirements for reference methods for measuring exposure covering inter alia, in addition to the definitions of limit values:

- definitions of suspended particulate matter
- methodology for assessment and a measuring strategy
- requirements concerning the qualities of those who do measurements and about the reliability of measurements.

It is in the Annex too that you will find the reference to CEN and its work on standards for measuring procedures. Although the Commission has never formally asked CEN to prepare standards, it is on the basis of this reference that the work has gone ahead at the request of Member States - which we shall hear more about later in the Session.

Two other Directives are of particular relevance to the need to measure - those covering lead (82/605/EEC) and asbestos (83/477/EEC).

These, like the VCM Directive, are very specific on what and how to measure. They include a European Reference Method of Measurement, which if my understanding is right, may well need to be modified if the principles laid down in the CEN standards are to be adopted. If you ask why the three Directives on individual substances should go into such technical detail it is because the exposure limit values laid down in these Directives are binding on Member States. The prescription for Methods avoids any distortions being introduced on their application as a result of different, perhaps less well tried and tested, methodology.

What I have said so far provides an outline of the situations in which measurement of exposure at the workplace is expressly covered in Directives - ie binding legal instruments from the European Community. From the point of view of this conference it is relevant to consider how you, as practising professionals involved in measurement at the workplace see this legal background in terms of its applicability, its benefits, its problems and what messages we can put forward to guide future developments.

I see reliability of measurement as a key point, as mentioned by Dr Hunter in the opening address. Do you as delegates to this Conference have views on the approach taken in those Directives which prescribe a European Reference Method? Or does the use of standards, referenced in legislation, provide a more practical way of keeping methodology up-to-date? Perhaps I should mention here that there is a mechanism written into most directives for a simplified procedure for updating purely technical matters, of which technical measurement methods could be one example.

I think I have said enough about the legal position on measurement to stimulate some thought. I don't want to trespass into areas to be covered by other speakers later on.

Measurement of pollutants in the air at the workplace may of course have important uses other than purely for checking compliance with legal limits. It may, for example, be used for:

- evaluating process design/controls
- checking the performance of control systems
- evaluating the effectiveness of building ventilation
- epidemiological purposes
- data gathering to aid the setting of limit values
- assessment of compliance with environmental/emission limits.

All of these uses demand methods that are suited to the particular need in terms of specificity, practicability, reliability and detection limits. All of these points will come up for discussion today, and later in the week. Failure to satisfy any one of these criteria will render the method at best of low utility, and at worst dangerous to the health of workers. The moral, professional and legal responsibilities on those who measure exposure are not to be overlooked. This underlines the importance of training and quality schemes about which I am sure we shall hear more in the paper from WHO.

Before I close this Chairman's introduction I should like to mention one more important point which is, in my view, crucial to the overall effectiveness of workplace monitoring - the question of cooperation between the analyst, the person carrying out the sampling (if different) the worker whose exposure is to be measured and the person who will use the result. Too often, measurements go wrong or the results are of little use because of a failure to understand the needs of one or other of these people. Before starting it is always important to ask "Why are we measuring and how will the results be used?". And also to be sure that the form in which a sample is to be presented to the laboratory will be usable by the analyst.

I hope that I have given a sufficiently clear outline of the legal background to measurement in Community law, and have stimulated you to think about some of the broader questions you need to address in the course of the session, and, later, if the Community law to come in the future is to provide an effective basis for measurement in maintaining "Clean Air at Work".

10

1344

THE DEVELOPMENT OF OCCUPATIONAL HYGIENE PRACTICE IN EUROPE: A WHO PERSPECTIVE

BERENICE I.F. GOELZER

Office of Occupational Health
World Health Organization, 1211 Geneva 27, Switzerland

1. WHO OBJECTIVES AND FUNCTIONS

According to its Constitution, "... the objective of the World Health Organization is the attainment by all peoples of the highest possible level of health". The World Health Organization and Member States are committed to the goal of Health for All, which can only be achieved if the health of the working populations is also protected and promoted.

Workers constitute an important and productive sector of any population. In addition, it is through work that development can be achieved, and development is indispensable, including to ensure the protection and promotion of health, as well as good quality of life. Industrialization, modern agriculture, energy production, provision of services such as transport and telecommunications, are among the essential components of development; their benefits should be enhanced and their negative impacts, if not completely prevented, at least minimized. These activities do not need to be associated with attacks on the health and well-being of workers and surrounding communities, as well as on the environment. It is necessary, and it is possible with technology available today, to find a formula for a 'healthy' and sustainable development.

For the protection of workers' health, it is necessary to anticipate or to identify the possible health hazards in the work environment, to carry out their evaluation based on correct exposure assessments and finally, to control them within the context of an adequate risk management approach.

The protection of communities and the general environment from the negative impacts of industrial activities cannot be dissociated from control strategies for airborne contaminants and other hazards in the work environment, as such strategies must also account for the prevention of environmental pollution, including adequate waste management.

One of the functions specified in the Constitution of the World Health Organization is "... to promote, in cooperation with other specialized agencies where necessary, the improvement of nutrition, housing, sanitation, recreation, economic and working conditions and other aspects of environmental hygiene".

2. SPECIFIC RESOLUTION AND PROGRAMMES

Throughout the years, the Director General of WHO has been requested and Member States have been urged to give attention to the health of working populations, including by actions on the work environment, as can be exemplified by the following extracts from resolutions:

"... wherever possible, to undertake **monitoring of the work environment** and workers' health, with a view to instituting **control measures** and evaluating the effectiveness of such measures ..." (Resolution WHA 29.57, 1976)

"... to develop and strengthen occupations health institutions and to provide **measures for preventing hazards in workplaces**, for the setting of standards and for research and **training** in occupational health ..." (WHA 32.14, 1979)

"... to support the developing countries in ensuring safe working conditions and effective protective measures for workers' health in agriculture, in mining and in industrial enterprises which already exist or which will be set up in the process of industrialization, by using the experience available in this field by both industrialized and developing countries, ..." (Resolution WHA 33.31, 1980)

As to the Eighth General Programme of Work Covering the Period 1990–1995, the six areas of activities, presented in the Global Medium Term Programme for Workers' Health, involve actions related to the work environment. Examples of such areas are the following:

- Development and application of a standard reporting system on the health status of workers and working conditions, aiming at the establishment of a nation-wide database and evaluation of workers' health programmes at both the country and the workplace level.
- Development and consolidation of the infrastructure for workers' health in national health systems: (a) to introduce appropriate preventive measures in workplaces, and (b) to support primary health care workers in occupational health practice.
- Development and application of guidelines for training and education of personnel in various disciplines to serve workers' health.
- Development and application of guidelines on health promotion in the workplace.

The Division of Health Protection and Promotion, where the Office of Occupational Health is located in WHO, Geneva, aims at the application of public health principles, for which the multidisciplinary approach is necessary. This is particularly relevant for the protection and promotion of workers' health, which require, among others, actions on the work environment. The importance of the multidisciplinary approach is also emphasized in programmes concerning workers' health developed by the WHO Regional Office for Europe[1].

3. THE RELEVANCE OF OCCUPATIONAL HYGIENE

From the previous considerations, it is evident that preventive actions on the work environment are needed, directly or indirectly, to achieve many of the WHO's goals. The profession responsible for a comprehensive approach to such actions is occupational hygiene, which can be defined as follows:

"Occupational Hygiene is the science of the anticipation, recognition, evaluation and control of health hazards arising in or from the work environment with the objectives of protecting the health and well-being of workers, the surrounding communities and the general environment".

Therefore occupational hygiene is one link of the chain necessary to attain WHO's objective of health protection and promotion for all, and its development on a worldwide basis is considered essential.

Unfortunately, occupational hygiene is not yet universally applied as it should be; 'healthy' work environments are still the privilege of a few, and too many workers continue to be exposed to hazards, often very serious, which could be prevented through the application of control technology known today. In many countries, occupational health programmes are very weak, in others their strength relies practically only on the medical component which, although essential, cannot by itself solve the problems of the work environment. The diagnosis of an occupational disease cannot prevent its further occurrence. For prevention, multidisciplinary

---

[1] Regional Adviser for Occupational Health: Dr F. La Ferla
World Health Organization, Regional Office for Europe
8 Scherfigsvej, DK-2100 Copenhagen, Denmark. Fax: 45-31.18.11.20

control strategy is indispensable, which must necessarily include interventions on the work environment.

4. OCCUPATIONAL HYGIENE IN WHO

Global

Observations in many countries reveal that common constraints to the effective development of occupational health programmes, including the occupational hygiene component, are insufficient political will, shortage of adequate manpower, and deficiencies in information. Financial problems may constitute serious constraints on some parts of the world, but these are beyond the scope of WHO's direct action.

First of all, **political will** to support and promote occupational health in general, through adequate legislation and effective enforcement, as well as through relevant programmes, requires awareness of the problems and knowledge of the possibilities and benefits resulting from their solution.

The shortage of **manpower** in the field of occupational hygiene is often acute, particularly in developing countries but also in some industrialized countries. The lack of adequately trained occupational hygienists may lead to undesirable situations such as when obvious and serious occupational hazards are completely overlooked, hence uncontrolled; when funds are wasted on extremely expensive equipment, purchased without sound professional judgement and left to rust even before being unpacked, and so on.

Even countries with adequate legislation for the protection of workers' health often do not have it implemented because this requires professionals, including occupational hygienists, who do not exist, at the country level, at least in enough numbers and with the desired level of competence.

Another point to consider is that critical and often 'life-saving' **information**, easily accessible to the scientific community, is not always available when it is needed, where it is needed, and in a suitable form. This is usually the result of deficiencies in information and communication systems. It may also happen that information on control options and their cost-benefit is not available to decision-makers, with the result that controls are not considered in development projects.

Therefore the areas selected as a focus for the present activities in the field of occupational hygiene in WHO give strong emphasis to education and training, as well as to the improvement of information systems and communication among professionals in this field in different countries.

The activities involved include:

- Preparation of educational materials for the different areas of knowledge required for training in occupational hygiene (e.g. Recognition of Health Hazards in the Work Environment; Evaluation of Airborne Contaminants in the Work Environment; Control Technology for Health Hazards in the Work Environment), utilizing knowledge available in different parts of the world and involving as much as possible WHO Collaborating Centers for Occupational Health.
- Identification of professionals and establishment of networks. For example, in order to have worldwide information on occupational hygienists and their qualifications, a computerized roster has been prepared, for easy retrieval of information and continuous updating[2].
- Technical cooperation with countries for the development or strengthening of programmes, services and training activities; when necessary, also promoting the essential role of occupational hygiene.

---

[2] The relevant forms can be obtained from The Office of Occupational Health, WHO, Geneva.

European Region

The development of occupational hygiene is unequal in different European Member States of WHO, and there is still much to be done in areas such as human resources and comprehensive services.

The need for the establishment of a clear profile for the occupational hygienist, as well as for formal training in occupational hygiene, exists in many countries. This also includes a number of European countries where expertise on the different aspects of exposure assessment and risk management in the work environment is already available, but where the profession of occupational hygienist is not yet officially recognized. Therefore, in these cases, rather than a technical support role, what is required from international organizations (such as WHO and the Commission of the European Communities (CEC)) is that they act as 'driving forces' towards the integration of the available specialized knowledge into a global approach for the provision of 'healthy' work environments, through the comprehensive practice of occupational hygiene.

It is important to mention the meeting on 'The Development of the Occupational Hygienist in Europe', which was a joint activity between WHO, Geneva, and the Regional Office for Europe, and took place in Geneva, 1–4 July 1991, with 38 participants.

The objective was to continue the work started in the international workshop and conference on 'Training and Education in Occupational Hygiene: An International Perspective', held in the CEC, Luxembourg, 1986, and to provide more detailed suggestions and guidelines for the occupational hygiene profession in Europe.

This meeting aimed at the harmonization of concepts and professional roles, and at the definition of minimum requirements for the development of adequate human resources in occupational hygiene, in Europe.

It was agreed that the occupational hygienist is an indispensable component of the occupational health and safety multidisciplinary team; and that this requires professionals applying special advanced skills and knowledge. Therefore, there is need for the establishment of the profession of occupational hygiene in Europe; for which education and training are of fundamental importance.

There was consensus about the detailed roles of the occupational hygienist who should be a professional able to:

- Anticipate the health hazards that may result from work processes, operations and equipment, and accordingly advise on their planning and design.
- Recognize and understand, in the work environment, the occurrence (real or potential) of chemical, physical and biological agents, as well as other stresses, and their interactions with other factors which may affect the health and well-being of workers.
- Understand the possible routes of agent entry into the human body, and the effects that such agents and other factors may have on target organs.
- Assess worker exposure to potentially harmful agents and factors and to evaluate the results.
- Understand the legal framework for occupational hygiene practice in their own country.
- Evaluate work processes and methods, from the point of view of the possible generation and release/propagation of potentially harmful agents and other factors, with views to eliminating exposures, or reducing them to acceptable levels.
- Design, recommend for adoption and evaluate the effectiveness of control strategies, alone or in collaboration with other specialized professionals, to ensure effective control at least cost.

- Participate in overall risk analysis and management of an agent, process or workplace, and contribute to the establishment of priorities for risk management.
- Educate, train, inform and advise persons at all levels, as well as participate in all aspects of hazard communication.
- Work effectively in a multidisciplinary team involving other occupational health and safety professionals.
- Recognize agents and factors that may have environmental impact, and understand the need to integrate occupational hygiene practice with environmental protection.

There was also consensus that, in principle, the profession of occupational hygiene has to be learned at the academic level in a university or similar higher education institution. Establishment of educational programmes, including design of appropriate curricula (covering a core body of knowledge to prepare for the above roles), selection of educational methodologies, and appointment of faculty, were considered to be of vital importance in promoting the professional development and ensuring that the high level requirements for adequate occupational hygiene practice are met.

The target should be for professional occupational hygienists to have studied at degree level (preferably at Master's level); even while this target is not yet possible, it should still be considered as a goal. A certified professional hygienist should in addition have several years experience, and subscribe to a code of ethics. It was pointed out that the expertise and responsibilities of the professional occupational hygienist often include general environmental matters, especially with reference to the effects of industrial pollutants on humans.

Other conclusions and recommendations of the meeting on 'The Development of the Occupational Hygienist in Europe' include the following:

- Professional associations should press for recognition of occupational hygiene at the national level.
- The European Community and national institutions should define and use the terms 'professional occupational hygienist' (as well as occupational hygiene technician'), and the term 'occupational hygiene', in a way consistent with the definitions and with the required body of knowledge as specified in the guideline document prepared in this meeting[3].
- The European network of occupational hygienists and occupational hygiene and related organizations should be continued and further developed; WHO/EURO should play a role in this respect.
- European collaboration in occupational hygiene, particularly concerning education and training activities, should be increased.
- A continuing dialogue should be kept among WHO, International Labour Organisation (ILO), CEC, International Occupational Hygiene Association (IOHA), and national associations, with a view to cooperating in joint efforts, promoting mutual support and avoiding duplication.

It was considered desirable to promote a European body to establish or coordinate matters such as codes of ethics, accreditation of courses and certification of professionals. If and when this is done, it should be in collaboration with national bodies and the International Occupational Hygiene Association.

---

[3] This document, entitled 'The Occupational Hygienist in Europe – Development of the Profession', is under preparation and will be available soon, upon request, from the Office of Occupational Health, WHO, Geneva, or EURO, Copenhagen.

It was also proposed to recommend that nations establish guidelines for state employment of professional occupational hygienists, according to clearly specified professional qualifications.

It can be said that this was an 'historic' occasion because it was the first time that, at a European level, agreement was reached on detailed roles and functions for the occupational hygienist, as well as on the required areas of knowledge for this profession. This meeting, which received very active collaboration from the CEC and the IOHA, can be considered as a landmark in the development of a harmonized occupational hygiene profession in Europe, and its results will hopefully have a positive influence on similar developments in other regions where it is very much needed, particularly in developing countries.

Future WHO Occupational Hygiene Activities in Europe

Concerning activities in the field of occupational hygiene in Europe, there is close collaboration between the Office of Occupational Health, WHO, Geneva, and the WHO Regional Office for Europe, Copenhagen. In early 1992, both offices will jointly develop a comprehensive plan of action for the next years. The general lines of action will include the following:

- development of the network of professionals and institutions dealing with occupational hygiene;
- promotion of and collaboration with education and training activities;
- promotion of the harmonization of techniques for exposure assessment, including both air and biological monitoring, and promotion of a regional (eventually global) quality assurance programme for analytical laboratories, in collaboration with a number of agencies and institutions;
- preparation of practical guidelines for the organization of occupational hygiene services;
- promotion of a better integration of occupational medicine and occupational hygiene services, including medical and environmental data;
- promotion of technical cooperation among European countries, strengthening the collaboration between Western and Eastern European countries.

5. COLLABORATION AMONG INTERNATIONAL ORGANIZATIONS

According to its Constitution, the World Health Organization has the function "... to establish and maintain effective collaboration with the United Nations, specialized agencies, governmental health administrations, professional groups and other such organizations as it may deem appropriate".

It is important to envisage collaborative approaches by which international organizations, such as WHO, ILO and the CEC, as well as national institutions and professional associations, may effectively join their forces and complement one another towards the common goal of protecting and promoting the health of the working populations.

The CEC and WHO have on many occasions joined efforts towards the development of occupational hygiene in Europe. Continuing collaboration can only enhance and multiply the achievements of our organizations and their Member States.

ASSESSMENT OF EXPOSURE TO CHEMICAL AGENTS IN AIR AT THE WORKPLACE FOR COMPARISON WITH LIMIT VALUES AND MEASUREMENT STRATEGY

R. GROSJEAN

Convenor of CEN TC 137 WG1
Ministerie van Tewerkstelling en Arbeid
Laboratorium voor industriële toxicologie
B-1040 Brussels

SUMMARY

Working group 1 of CEN TC 137 has produced a draft proposal for assessment of exposure to chemical agents and measurement strategy. A review of the standard is given. The purpose is to give practical guidance to those who have to carry out these assessments. A systematic approach allows a reduction in the number of measurements. The report of the work allows us to communicate in an efficient way with interested parties: workers, occupational physicians, and the labour inspectorate.

1. INTRODUCTION

The use of limit values is an important tool for the protection of the health of workers exposed to chemical agents.

It is not practical to provide every worker with measuring equipment permanently in order to be able to guarantee that his exposure does not exceed the limit values. By using some formalised approaches and techniques it is possible to reduce drastically the measuring effort. In many cases it may not even be necessary to rely on measurements.

Basic requirements for such formalised approaches can be found in a legislation e.g. Directive 88/642/EEC amending Directive 80/1107/EEC.

In some countries some legislation, guidelines or standards already exist on this subject e.g. in the United Kingdom, Germany and Belgium.

Working group 1 of CEN TC 137 started its activities in November 1988. In May 1991 it was able to present a draft proposal to the Technical Committee. (1)

At the beginning of the activities of WG 1 it appeared that it was very difficult to find a direction which could meet the different expectations of the experts from different countries.

However, after each meeting mutual understanding between the experts increased and this has resulted in a document which gives practical guidance to those who have to carry out assessments of exposure at the workplace.

2. PROCEDURE

The procedure includes two phases:

- an occupational exposure assessment (OEA): the exposure is compared with the limit value;
- periodic measurements (PM): to regularly check if exposure conditions have changed.

3. TERMINOLOGY AND DEFINITIONS

A number of definitions are given. Some of them are the same definitions as those from the Directive 88/642/EEC. No attempt is made to define 'compliance' as this is a matter for legislation.

The topic of definitions is often controversial: the question is how far should one go in trying to define some terms.

At the level of the Technical Committee it was decided that a working group with Mr. Guichard as the convenor should look at the definitions of the various working groups in order to avoid contradictions.

There was a consensus that the different working groups would not be forced to adopt definitions which they did not like or did not need.

4. OCCUPATIONAL EXPOSURE ASSESSMENT (OEA)

The occupational exposure assessment is in three steps.

First, potential exposures are identified. An inventory of all products and chemicals in the workplace is made and information on these substances is collected (physico-chemical properties, health hazards, limit values etc).

In many countries such an inventory should already be available in order to comply with regulations on chemical hazards, medical surveillance etc.

In the second place, the work processes and procedures in which the chemical agents are involved are studied.

The combination of steps one and two allows to make an assessment of the exposure. A structured approach is required in this process which may be conducted in three stages: an initial appraisal; a basic survey and a detailed survey.

Data about the temporal and spatial distributions of the concentrations of the substances in the workplace have to be collected. The procedure which is proposed does not require that every stage of the assessment is used. If it is expected that the exposure exceeds the limit value or if it is clearly determined that exposure is well below the limit value then the occupational exposure assessment can be concluded and appropriate action taken.

Initial Appraisal

In this stage, the list of chemicals and the workplace factors yield a consideration of the likelihood of exposure. Variables which effect personal exposure have to be looked at. These variables are influenced by the emission sources, the dispersion of air pollutants, the actions and behaviour of the individuals. If this initial appraisal shows that the presence of an agent in the air at the workplace cannot for certain be ruled out it needs further consideration.

Basic Survey

The basic survey provides quantitative information about exposure of workers, taking into account tasks with high exposure.

Possible sources of information are earlier measurements, measurements from comparable work processes, reliable calculations based upon relevant quantitative data. When calculations are used, care should be taken not to overlook aspects which are difficult to quantify: concentration gradients, diffusive emissions from spills, waste tins, contaminated clothing etc.

If the information obtained is insufficient to enable valid comparison to be made with the limit values, it shall be supplemented by workplace measurements.

Detailed Survey

The detailed survey is aimed at providing validated and reliable information on exposure when it is close to the limit value.

5. MEASUREMENT STRATEGY

Bearing in mind that measurements are expensive it is important to take an approach which enables the most efficient use of resources.

This means that it should be possible to stop the procedure in an early stage by using techniques which are easily applied and which may be less accurate. Other possibilities may be worst case measurements, sampling near emission sources or screening measurements.

The requirements and measuring ranges for these different measurement tasks should be defined in such a way that they allow a complete occupational exposure assessment to be made without further investigation.

In other cases, where exposures are suspected to be close to the limit value, it will be necessary to undertake a more accurate investigation, making full use of the capabilities of instrumental and analytical techniques.

6. SELECTION OF WORKERS FOR EXPOSURE MEASUREMENTS

Some general guidelines are given for selection of workers for exposure measurements.

If workers are sampled purely on an at random basis large numbers of samples are needed. Some techniques such as subdividing the exposed population into homogeneous groups, critical examination of the work pattern, examination of preliminary measurement results allow a reduction in the sampling effort compared to sampling on a 'blind' purely statistical basis.

7. REPRESENTATIVE MEASUREMENTS

Measurement conditions should be selected in such a way that measurement results give a representative view of exposure under working conditions. This means that personal sampling should be carried out for an entire working period, or a time period which is representative of it.

Fixed point measurements may be used when they allow an assessment of personal exposure. In some cases no personal measuring or sampling device is available.

Measurements should be carried out on sufficient days and during specific operations. It is important to consider different episodes during which exposure conditions may vary (night and day cycles, seasonal variations).

8. WORST CASE MEASUREMENTS

When it is possible to identify clearly episodes where higher exposures occur e.g. a high emission due to certain working activities, one can select sampling periods containing these episodes. Sampling efforts can be concentrated on these periods. If the concentrations measured this way are presumed to apply for the whole of the working period, a safety factor has to be built in.

9. MEASUREMENT PATTERN

A number of practical issues play an important role in the pattern of sampling: frequency and duration of a particular task, optimal use of occupational hygiene and analytical resources. Representativity of the data for the identified tasks and periods is essential.

In many workplaces work is varied throughout the working period which itself may be interrupted. The duration of an individual sample is often dictated by constraints of the method of sampling and analysis in practice. Unsampled time remains a serious weakness is the credibility of any exposure measurement. During this time careful observation is always necessary. The assumption that changes have not occurred in the unsampled period must always be critically examined.

Annex 1 of the draft standard contains a table which can be used as a guide for the minimum number of samples to be taken as a function of sampling duration in the case of a homogeneous working period. The table is a combination of practical experience and statistical arguments.

10. MEASUREMENT PROCEDURE

The measurement procedure shall give results representative of worker exposure. The measurement procedure contains: the agents, the procedure for sampling and analysis, the sampling location(s), the jobs to be monitored, the duration of sampling; the timing and duration between measurements, technical instructions concerning the measurements and the jobs to be monitored.

Annex 2 of the draft standard contains examples of calculations of the occupational exposure concentration from individual analytical values. The occupational exposure concentration is the arithmetic mean of the measurements in the same shift with respect to the appropriate period of the limit value of the agent under consideration. Working group 2 has already proposed general requirements for measurement procedures. (2)

11. CONCLUSION OF THE OCCUPATIONAL EXPOSURE ASSESSMENT

The prescribed procedure has to come to a conclusion. No uniquc formal scheme is proposed in the standard. Annex 3 and 4 give examples of formal schemes which might be used. There may be others. In Belgium a scheme proposed by Tuggle based on one sided tolerance limits is used. (3) (4)

Whatever scheme is used one of three conclusions should be made.

1. The exposure is above the limit value: the reasons for the overexposure should be determined and remedial action taken. The occupational exposure assessment should be repeated.
2. The exposure is well below the limit value and is likely to remain so on a long-term basis, due to the stability of conditions at the workplace and the arrangement of the work process. In this case periodic measurement are not needed. However it must be regularly checked that the occupational exposure assessment leading to that conclusion is still applicable.
3. The exposure does not fit into categories 1 or 2. Although the exposure may be below the limit value periodic measurements are still required. In certain cases, the periodic measurements can be omitted, depending on the properties of the agents in

the work process. Technical guidelines can provide criteria for deciding whether or not to carry out periodic measurements Annex 5 of the draft proposal gives an example of a procedure for considering if and when periodic measurements are required.

The purpose of the periodic measurement is to check the validity of the occupational exposure assessment and to recognize changes of exposure with time.

The occupational exposure assessment is only concluded when a report has been made of work done.

12. PERIODIC MEASUREMENTS (PM)

The objective of periodic measurements is to check that control measures remain effective. Information is likely to be obtained on trends or changes in pattern of exposure so that action can be taken before excessive exposures occur.

For the results of a periodic sampling programme to be of real use it must be possible to compare consecutive sets. This implies that the methodology used for collecting the samples needs to be rigorously planned to ensure that the overall error can be estimated and that genuine change in the exposure pattern can be recognised.

Annex 7 gives two examples of statistical analyses of data: the moving weight average and the probability plot.

The interval between measurements has to be established after consideration of a number of factors such as the process cycles, closeness to the limit value and the temporal variability of the results.

Together with other considerations this may lead to intervals between periodic measurements varying from less than a week to more than a year.

Annex 6 gives an example of a scheme for determining intervals between periodic measurements.

13. REPORTING

Reports have to be written of the occupational exposure assessment and of any periodic measurement. Each report should give reasons for the procedures adopted in the particular workplace. It is clear that such a report is very valuable for the labour inspectorate, for the workers who should have information on their exposure and the occupational physician.

14. HANDLING OF DATA

It is evident that an important quantity of data will be generated during the OEAs and the PMs. In order to prevent accumulation of a large quantity of unused data and in order to obtain a high return from the invested time and effort it is necessary to use appropriate statistical techniques. The data can be useful for epidemiological studies and the evaluation of occupational exposure limits.

15. AUTOMATED MEASURING SYSTEMS

During the discussions of working group 1 some participants felt it was necessary to deal with automated measuring systems. Such measuring systems are used in vinylchloride production and polymerisation plants, and sterilisation processes with ethylene oxide etc. It proved to be difficult to insert this concept into the working document in preparation without lengthy discussions. At the Technical Committee meeting in London in May 1991 it was decided to establish a joint *ad hoc* group from WG 1 an WG 2 on this subject.

16. CONCLUSION

The draft standard prepared by working group 1 is intended as a practical guide to help those who have to carry out assessments of exposure and workplace measurements. It makes proposals to provide basic requirements of a legislation which is sometimes too general and too vague. It cannot give solutions to problems which are clearly a matter for legislation such as exposure to mixtures and definition of 'compliance'.

ACKNOWLEDGEMENT

I am greatly indebted to all the members who participated actively to the discussions and the finding of solutions and compromises.

With the risk of forgetting important contributors, I would like to thank especially Messrs J. Auffarth, B. Bord, D. Sowerby, W. Ernst, T. Hafkenscheid, B. Hervé-Bazin, M. Molyneux, R. Lippi, P. Macchi, P. Rocchi and of course Dr. H. Blome, the chairman of CEN TC/137.

REFERENCES

(1) Document CEN TC 137 N48.

(2) prEN 482: General requirements for the performance of procedures for workplace measurements.

(3) NBN T 96-002 Workplace atmospheres-Monitoring strategy (June 1987) (only in Dutch and French).

(4) Assessment of occupational exposure using one sided tolerance limits. R.M. Tuggle. Am. Ind. Hyg. Assoc. J. 43: 338-346 (1982).

PERFORMANCE REQUIREMENTS FOR MEASURING METHODS

S. ZLOCZYSTI

Auergesellschaft GmbH
Thiemannstrasse 1, 1000 Berlin 44, Germany

SUMMARY

Programme and status of work of CEN/TC 137/WG 2 are described including history and background of the standardization work in this field. The first Draft European Standard issued by this Working Group is prEN 482 (March 1991). This is a General Standard which shall be used as an umbrella-specification for all subsequent Standards for workplace air measurements. The main ideas of the General Standard are described followed by a view on further activities in preparing European Standards for Diffusive Samplers, Sampling Pumps, Sampling Tubes and Detector Tubes.

1. INTRODUCTION

Working group 2 of CEN/TC 137 deals with the standardization of measuring procedures used for workplace measurements. The title of Working Group 2 is similar to the title of this presentation 'General Requirements for Measuring Procedures'. It may be of interest to see how CEN/TC 137 is organized and which tasks were assigned to the individual working groups.

CEN/TC 137

Title: Assessment of workplace exposure

Scope: Standardization in the field of assessment of exposure to agents at the workplace including the planning and performing of measurement but excluding the establishment of limit values

- WG 1: Monitoring Strategy
- WG 2: General Requirements for Measuring Procedures
- WG 3: Particulate Matter
- WG 4: Terminology and Definitions

CEN/TC 137 was founded in 1988 with a very clear intention. Based on European Directives, a need was identified for the preparation of Standards in the field of workplace air measurements. The following European Directives were understood to be the background for our standardization work:

- Council Directive of November 27, 1980 on the protection of workers from the risks related to exposure to chemical, physical and biological agents at work (80/1107/EEC)

- Council Directive of December 16, 1988 amending Directive 80/1107/EEC on the protection of workers from the risks related to exposure to chemical, physical and biological agents at work
(88/642/EEC).

With regard to the tasks of Working Group 2 of CEN/TC 137 the following excerpts from the Directive 88/642/EEC are more specific:

Annex II a

- Requirement for persons who carry out measurements:
Those carrying out measurements must possess the necessary expertise and facilities.
- Requirements for measuring procedures:
  a) The measuring procedure must give results representative of worker exposure.
  b) To ascertain the exposure of worker at the workplace, where possible personal sampling devices should be used attached to the workers' bodies.
  d) If the measuring procedure is not specific to the agent to be measured, the full value recorded must be counted as applying to the agent to be measured.
  h) If the European Committee for Standardization (CEN) publishes general requirements for the performance of measuring procedures and devices for workplace measurements together with provisions on testing, they should be referred to when selecting appropriate measuring procedures.

This was and is understood to be the justification of our work.

Prior to summarizing the status of our work I would like to report on the different approach of our Working Group compared with the activities of ISO/TC 146 'Air Quality'. Within ISO/TC 146/SC 2/WG 4 carries the title 'organic vapours'. This Working Group produces International Standards also under the heading 'workplace air'.

However, these Standards are very specific covering only one compound like vinyl chloride or groups of homologues like the determination of concentrations of C_3-C_{10} hydrocarbons in air. In addition, these ISO-Standards very often differentiate between specific sampling and analyzing procedures for the individual compound of interest.

To avoid any confusion or misinterpretation this description is not intended as a criticism of ISO/TC 146 or the work this committee has done. Many experts of CEN/TC 137 were or still are also members of ISO/TC 146. The only criticism which is more or less accepted by all experts involved is that the ISO-work is very slow due to a variety of reasons.

Therefore, a few years ago experts from several countries started a discussion in order to speed up standardization work. Most of these experts are now active members of CEN/TC 137 and are also actively involved in this Symposium.

The result of the discussion was our so-called new approach which was a new concept for standardization work in the field of workplace air measurements. The new concept was presented to the first plenary meeting of CEN/TC 137 in Berlin in November 1988 and the following working programme of WG 2 was accepted:

"The task of this Working Group shall be the preparation of general requirements for measuring procedures used in the assessment of workplace exposure and the necessary definition of terms.

The Working Group shall also specify evaluation procedures to determine the performance characteristics of the methods.

The preparation of test methods for individual substances or groups of substances shall be avoided and only carried out if this is necessary for political or other reasons. Other national or international organizations may prepare such Standards.

Standards with requirements for measuring equipment including diffusive samplers shall be in line with the general requirements."

2. STATUS OF WORK

CEN/TC 137/WG 2 started its work in November 1988 and the first meeting was needed to reach consensus on how to proceed. Finally, it was agreed to prepare a General Standard first which should work as an 'umbrella Standard' for all the subsequent work of WG 2.

The title of this Standard is: 'General requirements for the performance of procedures for workplace measurement' with the following scope:

> "This Standard specifies general requirements for the performance and validation of procedures for workplace measurements in line with the basic requirements of the Directive 80/1107/EEC as amended by the Directive 88/642/EEC."

This Draft European Standard was issued for public inquiry as prEN 482 dated March 1991 . Without presenting every detail of prEN 482 in this paper I would like to draw your attention to two major items which can be regarded as the highlights of this Standard.

First of all, accuracy, precision and bias are covered by a simple formula:

$$\text{Overall uncertainty} = \frac{|x - x_{ref}| + 2s}{x_{ref}}$$

where

x is the mean value of results of a number of repeated measurements

x_{ref} is the true or accepted reference value of concentration

s is the standard deviation of n measurements.

The expression 'overall uncertainty' was created to avoid misunderstandings with well-introduced and well-defined definitions like accuracy or precision. WG 4 of CEN/TC 137 has elaborated the following definition for overall uncertainty of a measuring procedure or of an instrument:

> "Quantity used to characterize as a whole the uncertainty of the result given by an apparatus or a measuring procedure. It is expressed on a relative basis by a combination of bias and precision".

During the discussion of this formula the group received some criticism concerning the inadequate use or understanding of statistics and the pertinent mathematical background.

The crucial point in the Standard is that the minimum number of repeated measurements is specified to be $n \geq 6$. WG 2 is fully aware of the shortcomings related to the minimum quantity of only six repeated measurements. Probably, the acceptance of this formula could have improved if the group would have agreed on a minimum quantity of 20 measurements instead of only six. However, it was agreed that 20 repeated measurements are not realistic. In addition, for type testing procedures in a test house a series of data will be obtained when testing the performance characteristics for the specified variety of parameters of instruments or measuring procedures.

The combination of these measurements will definitely increase the number of results. This perhaps needs to be clarified when revising prEN 482 as a result of the public inquiry.

The second major item of prEN 482 is a table of different measurement tasks combined with minimum requirements for measuring range and overall uncertainty.

Specification of performance requirements depending on the measurement task (LV = Limit Value)

Measurement task	Relative overall uncertainty	Minimum specified measuring range	Averaging time
Screening measurement of time weighted average concentration	≤ 50%	0.1 to 2 LV	≤ 8 h
Screening measurement of variation of concentration a) in time b) in space	 ≤ 20% ≤ 40%	dynamic range ≥ 10:1 dynamic range ≥ 10:1	≤ 5 min ≤ 15 min
Measurement near emission source	≤ 30%	0.5 to 2 LV	Source dependent
	≤ 50%	2 to 10 LV	
Measurement for comparison with limit values	≤ 50%	0.1 to 0.5 LV	≤ 8 h
	≤ 30%	0.5 to 2 LV	
Periodic measurements	≤ 50%	0.1 to 0.5 LV	≤ 8 h
	≤ 30%	0.5 to 2 LV	

First comments show that this table appears to be quite acceptable to the readers of the Draft European Standard with some suggested improvements. However, the terminology of the measurement tasks has not yet been finalized, because this has to be consistent with the Draft European Standard of WG 1 'Monitoring Strategy'. In particular, the terminology should be confined to technical definitions and should not have any socio-economic implications, which are the concern of the Commission.

It is hoped that the final Standard will be referred to specifically in new EC Directives, in which case its nature would be changed towards a more compulsory regulation.

Further comments related to having more entries to the table which is equivalent to additional different measurement tasks. Again, this will be brought in line with WG 1 and the 'Monitoring Strategy'.

As a conclusion of my report on prEN 482 it should be mentioned that type testing of measuring procedures or instruments in a test house shall be carried out under well controlled laboratory conditions only although WG 2 agrees that it is highly desirable also to carry out field tests. However, for the time being this could not be specified properly when preparing this Standard.

3. FUTURE WORK

The following projects to be addressed by Working Group 2 were agreed upon with CEN/TC 137.

- Diffusive Samplers
- Sampling Pumps
- Sampling Tubes
- Detector Tubes

The above four Standards are under preparation by the Working Group but with different priorities. Work is most advanced for the Diffusive Samplers as it was recognized that such a Standard is urgently needed.

The title is: 'Workplace atmospheres – Requirements and test methods for diffusive samplers for the determination of gases and vapours'.

This standardization work is a continuation of the Symposium on diffusive samplers which was held under the guidance of the Commission of the European Communities in Luxembourg, in September 1986. Our standardization work in this field includes European and NIOSH activities (United States) in order to harmonize the requirements for diffusive samplers on a wider basis.

The document is almost completed, the equivalent versions in French and German are currently under preparation. It is planned to submit the Draft European Standard on diffusive samplers in 1992 for public inquiry.

The work on sampling pumps and tubes and on detector tubes was started earlier this year. All three projects will be worked on more or less in parallel utilizing existing Standards from many countries. The Standard on sampling tubes will show many similarities with the diffusive sampler whilst the document on sampling pumps will be prepared in close cooperation with Working Group 3, 'Particulate Matter'. The reason is that we hope to prepare a basic document for sampling pumps which can be used for dust sampling as well as for gases and vapours.

The Standard on detector tubes will take into consideration existing Standards in several countries as a basis for discussion. The major part of our work for all subsequent Standards will be to harmonize the minimum requirements for individual procedures with our 'umbrella-specification' the 'General Requirements for Measuring Methods'.

1347

LEGAL BACKGROUND AND STANDARDS: PARTICULATE MATTER

T L OGDEN

Health and Safety Executive
1 Chepstow Place, London W2 4TF

SUMMARY

This paper summarises the legislative development of particle size fractions applied to exposure limits, both in national and European regulation. The parallel development of scientific work and standardisation is also considered. Important steps have been the adoption of the Johannesburg and American respirable fraction conventions, the specification for larger particles of a sampler entry velocity of 1.25 m/s, and the development of the inspirable (or inhalable) convention based on human head entry characteristics. The paper traces the influence of these on the measurement methods in the lead directive (82/605/EEC)[1] and the agents directive (80/1107/EEC, amended by 88/642/EEC)[2],[3]. The asbestos directive (83/477/EEC)[4] illustrates the problems of allowing 'equivalent' methods without defining equivalence; the approach of 88/642/EEC of defining this in terms of compliance decision is much better. The new draft European standard, stimulated by 88/642/EEC, is the culmination of this work, and includes specifications also accepted by ISO and ACGIH committees.

1. INTRODUCTION

This paper discusses the historical and legislative background of the definitions of respirable and inhalable (inspirable) dust, leading up to those recently agreed by committees of the European Standards Organisation (ISO) and the American Conference of Governmental Industrial Hygienists (ACGIH).

2. HISTORY OF CONVENTIONS TO 1988

Respirable conventions

It has been known for at least 50 years that only the smaller particles reach the lung, so that not all sizes can be responsible for respiratory diseases. The British Medical Research Council in 1952 agreed size-selection criteria for instruments to measure dusts associated with lung fibroses[5]. In 1959, the same selection curve was ratified by an international conference in Johannesburg, and is consequently often known as the Johannesburg Curve[6]. The Johannesburg Curve has remained the norm for measuring 'respirable dust' in most of northern Europe, and much of the rest of the world. In the US, a different convention was adopted[7], which, with slight variations, remains in use in North America and elsewhere, including some European countries. A few other respirable conventions have been used for special purposes, but have failed to gain general use in occupational hygiene.

Inhalable conventions

As increasing attention was paid to exposure limits for particulate materials which could cause health problems through deposition other than in the deep lung, it became apparent that some standardisation was necessary in this case also. By the 1970s, many experiments had shown that different sampler designs, not intentionally size-selective, would collect different amounts of dust in the same atmosphere. This was because inertia and sedimentation, especially of larger particles in moving air, could cause them to be preferentially collected or to miss the sampler. As samplers were generally not specified, the apparent precision of exposure limits for such materials was illusory. The first attempt to overcome this problem for particulate materials in general was for the German MAK List, which in the early 1970s specified a sampler intake velocity of 1.25 m/s (± 10%). This became widely used in continental Europe. Later work showed that specifying intake velocity was insufficient: indeed, it is one of the less important variables[8], but the German convention was a considerable advance because in practice only a small range of samplers was used. In the late 1970s this convention became the basis of a draft ISO 'total dust' method, which was overtaken by the improved ISO conventions described below.

Research groups in Britain and Germany studied the problem, with a view to deriving a specification for sampling large particles for health-related purposes. It was realised that the human head responded to external winds just as samplers did, and this led to the idea of making samplers which imitated the head's characteristics. Particles collected by such samplers would be those available for deposition in the respiratory tract, which might then be associated with disease. An 'inhalable fraction' convention was proposed, based on the directionally-averaged intake efficiency of the human head. (This fraction was later called 'inspirable', but ISO, CEN and ACGIH have all reverted to the original name 'inhalable'.)

The ISO and ACGIH reports

An ISO Technical Report was published in 1983 incorporating an inhalable fraction definition, a choice of respirable fractions based (for the workplace) on the Johannesburg and American conventions, and an intermediate thoracic fraction[9]. Because the report also applied to the general environment, an additional, finer respirable convention was included. Although the report was published in 1983, it only took into account scientific results available at the final drafting in early 1980. The main importance of the report was perhaps as an international basis for subsequent discussion, and improved proposals soon followed. One set was produced by an ACGIH committee[10], for possible application to TLVs. Because of the rapid improvements, a new ISO working group (under Technical Committee 146) was convened in 1987 to revise the Technical Report into a draft international standard. This work had not proceeded far before it was profoundly affected by agreement on European Directive 88/642/EEC.

3. LEGISLATIVE BACKGROUND

National positions

In the UK and Germany, the published sets of exposure limits have for many years specified the Johannesburg respirable dust definition. As already mentioned, for larger particles the MAK list has for almost 20 years specified a sampler entry velocity of 1.25 m/s. In Britain, official guidance for measuring the inhalable fraction has since 1983 laid down a particular personal sampler, and has referred to the ISO definition (also recently the ACGIH definition)(11). Official lists or methods in other countries have ignored the need for conventions or have used some combination of those mentioned above.

Amended directive 80/1107/EEC

The European situation was transformed by Council Directive 88/642/EEC, amending 80/1107/EEC. This requires Member States 'when they adopt provisions for the protection of workers, concerning an agent' to establish sampling and measuring procedures laid down in an Annex, 'or a method yielding equivalent results'. For airborne particles, the Annex (IIa) pays special attention to the inspirable (inhalable) and respirable fractions. The inspirable is defined in clause B4b as 'the fraction of suspended matter which can be breathed in by a worker through the mouth and/or nose', and the respirable fraction as (clause AI2) 'the fraction of the inspirable fraction reaching the alveoli'.

Some guidance is given on measurement. For the inspirable fraction, 'By way of example, in measurement practice, devices with an inspiration rate of 1.25 m/s +/- 10% or devices in conformity with ISO/TR 7708 1983 are used for sampling'. A few details are given for the 1.25 m/s example, but, as already mentioned, these two measurement approaches will only exceptionally agree with one another, because entry velocity is not the most important variable in determining entry efficiency. For the respirable fraction, the Annex specifies the Johannesburg convention. The Annex concludes (clause B4d) with two key codicils. The first is:

> If the CEN establishes specifications for the collection of suspended material at the workplace, they should be applied, by way of preference.

Thus specifications agreed by CEN appear to displace those in the Annex for the purposes of the Directive. The second codicil is:

> Other methods may be used provided that they yield the same conclusion or a stricter conclusion in relation to compliance with the limit values.

General discussion of equivalent methods is beyond the scope of this paper, but we may note that this clause is much easier to apply than a simple permission to use equivalent methods. Measurements always show statistical variation: simultaneous measurements in the same atmosphere by the same method will yield a range of results. Similarly, of ccurse, simultaneous measurements by different methods will yield a range of results, and some kind of statistical or other definition of equivalence is required before one can say whether or not methods are equivalent. The amended 80/1107/EEC solves the problem logically, by looking beyond the measurements to their purpose: other methods may be used if the results obtained lead to the same (or stricter) decision about what should be done.

Lead and asbestos

The measurement methods in two earlier directives illustrate the advance brought about by the amendment to 80/1107/EEC. Directive 82/605/EEC, on the protection of workers against lead, was negotiated about the same time as ISO report TR 7708, but, not surprisingly, the method in the directive is a compromise of contemporary national practices, and ignores the (then) new concept of the inhalable fraction. The sampler to be used under the directive must be 'closed face', have an entry velocity of 1.25 m/s, an entry orifice diameter of at least 4mm, and an entry flowrate of at least 1 l/min. This derived from the draft ISO method for 'total dust', mentioned above, which was itself an elaboration of the MAK method already discussed. The method has two shortcomings: it would be possible to make different samplers meeting the specification which collected different fractions - the specification is insufficient - and at least one such sampler has been shown to collect significantly less than the inspirable fraction. It is reasonable to infer that such a sampler could under-estimate biologically relevant intake of lead.

The second earlier directive is that on asbestos (83/477/EEC). This contains a reference method intended to estimate the respirable fibres, but in this case they are distinguished by microscope examination rather than by size-selective sampling. This is because particle shape, as well as respirability, is a factor in disease. In the context of this discussion, the directive is noteworthy as one which allows 'equivalent' methods to be used, but gives no definition of equivalent. In the UK, the enforcing authority (HSE) decided it would have to adopt the reference method despite perceived shortcomings, because in the absence of a definition HSE would find it difficult to prove if challenged that any other method was equivalent.

4. CEN WORK

Commencement and conclusion

Following agreement in 1988 of the amendments to 80/1107/EEC, CEN set up Working Group (WG) 3 of Technical Committee 137 to draw up definitions of respirable and inspirable fractions, and specifications for their measurement. The work was very similar to the revision by ISO of TR 7708, which had already started. The only difference was that the proposed ISO standard was intended to cover the general (non-occupational) environment, as well as the workplace. To ensure that the two standards were consistent, the same convenor was appointed for the CEN and ISO WGs, which then worked as closely together as the procedures of the parent bodies would allow. Contact was established with an ACGIH committee progressing the 1985 report referred to above.

The outcome has been a degree of agreement which probably exceeds the original expectation of the most optimistic participant. All three committees have agreed the same numerical definitions and the same nomenclature, removing, for example, the difference between the European and North American respirable dust definitions which have existed for about 30 years, and, consequently the need for two respirable definitions for the workplace in TR 7708. (The ISO report contains two supplementary specifications, one for respirable and one for inhalable in high winds, to deal with special conditions outside the workplace.) The agreement reflects the willingness of all concerned to give up long-established positions in the interests of international agreement, and the far-sightedness of S C Soderholm, the chair of the ACGIH committee, whose proposals are the basis of the agreed conventions.

The agreed definitions

Main features of agreements are summarised here. The draft European Standard, which is available in English, French and German, should be consulted for details. The fractions are named (in the English text) inhalable, thoracic and respirable (with extrathoracic and tracheobronchial fractions derived from these by subtraction). 'Inhalable' is the name used for the fraction called 'inspirable' in the Directive. The terms are equivalent, but inhalable is the original scientific term and is more obvious in English. The defining curve falls from 100% at 0μm aerodynamic diameter, through about 65% at 20μm, and smooths out at about 50% beyond 50μm. The thoracic fraction (not used in the Directive) and respirable fraction are defined as sub-fractions of the inhalable, as cumulative log-probability functions, with $\sigma_g = 1.5$. The aerodynamic diameter at which 50% of the total airborne particles are included is 10μm for the thoracic and 4μm for the respirable fractions. The curves are illustrated in Fig 1.

FUTURE WORK

At the time of writing, the CEN and ISO draft standards are approaching the end of their periods for comment and balloting. Some important comments have been received, but both drafts are very likely to be approved in essentials. Then follows the important step of implementation into national law, insofar as the Directive requires or Member States choose. An important question is instrumentation, and a standard on the difficult problems of testing instruments and specifying their practical performance is being drafted. In the US, ACGIH is taking the definitions through the 'proposed changes' procedures necessary to apply them to TLVs. Of course, the US regulatory authorities (MSHA and OSHA) are not bound to accept the ACGIH definitions, so it is unclear what these authorities will do.

Despite these remaining uncertainties, it is clear that the agreements resulting from the Directive are historic in their field. It is interesting that the 1970s scientific work in Britain and Germany which led to the inhalable definition was financed by the European Communities under the Health in Mines programmes. The CEN standard and the Directive which led to it therefore complete a work in which the Community has had a leading part from the beginning.

REFERENCES

(1) Council Directive of 18 July 1982 on the protection of workers from the risks related to exposure to metallic lead and its ionic compounds at work (82/605/EEC). Official Journal of the European Communities. Vol. 25, No. L247, pp12-21.

(2) Council Directive of 27 November 1980 on the protection of workers from the risks related to exposure to chemical, physical and biological agents at work (80/1107/EEC). Official Journal of the European Communities. 3 Dec 1980, No. L327, pp8-13.

(3) Council Directive of 16 December 1988 amending Directive 80/1107/EEC. (88/642/EEC). Official Journal of the European Communities. 24 Dec 1988, No. L356, pp74-78.

(4) Council Directive of 19 September 1983 on the protection of workers from the risks related to exposure to asbestos at work. (83/477/EEC). Official Journal of the European Communities. 24 Sept 1983. No. L263, pp25-32.

(5) HAMILTON, R. J. and WALTON, W.H. (1961). The selective sampling of respirable dust. In Inhaled Particles and Vapours, pp455-481. Pergamon, Oxford.

(6) ORENSTEIN, A. J. (ed.) (1960). Recommendations adopted by the Pneumoconiosis Conference. In Proceedings of the Pneumoconiosis Conference, pp619-621. Churchill, London.

(7) LIPMANN, M. and HARRIS, W. B. (1962). Size-selective samplers for estimating 'respirable' dust concentrations. Health Physics Vol. 8, pp155-163.

(8) OGDEN, T. L. (1983). Inhalable, inspirable and total dust. In Aerosols in the Mining and Industrial Work Environments, Vol. 1, pp185-304. Ann Arbor Science, Ann Arbor, Michigan (ISBN 0-250-40531-8).

(9) INTERNATIONAL STANDARDS ORGANISATION (1983). Air quality - particle size fraction definitions for health-related sampling. Technical Report ISO/TR 7708 - 1983 (E). ISO, Geneva.

(10) AMERICAN CONFERENCE OF GOVERNMENTAL INDUSTRIAL HYGIENISTS (1985). Particle size-selective sampling in the workplace. Report of ACGIH Technical Committee on Air Sampling Procedures. ACGIH, Cincinnati, Ohio.

(11) HEALTH AND SAFETY EXECUTIVE (1983). General methods for the gravimetric determination of respirable and total inhalable dust. Methods for the Determination of Hazardous Substances No. 14. HSE, Bootle. (ISBN 0 7176 0343 1).

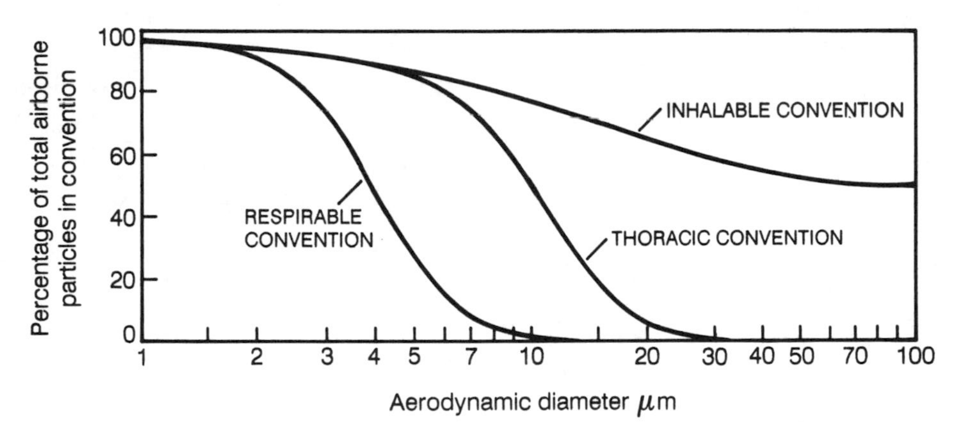

Fig.1 The size-fractions agreed for the workplace by committees of CEN, ISO and ACGIH

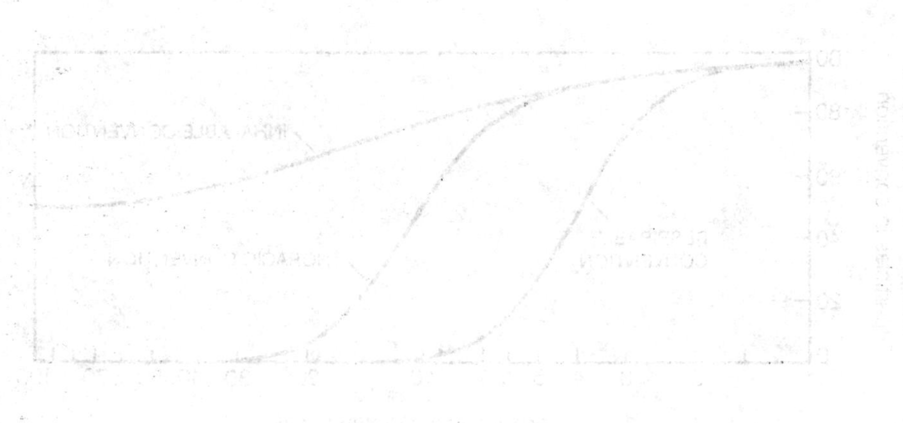

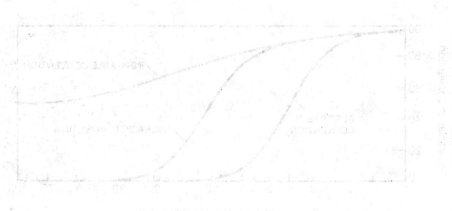

Session II

MEASUREMENT METHODOLOGY
Practical and Changing Needs

KEYNOTE PAPERS

POSTERS

1348

PRACTICAL AND CHANGING NEEDS

H. BLOME
Berufsgenossenschaftliches Institut für Arbeitssicherheit
Alte Heerstrasse 111, Sankt Augustin, Germany

1. INTRODUCTION

In accordance with the Council's Directive for the protection of workers against risks due to chemical, physical and biological agents at the workplace, it is essential to determine the nature and extent of exposure an employee might be subjected to when pursuing a certain activity. This has to be done to estimate if the health and safety of employees is endangered and to take suitable remedial measures.

2. DIFFICULTIES AND PROBLEMS IN THE FIELD

In implementing the Council Directive 80/1107/EEC and the amending Directive 88/642/EEC or corresponding national guidelines and regulations, enterprises, supervisory authorities and measuring services find themselves confronted with a number of difficulties which are mainly due to the following reasons.

1. In many cases, there is a lack of information concerning the hazardous substances used. Consequently, manufacturers and importers should be obliged to identify their preparations and products. It would also be a positive approach to improve the quality of the accompanying safety data sheets.
2. In numerous firms, know-how and technical instrumentation necessary for carrying out measurements is often lacking or incomplete. Even large enterprises are still developing concepts for the implementation of their monitoring obligation. For the majority of small and medium-size enterprises, this task is too much so that only a partial approach to the problem is possible.
3. Time expenditure and costs of monitoring are often considerable.
4. The complexity of guidelines and regulations, implying a variety of single aspects and mutual inter-connections, represents another problem. This is not only true for the different fields of work safety, but also for the interaction existing between work safety and environmental protection. The formerly neat borderline between work safety and environmental protection is becoming more and more indistinct and, very often, it is impossible to take measures on work safety without having to handle environmental problems at the same time.

   Technical processes should be designed so as to take into account the requirements and needs arising from environmental protection and to ensure an integral risk prevention approach. It would also be useful to consider – together with the idea of protecting exposed workers – the needs of all those who live in exposed areas (the community at large).
5. Particular problems are the monitoring of
   - 'mobile workplaces', i.e. workplaces with continuously changing exposure situations (e.g. construction sites, repair work).
   - workplaces with frequently changing procedures and activities (e.g. carpentry).
   - workplaces with frequently changing working materials:

- the waste disposal sector (e.g. lacquer, galvanic and petrol deposits, electric waste products, PCB-contaminated used oil, as well as solvent residues from the computer industry);
- batch processes (e.g. laboratories, preparation of chemicals, preparation of different lacquer qualities in the same installation).

- workplaces where multi-component mixtures occur (e.g. rubber industry, foundries, handling of cooling lubricants, fibres).

6. Many problems require interdisciplinary cooperation before they can be solved (analysts, technicians, hygienists, chemical engineers, medicines, toxicologists). Apart from the problems already specified, particular attention needs to be paid to difficulties encountered with cooling lubricants, hydrocarbon mixtures, welding fumes, fibre mixtures, as well as with the rubber industry in general and the treatment of epoxy resins.

3. REQUIREMENTS AND APPROACHES

These problems can only be solved on the basis of approaches taking into account preventive and scientific aspects as well as feasibility criteria. Such approaches could include the following.

Provision of Substance and Product Data

Work safety experts working in the field of hazardous substances have difficulties in obtaining clear, precise information on the substances under consideration. Various aspects must be considered, *viz.* the variety of substances, their risks, their storage and disposal, as well as their classification, their assessment and requirements in accordance with the wide spectrum of legal regulations, the complexity and great number of possible technical solutions. Consequently, the development of appropriate modern communication means is of paramount importance.

Elaboration of Recommendations or Guidelines

Recommendations containing information on, for example, measuring techniques, on the implementation of limit values, or on technical processes, may represent an aid for firms and supervisory authorities charged with the task of risk assessment. There are certain conditions (e.g. repair work or work with continuously changing exposure situations) where monitoring is not feasible. Other approaches have to be developed such as the specification of technical, organisational and personal protective measures. The efficiency of these measures can be easily verified.

The definition of framework conditions for the use of less expensive and easily applicable measuring techniques is of particular interest. On the one hand, there is a series of frequently used measuring techniques where - due to a lack of selectivity - no precise assessment is possible. On the other hand, these techniques may serve to estimate exposure situations under worst-case conditions. Clear instructions for the application of these measuring techniques would permit exploitation of their advantages; false interpretations could be avoided.

It must always be proved that high-technological measuring techniques are definitely applicable under field conditions.

Consideration should be given to those charged with the implementation of the complex regulation work by identifying certain substances, technical equipment and processes which allow the user to do without monitoring provided that there is no risk that the concentration remains below a certain limit value.

Three examples will be given to illustrate what has been stated so far: application of a measuring procedure (sorption tube), suggestion for the monitoring of a complex chemical mixture (cooling lubricants) and elaboration of two workplace analyses (waste disposal industry and ethylene oxide sterilization).

Recommendation for the use of sorbent tubes

The different parts of the measuring device (suitable sorption tube and corresponding sampling pump) should be selected, combined and used to fulfil the requirements.

Recommendations for practical use should mention possible non-selectivity and aim at a product oriented use of the sorbent tubes. Manufacturers and users should refer to existing standards. Instructions for correct use are of great importance.

Possible assessment in the case of cooling lubricants

The first step could consist in determining the nitrite concentration of a material sample by means of a simple procedure. Nitrite being an indicator for nitrosamines, the latter are often detected by measuring the nitrite concentration - in most cases n-nitrosodiethanolamine. Under certain circumstances, the mineral oil concentration could also function as a key component.

Workplace analysis in waste industry

Workplaces in waste industry are characterized by frequently changing substances, varying exposure situations and the presence of chemical mixtures. Preliminary investigations should also take into account experience related to branch specific ground contamination. Key components are determined on the basis of selected sanitation projects.

The assessment is effected using simple direct-reading measuring techniques (e.g. sorbent tubes, FID, PID, light scattering photometers).

Special cases (e.g. dioxine) may require separate consideration.

Recommendations for assessing workplaces in the case of ethylene oxide sterilization

Under certain circumstances, ethylene oxide concentrations may remain largely below the limit values in force. These conditions depend on the technical requirements applying to the device, on the correct handling in accordance with the instructions, as well as on the technical know-how. If all predefined criteria are fulfilled, measurements may - in certain cases - become superfluous.

Determination of requirements

The definition of testable requirements for measuring procedures is of great importance if it is to be established that only suitable measuring techniques are used in the field. The user needs assistance in selecting the right measuring procedure for his very specific application. The manufacturer of measuring devices needs concrete requirement profiles, preferably in the form of harmonized international standards. Within the framework of the European Standardization Organisation (CEN), the determination of requirements is mainly a task of the Technical Committee 137 'Assessment of Workplace Exposure'.

Exchange of experiences

The exchange of experiences between measuring services of firms and supervisory authorities has a marked influence on the elaboration of harmonized, technically irreproachable solutions. Up to this point in time, this has been realized only on a national level. International harmonization is still to be intensified. In this context, expert institutes, as well as international standardization activities, play an important role, the latter representing the framework in which different approaches can be discussed and the best solution (which has turned out to be the most effective under varying conditions) can be chosen.

Outlook

The implementation of Directives and standards requires the elaboration of recommendations, concepts and pragmatic agreements to ensure that work safety regulations are identically applied and satisfy the same qualitative requirements all over Europe.

1349

AIR MEASUREMENT METHODOLOGY: PRACTICAL AND CHANGING NEEDS FROM THE WORKERS' POINT OF VIEW

M SAPIR* and K. VAN DAMME**

*Director, **Research Officer
European Trade Union Technical Bureau for Health and Safety
27 rue Léopold, 1000 Brussels, Belgium

The priority for workers is the protection of their health and safety. Many pollutants present at the workshop can be dangerous for their health and safety. Limit values exist for a limited number of pollutants. Despite the use of safety margins - which can considerably raise the protection - in establishing limit values, the latter will only guarantee the protection of workers against those specific risks taken into consideration. Moreover, this will only be the case in so far as the biological and clinical detecting methods used in the epidemiological studies considered are sensitive to detect those risks. As a consequence, for the above mentioned reasons and others, respecting limit values will not necessarily be sufficient to protect the workers' health. Hygienic strategies must be the starting point in analysing and adapting the workshop. Air sampling methods must be used to evaluate those measures which are taken. They are necessary whenever new dangerous products or installations are used or new working procedures are adopted, and at regular intervals with a worst-case approach. Employers must guarantee transparency and competence. Workers must be fully informed about potential risks and must have the right to indicate times, places and circumstances of sample taking.

Our society at the present time is socially characterised by a permanent, quite considerable degree of unemployment, accompanied by continuous attempts to reduce social security. Finding or not losing a job is a permanent concern for a lot of workers. Economically, the industrial landscape is changing at a much faster rate than it has ever done before. Changing job or training is becoming a more and more common part of a worker's career.

Because of these social and economic circumstances, the preservation of good health becomes a permanent concern for a worker who wants to anticipate the risk of social marginalisation. It is rather exceptional that compensation will fully cover the victim's economic loss. Workers do realise this.

New technologies have made it necessary to adapt the cleanliness and the safety of workplaces. This has contributed to the worker's awareness of the possibility to realise more hygienic conditions at the workplace. What can be done for the product, can be done for the protection of a worker's health too! At the same time, studies in the field of occupational health and the development of occupational epidemiology have directly or indirectly contributed to a growing insight into the possibility to avoid a lot of health risks.

All this has served as a solid basis for trade unions' actions to make working conditions less painful and less harmful. Nowadays, health damage is no longer considered an inevitability and health protection is considered a worker's right.

Clean air at work is a very important aspect of health protection. Measurement methodologies will have to teach us whether the air at the workplace is 'clean' or not! But what does 'clean air' mean?

Should we understand by that, that the air is so pure that no individual worker will ever experience any additional short or long term health damage that can be attributed to the air he will breathe at the workplace, whatever his workload might be? Unfortunately, in a lot of areas where people live and work, such clean air can hardly be found: industrial and non-industrial chimney emissions, car exhaust, fire smoke, pesticide spraying etc...are omnipresent and can lead to health damage of at least part of the population breathing this polluted air.

So, should we understand by clean air at the workplace that the air quality is the same inside and outside the workplace, when at the workplace no pollution is added to the general pollution? (This would imply that, although workers may become sick when breathing air during working hours, their risk of sickness would have been quite the same if they had not been employed).

We could have agreed on that definition of clean air in the workplace. But apparently, society does not insist that 'clean' limit values of exposure are adopted. These limit values are generally much higher than the limit values adopted by public health authorities for the same substances in the general living environment. And so the question arises upon what are those industrial limit values based, how those differences can be explained, and if they are acceptable!

Above all, one should always keep in mind that limit values for exposure to chemical substances at the workplace are accepted levels rather than acceptable levels.

The differences between the limits in and out of the workplace adopted by public health authorities are being justified to a certain extent by the fact that the exposure period is limited to adult people, eight hours a day, five days a week and during a 40-year career, whilst the general population is continuously exposed since conception. They are also being justified to a certain extent by the fact that the workers may benefit from a well organised medical surveillance. This justification is acceptable if the medical surveillance allows for the detection of unknown adverse health effects at an early and reversible stage, and if it is not used as a substitute for other means to learn about the harmfulness of a substance.

However, the controversy about the TLV values of the ACGIH in the recent past in scientific magazines has made clear that a lot of the TLV values upon which most of the existing national legal values are based, can in no way be considered as purely health based. Some of the adverse health effects are apparently not even considered as such by the authorities that set the rules. Many of those values could lead and do lead to adverse health effects among exposed workers.

Moreover, the relative risk for exposed workers of getting a disease is very often calculated by comparing the frequency of the disease in an exposed population to its frequency in a reference group among which are persons who have been exposed to the same substances or to other substances causing the same disease. A relative risk calculated in such a way will always be an underestimation of the real risk attributable to the occupational exposure. This is particularly true for the risk of occupational cancer!

In other words, the best possible measurement strategy for comparison with limit values will in no way guarantee the protection of the workers' health if the limit values are set too high. The relevance to health of the limit value being used as a reference should therefore always remain the primary concern. For instance when discussing the problems in assessing the mean exposure to benzene of a single worker who is employed in a petroleum refinery, experts in environmental monitoring could focus on considerations of the confidence intervals of their measurements when reducing the sampling period. Of course we have no objection to the assessment of the benzene concentration using only four air samples of 15 minutes, or only ten momentary air samples with a colorimetric method, if the result provided is as reliable and less time and money consuming than continuous sampling throughout the entire working day. However, when the aim is to evaluate whether the benzene concentration in the air exceeds 10ppm as a time-weighted average limit value, then all discussions on reduced sampling time become irrelevant, since the adverse health effects on the bone marrow and the cancer risk are still considerable at that level of exposure. Equally, if the method used has its lowest overall uncertainty at around the concentration level of 10ppm, then that method is unsatisfactory! In this case, it is preferable that the method should have its lowest overall uncertainty at a much lower and safer level!

Moreover, if the same evaluation of the benzene concentration in the air were carried out for an industrial application in which benzene could easily be substituted, it would imply the acceptance of an avoidable risk, whatever the concentration might be, because a no-effect level probably does not exist.

It is obvious that even the best possible measurement strategy can in no way serve as an alibi to reduce the attention paid to hygienic measures aiming at risk elimination whenever possible or at risk reduction wherever possible, even if limit values in force are respected.

Experts in the field should be aware of all this, and should never be reticent to point it out. They should also be aware that it is not correct to make oral or written statements that 'the air is clean', or that the results show that there is 'no risk for health' or that 'the working conditions are OK', when in fact the only statement to make is that for a particular substance a defined value has not been exceeded.

Measuring the concentration of toxic substances in the air is part of a hygienic strategy. It is, of course, not a part of hygienic measures, because it does not in itself improve working conditions. It can either constitute an incentive to improve them, or be a means to verify the efficiency of the hygienic measures taken. Of course they can only have that meaning in so far as the risks are known, and in so far as the substances have been identified. The importance of air measurements for risk identification itself is rather limited. Nevertheless, risk identification - usually done by other means - will always have to be the first step in a strategy for occupational hygiene!

The next step will quite often have to be a systematic inventory of the measures of primary prevention. A systematic check-list seems to us to be the best practice here. There is no need to carry out measurements of the air quality before deciding upon an improvement of working conditions whenever this can be done easily, or whenever the risk is so serious that any improvement is welcome. In such cases, the only reason to carry out measurements is in order to be able to calculate the degree of improvement later on.

Once the risk is identified and the evident hygienic improvements are made, the right time has come to control the concentration of the substance in the air. Under these conditions, trade unions are very much in favour of measurements, and are the first to be interested in the improvement of their quality.

There may be several motives in asking for the improvement of measurement methodologies. One motive could be to reduce their cost. Trade unions are not at all opposed to that concern, but it has to take into account other concerns. These concerns can be related to the problems of sample taking, sample analysis, or both.

The first concern is that the measurement is precise, which simply means that the concentration that is measured is the real concentration in the air. This requirement refers to the limitation of the overall uncertainty as well as to the limitation of systematic errors. (In our opinion it is always preferable to use a method with a bigger systematic error rather than a method with a bigger overall uncertainty, conditionally on the fact that limit values also refer to the first method.)

The second concern is that the measurement is representative, which could mean two things:

- that it accurately reflects the process that has to be evaluated, which always implies a worst-case approach;
- that it accurately reflects the exposure of the workers, allowing for an accurate estimation of the inhaled dose when relating exposure to work load.

The third concern is that the measurements are carried out under standardised circumstances, so that a comparison is possible.

The reasons why we would like the measurements to be that precise, representative and comparable, are related to the aims that we see for carrying them out:

- The results should be used in epidemiological research and other research in the field of occupational health. This research is necessary for instance to evaluate whether a limit value is offering sufficient protection or not. To fulfil that aim, the relationship between exposure data and possible adverse health effects has to be established. The higher the precision of exposure data, the better the quality of that research! An underestimation of the exposure could lead to an overestimation of the risk of an adverse health effect at a specific concentration. An overestimation of the exposure could lead to an underestimation of the risk at the same concentration.
- The most common reason for carrying out measurements is the evaluation of altered procedures or working conditions, and the evaluation of the efficiency of hygienic measures. Imagine that using another working procedure could reduce exposure to such an extent that the frequency of adverse health effects would be considerably decreased. Imagine also that such a procedure is tried out, and although we know that it will surely lead to a reduction of exposure of about 40%, a measurement shows a concentration that is even a little bit worse, due to a high overall uncertainty of 50% of the methodology used! How can an employer then still be convinced, or the workers be motivated to apply that new procedure?

This last example illustrates the importance of using precise methods, and also that the degree of precision of the measurement methodology has to be an important element in deciding upon a measurement strategy. This at least is true when we agree that the aim of measurements is to improve working conditions.

The precision of the methodology being a very complicated and technical matter, we are pleased that the Commission focuses a lot of attention on it. The workers and employees should be fully informed about the possible lack of precision of the methods used.

The representativity of the air samples is a less technical matter and should always be discussed with the workers and the employees or their representatives. Those who are directly concerned have the evident right to participate in or to be informed about the process of their health protection. This is a matter of civilisation: it is perverse to treat data of environmental monitoring as if it were a military secret. It is very often an illusion and as a consequence a waste of money to carry out representative air sampling without knowing exactly the usual working procedure, the usual production intensity, the usual work organisation, etc. Who other than the people directly concerned can give precise information on this matter?

Assessing the cleanliness of the air at work has been a subject in several EEC Directives. Consulting and informing the workers and/or their representatives during this process has been imposed by them. More has to be done to enforce their full application. A process of elaboration of European limit values has started. We insist on the establishment of European limit values that offer much more guarantee for the prevention of adverse health effects than most of the current limit values do. We also insist on more compulsory Directives, in order to extend the field of obligatory application of hygienic measures and the air measurements necessary to evaluate their effectiveness, and in order to enforce clear rules concerning measurement strategies.

MEASUREMENT METHODOLOGY: NEEDS FROM THE OCCUPATIONAL HYGIENIST'S POINT OF VIEW

TH. L. HAFKENSCHEID

Directorate-General of Labour
The Hague, The Netherlands

SUMMARY

The occupational hygienist may have various objectives for performing measurements of concentrations of chemical agents in workplace air. Each objective will set characteristic requirements of the methods to be used.
By reviewing objectives, principal requirements and current measurement practice, areas in which developments and improvements of measurement methods are needed may be identified. From a strategic viewpoint the development and improvement of methods for identification of sources, processes, operations and activities leading to, or contributing to worker exposure, and of methods for the evaluation of the resulting exposure, are of major importance. Furthermore, some needs based on possible future trends within the field of occupational hygiene are (tentatively) recognized.

1. INTRODUCTION

Occupational hygiene may, in part, be approximately described as "The science or art aimed at the recognition, evaluation and control of ... chemical ... factors ... that originate from the occupational environment and may impair the health of the worker and his or her posterity". Hence, the work of the occupational hygienist contains activities aimed at the recognition, evaluation and control of exposure to chemical agents.

Measurement of concentrations of these agents in the air at the workplace are often indispensable elements. Such measurements may be performed with various objectives, each of which will set its own demands for the measurement methods to be used.

With reference to the possible requirements for a number of possible objectives, the present situation, with respect to the current practice of measuring concentrations of chemical agents in workplace air is reviewed in order to identify those areas in which developments and/or improvements may be useful.

Furthermore, on the basis of some expected developments within the field of occupational hygiene, possible future needs are (tentatively) predicted.

2. OBJECTIVES OF AIR MEASUREMENTS

The primary objectives of the occupational hygienist who takes measurements of concentrations of chemical agents in the air at the workplace are:

- confirmation of the presence or absence of agents;
- identification of sources, processes, operations and/or activities that may lead to, or contribute to exposure;
- evaluation of the resulting exposures;

- checking that exposure remains within acceptable limits, and that control measures remain effective.

Furthermore, a primary objective may be the determination of conformity with regulatory requirements.

Other, more or less secondary objectives include measurements as part of performance studies (e.g. validation) of newly developed methods, or, in retrospect, the use of results of measurements as data in epidemiological studies.

Although above a clear distinction has been made between the various measurement objectives, such sharp boundaries will not always exist in practice. Depending upon the nature of the exposure problem encountered, measurements may be planned in such a way, that some of the above purposes are combined. However, some objectives will always be more or less 'incompatible' due to specific (regulatory) requirements for specific measurement objectives.

Furthermore, the subsequent enumeration of the first four purposes does not imply that these are all indispensable measurement steps within the process of recognition, evaluation and control of exposure. In practice, measurement strategies can be such that there is no need to perform measurements for each separate objective.

3. MEASUREMENT METHODS

General: Current Scope

Estimates of the number of chemicals that are used and/or produced range from 40,00–60,000. Of these, some 2000 comprise the top 99.9% in production volume, and some 700 the top 99%. Most lists of occupational exposure limit concentrations also comprise some 700 agents.

Currently, methods are available for measuring the concentrations in the workplace air of some 450–500 chemical agents.

In practice therefore, there is a basic need for the development of measurement methods for those agents of prior interest for which no method is currently available. Due to the dynamic character of the 'chemical market' there will always be a practical need for measurement methods for new agents.

Measurement Methods by Objective

Each measurement objective described above may be characterized by a number of demands which the 'ideal' measurement method should fulfil. These requirements may differ between objectives.

In the following paragraphs the principal requirements as well as the current practice of measuring are 'screened' for each of the primary measurement objectives in order to identify the needs for development, improvement or adjustment of measurement methods.

Confirmation measurements

Confirmation of the presence or absence of an agent in the workplace air may be an essential step in the process of assessment of exposure (1). Methods for confirmation measurements need not always supply highly accurate estimates of concentrations. Such methods are particularly required to be selective, able to detect low concentrations of agents, and preferably rapid (e.g. directly readable) and easily manageable.

In current practice confirmation measurements will generally be conducted by direct-reading techniques, using flow-through detectors based on, e.g. IR-spectrometry, ionization, light dispersion, or using colour-reagent tubes, of which many applications exist. Nevertheless, there are some areas identifiable in which these techniques may be improved. The main areas are:

- selectivity/specificity for agents under investigation, with particular reference to (types of) particles (e.g. asbestos fibres) and constituents of aerosols;
- universality/versatility of application with respect to the (types of) agents measurable;
- ease of handling, particularly through further size reduction of instruments.

Future developments and improvements of techniques based on Fourier-transform IR–spectrometry and mass spectrometry may, for example, be promising in this respect (2).

For specific agents, such as carcinogens without a no-adverse effect level, these methods should, in addition, be applicable to measure low concentrations (sub-ppm to ppb range).

Identification measurements

In order to identify sources, processes, operations, activities etc. that may lead to or contribute to exposure, information needs to be obtained about the spatial and temporal distributions of concentrations of agents in the workplace air (1). Ideally, for this purpose 'near real-time' multi-dimensional scans should be made within a specific workplace of the concentration of a specific agent in order to visualize exposure sources/operations etc. and in order to 'map' the possible exposure situations. To date there are developments into this field (3–5), which should be continued and enhanced.

Presently, identification is mostly done using the methods described above, with the drawback of a relatively low resolution with respect to spatial concentration distributions, or by 'static' sampling using separate sample analysis, with additional low temporal resolution.

However, the application of these techniques for multiple purposes offers the possibility for a more efficient use of instrumentation.

Evaluation measurements

Decisions on the need for measures to reduce exposures are generally based on the evaluation of the exposures likely to be hazardous to worker health. This evaluation is aimed at:

- obtaining quantitative information on workers' exposure in general;
- obtaining information on the extent to which exposure is determined by certain workplace factors.

Methods suiting the former aim need in principle give results that are:

- representative of exposure;
- selective;
- accurate and reproducible.

(However, in order to achieve optimal utilization of available resources, a first criterion for selecting a method suiting the former aim may be the expected level of exposure in relation to proper exposure-limit concentration values. If it is expected, that levels are well above or below a limit value, this may be confirmed using less accurate techniques (1).

The latter aim is suited by methods that are:

- able to quantitate sufficiently low concentrations of the agent(s) under consideration;
- able to give exposure concentrations averaged over specific periods during which specific operations/activities take place ('task-specific sampling' (1)); such periods may vary from minutes to several hours.

Lastly, evaluation methods need to be reliable in operation.

The first three demands make methods based on personal air sampling the prime candidates for this purpose, although even these methods may sometimes lack adequate representativity with respect to the constitution of the air inhaled by the worker. Ideally, satisfaction of the other demands would require the use of selective personal samplers capable of integrating concentrations over relatively short periods, possessing data-logging facilities for off-line evaluation. Such instruments would in practice be applicable to most of the other objectives of measurements. This may therefore be a direction in which technical developments should be encouraged.

In the absence of such samplers, many of the methods used involve separate sample analysis, necessitating the use of fresh samplers for each specific period to be measured, rendering their application relatively costly and tedious.

Furthermore, specific demands with respect to their accuracy, such as laid down in a proposed EN-standard (6), will restrict the applicability of these methods that were mostly designed and validated for the purpose of determining conformity with (predominantly 8-h) TWA limit values, and not for relatively short sampling periods.

On the other hand it may be argued that, since the accuracy of the measurement method is only a minor constituent of the total uncertainty of the measurement, which is predominated by the representativity of measurements with respect to exposure, it should be the subject of disproportionate attention. A hypothetical example may illustrate this.

For The Netherlands, Buringh and Lanting (8) have estimated the 'average' day-to-day variation in personal exposure to be approximated by a lognormal distribution with a geometric standard deviation of ≈ 2.7. The result of taking a sample randomly (i.e. without any prior knowledge of exposure levels to be expected) from this distribution would have a 95% confidence interval of 0.14 to 7 times its value, while a typical analytical 95% confidence interval would be ≈ 0.7 to 1.3 times the measurement value.

It is evident therefore, that the reduction of 'strategic' measurement errors should have an ample share of the available attention for improvements within the field of workplace measurements.

Finally, the reliability and manageability of measurement methods for evaluation remains a topic of discussion. For example, some methods are still rather impractical in their use due to:

- the application of glass samplers filled with (sometimes noxious or reactive) liquids (e.g. of the methods listed in ref. 7, sampling in approximately 8% is still based on the use of bubblers or impingers);
- the application of sampling media with a relatively low stability or integrity on storage or transport.

There is definitely a need for the development of methods which can replace those methods with the above characteristics.

But also the reliability of well established methods may leave some room for improvement. For methods based on pumped sampling, reliability may be greatly influenced by improper performance of sampling pumps, although this feature may be most prominent when sampling during prolonged periods. For diffusive sampling methods, uptake rates may vary during sampling due to the non-ideal behaviour of the samplers. For both types of method, the availability of reliable 'materials' (pumps, 'ideal' samplers) is and will be a desire of the occupational hygienist. In this respect, the activities of Working Group 2 of CEN Technical Committee 137 (e.g. (9)) may be of significant interest.

Measurements with longer term objectives

Longer term objectives of measurements are to check periodically whether exposure situations which are believed to be 'in control', i.e. in which exposures are below acceptable limits, still prevail (1). These measurements are predominantly based on personal air sampling for prolonged periods (up to working shift length). Hence, the main requirements for measurement methods are the same as those for evaluation measurements with the exception of the need to be able to give average results over relatively short measurement periods. Instead, the reliability of these methods now becomes an important prerequisite, as (preferably) measurements should be performed unattended for longer periods. In this respect, the remarks made and needs established above are unabatedly, and even increasingly, valid here.

Measurements for determining conformity with regulatory requirements

Generally, measurements performed for the evaluation of exposures or measurements aimed at longer term objectives may be planned in such a way that no separate measurements will be needed to determine conformity with regulatory requirements. The results necessary for these determinations may be 'spins-off' of the other measurements performed. However, proper care should be taken to select measurement methods which fulfil contingent demands set by regulatory bodies, such as the use of (standard) reference methods, or the use of methods satisfying certain more general performance criteria (6).

Future Developments

Although the identification of possible future needs with respect to methods for measurement of concentrations of chemical agents in workplace air will have a somewhat speculative nature, there are likely developments that may lead to new demands. One of these is, of course, the introduction of new, and the steady lowering of health-based as well as regulatory limit values for worker exposure. This subject will be treated in more detail in a separate paper by Kennedy (10).

Another interesting development is the result of the consequent improvement of control measures for exposure, and the subsequent reduction thereof. For some substances worker exposure in the near future may not be predominated by occupation. Instead, environmental or domestic exposures may contribute to a considerable extent, current examples being asbestos fibres and benzene.

The protection of worker health may, as a result, require consideration of 'extra-occupational' exposures, and hence, measurements hereof. Whether or not this will be a task for the occupational hygienist may be a point of discussion. However, evaluation of extra-occupational exposures will especially require the availability of manageable, easy-to-handle, reliable and safe methods. For these reasons, methods based upon diffusive sampling may be potential candidates for such objectives.

4. CONCLUSIONS

Examination of the present situation with respect to measurement of concentrations of chemical agents in the air at the workplace has revealed that in several areas, developments and/or improvements of measurement methods are needed.

From a strategic viewpoint, the identification of sources, operations, activities etc. that may lead to or contribute to exposure, and the evaluation of the resulting exposures are measurement objectives of major interest to the occupational hygienist. These measurements will normally cover the major part of the uncertainty with respect to the 'where' and 'when' of exposure. For this reason, the development and improvement of methods as described above, i.e. of:

- techniques for near real-time scanning of workplaces in order to obtain information on temporal and spatial distributions of concentrations of agents;
- miniature personal samplers capable of giving average concentrations over varying periods with data-logging facilitics;

deserves ample attention.

Furthermore, improvements of currently applied measurement methods that may lead to a more effective and reliable application are desirable. Examples hereof include:

- enhancement of the selectivity (particularly for aerosols and constituents), universality/versatility of application and manageability of direct-reading methods;
- extension of the scope of 'off-line' personal sampling methods to relatively short sampling periods;
- replacement of methods based on liquid sampling media in glass samplers, or the use of sampling media with low stability or integrity;
- improvement of the reliability of personal sampling methods when used for prolonged periods.

Finally, the steady lowering of occupational limit values, and the subsequent control of exposure to relatively low levels may lead to future needs for developments of measurement methods. With respect to the latter development, measurement of extra-occupational exposure may become a potential feature of the health protection of workers.

REFERENCES

(1) EUROPEAN COMMITTEE FOR STANDARDIZATION, TECHNICAL COMMITTEE 137, WORKING GROUP 1 (1990). Assessment of exposure to chemical agents in air at the workplace for comparison with limit values and measurement strategy. Document CEN/TC 137/WG 1 N54.

(2) DAISEY, J.M. (1988). Real-time portable organic vapor sampling systems: status and needs. In: Advances in Air Sampling, pp. 225-241. ACGIH Edn. Lewis Publishers Inc., Chelsea MI.

(3) ROSEN, G. and LUNDSTROM, S. (1987). Concurrent video filming and measuring for visualization of exposure. Am. Ind. Hyg. Assoc. J. 48(8), 668-692.

(4) TODD, L. and LEITH, D. (1990). Remote sensing and computed tomography in industrial hygiene. Am. Ind. Hyg. Assoc. J. 51(4), 224-233.

(5) TER KUILE, W.M. and KNOLL, B. Specific and quantitative measurement and imaging of gases in industrial environments. Appl. Occup. Environ. Hyg., submitted.

(6) EUROPEAN COMMITTEE FOR STANDARDIZATION (1991). PrEN 482: General requirements for the performance of procedures for workplace measurements.

(7) NATIONAL INSTITUTE FOR OCCUPATIONAL SAFETY AND HEALTH (1984–1990). Manual of Analytical Methods, 3rd edn.

(8) BURINGH, E. and LANTING, R. (1991). Exposure variability in the workplace. Its implications for the assessment of compliance. Am. Ind. Hyg. Assoc. J. 52(1), 6-13.

(9) EUROPEAN COMMITTEE FOR STANDARDIZATION, TECHNICAL COMMITTEE 137, Working Group 2 (1991). Requirements and test methods for diffusive samplers for the determination of gases and vapours. Document CEN/TC 137/WG 2 N70.

(10) KENNEDY, E. (1991). Practical and changing needs in measurement methods. Needs arising from lowering of limit values. Paper presented at this symposium.

1351

NEEDS ARISING FROM LOWERING THE LIMIT VALUES

EUGENE R. KENNEDY, PH.D. and MARTIN T. ABELL

Methods Research Branch, Division of Physical Sciences and Engineering
Centers for Disease Control
National Institute for Occupational Safety and Health
4676 Columbia Parkway,
Cincinnati, Ohio, U.S.A. 45226

SUMMARY

Over the past years, the needs of the occupational safety and health community in the United States have been changing. After the passage of the Occupational Safety and Health Act of 1970, needs for sampling and analytical methods and standardization in procedures were emphasized. A major effort to address these needs was embodied by the joint National Institute for Occupational Safety and Health/Occupational Safety and Health Administration Standards Completion Program. Under this program, sampling and analytical methods for nearly 400 compounds were developed, as well as an experimental protocol for methods evaluation that is still in use today. Several years ago, the evaluation protocol was supplemented with specific evaluation experiments to address the performance characteristics of passive sampling devices. Recently, the trend toward lower exposure limit values has necessitated the evaluation of existing sampling and analytical methods at lower concentrations and the development of new methods. To address this need, the original sampling and analytical method evaluation protocol is being revised. In anticipation of further reduction of the exposure limit values, future needs are being projected.

1. INTRODUCTION

In the past years many changes have taken place in the occupational safety and health arena in the United States. The most notable event was the passage of the Williams-Steiger Occupational Safety and Health (OSH) Act of 1970 that established legal authority to promulgate occupational safety and health standards to "assure so far as possible every working man and woman in the nation safe and healthful working conditions and to preserve our human resources." This Act also created the Occupational Safety and Health Administration (OSHA) and the National Institute for Occupational Safety and Health (NIOSH). Under the provisions of the Act, NIOSH was instructed to develop and establish recommended occupational safety and health standards, and to perform all the functions assigned to the Secretary of Health, Education and Welfare (now the Secretary of Health and Human Services) under Sections 20 (Research and Related Activities) and 21 (Training and Employee Education).

As part of its duties under the Act, NIOSH was given the responsibility for performing research in support of occupational safety and health recommendations, including sampling and analytical methodology. In addressing this responsibility, one of the major functions of the scientists in the Methods Research Branch, Division of Physical Sciences and Engineering, NIOSH is the development and evaluation of new sampling and analytical methods for workplace contaminants in air. The research efforts of these scientists are guided in part by the analytical needs of Institute industrial hygienists and by the NIOSH Recommended Exposure Limits (RELs) (1). Other sources for direction in this area are the OSHA Permissible Exposure Limits (PELs) and the American Conference of

Governmental Industrial Hygienists (ACGIH) Threshold Limit Values (TLVs), Mine Safety and Health Act, as well as requests from other federal agencies and state and local agencies.

2. DISCUSSION

Past needs

After its formation in the OSH Act, NIOSH maintained laboratories in Cincinnati, Ohio and Salt Lake City, Utah. During this time of rapid development and evaluation of sampling and analytical methods, a means was needed to record sampling and analytical methods in a set format so that each laboratory could perform the same analysis in the same way. Based on this need, an informal methods manual was developed for the two laboratories. As other laboratories became interested in using the same methods as NIOSH and to insure consistency in the analyses performed at each laboratory, the first edition of the NIOSH Manual Of Analytical Methods (NMAM) (2) containing 39 methods covering 130 compounds was prepared and distributed to interested laboratories. Since then the NMAM has gone into its third edition containing 262 methods for 429 compounds (3). Currently, the NMAM is distributed to over 7,500 chemists and industrial hygiene professionals in 25 countries.

There was also a need to develop and evaluate methods quickly for about 400 compounds because many of the 1968 ACGIH TLVs were incorporated as OSHA PELs. In 1974, the NIOSH/OSHA Standards Completion Program (SCP) was begun to develop draft technical standards for these compounds. The methods evaluation portion of this work took over 5 years and resulted in all "S-" designated methods in the second edition of the NMAM (4).

One of the major accomplishments of the SCP, besides the development and evaluation of sampling and analytical methods for approximately 400 compounds, was the development of a method evaluation protocol to ensure the accuracy of reported exposure data. The basic premise of this protocol was that sampling and analytical methods must meet the NIOSH Accuracy Criterion, which requires that results are ± 25% of the "true" concentration 95% of the time. By experimentally evaluating method performance at 0.5, 1.0 and 2.0 times the OSHA PEL, the hypothesis that the method did not meet the accuracy criterion could easily be accepted or rejected. This protocol fulfilled a crucial need in the evaluation of sampling and analytical methodology by establishing performance specifications. NIOSH researchers have traditionally used the protocol and its associated performance specifications developed during the SCP for the evaluation of sampling and analytical methods (5,6). The protocol addressed the accuracy of the method in terms of its bias and precision. Performance specifications for capacity of the sampler for the analyte, the stability of the collected analyte in the sample, and the precision and accuracy of the determination of the analyte were developed for use during contract evaluation of methods included in the SCP (7,5), and are still in use by NIOSH researchers today with only minor modifications. Researchers at OSHA have developed a similar protocol for the evaluation of their sampling and analytical methods (8). Researchers associated with Technical Committee 137, Working Group 2 of the European Committee for Standardization (CEN) also are developing performance specifications for evaluation of methods for workplace air measurements (9).

Present needs

Sampling and analytical methods for compounds in the workplace serve as powerful tools for the industrial hygiene professional. These tools provide information that can be used to judge compliance with exposure limits, to assess exposure for health effects studies, and to verify engineering control effectiveness. The effectiveness of these tools to measure exposure at and below revised exposure limits is a topic of concern. In many ways the use of these tools is interrelated. Compliance monitoring data are often used in epidemiological studies to estimate exposure levels. Health effects information and engineering control effectiveness data often play key roles in the development of revised exposure limits. As exposure limits are revised (usually to lower levels), the verification of sampling and analytical method performance at the new limit becomes a key research area.

For health effects studies, workplace monitoring results provide important information about current and past exposures (10). Sometimes the exposures found in these studies may be below the exposure limits under which the sampling and analytical method was evaluated. In these instances, the accuracy of these measurements may be questioned, since these measurements fall outside the evaluation range of the method. Because of these situations, the evaluation of new, as well as revised, sampling and analytical methods should attempt to address accuracy at levels well below the current exposure limit.

For engineering control effectiveness, one aspect of the process of lowering limit values often involves the demonstration of the feasibility to control exposure below the new level with existing equipment. This demonstration relies on the ability of the sampling and analytical method to measure the analyte accurately in the area around the control device at levels below the current exposure limit. Again the accuracy of the method at levels below the current exposure limit comes into question.

For compliance monitoring, the sampling and analytical method is a key element in determining the compliance of an employer with established exposure limits. The accuracy of a given method has important legal aspects in terms of this type of monitoring. Therefore, it is vital that sampling and analytical methods developed for workplace exposure monitoring have defined accuracy, based on performance specifications used during the evaluation of the method. As uncertainty in the measurement of exposure is reduced by the use of precise and accurate methods, the magnitude of an apparent overexposure that can be considered as a true overexposure by the regulatory authority also is reduced.

In instances where an exposure limit for a compound has been revised downward or measurements have been made with a method at levels below the evaluation range, several questions arise. Does the method measure this compound accurately at these lower levels? Is the method applicable at the lower levels? Depending on what concentration levels are involved, the answers to both questions may be in doubt. The applicability of the method may be limited due to its ability to detect low levels. In instances where the method cannot accurately measure levels at the revised exposure limit, new analytical technology may have to be incorporated into the method or a new method may have to be developed and evaluated. The addition of new compounds to the exposure limit values also requires the development of new methods. In some instances where compounds have carcinogenic potential, NIOSH policy requires that exposure to potential occupational carcinogens to be controlled to the lowest feasible level.

The reduction in exposure limits has recently occurred in the United States with the 1989 OSHA Air Contaminants Final Rule (11). This legislation potentially has placed demands on the sensitivity and selectivity of existing sampling and analytical methods. The impact of these demands can only be determined by further evaluation of the methods. With the revision and further evaluation of existing methods, new sampling and analytical techniques can be incorporated into the methods to provide improved sensitivity, selectivity and sample stability.

When revising or modifying existing methods to measure lower concentrations, several aspects must be considered. Does the method meet performance specifications by additional evaluation at one or more lower concentration levels? Can the method be improved by changing only the sampler or the analysis or will an entirely different method have to be developed and evaluated? Often changes made to the sampler will require an evaluation of the sampler in terms of sample capacity, flow rate, recovery and precision, and bias.

Sometimes, modification of both the sampling and analytical aspects of the method is required to provide the necessary sensitivity and selectivity. For 1,3-butadiene, a field study conducted by NIOSH industrial hygienists required a method that could measure this compound at levels of 1/1000 of the OSHA PEL (1000 ppm). Modifications were required for both the sampling and analytical techniques. The analytical technique was changed from a packed gas chromatograph column to a capillary column. In the previous method for 1,3-butadiene (12), interferences were not a problem due to the fact that substantially higher concentration levels (500 ppm and above) needed to be determined. To address the interference problems at the lower levels (1-10 ppm), a backflushing procedure was incorporated into the analysis to remove some interferences in the analysis. The size of the charcoal beds in the sampler were increased from 100/50 mg to 400/200 mg to increase sampler capacity. The revised method was then evaluated over the range of 0.02 to 9 ppm (13,14) and was found to meet the NIOSH Accuracy Criterion.

In some instances, a method must be developed to address a newly established limit value. An example of this was the development of a sampling and analytical method for chlorodifluoromethane. When the OSHA Final Rule on Air Contaminants was published in the Federal Register (11), a suggested method for the determination of chlorodifluoromethane was NIOSH Method 1020 for 1,1,2-Trichloro-1,2,2-trifluoroethane (15). This method used a 100/50 mg charcoal tube for sample collection and carbon disulfide for solvent desorption. Preliminary investigation in our laboratory found that chlorodifluoromethane was not efficiently collected on the 100/50 mg charcoal tube and was not efficiently desorbed from the charcoal with carbon disulfide. Additional studies evaluated the use of Method 1018, Dichlorodifluoromethane and 1,2-Dichlorotetrafluoroethane, (16) for the determination of this chlorodifluoromethane (17). This method used a 400/200 mg charcoal tube with a similar charcoal tube as a backup and methylene chloride as a desorption solvent. To take advantage of improvements in analytical resolution, a capillary gas chromatograph was used for the analysis. The evaluation of this method for chlorodifluoromethane over the range of 0.5 to 2 times the PEL (1000 ppm) included capacity studies, recovery studies, sample stability studies, and precision and bias studies. The results of this study showed that this method did meet the NIOSH Accuracy Criterion using an evaluation protocol similar to that used in the SCP.

Revision of the evaluation protocol for sampling and analytical methods

One key element in the development of accurate methods is an experimental evaluation protocol that addresses all the important aspects of the performance of the sampling and analytical method. The sampling and analytical method evaluation may require specialized experiments, such as the case of the passive or diffusive monitor. Scientists at both NIOSH (18) and the United Kingdom Health and Safety Executive (19,20) have addressed the special evaluation requirements of this type of sampler.

Researchers at NIOSH are reexamining and revising the SCP evaluation protocol for sampling and analytical methods. A summary of some of the key experiments proposed for the revised protocol is presented in Appendix A. One of the most notable changes in this protocol is the reduced level of precision allowed for a method to meet the NIOSH accuracy criterion (result of a method is within ± 25% of the true value with probability of 0.95). Previously, the precision requirement for a method to meet the NIOSH accuracy criterion was a pooled coefficient of variation (CV) (standard deviation divided by the mean) of 0.105 for three concentration levels (0.5, 1.0 and 2.0 times the PEL) for an unbiased method. After rederiving the original statistical equations, a minor error was noted and the precision requirement was revised to 0.095 (18 samples total; three concentration levels) for an unbiased method with allowance for a 5% non-random pump CV factor (21). Figure 1 shows the previous and the revised CV limits and bias required to meet the NIOSH accuracy criterion. The treatment of bias addresses both positive and negative biases as shown in Figure 1. Along with this pass/fail approach of method evaluation, a statistical approach to estimate actual method accuracy is being investigated (21). This approach not only provides information on whether a method passes or fails the accuracy criterion, but also by how much. With the accuracy estimate of a method, the end user can choose a method that best fulfills his sampling and analysis needs.

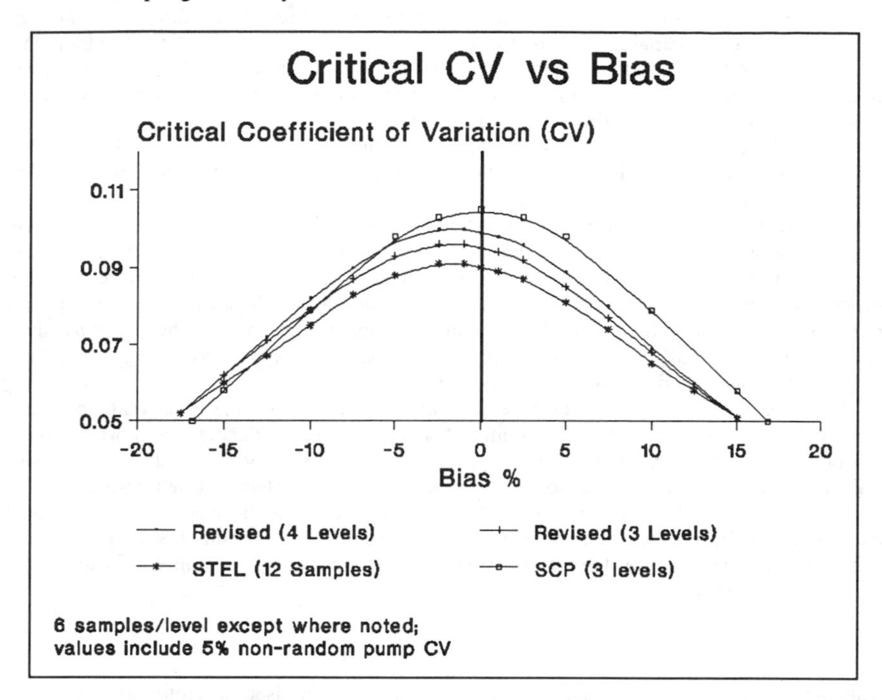

Figure 1 - Graph of critical coefficients of variation (CV) versus bias showing values used for Standards Completion Program and revised equations. A 5% non-random pump CV allowance has been incorporated into the calculations for the figure. CV must be less than the critical values for a sampling and analytical method to meet the accuracy criterion of the result being ± 25% of the true value with probability of 0.95.

The skewing of the new curves towards the negative bias side is due to the use of the estimated CV versus the true CV. To illustrate this point further, if two methods have the same standard deviation but one is positively biased and the other is negatively biased, then the CV of the positively biased method will be less than the negatively biased method. The shape of the new CV curve takes this fact into account and allows larger CVs for negatively-biased methods. The inclusion of a fourth evaluation level at 0.1 times the PEL is also proposed in this protocol so that further reduction in limit values may not require further evaluation of the method. Separate CV and bias limits have been developed for evaluations including four levels of concentration (0.1, 0.5, 1.0 and 2.0 times the PEL) and for short term exposure measurements (STEL) as well. Experiments have been expanded to include studies of the stability of desorbed samples overnight and the stability of collected samples for up to thirty days.

Future needs

As advances in sampling and analytical technology are made, revision of the sampling and analytical methods evaluation protocol to address new aspects of this technology will be needed. As always, there is the continual need for easy-to-use analytical methods to make the job of field sampling more convenient. Field readable methods and real-time monitoring are also desirable to allow industrial hygienists to make more timely recommendations for reducing worker exposure. An evaluation protocol is needed to address the performance of instruments that are capable of real-time or near real-time monitoring. New developments in sampling methodology also may require special evaluation criteria. With improvements in analytical methodology in terms of precision, accuracy requirements may be made more stringent to provide more reliable data on exposure.

Using the current methods evaluation protocol, the time required for the development and evaluation of a sampling and analytical method is from 1-2 years, depending on the complexity of the method. For an evaluation of a method at a lower exposure limit, 6 to 12 months are required, depending on the extent of the modifications in the method. Often, when exposure limit values are lowered, this type of lead time is not available. Therefore, automation of the methods development and evaluation process is needed to accelerate the process to provide accurate methods for revised exposure limit values. This facet also points out the need to have the methods development chemist involved in the exposure limit revision process as early as possible to identify research gaps and to maximize the lead time required for methods development and evaluation.

Short-term exposure limits require sampling and analytical methods that are more sensitive and measure the small amounts of analyte collected during the short sampling periods. To address this need, the application of thermal desorption will become more universally applied. Other methods to obtain the necessary level of sensitivity, such as immunochemical techniques, may be used, as well as state-of-the art classical analytical chemistry techniques, such as electrochemistry. Direct-reading instruments are particularly useful in this case because they can provide a measurement value at the end of any short term sampling period.

3. SYNOPSIS

The importance of accurate sampling and analytical methods to monitor at and below exposure limits is crucial to the reduction in worker exposure. This can only be accomplished with sampling and analytical methods that have been rigorously evaluated to ensure accurate results. As the field of analytical chemistry has grown, it has provided the industrial hygiene chemist with tools to solve complex problems and monitor workplace exposure with greater accuracy and at lower levels. Ion chromatography has been employed for the analysis of inorganic samples, providing better specificity and sensitivity than specific ion electrode analysis. As methods for volatile organic compounds have been changed from packed column gas chromatography to capillary column gas chromatography, the advantages of increased specificity and sensitivity have been incorporated into the methods. With the development of specific element detectors and the down sizing of the mass spectrometer, additional gains in sensitivity and specificity have been possible. Finally, thermal desorption has provided a means by which most of the sample can be analyzed for the utmost in method sensitivity. In the future, as in the past, the industrial hygiene chemist will have to take full advantage of the analytical methodology available to address the needs resulting from the lowering of the limit values.

ACKNOWLEDGEMENTS: We would like to acknowledge the technical assistance of John V. Crable, Thomas Fischbach and Dr. Stanley Shulman in the preparation of this manuscript.

REFERENCES

1. NIOSH Pocket Guide To Chemical Hazards. (1990). DHHS (NIOSH) Publ. No. 90-117, Cincinnati, Ohio.

2. CRABLE, J.V. and TAYLOR, D.G., Eds. (1974). NIOSH Manual Of Analytical Methods. 1st Ed. DHHS (NIOSH) Pub. No. 75-121, Cincinnati, Ohio.
3. ELLER, P.M., Ed. (1984). NIOSH Manual Of Analytical Methods. Third Ed. DHHS (NIOSH) Pub. No. 84-100, Cincinnati, Ohio.
4. TAYLOR, D.G. Ed. (1977). NIOSH Manual Of Analytical Methods. Vol. 1 through 7, 2nd Ed. DHHS (NIOSH) Pub. No. 77-157A, Cincinnati, Ohio.
5. BUSCH, K.A. and TAYLOR, D.G. (1981). Statistical Protocol for the NIOSH Validation Tests. In ACS Symposium Series 149, Choudhary, G., Ed. American Chemical Society, Washington, D.C., 503-517.
6. GUNDERSON, E.C., and ANDERSON, C.C. (1980). Development And Validation Of Methods For Sampling And Analysis Of Workplace Toxic Substances. DHHS (NIOSH) Pub. No. 80-133, Cincinnati, Ohio.
7. ANDERSON, C.C., GUNDERSON, E.C. and COULSON, D.M. (1981). Sampling and Analytical Methodology for Workplace Chemical Hazards. In ACS Symposium Series 149, Choudhary, G., Ed., American Chemical Society, Washington, D.C., 3-19.
8. SCHULTZ, G., BURRIGHT, D., CUMMINS, K., ELSKAMP, C.J. and HENDRICKS, W. (1987). An Outline for the Evaluation of Organic Sampling and Analytical Methods. Internal report, OSHA Analytical Laboratory, Salt Lake City, Utah.
9. EUROPEAN COMMITTEE FOR STANDARDIZATION, TECHNICAL COMMITTEE 137, WORKING GROUP 2. (June, 1991) General requirements for measuring procedures - Air quality - Workplace atmospheres. Draft proposal.
10. RAPPAPORT, S.M., and SMITH, T.J., EDS. (1990). Exposure Assessment For Epidemiology And Hazard Control. Lewis Publishers, Chelsea, Michigan.
11. FEDERAL REGISTER, Vol. 54, No. 12 (1989) Occupational Safety and Health Administration 29 CFR 1910 Air Contaminants Final Rule.
12. TAYLOR, D.G., Ed. (1977). Butadiene, Method S91. In NIOSH Manual Of Analytical Methods, Vol. 2, 2nd Ed., DHHS (NIOSH) Pub. No. 77-157B, Cincinnati, Ohio.
13. LUNSFORD, R.A., GAGNON, Y.T., and PALASSIS, J. (1987). 1,3-Butadiene, Method 1024. In NIOSH Manual of Analytical Methods, 3rd Ed., Eller, P., Ed., DHHS (NIOSH) Pub. No. 84-100, Cincinnati, Ohio.
14. LUNSFORD, R.A., GAGNON, Y.T., PALASSIS, J., FAJEN, J.M., ROBERTS, D.R., and ELLER, P.M. (1990). Determination of 1,3-Butadiene Down to Sub-part-per-million Levels in Air by Collection on Charcoal and High Resolution Gas Chromatography. Appl. Occup. Environ. Hyg. 5, 310-320.
15. PENDERGRASS. S. (1990). 1,1,2-Trichloro-1,2,2-trifluoroethane, Method 1020. In "NIOSH Manual of Analytical Methods," 3rd Ed., Eller, P., Ed., DHHS (NIOSH) Pub. No. 84-100, Cincinnati, Ohio.
16. GAGNON, Y.T. and WILLIAMS, K.J. (1987). Dichlorodifluoromethane and 1,2-Dichlorotetrafluoroethane, Method 1018. In NIOSH Manual of Analytical Methods, 3rd Ed., Eller, P., Ed., DHHS (NIOSH) Pub. No. 84-100, Cincinnati, Ohio.
17. LUCAS, M., and SEYMOUR, M.J. (1991). Internal report, NIOSH, Cincinnati, Ohio.
18. CASSINELLI, M.E., HULL, R.D., CRABLE, J.V. and TEASS, A.W. (1987). Protocol For the Evaluation of Passive Monitors. In Diffusive Sampling: An Alternative Approach to Workplace Air Monitoring, Berlin, A., Brown, R.H., and Saunders, K.J., Eds., Royal Society of Chemistry, Burlington House, London, 190-202.
19. FIRTH, J.G. (1987). HSE Protocol For Assessing Performance. In Diffusive Sampling: An Alternative Approach to Workplace Air Monitoring, Berlin, A., Brown, R.H., and Saunders, K.J., Eds., Royal Society of Chemistry, Burlington House, London, 177-184.
20. BROWN, R.H., HARVEY, R.P., PURNELL, C.J. and SAUNDERS, K.J. (1984). A diffusive sampler evaluation protocol. Am. Ind. Hyg. Assoc. J., 45, 67-75.
21. FISCHBACH, T., SHULMAN, S. and KENNEDY, E.R. (1991). Sampling and Analysis Method Development and Evaluation. Internal report, NIOSH, Cincinnati, Ohio.

Appendix A - Summary of key experiments proposed for the revised evaluation protocol for sampling and analytical methods.

1. **ANALYTICAL RECOVERY** Recovery of the analyte from the sampler should be complete and precise.

Experimental Design Fortify sets of 6 samplers with amounts of analyte equivalent to sampling concentrations of 0.1, 0.5, 1.0 and 2.0 times the exposure limit for a minimum of 4 h at 0.01 to 0.2 L/min for sorbent-based samplers and 1 to 4 L/min for filter-based samplers. If the analyte has a ceiling or short term exposure limit, the amount of analyte added should be adjusted for the shorter sampling time required for this type of exposure limit. If the sampler has a backup section, then a like number of separate backup sections should be fortified with amounts of analyte equivalent to 25% of the amount fortified on the front sections of the samplers. Analyze the samples and backup sections. Recap sample solutions and reanalyze with fresh standards after 1 day storage.

Interpretation of Results The recovery of the analyte from the sampler should be $\geq$ 75%. For a sorbent based sampler, the front section of the medium should be greater than 75% for levels equivalent to sampling 0.1, 0.5, 1.0 and 2.0 times the exposure limit. If recovery varies with analyte loading, results should be graphed as recovery vs loading, so that appropriate correction can be made to sample results. Recovery from the backup section of the sampler should be noted so that appropriate recovery corrections can be applied is there is breakthrough during sampling. The reanalysis of the sample solutions on the second day indicates whether immediate analyses after sample preparation is required. Results should agree within 5%. Often when processing many samples, it may be necessary to prepare the samples for analysis in a batch mode. In these instances, samples may not be analyzed for 24 h after preparation. If sample solutions are not stable prior to analysis, analysis must be scheduled as quickly after sample preparation as possible.

2. **SAMPLER CAPACITY/BREAKTHROUGH TIME** The capacity of the sampler for the analyte should be defined so that a maximum recommended sampling time and appropriate sampling rate can be specified.

Experimental Design Samples should be collected at the flow rate determined during the method development to provide for efficient collection of the analyte. If low humidity affects sampler capacity more than high humidity, generator humidity should be kept below 20% relative humidity. Otherwise, samples should be collected at 80% relative humidity. The generated concentration used for capacity determination should be at least 2 times the exposure limit. Duplicate samplers should be included to verify capacity. Samplers which use a primary and secondary bed of medium should have the secondary bed removed for this test. A means is required for the determination of the analyte in the effluent from the sampler. If the analyte is a particulate material and collected by filter, the capacity of the filter is defined by the pressure drop across the sampler or by the loading of the filter. If the particulate material has a measurable vapor pressure, a backup sampler should be included in the overall sampler and breakthrough of the filter/backup sampler should be determined.

Interpretation of Results When the concentration of the analyte in the effluent of the sampler exceeds 5% of the influent concentration, the capacity of the sampling medium has been reached. If a second backup sampler is used for this determination, capacity is exceeded when the amount of analyte collected on this sampler exceeds 5% of the mass collected on the front sampler. For filter-based samplers, the pressure drop should be less than 40 inches of water for less than 2 mg of total loading. With samplers that use reagents for collection of the analyte, the amount of the reagent in the sampler also will be a limiting factor in the capacity of the sampler, based on the stoichiometry of the reaction. Other factors, such as residence time in the sampler and the kinetics of the reaction of the analyte with the reagent, may affect the capacity of this type of sampler. The time at which any of these points is reached is the breakthrough time at a given sampling rate. To find the maximum recommended sampling time, the breakthrough time is multiplied by 0.667. The sampler capacity is defined as the volume of air sampled (sampling rate times breakthrough time) up to 5% breakthrough. If breakthrough is not detected after 10 h, a maximum recommended sampling time of 8 h should be used. If the capacity of the sampler is not sufficient to allow a reasonable sampling time, a lower flow rate should be used for this capacity study. Breakthrough times should always be reported along with the flow rate value at which the breakthrough was determined.

Appendix A - Summary of key experiments proposed for the revised evaluation protocol for sampling and analytical methods (cont.).

3. SAMPLE STABILITY Samples should be stable for at least 7 days after collection.

Experimental Design A concentration of 0.5 times the exposure limit should be sampled with 24 samplers for a minimum of 1/2 the recommended sampling time. The samplers should be divided into 2 groups of 6 and 4 groups of 3, with one group of 6 analyzed as soon after collection as possible. The other group of 6 should be analyzed after 7 days. The remaining groups of 3 should be analyzed after 10, 14, 21, and 30 days.

Interpretation of Results If the average analysis results of the group of 6 samplers analyzed on day 7 differs from the day 1 results by more than 10%, sample instability is a problem with the method. Either additional precautions may be required for storage or the method may have to be modified to address this problem. If a plot of recovery vs time (30 days) shows that recovery decreased by more than 10%, the samples are stable only for the amount of time where the recovery has not decreased by 10%. If samples need to be stored for longer periods, more restrictive storage conditions are required. If sample instability is still a problem after remedial actions have been attempted, the method does not meet the sample stability criterion and samples will require immediate analysis that will limit the utility of the method.

4. PRECISION AND ACCURACY (BIAS) Sampler results should be precise and accurate over a defined concentration range.

Experimental Design Concentrations corresponding to 0.1, 1.0, and 2.0 times the lowest exposure limit should be sampled with groups of 6 samplers for at least 1/2 the recommended sampling time. Results from samples collected for the sampler stability study (0.5 time the exposure limit) will also be used. If the analyte has a ceiling or short term exposure limit (STEL), twelve samples should be collected at the STEL. The concentrations of all levels should be verified by replicates of an independent method.

Interpretation of Results Before pooling the coefficients of variation (CV) (relative standard deviations) of the 4 sets of samplers, the homogeneity of the CVs should be checked using a test, such as Bartlett's test. If the CVs are not homogeneous, the sample set collected at 0.1 x exposure level should be removed and Bartlett's test recalculated. The pooled CV for the 3 groups of 6 samplers collected under this experiment and the group of 6 samplers collected under the Sampler Stability experiment should be less than or equal to 0.099 for a method known to be unbiased. For a method that has a known defined bias, Figure 1 shows the CV that must be achieved to meet the accuracy criterion that requires a single sample to be within ± 25% of the true value with probability of 0.95. If the CV exceeds 0.099, then the set of samples collected at 0.1 x exposure level should be excluded to determine if there is a point at which the method will meet the accuracy criterion. For the STEL measurements, the CV for the 12 samplers should be $\leq$ 0.09 for an unbiased method. For methods with a defined bias, Figure 1 shows the CV that must be achieved to meet the accuracy criterion. Results are incorporated in the estimation of bias. If the true values of the generated concentrations are known, then bias can be estimated by subtracting the mean of the method under study from the true concentration and dividing the result by the true concentration at each level. If the true values of the generated concentrations are only estimated by an unbiased independent method, bias is estimated by subtracting the mean of the method under study from the mean of the independent method and dividing the result by the independent method mean at each level. To account for the uncertainty in this calculation, the upper 95% confidence limit on the bias is used as the estimated bias value. This bias value should be used for the determination of the maximum CV allowed in Figure 1. The result yields 90% confidence that a passing method meets the accuracy criterion. Homogeneity of bias between levels should be tested using a one-way analysis of variance test. Estimated bias should be less than 10% to meet the accuracy criterion.

ALTERNATIVE APPROACHES FOR MEASUREMENTS

D. COTTICA

Laboratorio Igiene Industriale
Fondazione Clinica del Lavoro
Pavia, Italy

SUMMARY

The most frequent health requirement is to measure the dose of the pollutants absorbed by the workers at their workplace; industrial hygienists have many sampling and analytical methods to make these measurements but more and more often industrial hygienists need detailed data on human exposure versus time to identify the sources of contaminants, the circumstances and other information to reduce the exposure of the workers.

The use of direct and real-time monitoring instruments is a valuable aid for the continuous monitoring of the concentrations versus time.

More recently the combination of direct-reading instruments and video filming has been an alternative approach for measurements of exposure; not only to evaluate an average exposure but to identify the sources of the pollutants so as to take control measures, and as a training aid to show working practices; to let the workers know which stages of their job are critical and what they can do to reduce their exposure themselves.

1. INTRODUCTION

Sampling in the workplace is conducted to measure all the parameters that are needed to evaluate contaminants exposures. One of the most important health requirements is to measure the dose of the pollutants absorbed by the workers at their workplace and to determine compliance with reference standards (1).

To achieve this goal it is necessary to characterize the workplace air in the breathing zone of the individuals to evaluate their specific exposure.

Compact battery-operated personal sampling devices as well as diffusive samplers are useful to reach the aim without interfering with the worker's activity.

Industrial hygienists have many sampling and analytical methods to estimate the total amount of a pollutant during a work shift and to evaluate the average short or long term concentration (2).

The most used sampling devices for particulate matter, solids and liquids, are: filters (cellulose, glass fibre, PVC, silver, etc.); impactors; denuders; direct-reading instruments.

There are a lot of sampling devices for gases and vapours: plastic bags; pyrex tubes; absorbers (impingers, drechsel); adsorbers (activated charcoal; silica gel; molecular sieves; polymers; pre-impregnated substrates); diffusive samplers; detector tubes; direct-reading instruments (3).

The above mentioned methods are useful to satisfy the work physician's need to know the personal exposure at the workplace to evaluate health risks.

At the same time, industrial hygienists, plant engineers and workers must cooperate to reduce risks: they need detailed information on the factors which cause and influence the exposure: work methods, work organization, the use and the efficacy of exhaust ventilation and the effects of changes in work practice.

1352

To achieve this aim the industrial hygienist needs detailed data on workers' exposure versus time. They do not need to identify the contaminants and their specific concentration; for this there are a lot of conventional workplace monitoring techniques. Industrial hygienists need to correlate the fluctuations in exposure with the work activities in real time.

Industrial hygienists have to use alternative approaches for measurements. Have they the instruments and the techniques to reach the aims?

2. INSTRUMENTS

The instruments should be portable and/or enable sampling to be carried out in the breathing zone without interfering with the worker's activity. The response time must be as short as possible; continuous plotting of the concentration versus time is required.

For this kind of measurement the specificity is not the primary demand, the most important feature is that the instrument should show promptly the changes in the exposure.

Many of the direct and real-time, or near real-time, recently developed instruments for the monitoring of a workplace's contaminants seem to have such characteristics. The most used direct and real-time instruments for particulate matter are based on the light scattering principles (visible, infrared, laser) and particular scattering angles and response times in the range of seconds. They determine the total and the respirable dusts in fields such as: mining, cement processing, shipbuilding, mechanics industry, welding fumes, wood processing, mills, grain harvesting etc. (4) (5).

Numerous direct and real-time instruments for gases and vapours have been in use by industrial hygienists (4):

- Infrared spectrophotometers measure gases and vapours with characteristic absorption lines in the infrared region. They have seen the widest use in the paint industry, degreasing operations and in the measurement of anaesthetic gases in the operating theatre (6) (7).
- Photoionization (UV light) to measure organic compounds in the paints industry, printing industry, boat construction (8).
- Portable gas chromatograph - the flame ionization detector (GCFID) and gas chromatograph-mass spectrometer (GC-MS) are near real-time instruments; they are mainly used for ambient air monitoring organic vapours in the paints industry, printing industry, etc. Recent developments have worked on fast GC with a response time, on simple mixtures, of about ten seconds (9).
- The photoacoustic detection technique is a very recent detection principle: it is based on the fact that if a gas is irradiated with light of a wavelength coinciding with an absorption maxima of the gas, it will absorb the light and therefore increase in temperature. If the gas is in a chamber of constant volume the pressure will increase; if the light is modulated by a chopper there will be a pressure variation, and the sound wave can be detected by a condenser microphone. By using a selected wavelength of the light, a given gas can be selectively detected. The light source normally is in the IR band with an optical filter to restrict the wavelength. The response time is in the range of 30-100 sec for one to five gases and the technique presents very good specificity and sensitivity for a wide number of substances (10).

The use of direct and real-time monitoring instruments is a valuable aid for the continuous monitoring of the concentrations versus time, but very often it is impossible to correlate the cause of fluctuations of the concentrations with the work activities.

Video Filming and Measuring

In the past few years some researchers have developed methods for simultaneous monitoring of the personal exposure and video filming the worker and the workplace. The purpose of these methods is not to evaluate the personal exposure of the workers but to identify and reduce the sources and causes of exposure.

The video filming equipment consists of:

- a direct-reading instrument for measuring the pollutants of interest (for example, light scattering for dusts); infrared analyzer, photoacoustic infrared analyzers for gases and vapours;
- a video camera to film the workplace;
- a video mixer to amalgamate the signals from the measuring instruments and the camera;
- a video recorder and/or a TV monitor.

The combination of direct-reading instruments and video filming is an alternative approach for measurements of exposure: not only to evaluate an average exposure but to identify the sources of the pollutants. This allows control measures to be implemented and verified. It can be used as a training aid to show working practices; to let the workers know which stages of their job are critical and what they can do to reduce their exposure themselves.

Representative examples of the applications of these methods are the work of G.Rosén *et al* in Sweden and J. Unwin *et al* in the UK; scarce or no experiences are reported for the other countries of the EEC.

In their work, Rosén and Unwin demonstrate the applicability of video techniques in many working processes to correlate, on the video, the level of pollution with the critical stages of the job and/or the worker's activity.

By means of the video films they showed how exposure in: paint factories (batching, dilution); manufacture of furniture components (glue components); wood conversion plants (belt sander); GRP industry (styrene vapours); car servicing (gasoline vapours); degreasing operations (chlorinated hydrocarbons); solvent welding (dichloromethane); grinding grain feed; bakeries and mills (flour); operating theatre (anaesthetic gases and vapours) (7) (8) (11) (12) (13).

4. CONCLUSIONS

Traditional workplace monitoring techniques measure the average short or long term concentrations of the pollutants but they do not give any information on the cause of the fluctuations of the concentration; such information is necessary in order to establish control measures to reduce the exposure.

To obtain this information the industrial hygienist needs a strict correlation between the fluctuations of the concentration and work activities. The use of direct real-time instruments, such those mentioned above, coupled with equipment for video filming, seems to satisfy this need.

By the use of video techniques and direct real-time instruments the workers can immediately realize which stage of their job is critical and the best way to work and to minimize their exposure themselves. The industrial hygienist and the plant engineers can identify the sources, provide, realize and verify the control measures, and use the video technique to train new workers.

REFERENCES

(1) 1990-1991 Threshold Limit Values; ACGIH, Cincinnati, USA.

(2) NIOSH MANUAL OF ANALYTICAL METHODS, Third Edition; Vol. 1-2-3: Cincinnati (1984).

(3) POZZOLI, L. and MAUGERI U. (1986). Igiene Industriale. Campionamento Gas, Vapori, Polveri. Book; La Goliardica Pavese, pp 750 Pavia, Italy.

(4) LIOY, P.J. and LIOY, M.J.Y. Air Sampling Instruments; 6th Edition 1983. ACGIH, Cincinnati, USA.

(5) HARDCASTLE, S.G. and CAVAN, J. (1991). A portable, computer-controlled, multisize range, aerosol counting system for use in underground mining environments. Appl. Occup. Environ. Hyg. 6(3), 188-196.

(6) BURKHART, J.E. and STOBBE, T.J. (1990). Real-time measurement and control of waste anaesthetic gases during veterinary surgeries. Am. Ind. Hyg: Ass. J. (51) December 1990, 640-645.

(7) ROSÉN, G. (1986). Exposure to anaesthetic gas study using a direct-reading meter and video filming. International Congress on Industrial Hygiene, Rome, Italy, October 5-9, 1986.

(8) UNWIN, J. and WALSH, P.T. (1989). Visualisation of personal exposure using video techniques and fast response monitors. Visual Symposium, Lubeck 1989.

(9) MOURADIAN, R.F., LEVINE, S.P., SACKS, R.D. and SPENCE, M.W. (1990). Measurement of organic vapors at sub-TLV concentration using fast gas chromatography. Am. Ind. Hyg. Assoc. J. (51) 90-95.

(10) MULTI-GAS MONITOR TYPE 1302, Instruction Manual. Vol.1 Bruel & Kjaer, Denmark, May 1990.

(11) ROSÉN, G. and LUNDSTROM, S. (1987). Concurrent video filming and measuring for visualization of exposure. Am. Ind. Hyg. Assoc. J. 48 (8): 688-692.

(12) ROSÉN, G. (1988). Computers and video filming as aids in reducing exposure to air pollution. Scand. J. Work Environ. Health 14: suppl. 1, 37-39.

(13) ROSÉN, G. and Ing-Marie Andersson, (1989). Video filming and pollution measurement as a teaching aid in reducing exposure to airborne pollutants. Am. Occup. Hyg. Vol. 33, no.1, 137–144.

ALTERNATIVE APPROACHES FOR MONITORING BY MEANS OF SIMPLIFIED METHODS

D. BERGER* and H. KLEINE**

*Volkswagen AG, Wolfsburg, Germany
**Berufsgenossenschaftliches Institut für Arbeitssicherheit
Alte Heerstrasse 111, Sankt Augustin, Germany

1. LEGAL BASIS

Whenever hazardous substances are handled at a workplace or - due to the working procedure - are emitted into workplace air, suitable measures should be taken to protect employees against health hazards. Under certain circumstances, workplace atmospheres need to be monitored regularly. In the Federal Republic of Germany, these requirements are based on the ordinance on hazardous substances (Gefahrsstoffverordnung) in connection with the accident prevention regulations set up by the statutory accident prevention and insurance institutions in industry (gewerbliche Berufsgenossenschaften). In accordance with these regulations, the kind of protective measures to be taken as well as the nature of workplace monitoring to be provided, depend on the results of special investigations and examinations carried out by the employer.

Monitoring strategies and methods used in the Federal Republic of Germany basically comply with comparable existing European standards and draft standards. The modern state of standardization work was expounded in detail in the opening session of the conference. The present paper is meant to report on experience gathered in workplace monitoring in Germany. It is not only intended to give a brief survey of a limited field of the monitoring practice but to provide ideas for concrete monitoring activities in the European neighbour states.

2. MONITORING

If possible, investigations and examinations necessary within the framework of the monitoring of 'exposure-prone' workplaces are to be effected in a suitable way before or at least after the work has be taken up. Whenever the generation of atmospheric contaminants cannot be definitely excluded, these investigations are of particular interest because they aim at verifying whether the limit values in force are observed. In many cases, such investigations require exposure measurements. Exposure measurements in view of workplace monitoring are dealt with in a technical rule, determining the measuring strategy when controlling compliance with limit values (TRGS 402).

Since, in most cases, exposure measurements are complicated and lengthy, monitoring techniques ought to be preferred that permit the assessment of workplaces even without measuring pollution levels.

The following suitable instruments are referred to in pertinent regulations because they have proved effective in the field:

- procedure- and material-specific criteria, and
- use of tested tools, machines and installations.

To define criteria or approve tools, machines and installations, pollution levels have to remain continuously below the limit values, i.e. that the concentration of the atmospheric contaminants generated is low enough to exclude limit values to be exceeded. In the case of concrete workplace monitoring this means that the shift average must not exceed 25% of the

1353

limit value for random sample measurements or 100% of the limit value if the workplace atmosphere is monitored continuously.

Procedure- and material-specific criteria are described in the technical rule TRGS 420. They are generally elaborated within the framework of extensive field studies at comparable workplaces. So far, only one criterium concerning the decanting of solvents into 250 l casks has been established.

Other criteria for a series of different procedures could be imagined, e.g.

- use of certain copy machines under defined framework conditions, and
- chemical purifying installations.

We could also think of certain procedure-specific conditions which would allow the user to do without periodic measurements as, for example, in the case of closed installations (gas displacement devices).

As a matter of fact, material-specific criteria are particulary interesting in connection with hazardous substances, for example, the use of normally dust-producing materials in the form of:

- master batches,
- pastes (pigments),
- micro-encapsulated preparations (bonding agents).

Besides, measurements can be avoided or simplified by using type-tested and certified working means such as machines and installations. Type tests generally aim at determining whether a working means meets certain defined requirements which may concern, for example, the electrical or mechanical safety of the working means. In the case of technical working means emitting hazardous substances when correctly used, these requirements do also deal with the problem of pollutant emission. An example is captured air from working zones where carcinogenic substances are handled which may only be reintroduced into these working zones after elimination of the carcinogenic components by means of officially approved (supervisory authorities, statutory accident prevention and insurance institutions in industry) procedures and devices. Return air is considered sufficiently clean when the pollutant concentration is definitely below the exposure limit value. Regular modifications of limit values are to be considered. There are many working means where not only pollutant emission plays an important role but also the capture of hazardous substances. All these aspects have to be taken into account when determining requirements and procedures for type tests.

When using type-tested working means in accordance with the above mentioned requirements, permanent compliance with the limit values in force is however only guaranteed if it has been made sure that other sources of emission are of negligible importance, too.

Examples for type-tested devices can be found among devices for working asbestos/cement products. Among type-tested installations, those for decanting and dosing hydrazine should be included. In accordance with the technical rule TRGS 608 it must however first be checked whether the use of hydrazine cannot be avoided.

Cases where permanent compliance with the limit values is impossible are also frequently found in the field. Here, periodic control measurements according to the technical rule TRGS 402 are necessary. This is why we would also like to suggest some relatively simple measuring and test procedures that can be used for reliably monitoring workplace atmospheres. More complex workplace analyses may of course sometimes be necessary when using these simplified methods. We think that such measuring methods should meet the following requirements:

- immediate indication of the pollutant concentration, if possible,

- good profit/cost relation (moderately priced),
- precalibration or easy to calibrate,
- unambiguous result in terms of work safety,
- safe operation.

Some examples for the use of little complicated measuring methods:

- test tube,
- scattered light photometer,
- flame/photo ionization detector.

Test Tubes

The measurement of pollutant concentrations by use of test tubes normally consists in a reaction of the substance under consideration with a specially prepared carrier material which leads to the generation of a coloured reaction product. The carrier material is contained in a glass or plastic tube through which ambient air is sucked. The concentration level can be derived from the degree of discolouration of the indicating layer of the tube.

Test tubes are used to determine gas and vapour concentrations. The complete measuring device consists of the test tube and a corresponding sampling pump (passive sampling tubes (diffusive sampling tubes) where the contaminant is transported into the reaction zone by diffusion only, are less common).

There are two categories of test tubes: long term and short term sampling tubes. While short term sampling tubes are used for measurements of a few minutes duration only, long term sampling tubes have sampling times of several hours.

Whenever test tubes are used to measure workplace exposure, only such test tubes ought to be chosen that comply with the standardized requirements and that have been positively tested by an accredited testing laboratory. In addition, the user is obliged to ensure, within the framework of a workplace analysis, that the selected test tubes furnish correct results; this can be done, for example, by means of comparative measurements.

Particular attention is to be paid to climatic conditions or other interferences that might have modifying effects on the results. Consequently, detailed information has to be available on the environmental conditions and chemical components occurring at the measuring place. The conditions referred to in the instructions for use of the test tubes as well as known selectivities must be taken into account.

Experience related to the use of test tubes permits the following recommendations.

Test tubes can be used in cases where either pure substances are to be measured or mixtures of several substances for which the test tube does not show any selectivities.

If existing selectivities entail increased measuring results, it is sometimes possible to consider the measured value as indicating the correct pollutant concentration. As a matter of fact, this can only lead to an overestimation of the exposure situation which is acceptable in terms of work safety.

If comparative measurements by means of approved or recommended measuring techniques reveal decreased measuring results due to selectivities, it may be possible to take these selectivities into account by determining a correction factor (to be applied on the measuring result); it must, however, be made sure that the atmospheric composition has not changed after completion of the comparative measurement.

Whenever selectivities are known that lead to decreased measuring results, the use of test tubes is to be avoided. Further, users of test tubes should not forget that the different types of test tubes available on the market are of varying precision. The following types of test tubes turned out to be reliable: carbon monoxide, carbon dioxide, hydrogen sulphide, ozone, nitrogen monoxide, nitrogen dioxide, nitrous gases, ammonia, chlorine, phosphorus hydrogen, hydrocyanic acid and sulphur dioxide.

Scattered Light Photometer (Tyndallometer)

The measuring principle of this instrument relies on the fact that particles and aerosols exposed to IR light emitted under a certain angle generate a scattered light signal. The device is moderately priced, easy to handle and permits continuous measurement. Due to the above mentioned measuring principle it does not furnish any material-specific information. If scattered light photometers are used for control measurements, it is to be made sure that the correlation between the pollutant concentration and the outgoing signal of the measuring device has been determined within the framework of a workplace analysis.

A measuring facility for oil mists and oil vapours can be used for determining this correlation. It serves to determine both the oil mists by means of a tyndallometric analysis and separation on filters followed by an IR-analysis and the oil vapour content by use of downstream gas washing bottles filled with tetrachloroethylene.

If the analysis reveals a constant correlation, future control measurements can be carried out using the tyndallometer only.

Whenever the working conditions change, it is of course inevitable to have the conditions investigated again within the framework of another workplace analysis.

Flame Ionization Detector (FID) or Photo Ionization Detector (PID)

The measuring principle of these devices consists in measuring an ionic current produced either by burning organic compounds in a super clean hydrogen flame or by radiation from a high energetic light source.

If a single substance only is to be detected in workplace air, calibration of the measuring devices is possible. In the field, however, we often deal with chemical mixtures of changing concentration, the different components distinguishing themselves by varying response factors. Here, a calibration is not feasible.

In these cases, the determination of a key component represents a good approach. Principles for the selection of such key components are specified in the technical rule TRGS 404.

Carburettor fuels and other benzene-containing (more than 0.1% in weight) carbon hydrogen mixtures are excluded. On account of the contents of aromatic carbon hydrogens and n-hexane, limit values between 50ml/m^3 and 350 ml/m^3 were established for the different carbon hydrogen mixtures in accordance with the ACGIH procedure. Air quality control is effected by atmospheric sampling on activated carbon and gas chromatographic assessment within the framework of the workplace analysis, the total of carbon hydrogen components being referred to n-octane.

The systematic error for current mixtures comes up to 15%. If comparable results are obtained by use of direct-reading instruments (FID, PID), the latter can be used for periodic control measurements.

1354

AUTOMATED AND CONTINUOUS MONITORING OF ENVIRONMENTAL ANAESTHETIC GASES AND VAPOURS

V. COCHEO, C. BOARETTO and F. QUAGLIO

Fondazione Clinica del Lavoro
Centro di Ricerche Ambientali
6, via Tassoni
35125 PADOVA
ITALY

SUMMARY

The environmental anaesthetic gases and vapours concentration data, continuously acquired in surgical room by means of an automatic analyser, were submitted to graphic and statistical elaboration to allow accurate correlation with biological exposure indices. Correlation coefficients greater than 0.8000 were found between average environmental exposure and urinary concentrations of unmetabolised anaesthetics.

1. INTRODUCTION

The anaesthetic gas and vapour pollution level in surgical room can be very time variable. To establish the average exposure level during the whole surgical session, is very useful the emploing of continuous automated analyser. This kind of instrument can be also a power tool to show the pollution peaks and to localize leakage from anaesthesia equipment.

If the instrument is selective, if it ensures linear response for more analytes together in a wide range of concentration and if it is equipped with a memory for data storage, then the environmental concentration data can be *correctly* correlated with biological exposure indices after their elaboration by means of an appropriate soft-ware.

This approach is a general one.

2. EXPERIMENTAL

The environmental anaesthetic gases and vapours concentrations were continuously monitored by means of the photoacustical Brüel & Kjær *Multi-gas Monitor 1302*. To trust the results, the interference of chloroform, ethylether and *iso*-propanol, compounds absorbing in the same infrared region as the halogenated anaesthetics and likely found in surgical room, was removed by means of the procedure so-called *compensation for cross-interference*. The compensation was performed by using optical filters aving proper wavelenght window.

Analysed gases and vapours were N_2O and *ISOFLURANE* or *ENFLURANE* or *HALOTHANE*, both cross-compensated for above interfering compounds. Surgeon and anaesthetist were alternately monitored at the acquisition rate of one datum each 90 sec.

At the end of whole surgical session, urine samples were taken from the above operators. Those were analyzed by means of *head-space* method followed by capillary GC (*Pora-Plot-Q*, 27.5 m x 0.53 mm) and ECD.

3. DATA ELABORATION

The environmental data were stored by analytical instrument in *ASCII* format, as shown in fig. 1. The large amount of raw data is inconvenient to correlate the pollution level to the events occurred in surgical room. Furthermore, by this is very time consuming the evaluation of the average exposure level to be correlated to the biological exposure indices. To overcome these problems, the data were reelaborated by means of *STATGRAPHICS* soft-ware (STSC Inc. Rockville, USA) using a PC connected to the analyser *via* a RS232C interface, obtaining statistical values and graphic drawings, wich allow to display the more pollutant events (fig. 2-3).

Table 1: ASCII format of data stored in the photoacustical analyser *Multi-gas Monitor 1302*

Samples Measured From 1991-05-29 08:17

Samp. No.	Time hh:mm:ss	Gas A ppm	Gas B ppm	Gas C ppm	Gas D ppm	Gas E ppm	Water mg/m3
1	08:17:27	2.17E+00	-73E-03	. . .	. . .	1.51E+00	. . .
2	08:19:24	2.56E+00	-78E-03	. . .	. . .	1.49E+00	. . .
3	08:20:53	2.96E+00	-81E-03	. . .	. . .	1.51E+00	. . .
4	08:22:21	3.23E+00	-89E-03	. . .	. . .	1.52E+00	. . .
5	08:24:01	3.81E+00	-79E-03	. . .	. . .	1.53E+00	. . .
6	08:25:29	3.80E+00	-77E-03	. . .	. . .	1.52E+00	. . .
7	08:26:58	4.11E+00	-83E-03	. . .	. . .	2.17E+00	. . .
8	08:28:28	5.08E+00	-104E-03	. . .	. . .	1.98E+00	. . .
9	08:29:58	5.24E+00	-96E-03	. . .	. . .	2.20E+00	. . .
10	08:31:28	5.73E+00	-113E-03	. . .	. . .	3.05E+00	. . .
11	08:33:30	5.85E+00	-151E-03	. . .	. . .	4.01E+00	. . .
12	08:34:59	6.66E+00	-133E-03	. . .	. . .	4.14E+00	. . .
13	08:36:27	6.79E+00	-181E-03	. . .	. . .	4.94E+00	. . .
14	08:37:57	6.72E+00	-171E-03	. . .	. . .	5.68E+00	. . .
15	08:39:27	6.98E+00	-179E-03	. . .	. . .	6.28E+00	. . .
16	08:40:55	7.38E+00	-134E-03	. . .	. . .	7.85E+00	. . .
17	08:42:43	18.3E+00	-235E-03	. . .	. . .	7.81E+00	. . .
18	08:44:14	9.15E+00	-214E-03	. . .	. . .	8.11E+00	. . .
19	08:45:46	7.88E+00	-227E-03	. . .	. . .	9.02E+00	. . .
20	08:47:15	8.48E+00	-218E-03	. . .	. . .	8.76E+00	. . .
21	08:48:44	8.19E+00	-202E-03	. . .	. . .	9.03E+00	. . .
22	08:50:13	8.78E+00	-204E-03	. . .	. . .	9.70E+00	. . .
23	08:51:42	14.8E+00	-323E-03	. . .	. . .	11.3E+00	. . .
24	08:53:24	24.5E+00	-172E-03	. . .	. . .	11.1E+00	. . .
25	08:54:55	13.7E+00	-286E-03	. . .	. . .	11.7E+00	. . .
26	08:56:24	14.2E+00	-361E-03	. . .	. . .	13.2E+00	. . .
27	08:57:54	17.7E+00	-335E-03	. . .	. . .	350E+00	. . .
28	08:59:31	28.6E+00	-478E-03	. . .	. . .	49.1E+00	. . .
29	09:01:06	49.1E+00	-400E-03	. . .	. . .	27.3E+00	. . .
30	09:03:07	41.7E+00	-364E-03	. . .	. . .	34.5E+00	. . .
31	09:04:36	47.3E+00	-382E-03	. . .	. . .	34.0E+00	. . .
32	09:06:05	45.6E+00	-419E-03	. . .	. . .	32.8E+00	. . .
33	09:07:35	40.6E+00	-370E-03	. . .	. . .	28.5E+00	. . .
34	09:09:04	46.8E+00	-368E-03	. . .	. . .	27.1E+00	. . .
35	09:10:33	66.1E+00	-205E-03	. . .	. . .	24.6E+00	. . .

and more ...

Fig. 2: surgeon exposure to N_2O

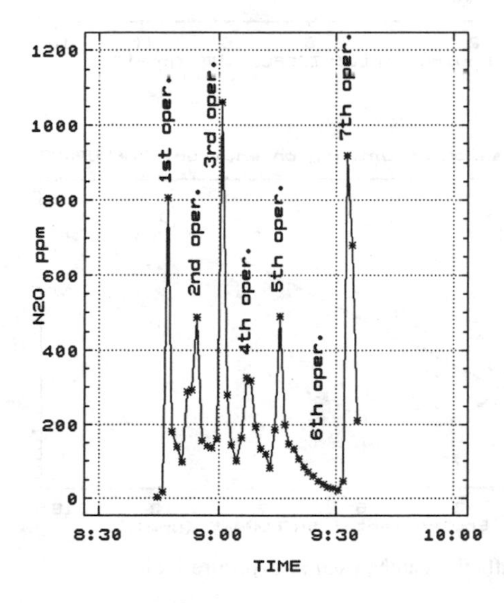

Fig. 3: surgeon exposure to HALOTHANE

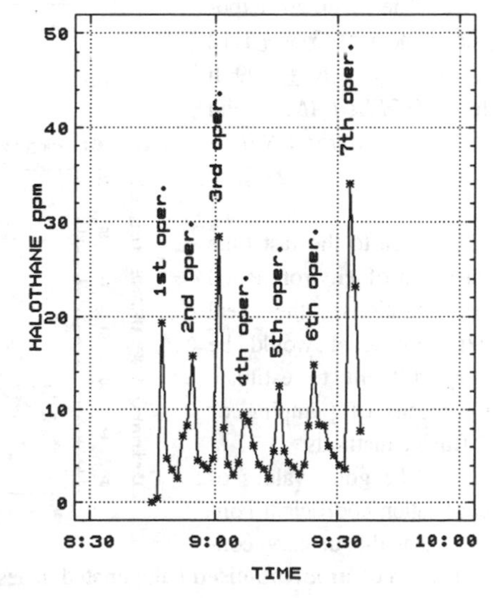

4. RESULTS AND DISCUSSION

The monitored surgical rooms were 27 for N_2O, 13 for *ISOFLURANE* and 12 for *ENFLURANE*. The correlation between environmental and biological data are shown in fig. 4-6 (the environmental values are the average of whole surgical session). The regression of urinary (C_U, μg/L) on environmental (C_E, ppm) anaesthetic concentrations resulted:

for **N_2O**:
$C_U = 0.435\ C_E + 4.629$
$r = 0.93796$, 77 samples

for ***ISOFLURANE***:
$C_U = 1.298\ C_E + 0.645$
$r = 0.91911$, 35 samples

for ***ENFLURANE***:
$C_U = 0.346\ C_E + 1.244$
$r = 0.80936$, 27 samples

The average exposure time was 3.36±1.12 h for N_2O, 3.64±1.09 h for *ISOFLURANE* and 2.83±1.18 h for *ENFLURANE*.

Due to the fast time variation of environmental concentrations, the real exposure level would be very difficult to estimate by means of pumping or diffusive methods.

The good values of correlation coefficient confirm that the urinary concentration of **unmetabolised** halogenated anaesthetics can be used as exposure indice.

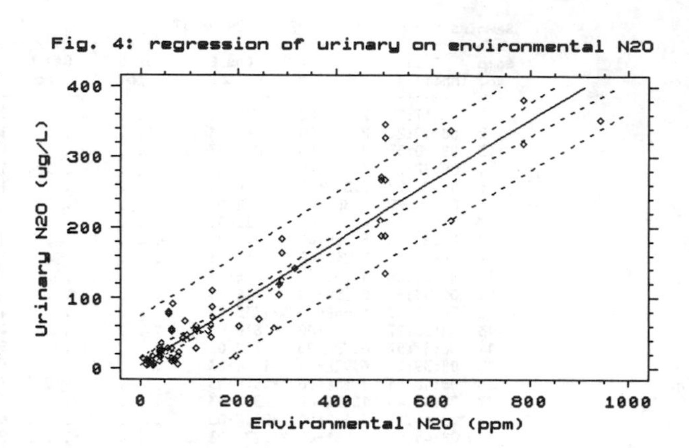

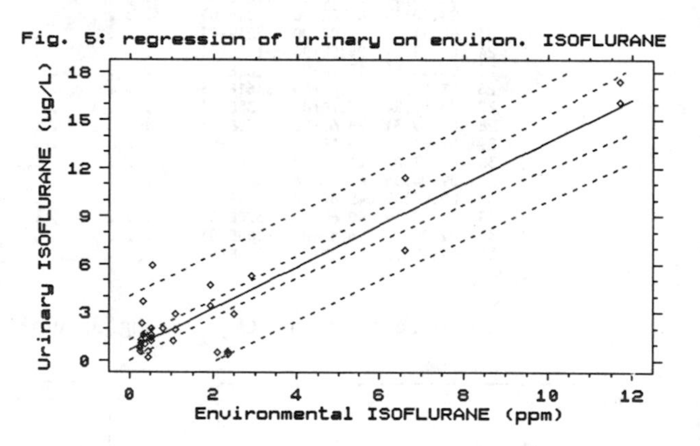

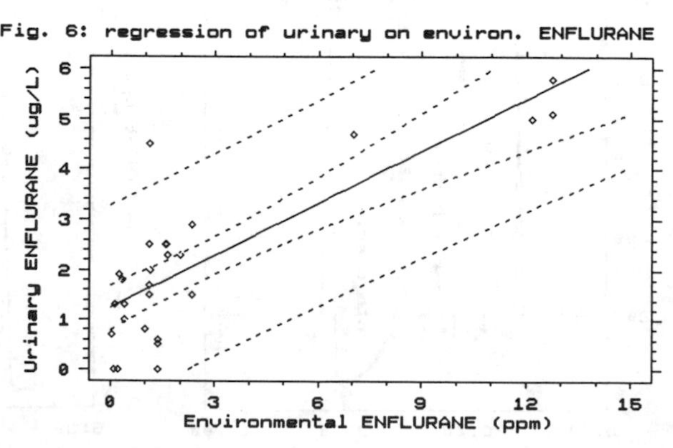

1355

EMISSION OF GASES DURING PROCESSING HEATSHRINKABLE SLEEVES ON TELEPHONE CABLES

J.J. HOOGESTEGER

Royal PTT Nederland N.V., Occupational Health Service
9726 CD Groningen, The Netherlands

SUMMARY

In a laboratory experiment we found that carbon monoxide, nitrogen oxides and formaldehyde were most important for investigating the exposure of workers processing heatshrinkable sleeves on telephone cables. In the field study we carried out next we observed a very low exposure with respect to the health limits. As we carried out this study in a worst-case situation we concluded that there are no health risks to be expected during the processing of heatshrinkable sleeves.

1. INTRODUCTION

The Dutch Telephone Company uses heatshrinkable sleeves for connecting telephone cables with copper conductors. These shrinkable sleeves are made of fibreglass strengthened polyethylene sheet with a polyamide glue layer on the inner side. The processing procedure takes place in a hole (1 x 1 metres and about 1 metre deep) dug in the ground over which a tent is placed. After making the connections the heatshrinkable sleeve is placed around the joint and closed using a propane torch. If the sleeve is to be removed, it is heated by the torch, cut open with a knife and pulled off.

Some workers, who process up to ten joints a day, complained about headache, irritation of the mucosal membranes and indicated that they were very thirsty. In one case a worker had to vomit. They were all convinced that their complaints were caused by their work with the heatshrinkable sleeves.

In a field study carried out with gas detection tubes we found relatively low concentrations of carbon monoxide (about 5ppm) and an indication of the presence of aldehydes.

In the literature we found that, when polyethylene is heated, emission of a large number of gases is to be expected (1) . Most important are the aldehydes. It is known that a (propane) torch produces carbon monoxide, nitrogen oxides and also aldehydes. The information that the manufacturer gave us about the emission of gases during the processing of sleeves was insufficient. Therefore we decided to measure the most important gases ourselves in an experimental situation in order to gather information for a field study, which had to be carried out next.

2. LABORATORY EXPERIMENT

Materials and Methods

The objective of the experiment was to quantify the emitted gases and to estimate their relative significance for exposure in field conditions. The gases measured were carbon monoxide, nitrogen oxides, phthalates and aldehydes.

An experimental situation was created in a fume cupboard (1.2 x 0.7 x 1.3 metres) that was not ventilated during the processing of the sleeves. Monitoring and air sample equipment was mounted right above the process. During the processing air was drawn from this fixed place in the cupboard. A qualified Telecom worker mounted or removed the sleeves. For this purpose the lower part of the fume cupboard was opened. The mounting and removing of the sleeves were measured separately, and both types of processing were measured three times. These processes took about 15 minutes each.

The same measurements were carried out twice while burning the propane torch for 15 minutes only.

Carbon monoxide was monitored with a Miran 1B infrared analyser, nitrogen oxides were monitored using a Monitor Labs. Inc. fluorometric nitrogen oxides analyser. Samples for phthalates were in each experiment gathered in duplicate on cellulose ester filters (0.8μ), using a personal air sampler. The analyses were carried out by gas chromatography. For sampling and analyses the instructions of NIOSH method 5020 were followed (2). Samples for aldehydes were also taken in duplicate in impingers containing a solution of 2.4-dinitrophenylhydrazin in iso-propanol, again using a personal air sampler. The analyses were carried out by liquid chromatography. Sampling and analyses were carried out according to the method of Hagenaars (3).

Results and Conclusion

The results of our measurements showed that there was an emission of carbon monoxide, nitrogen dioxide, nitrogen monoxide, formaldehyde and acetaldehyde in all the investigated situations. With respect to the Dutch Maximum Allowable Concentration (MAC) the concentration of nitrogen monoxide was very low, about 0.2ppm (MAC: 25ppm). The concentration of acetaldehyde was also very low, about $0.4mg/m^3$ (MAC: $180mg/m^3$). No phthalates could be detected and also no acrolein, propionaldehyde, buthyraldehyde, benzaldehyde, crotonic aldehyde (all $< 0.1mg/m^3$) and glutaraldehyde ($< 1mg/m^3$) was found. The average (15 minutes) concentrations for mounting, removing a sleeve and operating the torch only were respectively 2.6, 2.5 and 4ppm for nitrogen dioxide, 15, 13 and 5ppm for carbon monoxide and 5.1, 6.9 and 7.2 mg/m^3 for formaldehyde. It is likely that the carbon monoxide concentrations in this experimental situation were slightly heightened because of the possible lack of oxygen.

From the results it is obvious that the emission of the measured gases does not differ very much for mounting and removing the sleeves. By handling the torch alone the concentrations for nitrogen dioxide and formaldehyde were higher and for carbon monoxide lower. This indicates that the torch is an important source of the investigated gases.

It is concluded that nitrogen dioxide, carbon monoxide and formaldehyde are the most interesting gases to measure in field situations.

3. FIELD STUDY

Materials and Methods

We decided to measure the exposure of the worker to nitrogen oxides, carbon monoxide and formaldehyde in a worst-case situation. The conditions were: a normal hole covered with a one side closed tent (about 2 x 2 metres and 1.5 metres high) and poor ventilation. In normal practice the tent is only used when it is raining or cold. Since the ventilation depends on the breath of wind, the airflow during the investigations had to be low, below 1.5m/s. In Holland this airflow is normally exceeded for 90% of the time.

The measurement of nitrogen oxides and carbon monoxide were carried out with the same instruments as in the laboratory experiment. Since we wanted to measure formaldehyde and were not interested in the other aldehydes we used the more sensitive method of Meadows and Rusch (procedure B) for sampling and analyses (4). Air was drawn through impingers containing a solution of 1% sodium bisulfite in water. The analyses were carried out by spectrophotometry. For the nitrogen oxides and formaldehyde air was drawn through two tubes fixed to a helmet from the breathing zone of the worker. Carbon monoxide was measured stationary near by the breathing zone.

A qualified Telecom worker mounted and removed the sleeves in three different ways: using a soft blue flame, using a flame with a yellow top (according to the Telecom instructions) and paying attention to his position with respect to the emitting fumes. Mounting as well as removing a sleeve took about eight minutes. All experiments were carried out in duplicate.

Results and Conclusion

The results of these measurements showed that the exposure to all the investigated gases was very low with respect to the Dutch Maximum Allowable Concentration (MAC, t.w.a. eight hours) and to the Dutch Health Safety Limits (HSL).

The highest average (eight minutes) concentrations were about 0.3ppm nitrogen monoxide (MAC: 25ppm), 0.2ppm nitrogen dioxide (MAC: 2ppm, HSL: 0.5ppm t.w.a. 15 minutes), 3.6ppm carbon monoxide (MAC and HSL: 25ppm). No formaldehyde was detected so concentrations were below the detection limit which was 0.16mg/m^3 (MAC: 1.5mg/m^3, HSL: 1.5mg/m^3 t.w.a. 15 minutes).

There were only small differences between mounting and removing a sleeve. For mounting a sleeve we found that the blue flame gave a 40% higher exposure to nitrogen oxides than the flame with a yellow top. Paying attention to his position during the mounting of a sleeve the exposure dropped about 20% with respect to the exposure while using the instructed flame. During the removing of a sleeve these effects were not found, probably because this handling is less uniform than the mounting of sleeves.

The concentrations we found were of the same order of magnitude as those measured by Ulvarson and Ekholm, who investigated the same processes carried out on electric cable (5).

From the results we concluded that exposure to the investigated gases during processing heatshrinkable sleeves is below the Health Safety Limits and that a correct torch adjustment as well as a proper working position will result in an even lower exposure during mounting sleeves.

4. CONCLUSION

From literature study and our laboratory experiment we concluded that for health safety reasons the most important gases to which exposure during processing heatshrinkable sleeves is imaginable are: nitrogen dioxide, carbon monoxide and formaldehyde. Since we carried out our field study in a worst-case situation and found that the exposure to all three gases was far below the Health Safety Limits, we conclude that there are no health risks to be expected.

The results of our investigations however cannot explain the complaints of some workers. Because the number of complaining workers in an inquiry among our occupational doctors appeared to be small (five cases out of about 200 workers) there might be no relation with the exposure at all.

REFERENCES

(1) HOFF, A. et al. (1982). Degradation products of plastics. Polyethylene and styrene containing thermoplastics - Analytical and toxicological aspects. Scandinavian Journal of Work, Environment & Health, Supplement 2, 1-60.

(2) GROTE, A.A. (1985). Dibutyl phthalate and di(2-ethylhexyl) phthalate, method 5020. NIOSH Manual of Analytical Methods, third edition.

(3) HAGENAARS, A.F. (1986). Standardized sampling and measurement of aldehydes, method SD-86-04-1. The Dutch Directorate-General of Labour.

(4) MEADOWS, G.W. and RUSCH, G.M. (1983). The measuring and monitoring of formaldehyde in inhalation test atmospheres. American Industrial Hygiene Association Journal, 44(2), 71-77.

(5) ULFVARSON, U. and EKHOLM, U. (1982). Exposure measurement during installation of heat-shrinkable products. The National Swedish Board of Occupational Safety and Health, Research Department.

1356

APPLICATION OF FOURIER TRANSFORM INFRARED REMOTE SENSING TO AIR QUALITY MONITORING IN THE WORKPLACE

ROY W. BRANDON AND JOHN TRAUTWEIN, JR.
MDA SCIENTIFIC, INC.
405 BARCLAY BOULEVARD
LINCOLNSHIRE, ILLINOIS 60069 USA

SUMMARY: Fourier Transform Infrared Remote Sensing (FTIR-RS) can now be used to measure the quality of ambient air in workplaces. This paper briefly describes the technology used and presents the results of measurements made at two different sites, namely an electronics assembly area and an aluminum smelter.

INTRODUCTION: FTIR-RS is now commercially available as a field-mountable system capable of spectroscopically identifying and quantifying species in ambient atmosphere which absorb infrared radiation in the range of 450-4500 cm -1. The technique involves transmitting an infrared beam across the area to be measured and then spectroanalyzing the reflected light. The FTIR-RS is used in a unistatic configuration with a single telescope which acts as both the infrared transmitter and receiver. A corner cube retroreflector array is used to return the beam to the detector. Considerable installation flexibility is gained since one way path lengths of up to 650 meters are possible.

Single beam **transmission** spectra obtained during the measurement period are converted to **absorbance** spectra by comparison with a clean **reference** single beam spectrum, normally taken up-wind of the site being measured. The **absorbance** spectrum is then analyzed quantitatively for suspected contaminants by performing a least squares fit , over a selected spectral region, to **reference** spectra of pure gases for which the concentration path length products are precisely known. The method used is a Classical Least Squares algorithm with baseline correction, developed by Haaland and Easterling[1].

The number of scans per spectral analysis is dependent on the monitoring application. To allow the capture of **transient emissions**, using 5-10 scans (20-40 seconds) may be taken, and then averaged, for one measurement. Alternatively, **constant low level concentrations** can be sampled over longer periods of time, with a larger number of scans, to enhance the signal/noise ratio of the spectrum and gain greater sensitivity.

The FTIR-RS method has been used in many different applications in the United States, including typical chemical plants and Superfund sites. The two examples given in this paper are: a) a medium-sized electronics assembly area where a cleaning solvent and an aerosol propellant are present at low concentrations: and b) an aluminum smelter, which represents a potentially difficult situation because of the presence of particulates, heat, and strong magnetic fields.

RESULTS:

Figure 1 is a typical **absorbance** spectrum obtained using FTIR-RS over the spectral range of 450-4500 cm -1, showing strong residual water and CO2 peaks. Boxes A, B, C, and D indicate areas of spectral interest where most compounds have significant characteristic absorbances in the absence of strong water peaks.

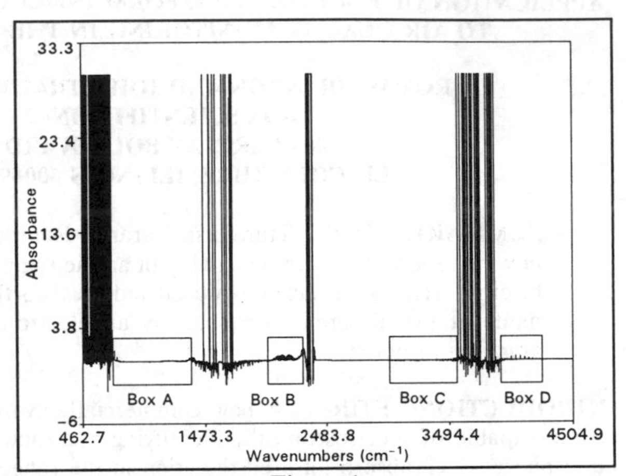

Figure 1

SITE A: ELECTRONICS ASSEMBLY AREA

The FTIR transmitter/receiver and retroreflector were set up on two balcony areas about 20 feet above, and on opposite sides of, the indoor manufacturing area. The open path distance was 50 meters. Figure 2 (a) shows the expanded region of Box A (Figure 1). It is a **workplace spectrum** recorded by taking 32 scans (2 minutes) at 1 cm-1 resolution. Visual comparison with the two **reference** spectra of Figure 2(b) and 2(c) confirms the presence of methanol and Freon 12 by matching significant peaks.

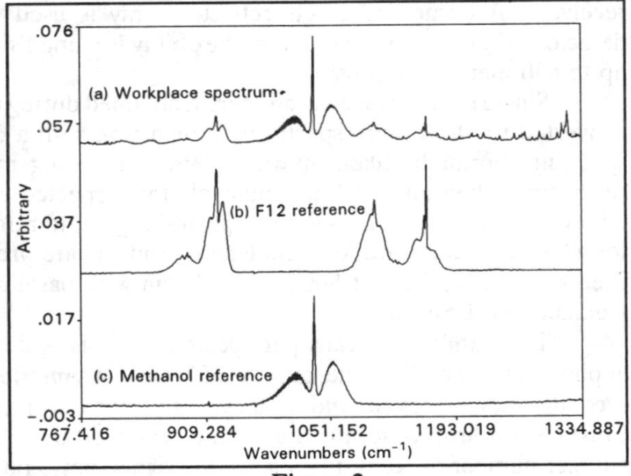

Figure 2

The MDA Scientific analysis package was used to quantify this measurement, and the concentrations of the two compounds were calculated as 0.135 (+/-0.002) ppm Freon 12 and 2.107 (+/-0.07) ppm methanol. These are the average concentrations present in the 50 meter open path. During the working day, these concentrations varied according to specific manufacturing activities in the area.

SITE B: ALUMINUM SMELTER

At this site the FTIRtransmitter/receiver and retroreflector were placed at several critical locations in the plant. Measurements were taken over 100 meters alongside the smelting pots as well as over 25 meter paths directly across and between individual smelting pots.

Figure 3 (a) shows just one area of interest in this application, the expanded region of Box D in Figure 1. This **workplace spectrum** was produced by taking 10 scans (40 seconds), across the top of one smelting pot, with an open path of 25 meters. The presence of HF is confirmed by comparison with the **reference spectrum** of Figure 3 (b), and quantitative analysis gives an average concentration of 5 ppm HF in the open path.

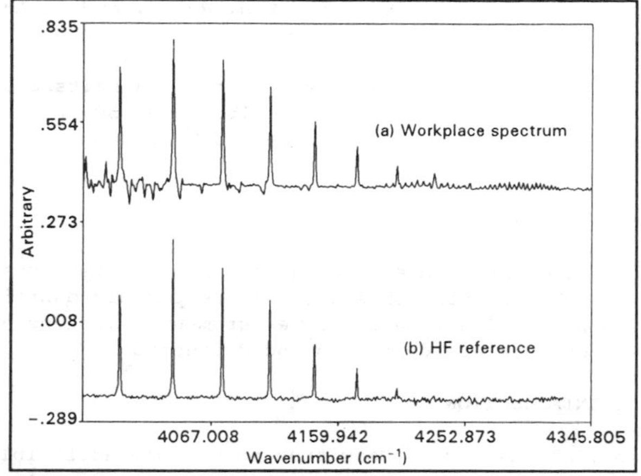

Figure 3

Detailed examination of workplace spectra obtained in different locations identified the presence of carbon tetrafluoride, carbon monoxide, carbonyl sulfide, ammonia, and silicon tetrafluoride, as well as the HF already described. Concentration variations could be correlated with plant operations, linking certain species with specific manufacturing activity.

The above examples illustrate the sensitivity and applicability of FTIR-RS to various manufacturing situations. In these examples the unit was used in a survey mode where individual spectra are recorded and stored for analysis at a later date. The unit can also be used in a continuous monitoring mode, in which the MDA analysis technique is applied immediately to each spectrum as it is recorded. The results are printed out in real time and action levels can be set.

References:

1. HALLAND, D. M. and EASTERLING, R. G. (1982). Application of New Least-Squares Methods for the Quantitative Infrared Analysis of Multicomponent Samples. Applied Spectroscopy 36, 665-673.

MEASUREMENT AND EVALUATION OF EXPOSURE TO DIESEL EXHAUST FUMES

E. Lehmann, K.H. Rentel and J. Auffarth

Bundesanstalt für Arbeitsschutz
Vogelpothsweg 50-52
D-4600 Dortmund 1

SUMMARY

Exposure to diesel exhaust fumes can be assessed by determining the total carbon content in the respirable dust in the particulate phase. The results of measurement campaigns at contaminated places of work are described and discussed.

1. INTRODUCTION

In 1987 the German Committee on Maximum Admissible Concentrations (MACs) classified the exhaust fumes from diesel engines as a carcinogenic substance in section III, Group A2 of the list of MACs[1]. This classification has subsequently been incorporated in the Order on dangerous substances[2]. The measures required for the protection of workers include both reducing emissions from diesel engines - for example, by developing less-polluting engines and suitable particle filters - and establishing a limit value for exposure at the place of work.

2. MEASUREMENT OF EMISSIONS FROM DIESEL ENGINES

Methods

According to present-day knowledge, the particulate phase of diesel exhaust fumes can be used to assess the carcinogenic potential[3]. The composition of the particulate phase is largely unkown. It mainly consists of graphite-like soot and organic compounds deposited on the soot particles. These include the polycyclic aromatic hydrocarbons and oxy- and nitro- derivates. The mean aerodynamic diameter of the particles is 0.1 to 0.2μm.

Because of the known properties of the particulate phase in diesel exhaust fumes, it was thought that determining the concentration of elemental carbon and the organic components in the respirable dust would be a simple method of measurement. Our laboratory experiments involved the combustion of particles collected on glass-fibre filters in muffle or annealing furnaces, plasma generators and coulometers. The influence of the material used on the results was very slight in the case of glass-fibre filters without binder. The values yielded by the individual methods were determined via the blank values in a number of test series with five or ten filters. Elemental carbon can be distinguished from the organic deposits by combustion of the filtrate in stages in the coulometer. The methods developed are summarized in Table 1. Details may be found in (4).

Method 1	Method 2
Collection on (binder-free) glass-fibre filter; respirable dust	
Incineration in furnace (500-550°C) or in plasma (200°C); Determining weight difference	Incineration in oxygen stream (650° C); Coulometric titration
Detection limit - 8-hour sampling period: 0.03-0.11 mg/m³	0.001-0.070 mg/m³

Table 1: Measurement methods for diesel exhaust fumes

The possibility of interference by other respirable organic dusts and carbonates must be taken into account with both these methods.

With a view to distinguishing between organic compounds bound to particles and elemental carbon, we investigated the possibility of determining the carbon content in stages in the coulometer, the main aim being to establish whether the organic components were completely incinerated at the copper oxide catalyst of the coulometer after desorption in the inert gas stream (He or N_2). The values determined with divided filters indicate that desorption in a stream of N_2 at 500°C and in a stream of He at 600°C will lead to comparable results.

Occupational exposure

Diesel-engined vehicles, such as fork-lift trucks, lorries and locomotives, are widely used in industry, and workers in production, storage and loading areas, repair and maintenance shops, tunnel construction, underground construction, customs points and transport are particularly exposed to exhaust fumes from diesel engines. Measurement campaigns at exposed places of work have yielded shift-averages of 0.003-1.3 mg/m³ total carbon in the respirable dust. On average, the measured values were higher than the known environmental absorption levels and were distributed as shown in Table 2.

Motor vehicle manufacture	0.010 - 0.194 mg/m³
Loading bays	0.003 - 0.136 mg/m³
Transport	0.008 - 0.230 mg/m³
Inland navigation	0.040 - 0.130 mg/m³
Internal transport	0.054 - 0.150 mg/m³
Tunnel construction	0.160 - 1.28 mg/m³
Non-coal mining	0.1 - 1.3 mg/m³

Table 2: Contamination by exhaust fumes from diesel engines (total C in respirable dust, method 2)

If the organically bound carbon is determined by thermal desorption of the deposited compounds and oxidation in stages in a coulometer, it is found to account for a considerable proportion of this diesel particle fraction (Table 3).

Respirable dust (mg/m^3)	C-total (mg/m^3)	C-elemental (mg/m^3)	C-organic (mg/m^3)	l/p *)
0.25	0.17	0.10	0.07	l
0.29	0.15	0.07	0.08	l
0.56	0.32	0.17	0.15	p
0.78	0.40	0.16	0.24	l

*) type of sampling: l = local p = personal

Table 3: Proportions of elemental and organic carbon in the respirable dust (bus station: method 2 with desorption stage)

3. CONCLUSIONS

The contamination of places of work exposed to diesel fumes can be measured by determining the total carbon in the respirable dust in the particulate phase. This can be recommended as a reference method for monitoring an MAC for diesel exhaust fumes based on such measurements.

REFERENCES

(1) DFG, Deutsche Forschungsgemeinschaft (1987), "Maximale Arbeitsplatzkonzentration und Biologische Arbeitsstofftoleranzwerte" 1987. VCH Verlagsgesellschaft Weinheim

(2) "Verordnung über gefährliche Stoffe (Gefahrstoffverordnung - GefStoffV)", 23 April 1990. Bundesgesetzblatt part 1, p.790.

(3) Pott, F. and Heinrich, U. (1988). "Neue Erkenntnisse über die krebserzeugende Wirkung von Dieselmotorabgas". Zeitschrift für die gesamte Hygiene. vol. 34, 686-689.

(4) Lehmann, E., Rentel. K.H., Allescher, W. and Hohmann, R. (1991). "Messung der beruflichen Exposition gegenüber Dieselabgas" Nr. GA 33 Schriftenreihe der Bundesanstalt für Arbeitsschutz. Dortmund. 2nd edition (printing)

1358

THE IMPACT OF HYGIENIC BEHAVIOUR ON THE RELATION BETWEEN ENVIRONMENTAL AND BIOLOGICAL MONITORING

M.E.G.L. LUMENS, P. ULENBELT, H.M.A. GéRON, R.F.M. HERBER

Research Group Work and Health,
Coronel Laboratory for Occupational and
Environmental Health, University of Amsterdam,
Meibergdreef 15,
1105 AZ Amsterdam,
The Netherlands

SUMMARY

The impact of hygienic behaviour on the relation between environmental (EM) and biological (BM) monitoring is studied in two lead and two chromium processing plants. Reason for conducting this study was the fact that the relation EM/BM is often not as high as expected. The study consisted of personal air sampling, biological monitoring and workplace observation and a questionnaire, in order to quantify hygienic behaviour. In these studies the impact of individual behaviour proved to be of major importance in explaining differences in BM.

1. INTRODUCTION

Working with toxic agents can be a threat to one's health. Usually environmental monitoring is used to assess the magnitude of this risk. In occupational field studies, however, the relation between environmental monitoring and biological monitoring often proves to be not as high as expected. Employees working at the same workplace, doing similar jobs prove to differ considerably in their uptake of toxic agents.

Although this may be due to differences in biological susceptibility and possible extra-occupational exposure, another possible cause of this discrepancy may be the fact that the worker is not a passive being, but can exert his own influence on the eventual uptake of chemicals from the workplace air, by behaving more or less hygienic.

Hygienic behaviour may be considered as all kinds of behaviour aimed at decreasing health risk, caused by exposure to toxic agents. This means that hygienic behaviour consists of both preventing uptake of chemicals into the body and preventing these chemicals to contaminate the workplace in the first place.

2. PURPOSE OF THE STUDY

The objective of this research is to study the impact of individual hygienic behaviour on the relation between environmental monitoring and biological monitoring.

3. MATERIAL AND METHODS

The studies were performed in lead and chromium processing industries: an electric accumulator factory (1), a secondary lead smelter (2) and two chromium plating industries. Each study consisted of the following parts:

Environmental monitoring

Personal air sampling (PAS) of all employees, willing to take part in the study, was performed during approximately 4 working days. Instruments used were duPont P-2500 pumps with custom-made PAS-6 devices. With this equipment dust, in accordance to the total dust definition of the German Staubforschungsinstitut, is sampled.

Chemical analysis of the filters was by A.A.S.

Biological monitoring

Lead: In the two lead factories all employees gave one sample of blood.

Chromium: In the chromium plating industries all worker gathered urine samples before and at the end of the working day, on which air samples were taken as well.

Chemical analysis of the biological samples was by A.A.S.

Observations

Workplace observations were performed to find out more about actual hygienic behaviour at the workplace. These observation consisted of describing the specific workplace workers occupied at the time of observation, the different tasks being performed, and whether they were wearing some sort of personal protection equipment. After that the frequency of specified actions that either positively or negatively influence the uptake of chemicals is counted. All workers were observed about 14 times, each observation lasted approximately 10 minutes.

Questionnaire

Because observation is a time-consuming activity and also while ideas, opinions and motivation cannot be observed directly, all workers were asked to complete a questionnaire. This questionnaire contained among others items concerning sanitary conditions, availability of protection devices and aspects of hygienic behaviour. The completed questionnaires were collected during the field study weeks. In total 57 out of 61 questionnaires were returned.

4. RESULTS

First the correlations between EM and BM were calculated for all four factories. In all cases the correlation between PAS and BM was low or even negative. In a next statistical analysis the impact of hygienic behaviour was recognized. Variables indicating hygienic behaviour were derived from the results of the observations and questionnaires. In a multiple regression analysis next to environmental monitoring measure, these variables were introduced in order to explain variance in biological monitoring measure, i.e. chromium in urine and lead in blood. The results of these calculations are described in Table 1. In order to facilitate a comparison, the results of the correlation between EM and BM are quadrated and multiplied so as to get % BM, explained by PAS.

Table 1. The results of fieldstudies in four factories

	N	r (EM/BM)	% BM explained by PAS	% BM explained by PAS+behaviour
accumulator factory	10	0.42	18	89
secondary leadsmelter	22	-0.45	21	53
chromium platery 1	20	0.67	46	94
chromium platery 2	9	-0.64	41	75

All correlations in Table 1 are significant ($p<0.05$). At all workplaces the correlation between PAS and BM was low or even negative. When however the impact of hygienic behaviour was recognized the variance in BM could be explained to a much larger extent. When comparing these results it is shown that in all cases the impact of air level at the workplace is of minor importance compared to the impact of hygienic behaviour.

The aspects of hygienic behaviour which need to be considered however differ considerably between the four factories. These differences may be due to several causes: Because of different routes of uptake of lead and chromium, different expressions of hygienic behaviour may increase or decrease the uptake of the chemicals. Between the factories general hygienic conditions varied considerably, for instance in the availability and type of personal protection devices and general workplace layout.

Due to these differences, different aspects of hygienic behaviour have their impact in the respective factories:

Electric accumulator factory

A higher frequency of putting the working gloves on and off, next to a higher occurrence of hand-mouth shunt increased the uptake of blood.

Secondary lead smelter

The lower the percentage of time the air stream helmet was worn correctly, i.e. with the face screen down, the lower the frequency of spitting during observation, and the higher the frequency of smoking at the workplace, the higher the Pb-B level was in this factory.

Chromium platery 1

A higher frequency of smoking, putting on and off gloves and having recent skin injuries, next to less washing hands before going to the toilet caused a higher Cr-U.

Chromium platery 2

A higher occurrence of skin injuries is related with high Cr-U. These injuries may be caused by dermal contact and subsequent absorption of chromium. Hygienic behaviour aimed at preventing dermal absorbance may cause lower Cr-U levels.

5. CONCLUSIONS

From the results of the studies in these four exemplary factories the following conclusions can be drawn:

* Hygienic behaviour is a major factor modifying the relation between environmental and biological monitoring measures

* Incorporation of hygienic behaviour in the EM/BM model improves the prediction of BM-levels

* However much stress has to be put on the necessity of reducing workplace concentrations, other variables, concerning individual differences in hygienic behaviour need to be considered as well in order to safeguard workers' health

6. REFERENCES

(1) ULENBELT, P. et al. (1990) Work hygienic behaviour as modifier of the lead air-lead blood relation. Int Arch Occup Environ Health 62:203-207

(2) ULENBELT, P. et al. (1991) A negative lead air lead blood relationship: the impact of air-stream helmets. Int Arch Occup Environ Health 63:89-95

EXPERIENCE IN THE USA WITH NOVEL PERSONAL MONITORING SYSTEMS

C.R. MANNING
Assay Technology, Inc.
Palo Alto, CA 94303 USA

SUMMARY

In response to lowered Exposure Limits for many chemicals and new regulatory requirements, several generations of Personal Exposure Monitors have evolved in the USA since 1970. A variety of monitors are in active use today, from historically important pump-and-tube samplers to diffusive badges incorporating reagent systems which provide immediate, on-site read-out. This presentation reviews product evolution as it responds to user needs and demonstrates that each type of monitor was developed to support certain applications which continue to the present time.

USA REGULATION DEFINES PELS AND STELS

Expanded efforts to monitor "personal exposure" to chemicals in the workplace followed the USA's Occupational Safety and Health Act (OSHA) of 1970 which defined Personal Exposure as the Time Weighted Average (TWA) level of chemical measured in an employee's "Breathing Zone". Since 1970, OSHA regulations have defined Permissible Exposure Limits (PELs) for more than 500 chemicals in USA workplaces and a Short Term Exposure Limits (STELs) for more than 100.

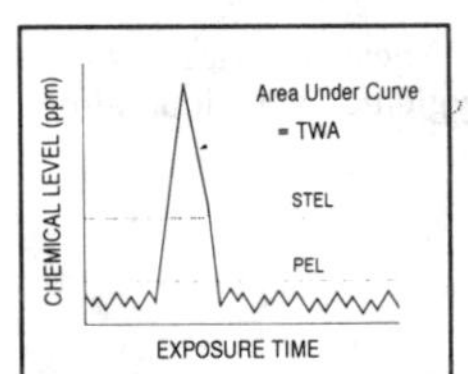

Fig 1: TWA Exposure

A PEL is a limit on the 8-hour TWA (day-long average) concentration of a chemical in each worker's breathing zone, while a STEL is a limit on the 15-minute, breathing zone TWA (measured during a 15-min peak exposure).

Once the existence of PELs and STELs defined Personal Exposure and set Limits, Personal Monitoring began to emerge as a standard methodology evolving in the USA in three waves or generations.

1ST GENERATION PERSONAL MONITORS

1st Generation Pump & Tube Monitors (Fig 2) were devised by adapting equipment available in the 1970's to meet the immediate needs of government industrial hygienists. In these devices, an air pump draws air at a metered rate through a trap (or "tube") filled with adsorbent. The tube's contents are then eluted and analyzed by conventional analytical methods, usually gas chromatography.

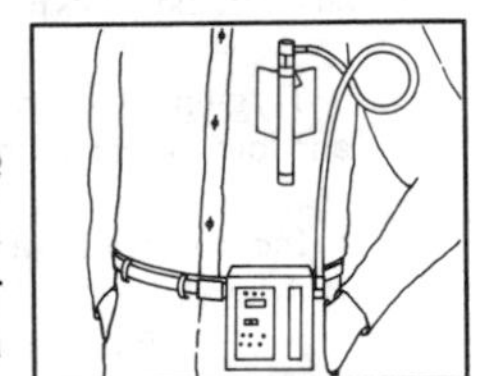
Fig 2: 1st Generation Monitor

2ND GENERATION PERSONAL MONITORS

2nd Generation Diffusive Monitors (Fig 3) consist of an adsorbent-loaded cartridge specially designed for Personal Monitoring incorporating a specially-designed "diffuser" opening capable of sampling solely via diffusion. Since molecular diffusion rates through a defined medium are a constant of nature, diffusional sampling has proved to be a highly-reliable method of sampling requiring no day-to-day calibration. Personal Monitoring "Badges", pioneered in the USA by the 3M Company and EI DuPont Company, have replaced the air pump with a more reliable, economical, and lighter method of sampling.

With 2nd Generation Personal Monitors, the contents of cartridges or badges were eluted and analyzed similarly to tubes so that existing analytical lab methods could be readily employed.

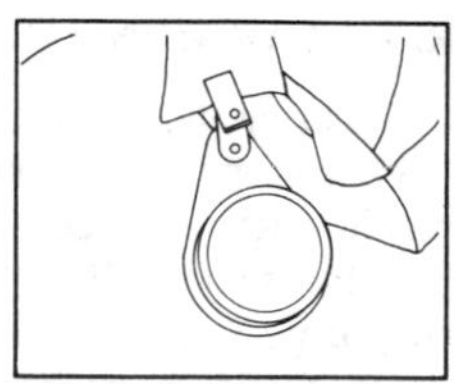
Fig 3: 2nd Generation Monitor

3RD GENERATION PERSONAL MONITORS

3rd Generation Monitors became available in the USA in the early 1980's (Advanced Chemical Sensors, Inc.) which **collected a sample via diffusion but which incorporated specific analytical reagents into the monitor**. One type of 3rd Generation Monitor had an appearance similar to a 2nd Generation Monitor (Fig 3) but could irreversibly bind sampled chemicals via incorporated analytical reagents, thus eliminating subsequent loss by "back diffusion". Incorporated reagents also facilitated subsequent analysis by making a stable chemical derivative amenable to gas chromatography or spectroscopy.

More advanced type of 3rd Generation Monitors (Fig 4) (Assay Technology, Inc.) made it possible to **obtain Personal Monitoring results immediately at the site of exposure**. In this case, the included analytical reagents formed a chemical derivative whose quantity (measured by color intensity or other means) was proportional to time-weighted-average (TWA) concentration of the analyte of interest. This system permitted, for the first time, immediate, quantitative, on-site measurement of PELs and STELs.

APPLICATIONS DRIVE PRODUCTS

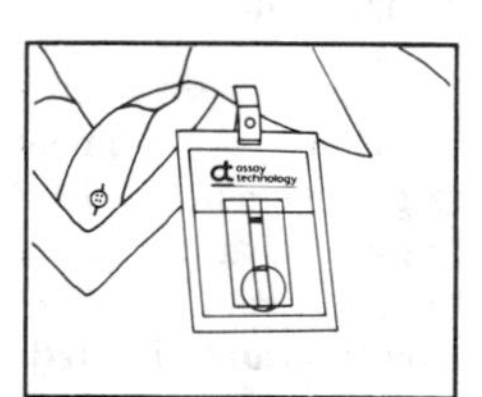
Fig 4: 3rd Generation Monitor

In the normal course of market evolution, customer needs drive technological developments. New products are designed to serve certain needs envisaged by a manufacturer but may fail to meet other needs. New "applications" developed by customers inform manufacturers of the new needs. Manufacturers respond by devising products for these "new applications", customers develop further applications, and so on as evolution proceeds. In this context, each "generation" of monitors is a response to evolving "applications" in industrial hygiene.

MONITORING "SURVEYS" CREATES DEMAND FOR 1ST GENERATION MONITORS

In its initial stages, Personal Monitoring was carried out by industrial hygienists in government and large corporations. With hundreds of chemicals to be monitored, there was a need to quickly develop and validate "applications" with simple, inexpensive equipment. Especially in government, there was a shortage of money for capital purchases but a relatively large quantity of human labor available.

Manufacturers responded by providing pump-and-tube technology certified by government industrial hygienists (Fig 2) which allowed an OSHA inspector or a corporate industrial hygienist to go into the field with portable equipment costing less than $1000 and monitor several hundred different chemicals.

Since the government or corporate hygienist's mission usually required their presence at a plant site, the technique-intensive and labor-intensive nature of the method was accepted. Since pump-and-tube tests were being used by government hygienists for enforcement, they took on a mystique of being "sanctioned by authority" leading conservative practitioners to continue to use them as one might use stone tablets for writing because Moses had used them in receiving the Ten Commandments.

REGULAR MONITORING CREATES DEMAND FOR 2ND GENERATION MONITORS

During the 1980's, studies by NIOSH and others showed that "a few samples" would not provide adequate data, and that **regular** monitoring would be required to demonstrate **statistical** compliance with PELs and STELs. As the quantity of testing increased, pragmatic **corporate hygienists sought cost-effective and convenient methods**.

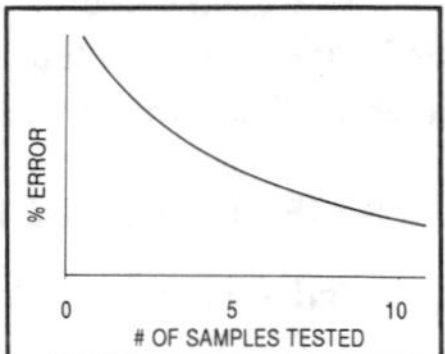

Fig 5: Accuracy vs # of Tests

Diffusional badges provided several advantages to corporate industrial hygienists. By **removing the uncertainty** and error associated with calibrating pumps, they enhanced the reliability of corporate-wide data bases while, at the same time, **eliminating the labor** associated with pumps. Diffusive samplers enabled a corporate hygienist to take hundreds of samples in one day (versus 10 or so possible with prior methods). Elimination of bulky pumps **made sampling less invasive** to workers, enhancing the likelihood that "typical" behavior would be exhibited during sampling leading to a sample which would be "representative" of real exposures.

These features combined to make to make 2nd Generation Diffusional Monitors (badges) the most pragmatic method for large corporations carrying out routine, regular Personal Monitoring programs.

EXPOSURE CONTROL PROGRAM CREATES NEED FOR 3RD GENERATION MONITORS

As the need for regular, cost-effective monitoring led to wide use of diffusive monitors in the 1980s, some limitations of collection by adsorbents were recognized. Certain chemicals (e.g., ethylene oxide, formaldehyde, isocyanates) were not well-retained or recovered from adsorbents at low levels. Further, there was growing interest in monitors which could provide on-site read-out of Exposure Levels creating the opportunity to **identify and correct overexposures immediately**.

Recognition of the dynamic (rather than static) nature of chemical exposure in the workplace led to an approach in which chemical exposure is viewed a parameter to be monitored and controlled much as a process quality attribute in a manufacturing plant. Such a model led investigators to begin to use Exposure Control Charts (Fig 6), long used in quality control, as a tool in Exposure Control Programs.

Medical institutions in the USA, already monitoring and controlling life signs of patients, followed this approach and were the first to practice Exposure Control Programs (ECPs). Making use of 3rd Generation Monitors capable of on-site read-out of ethylene oxide and formaldehyde, they were able to reduce typical exposures of ethylene oxide in hospitals from an estimated 10-15 ppm in 1983 to less than 0.3 ppm today as verified by more than 1 million data points generated in health care facilities by cost-effective, convenient, on-site Personal Monitors.

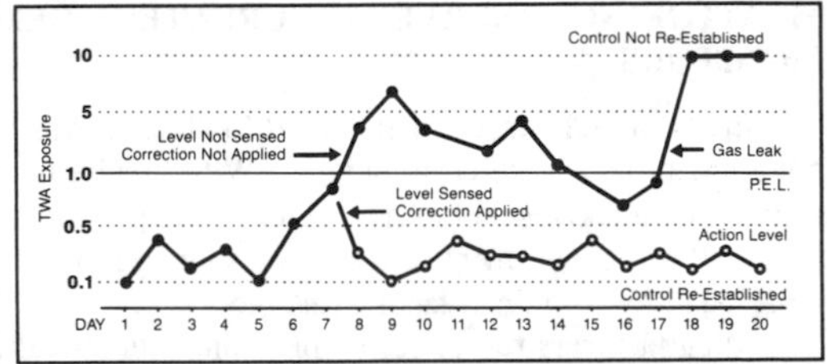

Fig 6: Exposure Control Program Monitored via Control Chart

EPILOGUE

1st Generation Monitors continue to be widely used by government hygienists due to their versatility and the insensitivity of governments to labor costs, while 2nd Generation Monitors have been adopted by corporate hygienists to obtain improved reliability and economy. For high tech and specialty firms focusing on controlling a few highly-toxic chemicals at low levels, 3rd Generation Monitors have become standard owing to the need for immediate, on-site results and their reliability in Exposure Control Programs. Fig 7 (below) summarizes practical economic characteristics of 1st, 2nd, and 3rd Generation Monitors. While results would vary, this analysis assumes "typical" industrial hygienist at cost of $100/hour (80 ECU/hour).

TYPE OF MONITOR	PUMP & TUBE	DIFFUSIVE MONITOR LAB ANALYZED	DIFFUSIVE MONITOR ON-SITE READ-OUT
GENERATION:	1ST	2ND	3RD
TESTS per MAN-DAY:	10	100	100
LABOR COST per TEST:	$ 80 = 64 ECU	$ 8 = 6 ECU	$ 8 = 6 ECU
MATL COST per TEST:	$ 3 = 2 ECU	$ 10 = 8 ECU	$ 20 = 16 ECU
LAB'TORY COST per TEST:	$ 30 = 24 ECU	$ 30 = 24 ECU	$ 00 = 00 ECU
TOTAL COST per Test:	$ 113 = 90 ECU	$ 48 = 38 ECU	$ 28 = 22 ECU
TIME TO RESULT:	7 days	7 days	20 minutes

Fig 7: Economic Characteristics of 1st, 2nd, and 3rd Generation Monitors

Two decades of product evolution have reduced the cost per test of Personal Monitoring more than four-fold while delivering a superior product. It is hoped that this evolution will continue so that future Personal Monitoring tests may be conducted at a cost comparable to complex blood tests which medical technologists have been able to deliver at a cost less than $10 (8 ECU) per test.

LEAD ASSESSMENT IN THE PRINTING AND PAINT INDUSTRIES

O. MAYAN, A. HENRIQUES, F. CAPELA, J. COELHO and J. CALHEIROS

Laboratory of Occupational Health, National Institute of Health
Largo 1° de Dezembro, 4100 Porto, Portugal

SUMMARY

In spite of the existing legislation, lead poisoning is a common public health problem. In practice, most of the people working with lead-based products are still unaware of the risks they are exposed to and the situation in Portugal is no exception. There is concern in the European Community and our laboratory has been deeply involved in the study of different professional groups working with lead.

MATERIALS AND METHODS

The present study, developed with two distinct risk groups from the printing and paint industries respectively, is based on both biological monitoring (lead in blood) and lead in air.

(i) Printing Industry (15 enterprises)

A case control study was conducted in the printing industry; the case group was composed of the typesetters (16 in number) who deal with lead printing types. The control group was formed by 42 other workers, lithographers and bookbinders whose activities do not directly involve lead handling.

(ii) Paint industry (5 enterprises)

This was a cross-sectional study involving 28 workers who develop paint production activities.

The results obtained were contrasted with the general population's values of lead in blood. For the three groups studied (printing industry workers, paint industry workers and general population) the subgroups of smokers and non-smokers were considered. The criteria adopted for the definition of these subgroups were the following: **smokers** - smoking a number equal to or higher than five cigarettes a day; **non-smokers** - those who had never smoked or had stopped doing it for more than one year.

The rates of lead in blood and in air were determined by atomic absorption spectrophotometry (AAS), using the Laboratory a PYE UNICAM SP–9 equipped with a Deuterium Background Correction System and Computer. Chelation - extraction by the APDC method - was employed for the biological samples and calibration solutions were prepared by additions of known amounts of lead to human blood. Air samples were analysed according to the NIOSH method.

RESULTS AND DISCUSSION

The results obtained are expressed in Tables 1 and 2.

Table 1. Lead in Air (microgr/m^3)

INDUSTRY	X	SD	LOW	HIGH
Printing	20.1	4.86	< 20	50.5
Paint				
- lead red oxide	660	245.52	400	1000
- other products	20	5.92	< 20	35

Table 2. Lead in Blood (microgr %)

GROUP	N	X	SD	LOW	HIGH	P
Typesetters	16	37.20	6.80	21.60	48.00	<0.001
Others	42	26.14	4.78	17.90	33.29	<0.001
Paint workers	28	29.56	9.78	12.50	53.00	<0.001
General population	51	14.49	5.15	6.00	28.00	-

N = number of subjects; X = mean; SD = standard deviation

Lead in air - The Portuguese legislation indicates that, in the case of lead, the admitted level of weighted average concentration is 150 microgr/m^3. In the printing industry we see that there are no lead concentrations in the air above the allowed level. In the paint industry, however, the production of lead red oxide is responsible for releasing large amounts of lead into the environment giving rise to risk exposures. The mean production of lead red oxide, at the different enterprises, is six hours a month. No risk situations are observed during the rest of the working time.

Lead in blood - After analysis of the results obtained we verified that the mean values of lead in the blood, in the groups under study, were significant. During our visit to different workplaces, it was often observed that among typesetters, handling lead printing types, food ingestion and smoking habits were frequent.

In the paint industry the same habits exist, aggravated by the handling of lead dusty material, freely expanded into the atmosphere. In the production of lead red oxide, due to the absence or inefficiency of prevention systems, there is a large liberation of lead dust producing hazardous working conditions.

Figure 1 shows the influence of smoking habits on the levels of lead observed in the blood.

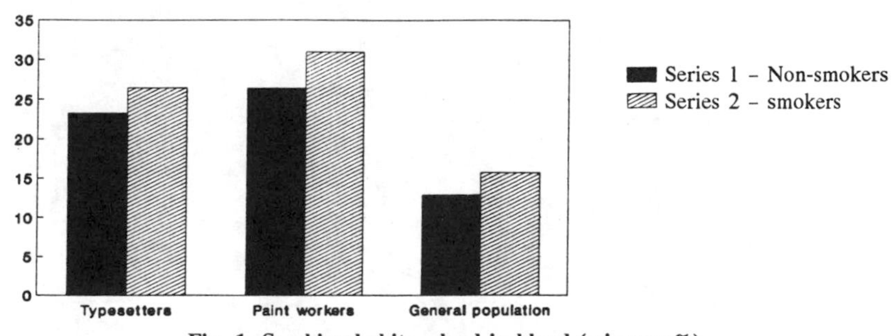

Fig. 1. Smoking habits - lead in blood (microgr %)

CONCLUSION

The results found in this study allow us to conclude that there is an urgent need for alerting both employers and employees against the risks they are exposed to and for promoting the right and safe utilization of lead-based materials. In the production of lead red oxide, measures to avoid the liberation of lead dust must be adopted. In some enterprises, it will be possible to automatize weighing, loading and unloading operations in closed circuits. Others will be provided with dust collection systems.

Finally, it is important to stress that environmental surveillance only, by means of evaluation and control of lead in air, is not enough. Food ingestion, smoking habits and lack of hygiene, so many times observed in these places of work, render easier the taking of lead into the worker's body. In control programmes relevance should be given to regular biological monitoring.

1361

BENZENE EXPOSURE AT SERVICE STATIONS, AN OCCUPATIONAL AND ENVIRONMENTAL PROBLEM

R. NORDLINDER and G. LJUNGKVIST
Department of Occupational Medicine,
University of Göteborg,
St Sigfridsgatan 85,
S-412 66 Göteborg, Sweden

SUMMARY

Measurements of the personal exposure to benzene during refuelling passenger cars at service stations showed that the exposure varied considerably. Wind speed, wind direction and type of car had the greatest influence on the level of exposure. Vapour recovery system reduces the exposure effectively for car drivers, service attendants and the emissions to the environment.

1. INTRODUCTION

Swedish motor petrol contains up to 5 % (v/v) benzene. The average benzene contents in petrol has increased during the last few years as a result of a legislative decrease of the lead contents. The National Chemicals Inspectorate in Sweden has decided that the petrol pumps at service stations should be marked with a scull and with cautions about the risk of benzene exposure. The emission of petrol vapour is an occupational risk for the service station attendants and also an environmental problem. The personal exposure to benzene during refuelling passenger cars at service stations have been measured.

2. METHOD

Samples were taken with personal sampling pumps in the refuellers breathing zone during the refuelling time (1-2 min). Hydrocarbons in the air was adsorbed on a porous polymer (Tenax TA) and thermally desorbed during the analysis with gas chromatography. Separation of benzene from other ambient hydrocarbons was done on a 60 m DB-1 column (i.d. 0.25 mm). Detection was accomplished using FID. The analytical system was calibrated with standard gaseous diluted hydrocarbon mixtures made at our laboratory. Temperature, wind speed, relative humidity and wind direction were registered continuously during sampling. More than 200 samples were taken at different service stations and at different seasons (winter and summer). The influence of wind direction in relation to the refueller was studied and also the type of car (van or sedan). The effect of a vapour recovery system upon the level of exposure was also studied.

3. RESULTS

The measurements show that there is a great variation in benzene exposure. Levels between 0.01 and 27.3 mg/m^3 has been measured. The mean exposure was 0.76 mg/m^3 (mean of geometric). The difference in exposure in the summer and the winter was small. The refuellers position in relation to wind direction had the greatest influence on exposure. Three cases were identified. Case A when the wind came from the side of the refueller, case B when the refueller had the wind coming from the front and case C when the refueller had the wind in his/her back. The results are shown in table 1.

Wind direction	Number of samples	Benzene mg/m^3	
Case A	112	0,51	(0,39-0,67)
Case B	17	0,49	(0,20-1,2)
Case C	46	2,4***	(1,5-3,7)

*** = $p<0,001$ (Case C versus case A or case B)

Table 1: Exposure to benzene at different wind directions, GM (95 % CI).

When the refueller has the wind coming from the back, the vapour-plume from the tank filling pipe is able to reach the breathing zone undisturbed.

The difference in exposure at different wind speeds is shown in table 2.

Wind speed m/s	Number of samples	Benzene mg/m^3	
<1,0	59	1,2*	(0,78-1,7)
>1,0	116	0,62	(0,45-0,86)

* = $p<0,05$

Table 2: Exposure to benzene at different wind speeds, GM (95 % CI).

Refuelling vans caused significantly higher exposure than refuelling sedan cars, table 3.

Type of car	Number of samples	Benzene mg/m^3	
Sedan	124	0,60	(0,45-0,81)
Van	51	1,3**	(0,85-2,1)

** = $p<0.01$

Table 3: Exposure to benzene when refuelling different types of cars, GM (95 % CI).

Vapour recovery systems decreases the exposure to benzene more than 80 % which is shown in table 4 below.

Type of stations	Number of samples	Benzene mg/m³	
With recovery	30	0,13	(0,08-0,21)
Without recovery	175	0,76***	(0,59-0,98)

*** = $p<0,001$

Table 4: Exposure to benzene at petrol stations with and without vapour recovery system, GM (95 % CI).

4. CONCLUSIONS

Benzene exposure during car refuelling at service stations varies very much due to several factors. Wind direction in relation to the refueller has the greatest influence on exposure. Vapour recovery systems considerably reduces both the exposure to the refueller, the people working at the service stations and the emission to the environment.

1362

ASBESTOS SEARCH IN BUILDING MATERIALS: A NEED TO LOWER THE ASBESTOS FIBRE EXPOSURE RISK

A. PASCUAL, A. FREIXA, A. CIRIA and M. J. BERENGUER
Instituto Nacional de Seguridad e Higiene en el Trabajo
Centro Nacional de Condiciones de Trabajo
Dulcet, 2-10 E-08034 Barcelona
SPAIN

SUMMARY

In order to identify asbestos presence in building materials, an specific methodology is developed. Analytical methods included: a) Light microscopy phase contrast, b) Microscopy, stereo and polarized light, with dispersion staining and c) X-ray diffraction. Using this methodologies during the last year asbestos fibres in fourteen building material have been analysed. The identification has been positive in seven materials that contained chrysotile. Five of them also contained amosite and one crocidolite and tremolite.

1. INTRODUCTION

Asbestos in air may be a serious health hazard for building occupants as it is well-known that air contamination may exist in indoor environments where asbestos-containing building materials have been used. Asbestos in buildings may be harmless if undisturbed but when maintenance, renovation or demolition operations are carried out, risk of both indoor and/or outdoor hazardous exposures comes up. So it is a major premise to avoid the utilisation of these materials in order to prevent the presence of asbestos fibres in the air and its of great importance to be able to identify its precence.

The aim of this work is to ascertain the presence or not of asbestos fibres in insulating materials, used in building industry, by the application of the analitycal methodology available in our laboratory that includes both optical and spectropic techniques.

In the Instituto Nacional de Seguridad e Higiene en el Trabajo we have succeded to identify asbestos fibres by means of a methodology that allow to caracterize their structure, their morphology and their optical properties.

2. METHODOLOGY AND INSTRUMENTAL CONDITIONS

Samples to be analysed in order to assess the presence of asbestos fibers arrive at our laboratory. The number of samples shows a trend towards continuous increase and last year fourteen samples were analysed.

The sampling method becomes critical owing to non-homogeneus composition of these materials. Several random samples of each material must be prepared and examined by stereoscopic microscope and light microscopy to assess the precence of fibres. To confirm the identification of any asbestos mineral, morphology and optical properties color, refractive indices, birefringence, extinction characteristics and dispersion staining characteristics are used. X-ray diffraction is then used for qualitative and semi-quantitative confirmation of chrysotile and qualitative confirmation of amosite, tremolite and crocidolite, the four major types of used asbestos.

The analytical methods we use to identify the most frequent asbestiform minerals in building materials have been: a) Stereoscopic Microscopy, b) Both phase contrast and polarized light microscopy and c) X-ray diffraction.

a) Stereoscopic microscopy:

Stereoscopic microscope, Nikon SMZ-2T, objective 1X ~ 6.3X, zoom ratio 6.3, eyepiece 10X.

b) Microscopy:

Microscope Nikon optiphot XFR, halogen lamp (100 W), photomicrographic system UFX - 35.

Light microscopy phase contrast:

Positive phase (dark) contrast, with green or blue filter, adjustable field iris, eyepiece 10X, and phase objective 40X (with adjustable piece for total magnification ca 500X), numerical aperture (NA) of 0.65, achromatic.

Polarized light microscopy, with dispersion staining objective:

Polarized light, with polarizer, analyzer, objetive lense 10X, ocular lense 10X, dispersion staining objective lens. Refractive index (RI) liquids for dispersion staining: 1) high- dispersion (HD) series, 1.550, 1.605, 1.620. 2) Refractive index liquids: 1.670, 1.680 and 1.700.

Asbestos Refractive Indices:

Chrysotile	n_{α}	1.537 - 1.553
	n_{γ}	1.545 - 1.557
Amosite	n_{α}	1.670 - 1.688
	n_{γ}	1.683 - 1.696
Crocidolite	n_{α}	1.680 - 1.700
	n_{γ}	1.684 - 1.708
Tremolite	n_{α}	1.599 - 1.620
	n_{γ}	1.622 - 1.639

c) X-ray diffraction:

X-ray diffractometers Philips PW1410 /10 and Siemens D-500, 40 kV, 20 mA equipped with broad focus copper target X-ray tube, proportionel detector, graphite monochromator and vertical goniometer.

The diffractometer is then scanned over the 2θ-range corresponding for a copper, $2\theta = 5^{\circ}$ to 40°. Presence of crystalline forms of asbestos is determined by the ocurrence of diffraction peak, as follows:

Asbestos	**d (most Intense)**	**d (second most Intense)**	**d (third most Intense)**
Chrysotile	7.31 Å	3.65 Å	4.57 Å
Amosite(Grunerite)	8.33 Å	2.77 Å	3.07 Å
Crocidolite(Riebeckite)	8.40 Å	3.11 Å	2.73 Å
Tremolite	8.38 Å	3.12 Å	2.71 Å

3. RESULTS AND DISCUSSION

Fourteen samples of building materials were analysed for asbestos fibres during the last year. Positive results were obtained from seven of these samples. Asbestos type identified are shown on TABLE 1. Photographies with images of stereoscopic microscopy, light microscopy with phase contrast and polarized light microscopy with dispersion staining that shows color change in across and along possition of fibres, are included in the poster.Diffacraction spectres of the most significant samples are also included.

TABLE 1.- Samples containing asbestos fibres

Asbestos Type	Sample No.						
	1	**2**	**3**	**4**	**5**	**6**	**7**
chrysotile	x	x	x	x	x	x	x
amosite	x	x	x		x		x
crocidolite	x						
tremolite	x						

In samples 3, 5 and 7, less than 1% of chrysotile has been detected in routine study by optical microscopy. In sample 1 tremolite was identified by a spectra registered on photographic film with camera Guinier with copper tube.

4. CONCLUSIONS

a)Owing to the very frequent presence of asbestos fibres in building materials, it is very important their analyse to ensure an indoor air free from asbestos fibres.

b)In spite of the need of more than one technique in this analysis, the methodology described has shown to be effective for asbestos identification.

c)Well experlenced staff in sample preparation as well as in analytical procedure and result interpretation is essential for good analythical performance.

d)This methodology is intended to be used for insulating materials but also for a number of the wide-ranging applications for which asbestos materials have been used.

5. REFERENCES

MICHAELS, L., CHISSICK, S.S. *Asbestos, Properties, Applications and Hazards.* Volume I, The Universities Press, Belfast, 1979.

NIOSH. *Analysis of Asbestos Fibres by Microscopy, Stereo and Polarized Light, with Dispersion Staining.* Method 9002.National Institute Occupational Safety and Health. 1989

NIOSH. *Analysis of Asbestos Fibres by Light Microscopy Phase Contrast.* Method 7400, National Institute Occupational Safety and Health. 1989.

J.C.P.D.S. *Selected Powder Diffraction Data for Minerals.* First Edition. Joint Committee and Powder Diffraction Standards. Philadelphia, 1974.

Acknowledgments: The authors wish to thank Dr. F. Plana and Dr. J. A. Montero, from the Instituto Jaime Almera of the CSIC, for their support in the study and valuable discussions.

1363

ASSESSING EXPOSURE TO ENVIRONMENTAL TOBACCO SMOKE IN THE NON-INDUSTRIAL WORK ENVIRONMENT

C. J. PROCTOR
Center for Environmental Health and Human Toxiocology
Washington, D.C., U.S.A.

SUMMARY

This paper discusses the most appropriate methods for assessing exposure to Environmental Tobacco Smoke (ETS) in the workplace. Data on nicotine, cotinine and respirable suspended particulates suggests that exposure to ETS at work is generally less than in other environments, including the home.

1. INTRODUCTION

Environmental Tobacco Smoke (ETS) is the complex mixture of substances found in indoor air as a result of tobacco smoking. Any assessment of non-smoker exposure to ETS is complicated by the dilute and unstable nature of ETS and by the lack of specificity to tobacco smoke of the majority of the component substances (1).

Because of this, exposure assessment of ETS is limited to the use of very few substances. Most commonly used is the measurement of personal exposure to airborne nicotine. This substance is specific in air to tobacco smoke. However, it should be noted that nicotine exists almost entirely in the vapor phase in ETS and exhibits slightly different decay characteristics than many other components of ETS, disappearing from the air rapidly but then remaining at low concentration for long periods as nicotine is re-emitted from surfaces (2). Even so, once the limitations of the method are accepted, nicotine can be a useful marker of ETS exposure. An advantage of using nicotine is that one of its metabolites, cotinine, can readily be measured in various body fluids, though research has shown that cotinine data also must be interpreted with care (3).

A further possible measure is respirable suspended particulates (RSP, generally defined as particles less than 5 micrometers in diameter). However, because a variety of sources may contribute to levels of RSP found in the workplace, it is essential that the analytical method used be capable of separating ETS-derived RSP from other sources. This can be achieved through several analytical methods, including measurement of the analyte solanesol, and comparison of ultra-violet and fluorescence emissions of extracts of the particulates against surrogate standards (4).

2. EXPOSURE DATA

Many studies have reported levels of nictoine and RSP in a variety of environments. The majority of these measurements have been made in the U.S., but some also have been made in Europe and Asia. The literature suggests nicotine levels in the air of offices where smoking is allowed of the order of 5 ug m-3 (4). In the U.S., the industrial 8-hour exposure limit is 500 ug m-3. Furthermore, chromatographic profiles of semi-volatile chemicals in the air of smoking and non-smoking offices show nicotine to be a minor component, and illustrate that there are few differences in both the type and amount of substances in air between smoking and non-smoking environments (5).

Studies that have measured respirable suspended particulates generally show a contribution from ETS of the order of 20 to 30 per cent in offices where smoking is permitted. This corresponds to levels of around 20 to 50 ug m-3 (4). These levels are somewhat dependant upon the manner by which the building is ventilated. Recent research has indicated that in office buildings that use adequate air handling systems, smoking and non-smoking offices could not be distinguished on the basis of airborne levels of RSP. Moreover, the most recent ventilation standard of the American Society of Heating, Refrigerating and Air-Conditioning Engineers (ASHRAE) recommends the same ventilation rate for smoking and non-smoking office areas.

A recent study performed in the United Kingdom measured exposures to ETS in non-smoking women (6). It compared daily personal exposures to airborne nicotine, salivary cotinine levels and time-activity information in groups of women who either lived with a smoker or not and that worked or not. The results indicated that exposure to ETS at work was considerably lower than that which might be expected at home. Women who lived with a smoker but did not work were exposed to a mean nicotine level of 7.4 ug m-3, while those who lived with a non-smoker and worked were exposed to a mean nicotine concentration of 0.8 ug m-3. Of the women living with a smoker, mean exposures were found to be considerably lower in the working group (1.6 ug m-3) over the non-working group (7.4 ug m-3). This data is consistent with the findings of the American Health Foundation, which used cotinine data to conclude that ETS exposure is lower in the workplace than in the home (7).

3. CONCLUSIONS

The data presented suggests that measurements of personal exposures to nicotine are useful in assessing exposure to ETS in the workplace. From the data that exists, exposures generally appear to be low. Further, such exposure is lower in the non-industrial workplace than in many other environments.

4. REFERENCES

(1) JENKINS, R., AND GUERIN, M. (1984). Analytical chemical methods for the detection of environmental tobacco smoke constituents. Eur. J. Resp. Dis. Suppl. 133, 33-46.
(2) BAKER, R., et al. (1990). The origins and properties of ETS, Environ. Int., 16, 231-245.
(3) ANDERSON, I., et al. (1991). Comparison of the measurement of serum cotinine levels by GC and RIA, Analyst, 116, 691-693.
(4) PROCTOR, C., et al. (1989). Measurements of ETS in an air-conditioned office building, Env. Tech. Lett., 10, 1003-1018.
(5) BAYER, C., AND BLACK, M. (1987). Capillary chromatographic analysis of VOCs in the indoor environment, J. Chrom. Sci., 25, 60-64.
(6) PROCTOR, C., et al. (1991). A comparison of methods of assessing exposure to ETS in non-smoking British women, Environ. Int., 17, 287-297.
(7) HALEY, N., et al. (1989). Elimination of cotinine from body fluids, Am. J. Pub. Hlth., 79, 1046-1048.

1364

EXPOSURE ASSESSMENT IN OCCUPATIONAL DISEASES

D. ROOSELS, J-M. BOSSIROY, J. BOULANGER and J. VANDERKEEL

Fund of Occupational Diseases (FOD)
Sterrenkundelaan 1, B-1030 Brussels

In Belgium the evaluation of occupational diseases is entrusted to the Fund of Occupational Diseases. This evaluation is not restricted to a medical examination but includes also the assessment of any occupational exposure explaining the defined illness. Such an exposure assessment (steps 6...) is preceded by a preliminary study (steps 1 to 5) including:

1. A description of the worker's environment including:
   - location on a map of the factory of
     - the machines
     - the operator
     - other activities in the work area
     - air streams including aspirations, windows, doors ...
   - registration of products used by the operator
     - their quantities
     - their frequency of use
   - registration of other products present in the work area.
2. A description of the worker's activities, movements, work rhythm and protective equipment including the composition of products used by the operator and in the neighbourhood.
3. A description of the analytical procedure(s) selected
   - method and instrumentation
   - sensitivity
   - reproducibility
   - detection limits on pure substances
   - interfering substances
4. A description of the sampling technique
   - volume
   - medium: solid adsorbent, coated filter, impinger, filter
   - rate
   - breakthrough limit
   - influence of humidity.
5. A test of the entire analytical procedure
   - on reference materials, and
   - on the products actually used at the factory
   - including the sampling technique.

The results of these steps 1 to 5 determine:

- the possibilities
- the shortcomings of the analytical procedure proposed.

6. From steps 1 and 2 a strategy is developed that defines:
   - the sampling times and location for specific work
     (short time operations)
     the number, and
     the characteristics of samplings:
   - personal samplings defining the exposure of the worker directly involved in the production,
   - stationary samplings defining:
     the exposure of workers not directly involved in the production
     contaminations from other activities (neighbouring machines)
     the production intensity,
   - repeatability as a function of the work description.

   The position of the samplings is drawn on a map of the factory.
7. As products used at the day of sampling can differ from those taken during the first visit (even when the same composition is mentioned), it is essential to take samples of the products actually used at the moment of the air samplings. Steps 1 and 2 must be repeated at the day of samplings and other observations such as temperature, air streams, productivity ... must be registered to compare the results obtained on other days.
8. The analytical results obtained are drawn on the map of the factory and discussed in relation to the observations registered.
9. The graphs are so designed that each product is addressed as a colour and expressed as a relative of its own TLV.
10. According to NBN T 96002 the samplings must cover four days of production representative of a normal working day.

1365

Detector Tubes in the Construction Industry

Reinhold Rühl and Michael Knoll
GISBAU
Arbeitsgemeinschaft der Bau-Berufsgenossenschaften
An der Festeburg 27 - 29
D 6000 Frankfurt 60

SUMMARY

The construction industry needs directly measurement techniques for the determination of hazardous material concentration on construction sites. By extensive comparison with usual measurements the usability of detector tubes are tested.

1. INDRODUCTION

Construction companies must fulfil the regulations laid down in the Hazardous Materials Decree, just as in every other industry. However, at present, there are several great difficulties primarily in the determination of hazardous material concentrations at construction sites which result from the impermanency of the sites:

- there is a multitude of work procedures, practically incalculable variations of the surrounding parameters and inexhaustible choice of building chemicals such as glues, strippers, thinners, etc.
- the measurement of the hazardous materials with the usual procedures is not possible on every construction site, e.g. because of the large number of companies, the listed range of parameters to be taken into consideration and the not existing measurement capacity
- in the usual procedures - sample taking on site and sample evaluation in laboratories - the result is only available after the site has been cleared
- the work procedures in question (undercoating, filling, glueing etc.) often only last minutes or hours
- control measurements are therefore not possible on the same sites.

The answer to these problems cannot be to exclude the building industry from the effectiveness of the Hazardous Materials Decree. In this way the projekt GISBAU has been started by the professional associations of the construction industry. The ministry of research and technology supportes GISBAU about the Research and Development Programm 'Work and Technology'. It is one of the aims of the GISBAU Project to fill up a large data pool of construction site measurements available to construction companies. This will make it possible to assess the dangers of a building chemical during work procedures with defined parameters by comparison with already existing measurements. In this way a determination is possible using comparison with existant data with a view to whether a situation is

- harmless (no protective measures necessary)
- hazardous (protective measures necessary)
- not comparable with existant data (measurements necessary).

Only in the last case a measurement on the building site would be necessary. This measurement must have immediate results, however, as a timely reaction would otherwise not be possible. We need measurement techniques, who allow directly references to the hazardous material concentration.

2. DETECTOR TUBES

At the moment, there are two measuring methods with the help of which the concentration of many materials can be determined:

- mobile laboratory units (e.g. gas chromatographs)
- direct result detector tubes.

As they cost more than DM 70,000.-- and require expert knowledge, the mobile gas chromatographs cannot be used in small and medium-sized building concerns.

However, the use of direct result detector tubes is possible in many cases and also economically reasonable (the hand pump required costs approximately DM 500.--, a detector tube between DM 5.-- and DM 10.--).

Two arguments are always being brought against the use of the test tubes; the transverse sensitivity (i.e., interference from different materials) and a great error of the mean. But, the usual measurements according to TRGS 402 (Technical rule of hazardous materials for measurements at work) too show an error of up to 30%. It is, however, most important in the building industry to have an immediate result (possibly with greater error) than after a period of six weeks, when the building site has been cleared. Additionally, the margin of error in the usual measurements can rise significantly as a result of implementation or documentation mistakes. It is safe to say that a well carried-out detector tube measurement is no less exact than a badly carried-out usual measurement.

The transverse sensitivity of the detector tubes must be taken seriously. The comparable measurements using the traditional methods make it possible for the GISBAU Project, however, to name precisely those products in which it is not possible to use the Detector tubes. GISBAU therefore chooses the products in which the transverse sensitivity is of importance and then recommends the use of the detector tubes only when reliable results are possible. As GISBAU gives information on hazardous materials both in respect to the product and the work procedure, the recommendation to use the detector tubes also takes into account that no other products containing interfering materials are being used alongside.

3. METHODS

All of the measurements taken by the twelve GISBAU technicians are made by decentralised sample taking and central analysis on the one hand and the direct result detector tubes on the other. While the Berufsgenossenschaftliche Institut für Arbeitssicherheit - BIA - is able to quantatively determine several materials found in the air on a building site[1], only one material we analysed with the detector tubes. Normally this substance is the most dangerous of the preparation and has been determined with the quotient resulting from the saturation concentration and the limit value.[2],[3] Table 1 shows the number of measurements for three substances.

In regard of the persons, who later have to use the detector tubes at construction sites, construction-engineers, -technicians, master masons and laboratory assistant carried out the measurements.

Table 1: Measurements of toluene, styrene and formaldehyde

	Number of Measurements		Detector Tubes
	Usual	Detector Tubes	every
Toluene	79	332	17 minutes
Styrene	35	279	15 minutes
Formaldehyde	65	373	20 minutes

4. RESULTS

The comparison of the results according to the usual measurements and the direct result detector tubes makes it possible to check the value of the detector tube measurements. Statements on the precision of the detector tube measurements are possible for the various building chemicals.

The results of the toluene, styrene and formaldehyde measurements show exemplary the usefullness of the detector tubes. If possible, all measurements have been evaluated according to

1) Intervals of the detctor tubes measurements (< 15, 15 - 30 and 30 - 60 minutes),
2) the measured substance has been the main part or a minor constitutent of the preparation,
3) hazardous material concentration over or under the limit value,
4) different groups of measuring technicians,
5) different producers of detector tubes.

at 1) Comparable results we have only obtained, if at least every 30 minutes a detector tube measurement has been taken. In this way it will be avoided, that we only measured the higest or lowest concentration during a work procedure.

at 2) Independent, wether toluene is the main part or not of the preparation, the results of the two measurement techniques agreed. In the case of toluene the detector tubes show no reliable results, if the sum of the concentrations of xylene and ethylbenzene is more than half of the toluene concentration.

at 3) The results of the two measurement techniques are comparable, wether the concentrations over or under the limit value. Table 2 show the comparison of the results of the two measurement technique (all measurements included).

at 4) The measurement with detector tubes needs the same accuracy and practice like the usual measurements.

at 5) There is no difference by using detector tubes of different producers.

Table 2: Results of usual measurements and Detector tubes* measurements

	Toluene				Styrene			
	usual	Detector tubes*			usual	Detector tubes*		
		A	B	C		A	B	C
< 1/4 MAK	28	12	14	2	1	1	-	-
1/4 MAK - MAK	43	4	33	6	7	-	4	3
> MAK	8	-	1	7	27	-	4	23

	Formaldehyde				
	usual	Detector tubes*			
		A	B	C	MAK : German TLV
< 1/4 MAK	38	37	1	-	A: < 1/4 MAK
1/4 MAK - MAK	22	15	7	-	B: 1/4 MAK - MAK
> MAK	5	-	2	3	C: > MAK

*mean values of the detector tubes measurements during a usual measurement

Simmilar results are expectet after the first evaluations for the following areas:

- dichloromethane used in strippers
- methyl methacrylate in floor coverings
- xylene as the main ingredient in certain solvent mixtures.

The first evaluations of the parallel measurements show, as expected, that it makes little sense to use benzinhydrocarbon tubes with complex solvent mixtures. These values seem to have no relation to those obtained by usual measurements.

The results confirm the useability of the detector tubes in the construction industry in many cases. In which way the construction companies have to use the detector tubes (for example, at the working step where the highest concentration is expected) an additionally evaluation of the GISBAU measurement data pool must show.

GISBAU will make these results available to experts for discussion and thereby create the basis for the recognisation of this practical measuring method in many areas.

REFERENCES

1) Coenen, W.: Berufsgenossenschaftliches System der meßtechnischen Überwachung; Staub-Reinh. der Luft 1988, S 51-56.
2) Piringer, R. and Leisser, H.: Strategie für die Beurteilung von chemischen Stoffen;
3) Technische Regel für Gefahrstoffe 420 - Verfahrens- und Stoffspezifische Kriterien; Dezember 1988, BArbBl 12/1989, S. 71

1366

SPECIFICITY OF CHEMCASSETTE DETECTION SYSTEMS TO WORKPLACE HAZARDS

Ron Walczak and John Trautwein Jr.
MDA Scientific, Inc.
405 Barclay Blvd.
Lincolnshire, IL 60069

SUMMARY: Toxic gas detection systems are commonly used to monitor for trace levels of hazardous chemicals in the process and storage areas of industrial facilities. The effectiveness of these systems may be evaluated by their resistance to potential cross-interference, response to low levels of gases or vapors and calibration requirements. The Chemcassette detection method provides unique advantages when evaluated for these characteristics. Many industrial gases may be monitored in a variety of applications.

INTRODUCTION: Based on the the existence of the various occupational exposure regulations, workers are expected to be protected from the toxic gases which may be present in the work environment. Process and effluent gases also require monitoring. Normally, levels in the ppm or ppb ranges are monitored. Typical methods of real-time monitoring include electrochemical sensors, solid state sensors, U.V. and I.R.absorbtion, gas chromatography, flame photometry, etc. Although these methods will measure some compounds well, there are some inherent disadvantages to each system (Fig. 1).

Detection System	Disadvantages of System	Chemcassette Advantages
Electrochemical	•Numerous Cross-Interferences •Frequent calibrations with toxic gases	•Gas Specific •No field gas calibration required
Solid State	•Numerous Cross-Interferences •Frequent calibrations with toxic gases •May "go to sleep"	•Gas Specific •No field gas calibration required •Fresh detector continuously used
Infra-Red	•Detection usually to only ppm levels	•Detection to ppb levels easily for many gases
Gas Chromatography	•Sample periods in minutes or hours for ppb concentrations	•Sample periods seconds or minutes for ppb concentrations
Flame Photometry	•Complex, expensive instrumentation	•Simple, relatively inexpensive instrumentation

Fig. 1: Chemcassette Detection System Advantages

CHEMCASSETTE DETECTION METHOD: The Chemcassette detection system integrates a dry color reaction, photo-optical sensors, sample flow system and microprocessor intelligence into a gas detection system. This method is useful for a variety of gases (Fig. 2). A wide range of applications are possible including ambient air, process streams and ventilation/stack monitoring (Fig. 3).

<u>Chemcassette Detectable Gases:</u>
Amines (NH_3,TDA,PPD)
Oxidizers/Corosives (CL_2,Br_2,NO_2,H_2O_2,O_3,SO_2,H_2S)
Isocyanates (TDI,MDI,CHDI,HDI,HMDI,IEM,IPDI,NDI, PCPI,PPDI,TMDI,TMXDI,XDI)
Toxics (HCN,$COCl_2$)
Hydrazines (MMH,UDMH,N_2H_4)
Hydrides (AsH_3,PH_3,B_2H_6,SiH_4,H_2Se, Si_2H_6,GeH_4,SbH_3,TBA,TBP)
Mineral Acids (HF,HCl,HNO_3,H_2SO_4)

Fig. 2: Chemcassette Detectable Gases (Partial List)

Gas	Arsine	Diisocyanates	Diisocyanates	Arsine	Hydrogen Sulfide	Hydrogen Fluoride
Application	Propylene Q.C.	Foaming: Insulation/Car Seats	Production & Storage	Semiconductor Production	Waste Water Treatment	Polymer Containers
Instrument Required	Continuous Single Point	Continuous Portable/Fixed Leak Detector	Continuous Rugged Single Point	Continuous Single/Multi Point Monitor	Wet Scrubber Efficiency Control System	Continuous Single/Multi Point
Range Of Detection	0-500ppb	0-200ppb	0-200ppb	0-500ppb[2]	0-5000ppb or 0-50ppm	0-30ppm
TLD	Not Applicable	ACGIH:5ppb (.036mg/m^3)[1] MAK:10ppb (.07mg/m^3)	ACGIH:5ppb (.036mg/m^3) MAK:10ppb (.07mg/m^3)	ACGIH/MAK: 50ppb[3] (.16mg/m^3)	Not Applicable	OSHA/MAK:[3] 3ppm (2mg/m^3)
Response Time	10 seconds at 50ppb	60 seconds at 5ppb	60 seconds at 5ppb	10 seconds at 50 ppb	10 seconds at 10ppm	30 seconds at 3ppm
Notes	Cut process time from hours to minutes	ppb levels documented in seconds	Early detection of leaks and emissions	Other semi gases also available in single/multi	Odor control: chemical dosing feedback	To produce corrosion-resist. surface

Fig. 3: Application Examples for Chemcassette-Based Monitors

Colorimetric Reaction: The actual sensor consists of a specially cut chromatography paper which is chemically treated with a proprietary formulation of reagents. These reagents are dissolved into volatile solvents, dispersed onto the filter paper; the solvents are removed by controlled evaporation to yield a dry, safe-to-handle substrate. This detector medium is called the Chemcassette. The reagents used are formulated for specificity to a target gas or a family of gases so that cross-interference is minimized (Fig. 4). This unique colorimetric method also provides physical evidence in the form of a well-defined stain that a gas has been detected.

Example Target Gas: Ammonia	
Responds to:	Ammonia and Aliphatic Amines
Will Not Respond To:	Acids, Alcohols, Amides, Aromatic Amines, Carbon Disulfide, Freons, Hydrazines, Hydrides, Hydrocarbons, Hydrogen, Hydrogen Cyanide, Hydrogen Sulfide, Isocyanates, Nitrogen, Nitrogen Dioxide, Ozone, Phosgene, Solvents and Sulfur Dioxide

Fig. 4: Gas Specificity of Chemcassette Method

Sample Flow and Optical Detection: The air sample is drawn through the Chemcassette at a constant rate of flow which is optimized for the gas species (Fig. 5). As the gas sample is drawn through the Chemcassette the target gas will cause a stain to develop in proportion to the concentration of gas present. The stain intensity and the speed of response are directly proportional to the concentration of the target gas (Fig. 6)[3].

A light source, such as a tungsten filament bulb or solid-state LED, is directed toward the Chemcassette. The reflected light is constantly measured by a photodiode and compared to an initial zero-gas reading. As the stain develops, the reflectance measurement is sent through an analog to digital converter and is then compared to look-up tables stored in ROM. The concentration is reported by digital display, 4-20 mA output or digital output, and alarm relays. After the stain reaches a preset intensity a new segment of Chemcassette is advanced into the flow/optics assembly and a new zero baseline is read. Ultra-low levels of gas are detectable by lengthening the amount of time which the sample is passed through a segment of Chemcassette, effectively concentrating a higher volume of sample.

Calibration: The need for dynamic calibration in the field is virtually eliminated with this method. Each Chemcassette is dynamically gas-tested and results are logged at the time of manufacture, assuring the accurate response of the Chemcassette. In the field the monitoring instrument is optically checked with a stain simulation card to ensure that a stain will be properly quantified. The flow system is also easily checked. All elements of the system are therefore verified without the use of calibration gas and gas dilution/mixing equipment.

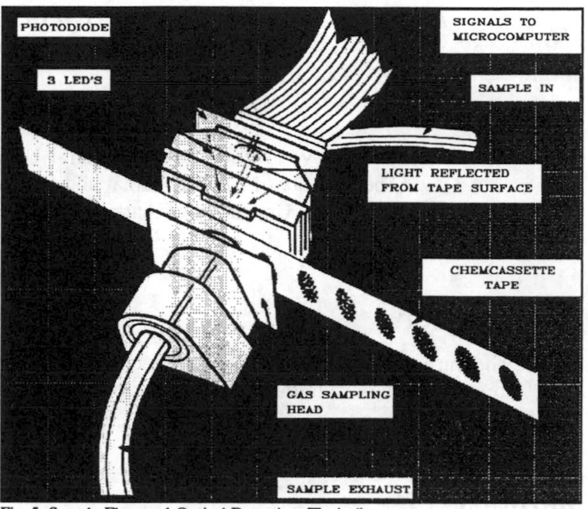

Fig. 5: Sample Flow and Optical Detection (Typical)

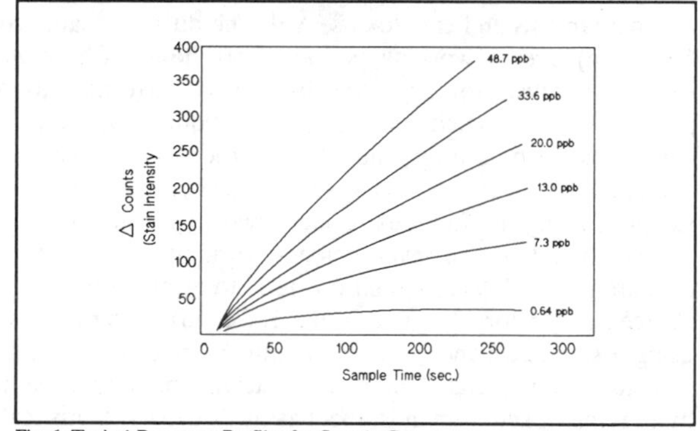

Fig. 6: Typical Response Profiles for Sample Gas

REFERENCES

1. American Conference of Government and Industrial Hygienists (1990). Guide to Occupational Exposure Values.

2. McMahon, R. (1989). Monitoring Hazardous Gases used in Semiconductor Fabrication. Microelectronics Manufacturing and Testing.

3. Rando, R.J. (May 12, 1989). Dynamic Calibration of the Model 7100 Monitor for Xylene Diisocyanate. Report submitted to MDA Scientific, Inc.

AIR QUALITY AND HUMAN RESPIRATORY HEALTH AT DANISH PIG FARMS

P. BÆKBO[1] and C. HJORT[2]

[1]The Royal Veterinary and Agricultural University, Denmark
[2]The Institute of Environmental and Occupational Health
University of Aarhus, Denmark

SUMMARY

The purpose of the present paper was to study the air quality at Danish pig farms and to examine the respiratory health of the persons working in the units.

METHODS

The fattening units of 44 farms were examined for dust, gases (CO_2 and NH_3), microorganisms and endotoxins. Airborne dust was sampled on a Millipore filter (0.8 μm; 1.9 l/min) over a working period of six hours. Gases were measured with the Dräger system. Analysis for microorganisms was performed by a small cyclon sampler where particles are impacted on a collection liquid. Different nutrient media and incubation conditions were used by which the microorganisms were differentiated to some extent.

Analysis for endotoxins was performed on the exposed collection liquid from the cyclon sampler, using the Limulus Amoebocyte Lysate-test (Coatest, Kabi-Vitrum).

One hundred and seven persons working in the units were interviewed and examined by spirometry. A Vitalograph was used to measure FVC (Forced Vital Capacity) and FEV_1 (Forced Expiratory Volume in the 1.second). Allergy was performed as a skin-prick test using a standard panel as well as antigens related to farming. Serum antibodies (IgG and IgA) to pig farm related antigens (dust, pig epithelium, pig urine, pig serum) were detected by enzyme linked immunosorbent assay (ELISA). Statistical calculations were done in SAS and BMDP.

RESULTS

Results from examination of 44 fattening units: Means: Dust (mg/m^3) – total: 2.11 ($\pm$1.77), respirable (<5 μm): 0.49 ($\pm$0.48), NH_3 (ppm): 8.7 ($\pm$5.1), CO_2 (ppm): 1381 ($\pm$598), Bacteria (x 10^5 CFU*/m^3): 20.1 ($\pm$31.7), Fungi (x10^4 CFU*/m^3): 2.96 ($\pm$7.22), Endotoxins (x10^3 EU/m^3): 38 ($\pm$37).

Based on the results from the 44 fattening units, multiple regression models (GLM-procedure) could be established, showing that all air contaminants were negatively correlated to the outdoor temperature and/or positively correlated to the CO_2 concentration indoors. The ventilation rate of the building is closely connected to both outdoor temperature and indoor $C0_2$ concentration: 64% of the persons reported at least one respiratory symptom (i.e. dry cough, cough and expectoration, wheezing, chest-tightness); 18% had lung function impairments of the obstructive type; 15% showed positive reactions in the skin-prick test to one or more allergens.

Most frequently reactions to storage mites were detected. Many persons had high levels of serum antibodies to different pig farm antigens and generally the antibody levels in farmers differed from the levels seen in blood donor controls. However, lung function impairments and occurrence of respiratory symptoms did not correlate. Neither did respiratory symptoms and lung function impairments correlate with antibody level or allergy status. Furthermore it was not possible to find any correlation between the occurrence of symptoms and measurements of environmental parameters.

CONCLUSION

From the results it seems possible to improve the air quality in every respect by increasing the air exchange rate. Many farmers have respiratory symptoms but the environmental measurements of this study cannot be related to the respiratory health of the persons working in the buildings. However, with other types of measurements this might be possible.

CFU: Colony Forming Units
EU : Endotoxin Units (25 EU = 1 ng; FDA-standard EC-5)

OCCUPATIONAL EXPOSURE TO CHEMICALS WITH ADDITIVE EFFECTS: A STATISTICAL PROCEDURE FOR EVALUATION

M. RE

Dipartimento Medicina Preventiva Occupazionale e di Comunità
Sezione Medicina Preventiva dei Lavoratori e Psicotecnica
Università degli Studi di Pavia
Via Severino Boezio, 24 – Pavia – Italia

SUMMARY

A method is described for the evaluation of occupational exposure to a mixture of two chemicals under the assumption of additive effects. The procedure consists in calculating statistical confidence limits for the sum of average work shift concentrations and is derived from the NIOSH method for assessment of compliance with TLV-TWA standards. The aspect of statistical correlation between concentrations of the two substances is considered by its effect on confidence limits. Results are reported of the application of the method to a case of additive exposure in a workplace where a mixture of ethylacetate and MIBK is used as paint solvent.

INTRODUCTION

Simultaneous exposure to more than one risk factor and, more specifically, to several hazardous substances occurs very frequently in workplaces. In most cases however single risk factors have quite specific and independent effects on health, so that exposure evaluation must be done separately for each factor: for example lead and sulfuric acid (1). However, it has been pointed out (1,2) that some activities often imply contemporary exposure to substances with similar health effects: e.g. painting, certain foundry operations etc. In such cases it has been suggested that risk evaluation should be referred not only to exposure to single substances, but also to cumulative exposure under the hypothesis of additive effects. It should be remarked that simultaneous additive exposure to several substances may occur systematically when a mixture of chemicals is used in a workplace and more or less occasionally when a workplace may be affected also by substances used in the surrounding ones. In the latter case concentrations of single substances are expected to be independent, whereas in the former case they are expected to be correlated.

A method is proposed for estimating average daily exposure in the simple case of mixture of two substances. It is derived from the statistical procedure for estimating daily exposure to one substance by confidence limits of the mean (3) and is based on the same assumptions.

METHODS

Some preliminary conditions for applicability of the estimating procedure of additive exposure to two substances by confidence limits are as follows:

(i) Measurements of concentration for both substances in air (preferably breathing zone) are available for the whole 8–hour workshift or at least for 6 hours in the workshift (in this case the sampling period should be conveniently determined to be representative of exposure during all the workshift). Sampling positions and intervals must be the same for both substances. Lenght of sampling intervals may be various, e.g. 8 sampling intervals of one hour in a workshift or a single 8–hour measurement.

(ii) The coefficient of variation (CV) for the sampling and analytical method is known for both substances. It should be remarked that this coefficient is often the same for both substances when their chemical structure is similar and therefore the analytical technique is the same, e.g. gaschromatography for solvents.
(iii) When several samples are available for both substances a statistical test for significance of correlation (4) between concentrations of the two chemicals should be applied; a statistically significant correlation coefficient R implies a wider confidence interval and therefore a greater uncertainty in estimating mean additive exposure. In fact, mean additive exposure $X_{a,m}$ to substances 1 and 2 in a workshift is given by:

$$(1) \quad X_{a,m} = X_{1,m} + X_{2,m}$$

where $X_{1,m} = \sum_{i=1}^{n} \frac{T_i}{T} X_{1,i}$ $\qquad X_{2,m} = \sum_{i=1}^{n} \frac{T_i}{T} X_{2,i}$

n = number of sampling intervals in a workshift
T_i = length of sampling intervals
$X_{1,i}$ = standardized concentration of substance 1 in interval i
$X_{2,i}$ = standardized concentration of substance 2 in interval i

It can be demonstrated (5) that the upper and lower confidence limits of $X_{a,m}$ are given by the expression:

$$UCL, LCL = X_{a,m} \pm 1{,}645 \frac{CV}{n} \left\{ \sum_{i=1}^{n} X_{1,i}^2 + \sum_{i=1}^{n} X_{2,i}^2 + 2R \left(\sum_{i=1}^{n} X_{1,i}^2 \right)^{\frac{1}{2}} \left(\sum_{i=1}^{n} X_{2,i}^2 \right)^{\frac{1}{2}} \right\}^{\frac{1}{2}}$$

in the common case that the coefficient of variation is the same $CV_1 = CV_2 = CV$ and that sampling intervals are all equal, i.e. $T_i = \frac{T}{n}$. In the above expression the term containing R needs to be considered only when R is statistically significant at high level.
The evaluation of additive exposure relies on the resulting value of the upper confidence limit UCL, that should be lower than unity in case of compliance additive exposure for the two substances.

RESULTS

The procedure reported above was applied for the evaluation of additive daily exposure in a workplace where a mixture of ethylacetate and methylisobutylketone is used as a paint solvent. Breathing zone air was sampled on activated charcoal tubes. After desorption with carbon disulfide samples were analyzed by FID–gaschromatography on Carbowax 20M packed column; the coefficient of variation was 0.1 for both chemicals.
Results are reported in table 1. It can be seen that concentrations of ethylacetate (TLV–TWA = 400 ppm) and of MIBK (TLV–TWA = 50 ppm) (1) are statistically correlated at 95% significance level (4); therefore upper and lower confidence limits UCL, LCL for additive exposure were calculated taking into account also the correlation coefficient; without the term containing R a smaller value for UCL would be obtained, i.e. UCL = 0.94 + 0.41. If the same average exposure values were obtained with one 8–hour measurement, resulting UCL would be higher than unity (0.94 + 0.11), indicating non–compliance additive exposure.

CONCLUSIONS

Evaluation of additive exposure to two chemicals by confidence limits causes, as for

exposure to a single substance, a more severe judgment than evaluation by average daily concentrations only. This effect increases as the coefficient of variation of the sampling and analytical procedure and the correlation coefficient between concentration values of the two chemicals increase; on the contrary the effect is reduced by increasing the number of samples in a workshift (n).
The procedure is easily applicable for its simplicity. A statistical aspect that might be considered by a more complex procedure is time autocorrelation of concentrations, that in evaluation of occupational exposure could have significant effects in some cases when sampling intervals are short, about 15 minutes or less (6).

Standardized	Concentrations			
ethylacetate	MIBK	$X_{a,m}$	R	UCL, LCL
0,24	0.18			
0.78	0.67			
0.26	0.23			
0.60	0.51			0.94 ± 0.054
0.43	0.36	0.94	0.76	
0.83	0.38			UCL = 0.994
0.51	0.44			
0.65	0.39			
$X_{1,m} = 0.54$	$X_{2,m} = 0.40$			

Table 1: Evaluation of daily additive exposure for two substances (8 hourly measurements in a workshift)

REFERENCES

(1) AMERICAN CONFERENCE OF GOVERNMENTAL INDUSTRIAL HYGIENISTS (1990). Threshold Limit Values and Biological Exposure Indices for 1990–91. ACGIH, Cincinnati.

(2) AMERICAN CONFERENCE OF GOVERNMENTAL INDUSTRIAL HYGIENISTS (1985). Proceedings of International Symposium on Occupational Exposure Limits. Annals ACGIH. Vol. 12, 111–118.

(3) LEIDEL, N.A. and BUSCH, K.A. (1985). Statistical Design and Data Analysis Requirements. Patty's Industrial Hygiene and Toxicology, 2nd ed. Vol. 3A. John Wiley and Sons, New York.

(4) SNEDECOR, G.W. and COCHRAN, W.G. (1978). Statistical Methods. 6th ed. Iowa State University Press, Ames.

(5) RE, M. A Generalized Procedure for Evaluation of Multiple Occupational Exposure. Manuscript in preparation.

(6) FRANCIS, M., SELVIN, S., SPEAR, R. and RAPPAPORT, S. (1989). The Effect of Autocorrelation on the Estimation of Workers' Daily Exposures. Am. Ind. Hyg. Assoc. J. Vol. 50, 37–43.

Session III

MEASUREMENT METHODOLOGY

Gases and Vapours

KEYNOTE PAPERS

POSTERS

1369

MEASUREMENT METHODOLOGY : GASES AND VAPOURS TECHNICAL REQUIREMENTS, RESPONSES AND MEASUREMENT

D.C.M. Squirrell

9 Graysfield, Welwyn Garden City,
Hertfordshire, AL7 4BL, England.

SUMMARY

The responses made to needs arising for improved measurement methodology in the control of harmful gases and vapours in workplace atmospheres are summarised. These include fixed plant monitors, portable instruments and instruments for leak seeking, miniaturised instrumentation and personal monitors, calibration and reference materials together with special requirements covered by innovative analytical chemistry.

THE FOUNDATION OF PROGRESS

One of my duties as Chairman of this session is to try to set the scene for the papers that are to follow. Hence the title of this introductory paper.

At the time of our 1986 symposium concerned with passive sampling, the main problems that industry faced in relation to workplace atmospheres and the environment were very apparent and action had resulted in significant progress being made in bringing these problems under control. It was also clear that further improvements would be required for the future and that these improvements could only result from better understanding, training, cooperation and innovation from all concerned.

Changes have therefore continued. Several known problem chemicals have been replaced by safer alternatives; composite formulations have been changed and manufacturing processes and working procedures have been modified. A great deal of work has gone into the containment of volatile materials during manufacture and use and these improvements carried over to transport and transfer procedures. All these actions have brought about improvements and reduced the levels of harmful and potentially harmful substances in workplace atmospheres. At the same time however the expected or allowable concentrations of atmospheric contaminants has also decreased and in some cases better specificity as well as sensitivity has made demands on our measurement technology and methodology. In this context I think it would be fair to say that the last 5 year period has been one of consolidation, improvement and more efficient use of our modern technology, rather than the discovery of new and revolutionary means of monitoring volatile solvents and chemicals in air.

Within our manufacturing and compounding areas we have had to improve our survey procedures and methods of leak detection. It has been necessary to provide more efficient means of fume extraction after a leak or spillage and reliable means of monitoring the "clean up" process. Scrubbing systems have had to be installed on vent gas installations to protect the outside environment. Continuous monitoring at several key areas within a manufacturing plant is sometimes required to give early warning of small leaks, provide data on trends and permit calculation of exposure levels for comparison with those obtained by other methods, for example, personal monitoring. Considerable attention has been paid to this latter area and the sampling systems for personal monitoring have been made more reliable and more comfortable and convenient to wear.

As previously indicated, success in any prevention and monitoring system can only be attained with the full cooperation of the workforce and all concerned with the over all efficiency of the plant and the welfare of those working in and near it. (some extraction systems can be noisy.) Communication skills and training are thus important and knowledge of the Why, When, Where and How of the safety regulations, methodology, measurement hardware together with its operation, capability and limitations are very important. The measures taken to preserve a safe working atmosphere in the workplace and the ways in which this is monitored are now more widely understood.

FIXED PLANT MONITORING SYSTEMS

The types of fixed plant monitoring systems together with a review of the whole field of workplace monitoring and including a useful bibliography is provided in a paper by Miller (1). For the systems relying on multiple point sampling and sequential analysis at a central analyser, it is very important that a fast loop sampling system be installed so that virtually real time samples are examined from each sampling point. The provision of a second analyser which can be locked on to any individual point to monitor "clean up" after a leak or spillage without interrupting the surveillance cycle of the other sampling points is also to be recommended. See fig. 1.

The deployment of a series of independent sensors at key points around a plant to relay information back to a central data handling or control point is a popular concept. Current efforts in the general sensor field are being directed towards providing better selectivity or specificity and to overcoming problems with reproducibility in manufacture and performance which in many cases is variable and necessitates a higher frequency of calibration and maintenance than is desirable for plant use. Multisensor units, each utilising an array of different sensors, the different responses from which can be processed to provide some specificity are under active investigation, for example for monitoring ammonia (2) and mine atmospheres (3). The advanced work of the Warwick Electronic Nose Group (University of Warwick, UK.) and similar advanced studies at the Manchester Institute of Science and Technology, UK. are very relevant in this area. A new report prepared for the UK. Department of Trade and Industry (4), indicates that there will be many opportunities for new optical and solid state sensor technology to meet measuring requirements.

Lasers, particularly infrared lasers, either in open path

or long path-length cell configurations have advantages for some applications. Use in conjunction with optoacoustic spectrometry has shewn promise for trace gas detection. (5). Infra-red gas analyser instrumentation has been refined and wide measuring ranges are now available for such gases as CO, CO_2, CH_4, and NO_2. Remote spectral data and high specificity can be obtained by use of Fourier transform. For example, one Commercial instrument (6) permits unattended operation and remote measurement from multiple sampling locations, quantitation of up to 20 analytes, and provides set-point alarms together with a warning if unexpected components are detected.

Paper-tape instruments, used as fixed or portable analysers continue to provide good service. In combination with optical fibre technology, potential has been demonstrated for possible use for continuous multi-site and multi-gas analysis. (7).

PORTABLE INSTRUMENTATION AND INSTRUMENTS FOR LEAK SEEKING

Portable instrumentation suitable for workplace air monitoring is in good supply. Instruments based on infra-red, flame ionisation, photoionisation, thermal conductivity, ion mobility, fuel cell technology and solid state sensors are frequently advertised. Multipurpose use and good selectivity are provided by the infrared instruments or by combining the measurement technology with gas chromatographic separation. For example, in one of the more sophisticated instruments (8), wide measuring ranges are facilitated for a variety of gases. Changing from the measurement of one gas to another is facilitated by replacing the calibration set with an alternative from a series of calibration sets each comprising an electronics module, infra-red filter and calibration scale. A large number of these sets is available.

A series of portable photoionisation gas chromatographs (9) which are fully self contained, capable of multicomponent analysis and high sensitivity is available with various options. With the computer in board, calculation of such figures as time weighted averages and maximum concentrations can be made and stored. Appropriate alarms can, of course, also be triggered.

A programmable vapour monitor, utilising ion mobility spectrometry is capable of monitoring and storing the concentration of a range of gases. The instrument is programmed for each gas by setting up three windows in the ion mobility spectrum and calibration with standard concentrations. Alarm levels can be set for immediate, 10 minute and 8 hour TWAs.

Detailed information about other direct reading instrumentation and on the development and use of detector tubes follows later in this session. Suffice it to say that significant advances have been made in response to needs in these areas. Mention must however be made here of miniaturised instrumentation and personal monitors.

MINIATURISED INSTRUMENTATION AND PERSONAL MONITORING

It is a usual requirement that this category of monitor be intrinsically safe for use in hazardous areas and this criteria is now normally satisfied. A commercially available triple gas alarm (10), is BASEEFA certified and can monitor three gases simultaneously. Three measuring cells include one which is a catalytic sensor to warn of explosive mixtures, whilst two separate electrochemical cells can operate for a toxic component and for oxygen level. Another range of self contained instruments (11), monitor and give audio and visual alarms for such gases as SO_2, H_2S, CO and O_2.

Sampling for direct readers, detector tube monitors and indeed almost all personal monitoring methods rely on pumps or diffusion principles and in the former case the efficiency and reliability of the pumping systems and associated valves, particularly of the miniaturised variety has improved. Various designs of passive samplers of both badge and tube type are in common use and several innovative samplers have been described. These include a novel design due to Fields (12) which departs from the normal pattern of having a diffusion gap over which the analyte is transported for sorption onto a solid or supported liquid phase/reagent. In his design the liquid reagent is not dispersed but is held by surface tension in a small cell which is an integral part of the sampler as shown in fig. 2. The advantage of this device is that it can be conveniently used for those gases for which there is no suitable solid sorbent and of course no desorption stage is required. In addition, because of the low cell volume, low detection levels are possible. The device has been used for amines, aldehydes and inorganic acids but wider applications are envisaged.

A useful review of the use of denuder tubes for sampling gaseous species has been published by Ali, Thomas and Alder (13) and deals with both cylindrical and annular designs.

Where desorption is required before the analytical stage this is still normally carried out either by solvent extraction or thermal desorption procedures. Improvements have been made in both simple and sophisticated thermal desorption units. The now widely used automated thermal desorber has been upgraded to a system capable of handling up to 50 tubes in one carousel with a desorption temperature range of -100 to +400 C. (14). This instrument can thus cope with mixtures ranging in volatility from C1 to n-C40 in a single analysis and hopefully overcomes some memory effects experienced by Alder et al. (15) with desorption of such low volatility compounds as methyl salicylate and dichlorodiethyl ether at lower temperatures. The same Company also markets a single shot, two stage desorber, based on similar principles, together with a fused silica heated transfer line for direct connection to a gas chromatograph. (16).

A combination of diffusive sampling with controlled flow thermal desorption on to a stain tube with subsequent quantitation by measurement of the stain length has now been evaluated for some 25 commonly encountered vapours. A commercial instrument assembly is now available (17).

CALIBRATION

All systems of course require calibration and the use of permeation tube and dilution devices is still popular. There is, however, urgent need for more reference materials and standardised mixtures if high accuracy is to be attained (18). Some volatile single substances and test mixtures are already available from commercial suppliers for the direct calibration of monitoring equipment as are ratio devices for mixing gases from separate supply lines. Of particular significance is the cooperative work of the BCR in Brussels with workers in the Netherlands and the UK. in the production and certification of a reference material for aromatic hydrocarbons in Tenax samplers (19). Certified values for benzene, toluene and xylene are provided.

SPECIAL REQUIREMENTS

There will always be special requirements or problems that cannot be solved by standard monitoring equipment or methodology and these are being addressed by innovative chemistry and use of main-line analytical instrumentation. This is demonstrated by a brief survey of publications since 1986, taken from just one international journal.

Phthalic anhydride has been determined in workplace atmospheres, in which workers were suspected of sensitisation, by gas chromatography with electron capture detection (20). Mass spectrometry, either directly computer integrated or coupled with a gas chromatograph has provided high sensitivity and specificity to trace gas analysis in many areas. Stopped flow and conventional flow injection procedures for the determination of formaldehyde in air has utilised pararosaniline in the presence of sulphite as reagent with spectrophotometric detection at 570nm. (21).

High performance liquid chromatography in conjunction with tryptamine as a derivitisation agent has been successfully used by Wu and his coworkers for the determination of various isocyanates, utilising fluorescence and amperometric methods of detection (22, 23). The fact that the fluorescence characteristics and amperometric behaviour of tryptamine are retained even after its reaction with isocyanates means that a single pure derivative, eg. of hexamethylene diisocyanate, can be used as a calibration standard for many isocyanates for which pure samples might not be available (24). Examples of other methods based on derivatisation and HPLC are as follows:- 2-methylaziridine as its 2,4,6,-trinitrobenzenesulphonic acid derivative (25): primary aliphatic amines as o-acetylsalicylamide derivatives (26): hexamethylenetetramine as its 2,4-dinitrophenyl-hydrazone derivative (27). Clearly there is a wide scope for such derivatisation/HPLC procedures. The importance of good chemistry to support the instrumentation and to meet special requirements must be emphasised.

I hope that in this brief survey I have selected sufficient examples to demonstrate that analytical scientists and instrument manufacturers have indeed responded, with considerable enthusiasm and success, to the requirements of the 1980s for measurement and control systems in workplace atmospheres and that this consolidation and innovation is continuing into the 1990s.

REFERENCES

(1) Miller, B., Anal. Proc., 1990, 27, 267.
(2) Fraser, S.M., Edmonds, T.E., West, T.S., Analyst, 1986, 111, 1183.
(3) Bott, B., Jones, T.A., Anal. Proc., 1986, 23, 61.
(4) Fleming, J., McCallum, J., The Government Chemist, 1991, 7, 5.
(5) Johnson, S.A., Anal. Proc., 1986, 23, 1.
(6) Equipment News, Anal. Proc., 1990, 27, 312.
(7) Narayanaswamy, R., Sevilla, F., Analyst, 1988, 113, 661.
(8) Lab. Products Tech. 1990, October, 12.
(9) Adams, M., Collins, M., Anal. Proc., 1988, 25, 190.
(10) Equipment News, Anal. Proc., 1990, 27, 286.
(11) Equipment News, Anal. Proc., 1989, 26, 297.
(12) Fields, B., The Diffusive Monitor, 1989, 3, 3.
(13) Ali, Z., Thomas, C.L., Alder, J.F., Analyst, 1989, 114, 759.
(14) Equipment News, Anal. Proc., 1990, 27, 346.
(15) Alder, J.F., Hildebrand, E.A., Sykes, J.A.W., Analyst, 1985, 110, 769.
(16) Equipment News, Anal. Proc. 1991, 28, 128.
(17) Dabill, D.W., The Diffusive Monitor, 1991, 4, 7.
(18) Maier, E.A., Anal. Proc., 1990, 27, 269.
(19) Vandendriessche, S., Griepink, B., Hollander, J.C.Th., Gielen, J.W.J., Langelaan, F.G.G.M., Saunders, K.J., Brown, R.H., Analyst, 1991, 116, 437.
(20) Pfaffli, P., Analyst, 1986, 111, 813.
(21) Munoz, M. P., Manuel de Villena Rueda, F.J., Polo Diez, L.M., Analyst, 1989, 114, 1469.
(22) Wu, W.S., Nazar, M.A., Gaind, V.S., Calovini, L., Analyst, 1987, 112, 863.
(23) Wu, W.S., Szklar, R.S., Gaind, V.S., Analyst, 1988, 113, 1209.
(24) Wu, W.S., Stoyanoff, R.E., Szklar, R.S., Gaind, V.S. Rakanovic, M., Analyst, 1990, 115, 801.
(25) Gaind V.S. Jedrzejczak, K., Huang, L., Vohra, K. Analyst 1990, 115, 925.
(26) Jedrzejczak, K., Gaind, V.S., Analyst, 1990, 115. 1359.
(27) Levin, J., Fangmark, I., Analyst, 1988, 113, 511.

1370

SORBENTS FOR ACTIVE SAMPLING

M.J. QUINTANA, B. URIBE and J.F. LOPEZ ARBELOA

Instituto Nacional de Seguridad e Higiene en el Trabajo
48903 Cruces-Baracaldo, Spain

SUMMARY

The use of sorbent tubes for active sampling of gases and vapours requires the careful consideration of the sorbent and contaminant nature comprising the sorbent-sorbate system and the factors affecting their performance in the sorption and desorption processes. The review of the experiments undertaken, based on the application of sorption theory allows some recommendations of practical use and the selection of the suitable sorbent.

1. INTRODUCTION

Solid sorbents, coated or not with liquids, packed in glass or metal tubes, are at present the most used system for sampling gases and vapours in workplace air. The use of the so-called sorbent tubes requires careful consideration of the sorbent and contaminant nature comprising the sorbent-sorbate system, and the factors affecting their performance both in the sorption and desorption processes. Main problems to be considered are the ubiquitous presence of water vapour in the air to be sampled as well as the presence of other contaminants that could invalidate the sampling and/or the analytical standard conditions by the methodology in use. The standardization of the demanded requirements to the sorbent tubes and test methods involved is set out in the CEN standard (1). Moreover, with the data obtained from these tests the sorbent behaviour can be derived under other conditions.

In this article a revision of the current situation of the solid sorbents is submitted under the two critical stages of their use, sampling and recovery of collected contaminant. Some practical guidelines based on the information about sorbent features, and about the sorbentsorbate system is given.

2. MAIN TYPES OF ADSORBENTS

Adsorption is a surface phenomenon, and therefore the 'available' surface is of prime consideration. The adsorbent surface area refers to the solid surface accessible to the adsorptive and the specific surface area to the area per adsorbent mass unit. Many of the adsorbents of great surface areas are porous. The porous structure can be developed by aggregation, or by elimination of some parts of the original solid. It can also be found naturally in crystalline solids (2).

Following IUPAC recommendations (3), it is convenient to classify the adsorbent pores according to their sizes: macropores, pores whose width exceeds 50nm (500 Å); mesopores, pores whose width is between 2nm and 50nm (20 Å - 500 Å); micropores, pores whose width does not exceed 2nm (20 Å). These limits are not very defined but they are not totally arbitrary. The micropore filling mechanisms depend on the pore shape though they are influenced by the adsorptive properties, and the adsorbent-adsorbate interactions.

Table 1. General Description of Some Types of Solid Sorbents

TYPE	GENERAL CHARACTERISTICS	EXAMPLES	ADSORPTION/DESORPTION
CARBONACEOUS SORBENTS			
Activated charcoals	High specific surface areas (800-1200 m2/g); basically microporous; wide distribution of pore sizes ; high adsorption capacities for most organics; partly polar surface .	Charcoal coconut. Charcoal petroleum	Solvent desorption. Polar compounds can not be quantitatively desorbed with a non polar solvent. Temperature can cause pyrolysis in organic adsorbates. Relatively high adsorption of water (HR>50%). High temperature stability.
Carbon molecular sieves	Medium-high specific surface areas (400-1000 m2/g); microporous; narrow distribution of pore sizes; similar adsorption capacities to activated charcoal; low polarity.	Carbosieve S-II, S-III,Spherocarb Ambersorb XE-340, XE-347, Carboxen569,564 563. Anasorb cms	To be used for thermal and solvent desorption. High temperatures can cause pyrolysis in organic adsorbates. Better desorption of polar compounds than activated charcoals. Low adsorption for water vapour High temperature stability (400 °C)
Graphitized carbon blacks	Low specific surface areas; non porous very homogeneous surfaces; non polar.	Carbotrap B Carbotrap C Carbotrap F	Thermal desorption. Low adsorption capacities for lighter compounds and water vapour. Appropiate for heavier compounds.High temperature stability
POROUS POLYMERS			
Styrene polymers and copolymers	Depending on monomer and cross-linked: specific surface areas; 300-800 m2/g;mesoporous/macroporous; variable surface polarity.	Chromosorb 102, 106, Poropak Q Amberlite XAD-2 Amberlite XAD-4	Thermal desorption. Not convenient for solvent desorption. Low affinity to water. Operational temperatures: 200°C - 250°C
Phenyl-phenilene oxide polymers	Low specific surface areas (20-35 m2/g); macroporous; low polarity.	Tenax GC Tenax TA	Thermal desorption. Not convenient for solvent desorption. Low affinity to water. Operational temperatures: 375°C

In Table 1 some features and properties of the carbonaceous adsorbents and porous polymers are collected among the ones in most use. Carbonaceous adsorbents can be gathered into *(a)* activated charcoals coming from charring of shell coconut, wood and also from petroleum and coals, *(b)* graphitized coals as the graphitized carbon black obtained by heating coals at very high temperatures in an inert atmosphere, and *(c)* carbon molecular sieves made from pitch and porous polymers by pyrolysis at lower temperatures than those for graphitization.

Within the porous organic polymers the most numerous group is made up by the styrene polymers and copolymers. The different types and denominations, which are available in the market, differ in their characteristics and properties due to the cross-linking grade, to the monomers present, and because they are usually subject to different treatments within each series.

In order to take advantage of the adsorbent properties of the different materials, sorbent tubes have been marketed having two or more types of different adsorbents. The goal is to collect compounds of different characteristics with the same tube. The first layer holds back the heavy compounds and the last one the lightest compounds.

In general porous polymers have lower adsorption capacities, in many cases much bigger pore sizes and lower thermal stability, in comparison with activated charcoals and carbon molecular sieves. Due to their lower adsorptive strength they cannot hold back efficiently the lightest compounds and the polar ones. Another item to be borne in mind is that porous polymers are swelled up with many solvents and water which may increase the back pressure through the tube during the sampling (4, 5). Regarding desorption, some of them such as Tenax or Amberlite XAD-2 cannot be used with solvents such as carbon disulphide. On the other hand, troubles can arise with the thermal desorption of heavier compounds when the maximum temperature of the polymer is exceeded.

Carbonaceous adsorbents are thermally more stable than porous polymers but in the microporous adsorbents, desorption is more difficult than it is in the macroporous or non-porous sorbents though the adsorption capacities are similar (6, 7). Activated charcoals and carbon molecular sieves can produce pyrolysis in some compounds during the thermal desorption process.

Both of them are suitable for solvent desorption. Carbon molecular sieves have the advantage of their lower affinity to the water, maintaining adsorption capacities similar to activated charcoals with regard to organic compounds. Graphitized carbon blacks have low surface areas because they are non-porous adsorbents, and with very homogeneous surfaces. These features, together with their high thermal stability, make them specially suitable for the collection and subsequent thermal desorption of the heaviest compounds.

3. ADSORPTION PROCESS

Active sampling using sorbent tubes is a dynamic adsorption process (8). The adsorption theory application under dynamic flow conditions has been used by several authors in order to predict the behaviour of the adsorbent-adsorbate system under different conditions of contaminant concentration, adsorbent amount, relative humidity, and presence of other contaminants. Adsorption equilibrium data allows us to predict the maximum capacity of an adsorbent, but contact time between the adsorbent and the adsorbate depends on the process kinetic. On the other hand, under the conditions in which the active sampling is held, usually it is not possible to saturate all the adsorbent without losing some adsorbate in the effluent stream. The adsorbent saturation level also depends on the process kinetic.

In studying a dynamic adsorption process, breakthrough curves are a useful tool. For different operation conditions breakthrough curves and therefore breakthrough times for a percent breakthrough can be obtained through theoretical calculations, or experimentally by direct or indirect methods. Indirect methods (9–12) have the advantage of speed but they have limitations which may lead to great differences with the actual results. Adsorption coefficients and maximum capacities which are obtained (12), would be applicable to smaller linear ranges of their adsorption isotherms at low concentrations where they are supposed to fulfil Henry's law.

Wheeler-Jonas (13–26) and Bergström (8, 27–29) equations provide an approximate mathematical description for breakthrough curves and they are the most employed equations for activated charcoal beds. Also, they are expected to be applied to other adsorbents. The Bergström equation reproduces better the breakthrough curve sigmoidal shape though for low values of percent breakthrough is the same for both equations. Following this author and from the breakthrough curve data experimentally obtained at several concentrations, the maximum adsorption capacity, W_e, and kinetic adsorption constant, k_v, can be calculated. From these the breakthrough time is obtained with the following equation:

$$t_b = \left[\frac{1}{C_o} \frac{W_e M}{Q} - \frac{1}{k_v} \log \left[\frac{C_o}{C_x} - 1 \right] \right] \quad \textbf{(1)}$$

where t_b is the breakthrough time for a breakthrough fraction C_x / C_o, C_o is the inlet concentration, C_x is the outlet concentration for a time t_b, M is the adsorbent mass and Q is the flow rate. From the maximum adsorption capacities and corresponding inlet concentrations, the constants of the adsorption isotherm which fit better into the experimental data, may be calculated (8).

In Table 2 maximum adsorption capacity values are collected for n-hexane and 2-propanol on activated charcoal, experimentally obtained and calculated from Langmuir's adsorption isotherm and from breakthrough curves. The joint use of equation **(1)** and the suitable adsorption isotherm allows us to calculate the breakthrough for any percent breakthrough and inlet concentration. Since W_e and k_v are constant values for each adsorbent-adsorbate system, the breakthrough time can also be obtained for different adsorbent masses and tube sizes for an equal lineal velocity value. Besides, useful conclusions can be obtained from a practical point of view in order to consider the effects of the presence of water vapour and other substances in the air.

Table 2. Maximum Adsorption Capacities for 2-propanol and n-hexane Individually and in Mixtures

ADSORBENT	COMPOUND	ATMOSPHERE CONCENTRATION (mmol/l)	MAXIMUM ADSORPTION CAPACITY (mmol/g)			
			Individually		In mixture	
			expected	founded	expected	founded
charcoal coconut	2-propanol	0.034	2.27	2.38	1.81	1.92 •
		0.017	1.90	1.89	1.56	1.51 ••
	n-hexane	0.020	1.67	1.91	0.95	0.82 •
		0.010	1.26	1.61	0.69	0.93 ••
charcoal JXC	2-propanol	0.034	2.91	2.89	1.85	1.63 •
		0.017	2.19	1.93	1.49	1.40 ••
	n-hexane	0.020	2.22	2.29	1.11	1.18 •
		0.010	1.64	1.78	0.88	0.88 ••

• Mixture 1 •• Mixture 2

Humidity Effect

The humidity effect over the behaviour of adsorbents on the adsorption of organic compounds has been studied by many authors, mainly about activated charcoals (18, 19, 21, 24, 28, 30, 31), though there are also studies about other carbonaceous adsorbents and porous polymers (5). Water vapour is always present in the air, and its concentration, even at low relative humidity values, is higher than any other substance present in the workplace air. Models based on the joint use of breakthrough curves and adsorption isotherms (18, 19, 21, 24, 28, 31) and others based on chemical kinetic models (30) have been developed in order to study humidity effects and predict the adsorbent-adsorbate system behaviour for activated charcoals. The common remarks in all the cases are the following:

(a) Maximum adsorption capacity, W_e, decreases at high relative humidity values.

(b) Kinetic adsorption coefficient, k_v, can increase (28) or decrease (21) depending on the adsorbate but in any case there can be a significant reduction in the breakthrough time for a given percent breakthrough.

(c) Humidity effect can be noticed at relative humidity values of 50%–60%, and the effect is higher as the concentration of the adsorbable substances in the air is reduced, and its solubility in the water is higher.

For other adsorbents such as graphitized carbon blacks, carbon molecular sieves (12), and porous polymers, the water is practically not adsorbed until reaching the relative humidity values higher than 90%. On porous polymers, however, we must bear in mind, the swelling effect because of the water vapour which can produce an increase of the back pressure through the tubes. In any case, direct methods have not been used to study the humidity effect over the breakthrough curve.

In the light of the available experimental evidence and from a practical point of view it could be suggested:

(a) To choose among the adsorbents having the suitable features, the one that is less affected by humidity, at least in those cases where relative humidities over 50%-60% could be expected.

(b) To use sampling times based on breakthrough times obtained with higher relative humidity than 80% with a direct method.

Multicomponent Systems

For the study of multicomponent systems similar equipment as is used for obtaining breakthrough curves by a direct method (a gas chromatograph with an analytical column) can be used. After passing through the sorbent tube, the gas goes into the gas chromatograph, and the mixture components are analysed for percent breakthrough. In this way, simultaneous breakthrough curves to all mixture components can be built up.

Despite its importance, the mutual effect of the presence of several components in the air has not been studied very much, and there are only some papers about activated charcoals (13-15, 25). Jonas and Col. (13-15), by using Wheeler-Jonas and D-R for the adsorption isotherm equations have found a good agreement with the experimental data for binary mixtures.

The theoretical equations, which would allow us to predict a multicomponent system behaviour have to be necessarily complex because not only the operation variables but the properties of their constituents and the competition among them to reach the active points of the adsorbent solid surface will influence the system. In spite of it, breakthrough curves maintain their usefulness in order to derive practical conclusions. Swearengen and Weaver (32), basing on their own remarks about the obtained breakthrough curves for binary systems, suggest the grouping of the molecules based on their molecular weights, boiling points, and vapour pressures in order to predict their behaviour in a multicomponent system. In accordance with the remarks made by these authors the tests made at our laboratory show that in an adsorbent-multiadsorbate system there is a decrease of the maximum adsorption capacity and breakthrough time at a certain percent breakthrough compared to the one which the components would show if they were alone in the same concentration as the mixture.

This effect can be clearly seen in Figures 1 and 2 for a mixture of seven components on different adsorbents obtained at our laboratory. At first, the most surprising phenomenon which can appear on these curves is that before reaching the equilibrium all breakthrough curves show regions over 100% breakthrough. Taking into account that breakthrough curves could be considered as a special case of the adsorption zone concentrations, it could be understood as the displacement of some components by others, as the adsorbent is being saturated with the compounds more strongly adsorbed.

The adsorption of each mixture component is affected by the presence of the other components. With the obtained concentration data and breakthrough time for the mixture of seven components we have drawn the 'mixture breakthrough curve' (Figures 1 and 2). It can be seen that while percent breakthrough of the individual compounds surpasses 100%, 'percent breakthrough of mixture' increases until reaching 100% without surpassing this value at any time during all time registered.

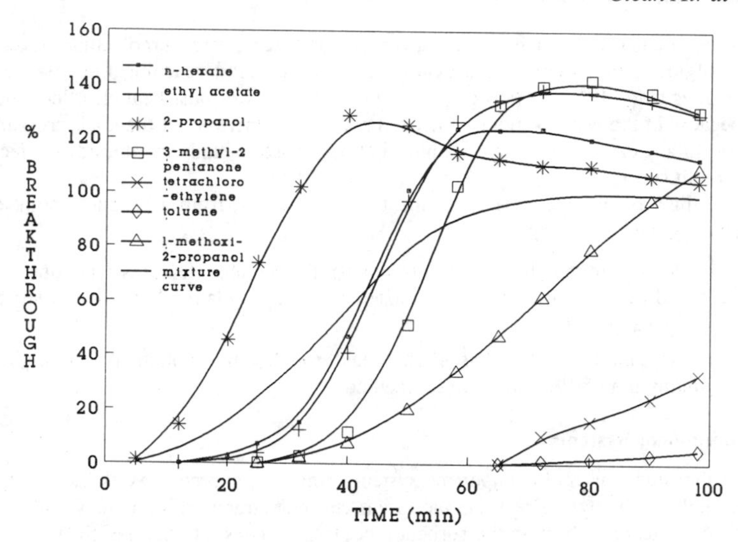

Fig. 1. Breakthrough curves for a seven component mixture on carbon molecular sieve (Carboxen 564)

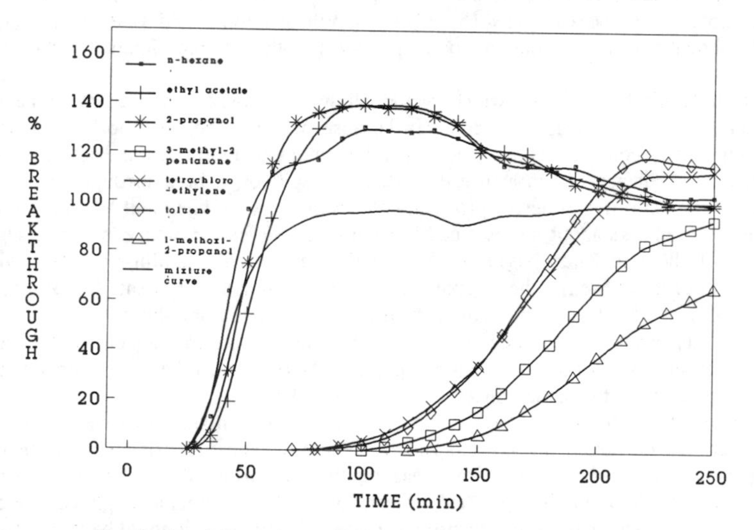

Fig. 2. Breakthrough curves for a seven component mixture on charcoal coconut (SKC)

Breakthrough curves suggest the following points of practical interest:

(a) Compounds, which are adsorbed less strongly, would be expected to appear in first place.

(b) The constituent in a mixture, which appears in first place when the adsorbent bed breakthrough is produced, is the determinant for the appraisal of the maximum allowed sampling time.

(c) The first component breakthrough time will decrease as the concentrations of the other components are higher. Our results suggest that this time might be appraised by considering a concentration, in number of moles, equal to that of all the other components in the mixture.

4. DESORPTION PROCESS

The procedures, which are basically used for the recovery of the retained substances into the sorbent tube, are solvent desorption and thermal desorption. Solvent desorption has the inconvenience of the use of solvents which are toxic in the analysis and dilute the sample (9). We can quote as main advantages the versatility with regard to the use of different desorption mixtures, possibility of analysing the sample with different analytical techniques, and that additional equipment is not necessary.

All the inconveniences of using solvents are eliminated in the thermal desorption, with the additional advantage that the adsorbents can be reutilized. As inconvenient we can quote that their use is practically restricted to a subsequent analysis by gas chromatography and their limitations with regard to desorption of high boiling point or strongly adsorbed compounds. Thermal desorption would be specially indicated for short sampling or low concentrations of contaminants in the air.

In general, it can be said that those compounds which are more strongly adsorbed, are desorbed with greater difficulty. This will lead to the need of using higher temperatures in thermal desorption and more powerful solvents in desorption with solvents. Desorption efficiency is calculated as the ratio of the recovered amount from a sorbent tube to that applied. It is desirable that the recovery is total, and does not vary with the concentration or with other operation factors.

Thermal Desorption

The increase of temperature makes easy the desorption as the sorbent-sorbate interaction strength decreases. Temperature and heating time are the two critical factors in the thermal desorption, and should be maintained as low as possible in order to avoid breakdowns or irreversible reactions as some authors have reported (5, 10). The breakdown during the thermal desorption process occurs with greater ease, and at lower temperatures with oxygen containing compounds.

On the adsorbents, irreversibility in the desorption will be foreseeably higher as long as the surface polarity is higher, and the adsorbent has greater adsorption power, and requires, thus, higher desorption temperatures. In this way, the least suitable ones for use in thermal desorption are the activated charcoals.

Light adsorbents such as Tenax and Chromosorb 106 would be suitable for medium or heavy compounds provided that the maximum polymer temperature is not exceeded. In this case, graphitized carbon blacks might be used. For the lighter compounds stronger adsorbents would be needed such as carbon molecular sieves.

Finally, just to outline with regard to the reutilization of the adsorbent, the recovery of the adsorbed compounds must be total.

Solvent Desorption

Solvent desorption is a static process where adsorption theoretical bases can also be applied.

Rudling (33) compares three models based on multicomponent adsorption isotherms to explain the mixture effects in the desorption process starting from the adsorption data obtained for the monocomponent systems.

The desorption efficiency will be determined by the compound equilibrium concentrations in the adsorbed phase and in solution. The lower the solubility of the compound of interest in the solvent, the greater is the affinity for the sorbent surface. Desorption efficiency is the result of the balance between the possibility of access of the solvent molecules to the absorbed molecules between the adsorbent pores and its capacity to displace them from the surface.

The suitable solvent characteristics, which displace the adsorption equilibrium towards the full adsorbate recovery, will depend on the sorbent and sorbed compound characteristics. In the carbon disulphide/activated charcoal system, as is well known, apolar compounds show total recoveries independently of the presence or absence of other polar or apolar compounds and of their own concentration. The reverse occurs for the polar compounds and the effect is greater as their polarity increases. The water also affects the recovery and stability of the polar adsorbates (34-36).

Recoveries also depend on the chosen procedure for their determination. Different methods are used: direct spiking method (NIOSH method), phase equilibrium method, direct spiking method with subsequent air circulation, and dynamic method.

When the recoveries are not total, the obtained values are different with the four methods (37). Therefore, the chosen method for the validation of a measurement procedure should be the dynamic method. Once the optimum desorption conditions are settled down and facing routine work the use of the remaining methods may be good alternatives as long as the actual environmental conditions are reproduced.

Table 3. Desorption Efficiencies of a Mixture of Compounds on Some Sorbents

DESORPTION EFFICIENCIES (SOLVENT: S_2C)

COMPOUND	Charcoal coconut		Charcoal JXC		Carboxen 564		Carbosphere 60/80	
	Average	CV	Average	CV	Average	CV	Average	CV
n-hexane	102.89	1.09	101.62	0.52	102.03	0.44	103.00	1.43
ethyl acetate	93.33	0.89	98.84	0.46	101.48	0.59	101.13	1.36
2-propanol	49.50	2.02	88.66	0.71	96.13	0.65	87.26	1.68
3-methyl-2-pentanone	91.57	1.79	10.11	0.36	102.93	1.02	101.41	2.31
tetrachloroethylene	100.00	2.41	99.78	1.47	100.00	2.12	100.65	2.16
toluene	98.89	0.99	99.25	0.44	100.04	0.59	100.55	1.27
1-metoxi-2-propanol	22.47	6.48	77.65	2.41	96.56	0.50	90.73	1.50

DESORPTION EFFICIENCIES (SOLVENT: S_2C + 5% 2-BUTANOL)

COMPOUND	Charcoal coconut		Charcoal JXC		Carboxen 564		Carbosphere 60/80	
	Average	CV	Average	CV	Average	CV	Average	CV
n-hexane	104.55	0.93	102.70	0.47	103.65	1.24	103.88	1.83
ethyl acetate	101.31	0.82	102.32	0.34	102.82	1.06	102.87	1.68
2-propanol	92.24	1.08	102.32	0.46	103.35	0.96	102.35	1.57
3-methyl-2-pentanone	101.39	1.68	103.29	1.06	103.43	1.48	101.99	1.95
tetrachloroethylene	100.91	2.68	100.67	1.30	101.79	1.80	98.21	2.65
toluene	100.58	0.67	101.20	0.48	100.52	1.16	99.64	2.45
1-metoxi-2-propanol	79.96	1.37	102.04	0.62	102.56	1.02	100.80	2.56

The recovery depends in an important way on the used adsorbent features. Desorption efficiencies, which have been obtained at our laboratory for a mixture of organic compounds, usually found in the solvents at the workplaces, adsorbed on several types of carbonaceous adsorbents, are shown in Table 3. Hydrocarbons (n-hexane and toluene) show in all cases total recoveries. However, the recoveries of the polar compounds, especially for 2-propanol and 1-methoxi-2-propanol, vary from one adsorbent to another. The lowest recoveries and higher variation coefficients correspond to the coconut charcoal.

In order to solve the problem of desorption and the stability of the polar compounds in activated charcoal, more polar solvents such as dichloromethane with 5%-methanol (34,35) have been used, or else another more polar compound as an alcohol in concentrations 1 to 5% has been added to the carbon disulphide. In previous works (38) we found that a suitable mixture for most organic compounds is the carbon disulphide + 5% 2-buthanol. Recently, we have tested this desorption mixture with other carbonaceous adsorbents as carbon molecular sieves (see Table 3). It may be seen that though the recoveries on these latter adsorbents are better than the coconut charcoal for the polar components, the recovery does not reach 100%. The use of carbon disulphide + 5% 2-buthanol, makes possible virtual total recoveries and lower variation coefficients from these adsorbents for all the compounds tested. This fact shows that, also for desorption, consideration of the sorbent characteristics is as important as the selection of the suitable solvent.

5. CONCLUSIONS

Choosing the suitable adsorbent is of prime importance for workplace air sampling and analysis of gases and vapours. On choosing the adsorbent its features should be considered (surface area, porosity, surface groups) and its properties determined with regard to adsorption and desorption processes. Therefore, it would be desirable for the information supplied by the manufacturers to be homogeneous and follow, if possible, IUPAC recommendations (3).

Breakthrough curves constitute the most suitable means to study the behaviour of the adsorbent-adsorbate system under dynamic flow conditions. The obtaining of breakthrough curves for different conditions in the CEN standard (1) the reliable use of the adsorbent tubes and allows the extrapolation of the data into other conditions.

In order to use sorbent tubes under high moisture conditions or in the presence of other organic vapours it would be useful to consider the recommendations included in Chapter 3.

On choosing the desorption system, adsorbent and adsorbate features should be considered as well as the type of sampling intended.

For the calculation of desorption efficiency during the validation process of a measuring procedure a dynamic method should be used. Once the suitable conditions are stated and in order to use them daily, static methods may be good alternatives.

6. REFERENCES

(1) EUROPEAN COMMITTEE FOR STANDARDIZATION (1991). General requirements for the performance of procedures for workplace measurements. prEN 482.

(2) GREGG, S.J. and SING, K.S.W. (1967). Adsorption, surface area and porosity. Academic Press, London.

(3) SING, K.S.W., EVERETT, D.H., HAVL, R.A.W., MOSCOU, L., PIEROTTO, R.A. ROUQUEROL, J. and SIEMIENIEWSKA, T. (1985). Presentación de datos de fisisorción en sistemas gas/sólido (Recomendaciones de la IUPAC de 1984). Spanish version: Ruiz Paniego, A. (1989). Anales de Química 85, 386-399. Original English text: Pure Applied Chem. 57, 603-619.

(4) BELYAKOVA, L.D. (1991). The porous structure of polymeric sorbents of different nature. In Characterization of porous solids II – Proceedings of the IUPAC Symposium (COPS II) by Rodríguez-Reinoso F., Rouquerol, J., Sing, K.S.W. and Unger K.K. (editors). Elsevier, Amsterdam. pp. 701-708.

(5) FABBRI, A., CRESCENTINI, G., MANGANI, F., MASTROGIACOMO, A.R. and BRUNER, F. (1987). Advances in the determination of volatile organic solvents and other organic pollutants by gas chromatography with thermal desorption sampling and injection. Chromatographia, 23, 856-860.

(6) SING, K.S.W. (1989). The use of physisorption for the characterization of microporous carbons. Carbon, 27, 5-11.

(7) KARTEL, N.T., PUZY, A.M. and STRELKO, V.V. (1991). Porous structure of synthetic active carbons. In Characterization of porous solids II – Proceedings of the IUPAC Symposium (COPS II) by Rodríguez-Reinoso, F., Rouquerol, J., Sing, K.S.W. and Unger, K.K. (editors). Elsevier, Amsterdam, pp. 439-447.

(8) BALIEU, E. and BJARNOV, E. (1990). Activated carbon filters in air cleaning processes – II. Prediction of breakthrough times and capacities from laboratory studies of model filters. Ann. occup. Hyg., 34,1-11.

(9) BROWN, R.H. and PURNELL, C.J. (1979). Collection and analysis of trace organic vapour pollutants in ambient atmospheres. The performance of a Tenax GC adsorbent tube. J. Chromatogr., 178, 79-90.

(10) BERTONI, G., BRUNER, F., LIBERTI, A. and PERRINO, C. (1981). Some critical parameters in collection, recovery and gas chromatographic analysis of organic pollutants in ambient air using light adsorbents J. Chromatogr., 203, 263-270.

(11) NAMIESNIK, J. and KOZLOWSKI, E. (1982). Comparative study of breakthrough volumes BTV on various sorbents. Fresenius Z. Anal. Chem. 311, 581-584.

(12) BETZ, W.R., MAROLDO, S.G., WACHOB, G.D. and FIRTH, M.C. (1989). Characterization of carbon molecular sieves and activated charcoal for use in airborne contaminant sampling. Am. Ind. Hyg. Assoc. J., 50,181-187.

(13) JONAS, L.A., SANSONE, E.B. and FARRIS, T.S. (1983). Prediction of activated carbon performance for binary vapor mixtures. Am. Ind. Hyg. Assoc. J., 44, 715-719.

(14) JONAS, L.A. and SANSONE, E.B. (1986). Prediction of activated carbon performance for sequential adsorbates. Am. Ind. Hyg. Assoc. J. 47, 509-511.

(15) SANSONE, E.B. and JONAS, L.A. (1981). Prediction of activated carbon performance for carcinogenic vapors. Am. Ind. Hyg. Assoc. J., 42, 688-691.

(16) YOON, Y.H. and NELSON, J.H. (1984). Application of gas adsorption kinetics – I.A theoretical model for respirator cartridge service life. Am. Ind. Hyg. Assoc. J., 45, 509-517.

(17) YOON, Y.H. and NELSON, J.H. (1984). Application of gas adsorption kinetics – II. A theoretical model for respirator cartridge service life and its practical applications. Am. Ind. Hyg. Assoc. J., 45, 517-524.

(18) YOON, Y.H. and NELSON, J.H. (1988). A theoretical study of the effect of humidity on respirator cartridge service life. Am. Ind. Hyg. Assoc. J., 49, 325-332.

(19) YOON, Y.H. and NELSON, J.H. (1990). Effects of humidity and contaminant concentration on respirator cartridge breakthrough. Am. Ind. Hyg. Assoc. J., 51, 202-209.

(20) YOON, Y.H. and NELSON, J.H. (1990). Contaminant breakthrough: A theoretical study of charcoal sampling tubes. Am. Ind. Hyg. Assoc. J. 51, 319-325.

(21) HALL,T., BREYSSE, P., CORN, M. and JONAS, L.A. (1988). Effects of adsorbed water vapor on the adsorption rate constant and the kinetic adsorption capacity of the Wheeler kinetic model. Am. Ind. Hyg. Assoc. J., 49, 461-465.

(22) ACKLEY, M.W. (1985). Residence time model for respirator sorbent beds. Am. Ind. Hyg. Assoc. J. 46, 679-689.

(23) COHEN, H.J. and GARRISON, R.P. (1989). Development of a field method for evaluating the service life of organic vapour cartridges: Results of laboratory testing using carbon tetrachloride. Am. Ind. Hyg. Assoc. J., 50, 486-495.

(24) COHEN, H.J., ZELLERS, E.T. and GARRISON, R.P. (1990). Development of a field method for evaluating the service lives of organic vapor cartridges: Results of laboratory testing using carbon tetrachloride. Part II: Humidity effects. Am. Ind. Hyg. Assoc. J., 51, 575-580.

(25) COHEN, H.J., BRIGGS, D.E. and GARRISON, R.P. (1991). Development of a field method for evaluating the service lives of organic vapor cartridges - Part III: Results of laboratory testing using binary organic vapor mixtures. Am. Ind. Hyg. Assoc. J., 52, 34-43.

(26) WOOD, G.O. and MOYER, E.S. (1989). A review of the Wheeler equation and comparison of its applications to organic vapor respirator cartridge breakthrough data. Am. Ind. Hyg. Assoc. J. 50, 400-407.

(27) BALIEU, E. (1976). Characterization of respirator adsorbent filters by means of penetration curve parameters. Ann. occup. Hyg., 19, 203-213.

(28) BALIEU, E. (1977). Characterization of respirator adsorbent filters by mean of penetralion curve parameters - II. Effect of relative humidity. Ann. occup. Hyg., 20, 375-384.

(29) BALIEU, E. (1989). Activated carbon filters in air cleaning processes - I. Introduccion and fundamental aspects. Ann. occup. Hyg., 33,181-195.

(30) WOOD, G.O. (1987). A model of adsorption capacities of charcoal beds. I. Relative humidity effects. Am. Ind. Hyg. Assoc. J., 48, 622-625.

(31) WERNER, M.D. (1985). The effects of relative humidity on the vapor phase adsorption of trichoroethylene by activated charcoal. Am. Ind. Hyg. Assoc. J., 46, 585-590.

(32) SWEARENGEN, P.M. and WEAVER, S.C. (1988). Respirator cartridge study using organic-vapor mixtures. Am. Ind. Hyg. Assoc. J., 49, 70-74.

(33) RUDLING, J. (1988). Multicomponent adsorption isotherms for determination of recoveries in liquid desorption of mixtures of polar solvents adsorbed on activated carbon. Am. Ind. Hyg. Assoc. J., 49, 95-100.

(34) RUDLING, J., BJÖRKHOLM, E. and LUNDMARK, B.O.(1986). Storage stability of organic solvents adsorbed on activated carbon. Ann. occup. Hyg. 30, 319-327.

(35) RUDLING, J. and BJÖRKHOLM, E. (1986). Effect of adsorbed water on solvent desorption of organic vapors collected on activated carbon. Am. Ind. Hyg. Assoc. J., 47, 615-620.

(36) URIBE, B., LOPEZ ARBELOA, J.F., ADRIAN, M. and TRANCHO, M.L. (1990). Estudio y evaluación de un método analítico para la determinación de una mezcla de alcoholes en aire. Presented at the 5th Jornadas de Análisis Instrumental, Grupo de Cromatografía y Técnicas afines. (Real Sociedad Española de Química), 7-9 November, Barcelona.

(37) LOPEZ ARBELOA, J.F., URIBE, B., ADRIAN, M. and TRANCHO, M.L. (1990). Análisis de compuestos orgánicos en aire. Estudio comparativo de los procedimientos de cálculo de la eficacia de desorción. Presented at the 5th Jornadas de Análisis Instrumental, Grupo de Croma tografía y Técnicas afines (Real Sociedad Española de Química), 7-9 November, Barcelona.

(38) LOPEZ ARBELOA, J.F., URIBE, B., ADRIAN, M. and TRANCHO, M.L. (1987). Propuesta de un nuevo procedimiento para la recuperación cuantitativa de compuestos orgánicos captados en carbón activo. Presented at the XI Congreso Nacional de Medicina, Higiene y Seguridad del Trabajo, Instituto Nacional de Seguridad e Higiene en el Trabajo, 1-4 December, Madrid.

1371

SAMPLING OF REACTIVE SPECIES

JAN-OLOF LEVIN

National Institute of Occupational Health
P. O. BOX 6104
S-90006 Umeå
Sweden

SUMMARY

Reactive organic compounds can often be sampled using solid sorbents coated with suitable reagents. This chemosorption technique is described for important classes of compounds like aldehydes, amines and diisocyanates. Requirements for chemosorbents are discussed and a number of examples are given.

1. INTRODUCTION

The use of solid sorbents is the most important method for workplace air monitoring. The most used sorbent is activated charcoal, a sorbent that can be used successively for the sampling of important unreactive compounds like aliphatic and armomatic hydrocarbons, chloroaliphatics, chloroaromatics etc. Unfortunately, compounds of moderate or high reactivity can not be sampled on charcoal owing to the high activity of this sorbent. Decomposition, polymerization or irreversible adsorption is frequently observed for reactive and high molecular compounds when sampled on activated charcoal.

For moderately reactive compounds, like certain organic monomers, ketones etc, a porous organic polymer like Tenax or Amberlite XAD can be used instead. For thermal desorption, where inertness of the sorbent is especially important, these organic polymers are often used. For highly reactive species, however, like the majority of low molecular aldehydes, amines, and diisocyanates, ordinary solid sorbents can not be used, even with solvent desorption.

Traditionally, methods employing bubbler collection have been used for the sampling of reactive compounds. Bubblers or impingers are not convenient in field investigations, especially not for personal monitoring of worker exposure, where breathing-zone sampling is required.

The introduction of reagent-coated sorbents for the sampling of reactive compounds has much simplified the measuring of these compounds. The technique of this methodology is to coat a suitable reagent onto a solid sorbent. During sampling a stable derivative is formed *in situ* on the adsorbent. The derivative is solvent desorbed and determined by a sensitive analytic technique like GC or HPLC. One of the first methods using a reagent-coated solid sorbent was reported in 1979 (1). The method utilized 2,4-dinitrophenylhydrazine (DNPH) coated on Amberlite XAD-2 for the sampling of formaldehyde. In this reaction, a stable hydrazone is formed, which is determined by HPLC and UV-detection. This is an example of *chemosorption*, and for this technique to be sucessful, the following criteria have to be met:

- the chemosorbent should be chemically stable
- the reaction should be rapid and quantitative
- the derivative should be chemically stable
- desorption of the derivative should be quantitative

In this work examples are given, illustrating the chemosorption technique for the measurement of reactive organic compounds in air.

2. CHEMOSORPTION OF REACTIVE COMPOUNDS

Aldehydes

From an occupational point of view, five aldehydes are of special interest, namely, formaldehyde, acrolein, glutaraldehyde, acetaldehyde and furfural. These are all reactive aldehydes, but furfural is stable enough to be sampled on an un-coated organic porous polymer like Amberlite XAD-2 (2).

Formaldehyde is the most important of the aldehydes, since it has an enormous industrial use. A large number of methods for the determination of formaldehyde in air have been published. In recent years several of the methods reported have utilized sampling on a reagent-coated sorbent. The reagent of choice is in most cases 2,4-dinitrophenylhydrazine (DNPH). The reaction between formaldehyde and DNPH is rapid and quantitative in the presence of acid, and the hydrazone formed is determined with excellent sensitivity by HPLC with UV detection. DNPH has been coated on various solid sorbents like XAD-2 (1), silica gel (3), octadecylsilane-bonded silica (4) and glass fiber filter (5, 6).

Aldehydes form oxazolidine derivatives with 2-(hydroxymethyl)piperidine (2-HMP), a reaction used by NIOSH and OSHA in the US for formaldehyde and several other aldehydes (7, 8). The 2-HMP is coated on XAD-2, and the derivative is determined by GC.

For the determination of formaldehyde in workplace as well as ambient air, the method using a 13-mm DNPH-coated glass fiber filter has several advantages:

- type AE glass fiber filters have extremely low formaldehyde background
- coated filters are easy to prepare and use in standard 13-mm filter cassettes
- sampling rates from 10 ml·min^{-1} to 1000 ml·min^{-1} can be used
- particulate formaldehyde (paraformaldehyde) is sampled as well
- simple desorption- and analysis steps, which can easily be automated

The DNPH-filter method has been validated in a number of laboratory and field studies against a number of other formaldehyde reference methods such as the chromotropic acid (CA) method, the hydrazinobenzothiazole (HBT) method, and the acetylacetone (AA) method. Some previously unpublished results are summarized in Tables 1 and 2.

Table 1. Comparison between pumped DNPH-coated 13-mm fliter (1.0 l·min^{-1}, 15 min, n=3) and continously reading acetylacetone method (SCALAR instrument) in exposure chamber experiments.

DNPH-coated filter	SCALAR instrument
7.0 μg·m^{-3}	6.5 μg·m^{-3}
14	7.2
23	25
40	32
72	69
123	128
193	243
437	441
725	744

Table 2. Comparison between pumped DNPH-coated 13-mm filter (0.2 $l\cdot min^{-1}$, 60 min, n=3) and impingers with HBT (0.4 $l\cdot min^{-1}$, 60 min, n=3) in exposure chamber experiments.

DNPH-coated filter	HBT impinger
0.93 $mg\cdot m^{-3}$	0.89 $mg\cdot m^{-3}$
0.94	1.06
3.63	3.73

One great advantage with the DNPH-coated filter is that it can be used for diffusive (passive) sampling of formaldehyde. Diffusive sampling has been recognized as an efficient alternative to pumped sampling in occupational hygiene (9). A diffusive sampler for formaldehyde consisting of a 37-mm, DNPH-coated glass fiber filter mounted in a cassette constructed from a a standard 37-mm filter holder has been used for both occupational and indoor air applications (6, 10). The sampling rate of the sampler is 61 $ml\cdot min^{-1}$. The sampler is shown in Figure 1.

Figure 1.

Another diffusive sampler especially designed for the sampling of reactive compounds using a reagent-coated filter is shown in Figure 2. This sampler consists of a 20 x 45 mm impregnated filter mounted in a polypropylene housing. For formaldehyde the DNPH-coated filter is used. The uptake rate of this sampler is 25 $ml\cdot min^{-1}$, and it has been extensively validated in laboratory as well as field studies (11). Since the chromatographic determination of the hydrazone is highly sensitive and selective, the sampler can be used for short-time sampling (15 min) in the ppm range. In an 8-h sample the sensitivity is about 5 ppb (0.006 $mg\cdot m^{-3}$).

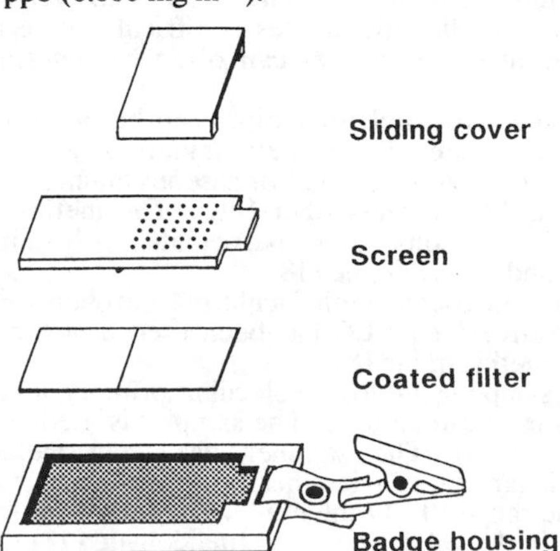

Figure 2.

It has been shown that the sampler performs well at extremely low wind velocities. Thus it can be used for static monitoring of low levels of formaldehyde in indoor air, as well as for personal monitoring. The sampler is commercially available and intended for one-time use (11).

A Round Robin interlaboratory exercise with the sampler has been carried out with a number of Swedish occupational health laboratories participating. Samplers were exposed for various levels of formaldehyde in a dynamic generation system (6). Formaldehyde concentrations in the chamber were determined using the pumped 13-mm DNPH-coated filters. Each laboratory received four samplers exposed at different levels. The results are shown in Table 3, and indicate that most laboratories performed quite well. The sampler has at the present time been in routine use in Sweden for about three years.

Table 3. Round Robin exercise with formaldehyde diffusive samplers exposed for 60 min at 0.59 (I), 0.96 (II), 1.37 (III) and 1.91 (IV) $mg{\cdot}m^{-3}$. Amount formaldehyde found corrected for blank, in μg.

Laboratory	I	II	III	IV
NOMINAL	0.90	1.44	2.06	2.87
LAB A	0.93	1.51	2.18	3.16
LAB B	0.87	1.41	2.11	2.83
LAB C	0.89	1.29	1.99	3.01
LAB D	0.80	1.18	1.74	2.46
LAB E	0.87	1.49	2.48	3.02
LAB F	0.80	1.31	1.88	2.64
MEAN	0.86	1.37	2.06	2.85

Amines

Both aliphatic and aromatic amines are reactive compounds. However, tertiary aliphatic amines are often stable enough to be sampled on uncoated solid sorbents, preferably organic polymers (12-14). For the more reactive primary and secondary amines chemosorption can be used, in order to avoid the cumbersome wet sampling methods. The simplest way is to use acid-coated sorbents or filters followed by GC determination. Since this procedure normally includes no derivatization, and since direct GC determination of free aliphatic amines is difficult, these methods are usually not very sensitive. Larger aromatic amines can often be determined by HPLC without derivatization.

Primary and secondary aliphatic amines can be sampled using a sorbent coated with 1-naphthylisothiocyanate. Substituted thioureas are formed, which can be determined by HPLC and UV-detection. For gaseous amines coated XAD-2 is used, and for particulates coated 13-mm glass fiber filter. The method has been validated for ethylenediamine (15), gaseous and particulate polyamines (16), mono- and diethanolamine (17), and diethylamine (18).

An XAD-7 sorbent coated with 7-chloro-4-nitrobenzo-2-oxa-1,3-diazole (NBD) for producing a derivative for HPLC has been used in a similar manner for methyl-, dimethyl-, ethyl-, and diethylamine (8).

For diffusive sampling of low molecular primary and secondary amines, the sampler shown in Figure 2 can be used. The sampler is used with a filter coated with 1-naphthylisothiocyanate (18). The sampler has been validated for methylamine, dimethylamine, diethylamine, allylamine, isopropylamine, and n-butylamine. The uptake rates are in the range 10 - 17 $ml{\cdot}min^{-1}$. The sensitivity for an 8-h sample is about 10 ppb. It was shown in the validation for formaldehyde (11) that the sampler performs well at extremely low wind velocities. Thus, the sampler can be used for static as well as personal monitoring of low levels of primary and secondary amines in air.

Diisocyanates

Diisocyanates are highly reactive monomers used in polyurethane production. For sampling, a number of amine reagents have been used, mostly in solution. Substituted ureas are formed with amines, which are conveniently determined by HPLC and UV or fluorescence detection, or a combination of UV and electrochemical detection. The most common reagents employed are 1-(2-methoxyphenyl)-piperazine (2-MP) (20), 1-(2-pyridyl)-piperazine (2-PP), and 9-(N-methylaminomethyl)-anthracene (MAMA). Some of these reagents have been coated on sorbents or filters (7, 8).

Sampling of diisocyanates is difficult mainly for two reasons. Firstly, the isocyanates represent a wide range of volatility, making it necessary to sample both vapour and particles. Secondly, the reactivity of the isocyanate monomers will produce dimers and other oligomers in industrial environments, as well as the corresponding amines. This raises the question whether the monomer or the isocyanate group should be determined in the analytical step.

Reagent-coated sorbents have been shown to work well with volatile isocyanates like TDI (19). MDI is one of the most important isocyanates. Because of its low vapour pressure it exists both as vapour and aerosol in the workplace atmosphere. Both MDI vapour and particles have been sampled in laboratory experiments with coated sorbents and filters. However, when sampling MDI in the field, there have been numerous observations of discrepancies between methods using impingers and methods using coated sorbents or filters. Generally the impinger methods have been reading higher than the chemosorption methods. This has been interpreted in terms of too low inlet capture velocities for the MDI particles when 25 mm or 37 mm coated filters have been used, and hence the use of coated 13-mm filters has been recommended (20).

Another explanation for the discrepancies observed in field investigations is that the MDI monomer forms some sort of complex, which has to be broken down in solution before reaction with the amine can occur. So far there is no strong field data supporting either hypothesis. Until this question has been addressed properly, MDI should be sampled using impingers.

Miscellaneous

In addition to the important classes of compounds mentioned above, there are several reactive organic compounds that can be sampled using reagent-coated sorbents. For example ethylene oxide is sampled on HBr-coated charcoal and hydrazine on benzaldehyde-coated XAD-2. Ethyleneoxide is converted to 2-bromoethanol, which is determined by GC (8). Hydrazine forms a stable aldazine derivative, which is determined by HPLC (21).

3. CONCLUSIONS

Sampling methods using impingers or gas dispersion bottles with liquids are not convenient in personal sampling. Chemosorption (sorption on a reagent-coated sorbent) offers a convenient alternative for those compounds that can not be sampled on ordinary solid sorbents because of their reactivity. The reactive compound is trapped on the sorbent by means of a suitable derivatizing agent. For aldehydes, amines and diisocyanates, the reaction between nitrogen of an amino group and carbon of a carbonyl, an isocyanate or a thioisocyanate is utilized. The derivatives formed are stable hydrazones, ureas or thioureas. By incorporating a group in the reagent molecule that is suitable for a specific detection system, the sensitivity of the method can be optimized.

REFERENCES

(1) ANDERSSON, K., ANDERSSON, G., NILSSON, C.A. and LEVIN, J.O. (1979). Chemosorption of formaldehyde on Amberlite XAD-2 coated with 2,4-dinitrophenylhydrazine. **Chemosphere 8,** 823-827.

(2) LEVIN, J.O., NILSSON, C.A. and ANDERSSON, K. (1977). Sampling of organic substances in workroom air using Amberlite XAD-2 resin. **Chemosphere 6,** 595-598.

(3) BEASLEY, R.K., HOFMANN, C.E., RUEPPEL, M.L. and WORLEY, J.W. (1980). Sampling of formaldehyde in air with coated solid sorbent and determination by high performance liquid chromatography. **Anal. Chem. 52,** 1110-1114.

(4) KUWATA, K., UEBORI, M., YAMASAKI, H., KUGE, Y. and KISO, YOSHIYUKI, Y. (1983). Determination of aliphatic aldehydes in air by liquid chromatography. **Anal. Chem.** 55, 2013-2016.

(5) LEVIN, J.O., ANDERSSON, K., LINDAHL, R. and NILSSON, C.A. (1985). Determination of sub-parts-per-million levels of formaldehyde in air using active or passive sampling on 2,4-dinitrophenylhydrazine-coated glass fiber filter and high performance liquid chromatography. **Anal. Chem. 57,** 1032-1035.

(6) LEVIN, J.O., LINDAHL, R. and ANDERSSON, K. (1986). A passive sampler for formaldehyde in air using 2,4-dinitrophenylhydrazine-coated glass fiber filters. **Environ. Sci. & Technol. 20,** 1273-1276.

(7) U.S. NATIONAL INSTITUTE FOR OCCUPATIONAL SAFETEY AND HEALTH. (1984). NIOSH Manual of analytical methods, 3rd ed, Cincinnati, USA.

(8) U.S. OCCUPATIONAL SAFETY AND HEALTH ADMINISTRATION. (1985). OSHA Manual of analytical methods, Salt Lake City, USA.

(9) BERLIN, A., BROWN, R.H. and SAUNDERS, K.J., eds. (1987). Diffusive sampling. An alternative approach to workplace air minitoring. Royal Society of Chemistry, London.

(10) LEVIN, J.O., LINDAHL, R. and ANDERSSON, K. Monitoring of part-per-billion levels of formaldehyde using a diffusive sampler. (1989) **J. Air Pollut. Control Assoc. 39,** 44-47.

(11) LEVIN, J.O., LINDAHL, R. and ANDERSSON, K. (1988). High performance liquid chromatographic determination of formaldehyde in air in the ppb to ppm range using diffusive sampling and hydrazone formation. **Environ. Technol. Lett. 9,** 1423-1430.

(12) ANDERSSON, B. and ANDERSSON, K. (1986). Air sampling of N-methylmorpholine on solid sorbent and determination by capillary gas chromatography and a nitrogen-phosphorus detector. **Anal. Chem. 48,** 1527-1529.

(13) ANDERSSON, B. and ANDERSSON, K. (1989). Determination of tertiary amines in air. **Appl. Ind. Hyg. 4,** 175-179.

(14) ANDERSSON, B. and ANDERSSON, K. (1991). Determination of heterocyclic tertiary amines in air. **Appl. Occup. Environ. Hyg. 6,** 40-43.

(15) ANDERSSON, K, LEVIN, J.O., HALLGREN, C. and NILSSON, C.A. (1985). Determination of ethylenediamine in air using reagent-coated adsorbent tubes and high performance liquid chromatography on the 1-naphthylisothiourea derivatives. **Am. Ind. Hyg. Assoc. J. 46,** 225-229.

(16) LEVIN, J.O., ANDERSSON, K., FÄNGMARK, I. and HALLGREN, C. (1989). Determination of gaseous and particulate polyamines in air using sorbent or filter coated with naphthylisothiocyanate. **Appl. Ind. Hyg. 4,** 98-100.

(17) LEVIN, J.O., ANDERSSON, K. and HALLGREN, C. (1989). Determination of monoethanolamine and diethanolamine in air. **Ann. Occup. Hyg. 33,** 175-180.

(18) LEVIN, J.O., LINDAHL, R., ANDERSSON, K and HALLGREN, C. (1989). High- performance liquid-chromatographic determination of diethylamine in air using diffusive sampling and thiourea formation. **Chemosphere 18,** 2121-2129.

(19) ANDERSSON, K., GUDHÉN, A., LEVIN, J.O. and NILSSON, C.A. (1982). Analysis of gaseous diisocyanates in air using chemosorption sampling. **Chemosphere 10,** 3-10.

(20) COYNE, L.B., KLINGER, T. and MOORE, G. (1989) The development and validation of a new air monitoring method for MDI vapor and aerosols in the workplace. **American Industrial Hygiene Conference,** St. Louis, USA.

(21) ANDERSSON, K., HALLGREN, C., LEVIN, J.O. and NILSSON, C.A. (1984). Liquid chromatographic determination of hydrazine at sub ppm levels in workroom air as benzaldazine with the use of chemosorption on benzaldehyde-coated Amberlite XAD-2. **Anal. Chem. 56,** 1730-1731.

DIFFUSIVE SAMPLING

R. BROWN

Health and Safety Executive,
Occupational Medicine and Hygiene Laboratory,
403, Edgware Road, London, NW2 6LN, UK

SUMMARY

The present state-of-the-art of diffusive sampling technology is reviewed in the light of the performance requirements and other desirable features that samplers must have to be of practical use in monitoring gases and vapours in workplace atmospheres. In general, diffusive samplers match up to these expectations, provided they are not regarded as a universal panacea. For samplers to become widely accepted, there is still a need for more research and sharing of validation information.

1. INTRODUCTION

Diffusive samplers, as a means of monitoring workplace atmospheres for toxic gases and vapours, have evolved over the last ten years or so in response to a growing demand for measurement. Early prototypes were largely experimental, and some which were found to be technically inadequate have not survived. The present generation of samplers, however, have been developed from a sound base of knowledge of the factors affecting performance, and have resulted from clearly understood technical requirements in relation to key application areas.

First and foremost, samplers must be reliable and give valid results. Reliability is not too easy to define, but includes freedom from interferences or artefacts and resistance to contamination or abuse. The measured result should be unambiguous. Validity is measured in CEN terms (1) as 'overall uncertainty'; a combination of precision or random error and bias or systematic error. Depending on the measurement task (2), overall uncertainty should be better than a specified amount, usually 30%, and this criterion should be met under conditions typical of the workplace (1). Validity is also linked to quality control, quality assurance and laboratory accreditation (3).

In addition, there is a convenience factor. Ideally, particularly if the sampler is to be used for personal monitoring, the sampler should be small, light, unobtrusive and convenient. It is then unlikely to be influenced by worker behaviour. It should be easily calibrated and compatible with automation. Depending on the specificity requirement, it may need to be capable of simultaneous measurement of more than one pollutant.

Depending on the measurement task (2), a sampler may be required for screening measurements, emission measurements or for comparison with limit values. The sampler may need to monitor continuously, measure short-term excursions, or give a long-term time-weighted average concentration. The sample location may be for personal monitoring, background (static) monitoring or boundary fence/ environmental monitoring.

2. VALIDATION

For a sample measurement to be valid, the sampler should meet the CEN criteria for performance requirements (1). These requirements include unambiguity, selectivity and overall uncertainty; the specific requirements for the last two depending on the measurement task. Not all samplers will meet these requirements, but they must a priori have been tested, and the corresponding CEN technical method of test (4) should be followed. This test procedure examines sampler performance in the laboratory when the sampler is exposed to atmospheres of gas or vapour of known concentration under a variety of environmental conditions, typical of the workplace. Environmental variables include time, concentration, humidity, temperature, interferences and storage.

The CEN recommendations are minimum requirements, and it is strongly advised that field tests are also undertaken (5) following the procedures in the annexes (4).

3. OVERVIEW OF SAMPLER TYPES

A variety of diffusive samplers have been developed (5) and only a selection of the major types manufactured can be described here. Diffusive equivalents to the more familiar pumped methods exist for nearly all types; the main exception being the direct collection of gas samples, where the nearest equivalent is an evacuated canister. Thus the diffusive equivalent of an impinger is a liquid-filled badge such as the Pro-Tek(TM) inorganic monitor or the SKC badge; the diffusive equivalent of the charcoal tube is the charcoal badge such as the 3M OVM or the MSA VaporGard organic; and the diffusive equivalent of the thermal desorption method is the Perkin-Elmer tube or the SKC thermal desorption badge. There are also diffusive devices based on reagent-impregnated solid supports. Some of these are direct-reading (the equivalent of pumped gas detector tubes) and some require a separate analytical work-up. Lastly, some direct-reading instruments use diffusive sampling to collect the sample, but these and detector tubes are dealt with in separate papers.

It is perhaps convenient to consider diffusive samplers as falling into eight categories, resulting from three variables; the geometry may be 'badge' or 'tube', the collection medium may be a chemical reagent or a non-specific (ad)sorbent, and the measurement may be direct read-out or separate analysis. Examples of nearly all these combinations are in (5). The main exception is the combination of non-specific sorbent with direct read-out.

4. CALIBRATION

The basic expression of Fick's Law is;

$$J = D/L(C_o - C_e) \tag{1}$$

and

$$Q = DA/L(C_o - C_e)t \tag{2}$$

where:
- J = diffusive flux (g/sec)
- D = coefficient of diffusion (cm^2/sec)
- A = cross sectional area of diffusion path (cm^2)
- L = length of diffusion path (cm)
- C_o = external concentration being sampled (g/cm^3)
- C_e = concentration at the interface of the sorbent (g/cm^3)

Q = mass uptake (g)
t = time of sampling (sec)

It is apparent from an inspection of these equations that the expression DA/L has units of cm^3/sec and therefore represents what can be considered as a "sampling rate" of the diffusive sampler when comparing to a pumped sampling system. This simple use of the sampling rate concept has proven of considerable value to users of the devices and is often expressed in the dimensionally equivalent un its of ml/min. Knowledge of the geometry of the sampler (which will be fixed for any given sampler type) permits the calculation of the sampling rate provided the diffusion coefficient is known. A number of manufacturers have published tables of sampling rates calculated in this way, most of whom have used the same source of published diffusion coefficients (6). Diffusion coefficients that are not in this list can be calculated theoretically (7).

5. ENVIRONMENTAL FACTORS AFFECTING SAMPLER PERFORMANCE

Temperature and pressure

From Maxwell's equation, the diffusion coefficient, D, is a function of absolute temperature and pressure;

$$D = f(T^{3/2}, P^{-1}) \quad (3)$$

But from the general gas law;

$$PV = nRT$$

$$C = n/V = P/RT \quad (4)$$

Substituting (3) and (4) in (2), we get;

$$Q = f(P/T,\ T^{3/2}/P)$$

$$= f(T)^{1/2} \quad (5)$$

Thus Q is independent of pressure, P, but dependent on the square root of absolute temperature, T. In practice, the temperature dependence of the sampling rate at ambient temperature levels (about 0.2% per ^{o}C) may be ignored. However, temperature may affect the sorption capacity of a sampling medium adversely.

Humidity

High humidity can affect charcoal (ad)sorption adversely, resulting in a reduction in the saturation capacity of charcoal badges. If the sampler becomes saturated, C_e in Equation 1 is no longer zero, and the sampling rate becomes non-linear. Porous polymers used for thermal desorption are relatively unaffected by humidity.

Transients

Simple derivations of Fick's Law assume steady-state conditions, but in the practical use of such samplers, of course, the ambient concentrations of pollutants are likely to vary widely. The question then arises whether a diffusive sampler will give a truly integrated response, or will "miss" short-lived transients before they have had a chance to diffuse into the sampler. The problem has been discussed theoretically and practically. Generally, transients do not present a significant problem provided the total sampling time is well in excess of the time constant of the sampler, i.e. the time a molecule takes to diffuse into the sampler under steady-state conditions. The time constant of most commercial samplers is between 1 and 10 secs.

Sorbent factors

All diffusive samplers rely on sorbents having a high affinity for the contaminant being sampled, i.e. C_e = zero in Equation 1 and uptake is linearly proportional to concentration and time of exposure. Useful checks on sorbent suitability are a back-diffusion test given in Bartley (9) and used in CEN (4) and the measurement of adsorption isotherms (10).

6. CALCULATIONS

The method of calculation of atmospheric concentrations is essentially the same as for pumped samplers, i.e. the collected sample is analysed and the total weight of analyte on the sampler determined.

Then,

$$C = \frac{(m_1 + m_2 - m_{blank})}{DE \times V} \quad (7)$$

where: m_1 = weight of analyte on first tube section (ug)
m_2 = weight of analyte on back-up tube section (if used) (ug)
DE = desorption efficiency corresponding to m_1

(m_2 and DE are ignored for liquid sorbent badges)

V, the total sample volume, is calculated from the effective sampling rate (litres/min) and the time of exposure (min).

This calculation gives C in mg/m^3, and strictly speaking, a appropriate sampling rate for the ambient temperature and pressure should be made, as Equation 4 assumes C is in ppm.

Alternatively, sampling rates can be expressed in units such as ng/ppm/min (dimensionally equivalent to cm^3/min), when C′ is calculated directly in ppm;

$$C' = \frac{(m_1 + m_2 - m_{blank})}{DE \times U \times t'} \times 1000 \quad (8)$$

where: U = sampling rate (ng/ppm/min)
t′ = sampling time (min)

Note

m_2 is relevant only to samplers with a back-up section, and an additional multiplication factor may be needed to account for differing diffusion path lengths to primary and back-up sections.

7. EVALUATION OF DIFFUSIVE SAMPLERS

Validation of diffusive monitors

Brown (11) examined the Perkin-Elmer tube for acrylonitrile, benzene, butadiene, carbon disulphide and styrene and found the sampler to be at least as accurate as the equivalent pumped method. Laboratory precision was, on average, 10% for the diffusive sampler. Field precision was 12% for the diffusive sampler and 13% for the pumped sampler.

Kennedy (12) evaluated a range of inorganic samplers, including 3M, DuPont, MSA, REAL and SKC samplers, and found they generally met NIOSH criteria (Cassinelli, 13).

A European interlaboratory comparison (14) of the 3M badge exposed to butanol, pentanal, trichloroethane, octane, butyl acetate, 3-heptanone, xylene, a-pinene and decane showed generally good agreement with the charcoal tube, except for butanol and pentanal, where the diffusive samplers read low. Laboratory precision was 13%, excluding pentanal, largely arising from inter-laboratory error.

Generally speaking, the laboratory precision of diffusive samplers is about 10% with field precision a little greater. Systematic bias, measured against a test gas of constant composition, is generally zero, as samplers are calibrated against such a gas. However, some bias can result from measuring non-constant concentrations with samplers containing non-ideal sorbents (see 'Transients' above). For example, if the CEN back-diffusion test (3) is applied to benzene on the Perkin-Elmer tube with Tenax sorbent, the difference between samplers exposed or not-exposed to clean air is 13% (B Fields, private communication). This represents the maximum bias that might be encountered in the field. The CEN calculation of overall uncertainty (1) is then bias + 2x precision (CV), i.e. normally 20% but 33% in the benzene example. The latter would then fail the CEN 30% criterion (but would be useful for screening for example). Precision can be improved by careful attention to detail, and the back-diffusion bias can be eliminated by using stronger sorbents (Chromosorb 106 in the benzene example).

Certified Reference Materials

A CRM for benzene, toluene and xylene in Tenax samplers has been prepared by BCR (15). The sample loading is appropriate for either pumped or diffusive sampling.

Other technical requirements

Diffusive samplers are generally free from artefacts and resistant to contamination. Exceptions are some early samplers with carbon sorbents, which had relatively high levels of blank hydrocarbons, but these problems have now been overcome. If Tenax is used as sorbent, artefacts can be caused by oxidising gases (ozone and NO_2) but levels are not significant for workplace air monitoring. Thermal desorption samplers must be carefully sealed because of the very high sensitivity of the analysis, and compression fittings with PTFE seals are preferred if contamination is to be minimised.

There is little doubt that diffusive samplers are generally convenient and easy to use. Just a few badge-type samplers, particularly those with liquid sorbents, are a little difficult to assemble. For monitoring sub-ppm levels, it may be necessary to assemble samplers in specially clean areas to reduce blank levels. Calibration is generally straightforward, although it is of course necessary to know the sampling rate (except for some pre-calibrated direct-reading devices) and this information may not be available. A preliminary study (16) has suggested that diffusive samplers affect worker performance less than conventional samplers.

A wide variety of diffusive samplers is available, which meet most of the needs of the occupational hygienist in his measurement tasks (1). However, because of face velocity effects, some designs of diffusive sampler are less suitable for monitoring in still air conditions.

8. DOES ANYONE USE THEM?

In a recent aromatic hydrocarbons sampling exercise organised by BCR (March 1991) only 14% of the sampling methods used by the participants were diffusive. The proportion of national standards (for gases and vapours) using diffusive samplers is somewhat higher at 25% and a selection of CA Selects abstracts for 1990/1 rate 31% diffusive. Overall, therefore, only about a quarter of users prefer diffusive samplers, in spite of the apparent advantages.

The reasons for this are not obvious, and are probably a combination of factors. National authorities (particularly in the US) have been slow to 'approve' diffusive samplers. Validation data, and sampling rates, are difficult to find and are not always reliable. But basically, I believe there is still a distrust of diffusive samplers. They are still being discredited because some of the earliest models were poorly designed, because of a poor understanding of the (real) factors affecting performance, and because of poorly designed field trials.

What is needed is more research, more collaboration on validation and more sharing of information.

REFERENCES

(1) CEN/TC137 General requirements for the performance of procedures for workplace measurements (PrEN 482).

(2) CEN/TC137 Guidance on the assessment of exposure to chemical agents in air at the workplace for comparison with limit values and measurement strategy (draft standard, doc N48, March 1991).

(3) HEALTH AND SAFETY EXECUTIVE (1991). Methods for the Determination of Hazardous Substances. Analytical quality in workplace air monitoring. MDHS 71. HSE/OMHL, London, UK.

(4) CEN/TC137 Workplace atmospheres - Requirements and test methods for diffusive samplers for the determination of gases or vapours (draft standard, doc N55, May 1991; accepted as PrEN).

(5) SQUIRRELL, D.C.M. (1987). Diffusive sampling - an overview. In: Diffusive Sampling; an Alternative Approach to Workplace Air Monitoring, A. Berlin, R.H. Brown and K.J. Saunders, Eds. CEC Pub. No. 10555EN, Bruissels-Luxembourg.

(6) LUGG, G.H. (1968). Diffusion coefficients of some organic and other vapours in air. Anal. Chem. 40:1072.

(7) PANNWITZ, K.-H. (1984). Diffusion coefficients. Draeger Review 52:1.

(8) BROWN, R.H. and WOEBKENBERG, M.L. (1989). Gas and vapor sample collectors. In: Air Sampling Instruments, 7th Ed, S.V. Hering, Ed., ACGIH, Cincinnati, OH, US.

(9) BARTLEY, D.L., DEYE, G.J. and WOEBKENBERG, M.L. (1987). Diffusive monitor test: performance under transient conditions. Appl. Ind. Hyg. 42:119.

(10) VAN DEN HOED, N. and HALMANS, M.T.H. (1987). Sampling and thermal desorption efficiency of tube-type diffusive samplers: Selection and performance of adsorbents. Am. Ind. Hyg. Assoc J. 48:364.

(11) BROWN, R.H. (1987). Applications of the HSE diffusive sampler protocol. In: Diffusive Sampling; an Alternative Approach to Workplace Air Monitoring, A. Berlin, R.H. Brown and K.J. Saunders, Eds. CEC Pub. No. 10555EN, Brussels-Luxembourg.

(12) KENNEDY, E.R., CASSINELLI, M.E. and HULL, R.D. (1987). Verification of passive monitor performance: applications. In: Diffusive Sampling; an Alternative Approach to Workplace Air Monitoring, A. Berlin, R.H. Brown and K.J. Saunders, Eds. CEC Pub. No. 10555EN, Brussels-Luxembourg.

(13) CASSINELLI, M.E., HULL, D., CRABLE, J.V. and TEASS, A.W. (1987). Protocol for the evaluation of passive monitors. In: Diffusive Sampling; an Alternative Approach to Workplace Air Monitoring, A. Berlin, R.H. Brown and K.J. Saunders, Eds. CEC Pub. No. 10555EN, Brussels-Luxembourg.

(14) DEBORTOLI, M., LOLHAVE, L. and ULLRICH, D. (1987). European interlaboratory comparison of passive samplers for organic vapour monitoring in indoor air. In: Diffusive Sampling; an Alternative Approach to Workplace Air Monitoring, A. Berlin, R.H. Brown and K.J. Saunders, Eds. CEC Pub. No. 10555EN, Brussels-Luxembourg.

(15) VANDENDRIESSCHE, S., GRIEPINK, B., HOLLANDER, J.C.Th., GIELEN, J.W.J., LANGELAAN, F.G.G.M, SAUNDERS, K.J. and BROWN, R.H. (1991). Certification of a reference material for aromatic hydrocarbons in Tenax samplers. Analyst 1 16:437.

(16) CHERRIE, J.W., LYNCH, G., CORFIELD, M., COWIE, H. and ROBERTSON, A. (1991). Effects of workers behaviour on measurements made using pumped and diffusive sampling techniques. Report TM/91/04. Institute of Occupational Medicine, Edinburgh, UK.

1373

DETECTOR TUBES

K. LEICHNITZ

Drägerwerk AG
Moislinger Allee 53-55, D-2400 Lübeck

SUMMARY

Detector tubes are laboratory methods simplified to the extent that they could be employed as a rapid test in the field. Independent institutes have assessed the detector tube method and confirmed its reliability. The user of detector tubes has various possibilities for accepting this measuring system for analysis. He can accept the manufacturer's calibration, check his quality control by a visit to the manufacturing facility, or the user can recalibrate the detector tubes by checking a representative sample of the detector tube shipment by means of test gases. Regardless of the way in which the quality of detector tubes is checked, the usage of this system has to take place within a carefully prepared sampling strategy.

1. INTRODUCTION

Measuring air contaminants is an essential requirement with regard to effective occupational hygiene and safety. Only by measurement is it possible to detect critical cases and to take appropriate measures.

The number of different measurement tasks to be performed in order to determine the contaminant concentration needs to be minimised, as the task is enormous. However, the experts' view concerning the feasibility of analysis methods is not as easily reduced. The 'great analysis' is still treated as equivalent to 'great accuracy'; whereas one tends to classify simple measuring methods as 'rough' measurements. The very fact that even the 'great analysis' – as far as accuracy is concerned – is limited, will not be discussed in detail. The scope of this paper is to give information concerning a simple measuring method, which allows the determination of contaminants in the air at workplaces. Detector tubes can be used as a measuring method for this type of analysis. However, as with any other kind of measuring method, it is necessary to use them within the framework of an efficient measuring plan; i.e. even a simple measuring method is a matter for experts.

Figure 1 shows detector tube measuring equipment consisting of a glass tube filled with a gas-sensitive reagent preparation and a delivery pump. The air sample to be analysed is drawn through the detector tube by means of the pump. The reagent in the detector tube undergoes a colour reaction with the gas to be measured. The length or intensity of the resulting stain is a measure of the gas concentration.

Detector tubes are described as a ready-to-use analysis. Manufacturers of detector tubes provide a defined composition of their products; the user may – after having established a measuring plan – use the detector tubes to measure gaseous contaminants without a specialist analytical effort.

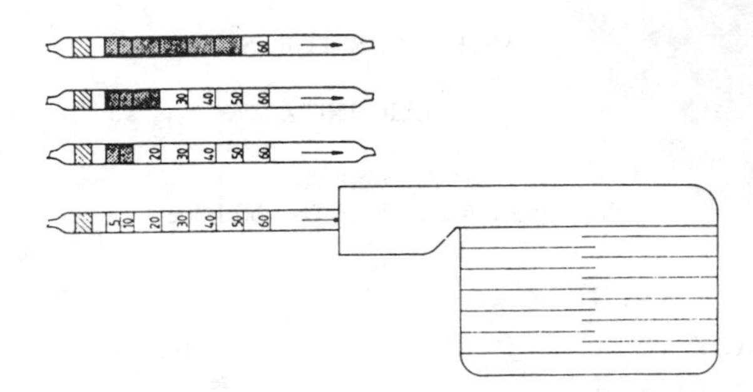

Fig. 1: Detector tube with hand operated pump for the determination of gases
- Detector tube connected to the pump
- After the test part of the indicating layer is discoloured (stain reaches scale mark '10')
- Test conducted at higher concentration (stain reaches scale mark '20')
- Test conducted at higher concentration (stain reaches scale mark '50')

Measurements of contaminants in workplace air by means of detector tubes have been carried out for more than 70 years. A wide range of detector tubes for the measurement of hundreds of gases and vapours, together with the necessary pumps, are available on the market. This range includes short term tubes which deliver the measurement results within a few minutes and long term tubes for the measurement of average concentration over a period of several hours.

Official institutions both on national and international level have issued performance standards on detector tubes (1, 2, 3). Many reports have been published, which confirm the reliability of the detector tube method (4, 5).

2. COURSE OF REACTION BETWEEN GAS AND DETECTOR TUBE

The reaction between gas and reagent preparation of the detector tube is clearly defined and can be described by chemical equations and mathematical formulas (6).

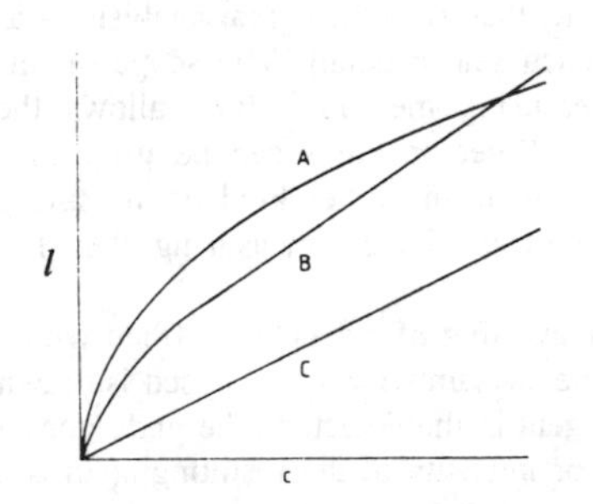

Fig. 2. Calibration curves of detector tubes

l = length of the indication
c = gas concentration

Curve A: curved throughout its whole length
Curve B: curved initially and then linear
Curve C: linear throughout

The indication obtained with detector tubes is a measure of the gas concentration. For scale tubes, the relationship between indication (length of the discoloration) and gas concentration can be described by curves like those shown in Figure 2.

On evaluation of such calibration curves, it can be seen that there is a linear relationship between indication and concentration for some detector tube types.

However, in the majority of cases, a curved line is observed. It will be shown, however, that – regardless of shape – all calibration curves do follow mathematical relationships. The following relationship can be drawn between the length of the indication and the mass of gas (6):

$$l = \frac{m}{c} + \frac{v}{A}\left[1 - e^{-\frac{lA}{v}}\right]$$

l = length of the indication

v = flow rate of the gas sample in the tube

A = a constant (measure of the rate at which the gas component to be determined reacts with one layer element)

e = 2.718

m = the total mass of the gas component absorbed

C = the absorption capacity of one layer element with respect to the gas component

This equation is not valid for all detector tube calibration curves. There are a number of causes for deviations:

a. The absorption capacity C of a layer element is not constant, but depends on the gas concentration c of the gas component to be determined. This dependence is always present when the reaction of the gas component with the reagent is not quantitative, but takes place with the formation of an equilibrium.

b. The mass of gas m reacting with one layer element is not constant, since part of the gas component is adsorbed by the carrier material. Under these conditions, two separate sorption processes take place in a layer element, one being chemical reaction with the reagent and the other adsorption by the large surface of the carrier material.

c. If the reagent system of a detector tube consists of several components, it is possible for the gas to react only with one reagent component. Under these conditions different courses of reaction can take place and the total absorption capacity C depends on the gas concentration c.

d. Finally, we must consider those reactions in which transfer from the gaseous phase to the indicating preparation is so slow that a delimitation of the discoloured zone can barely be detected. Here the entire indicating layer changes colour even after sucking through small gas volumes. The intensity of the colour indication then increases with increasing volume, again over the entire indicating layer. Evaluation of the indication of such detector tubes cannot take place by a scale but by comparison with a colour standard.

The equation shows that relatively simple interrelationships are valid in a summary consideration of the processes which come into play during the reaction of the gas sample with the detector tube preparation. Thereby, the course of a calibration curve can at least be referred to as covered by chemical and physical laws.

3. RELIABILITY OF DETECTOR TUBES

There is no analytical method with zero error. The results of measurements obtained with detector tubes are also beset with errors. In their literature (7), detector tube manufacturers give data as to the magnitude of the error, as a relative standard deviation (coefficient of variation).

Random Errors (reproducibility)

If several measurements are made using gas samples of precisely defined concentration, the results fluctuate around a mean value. If, for example, the indication of one detector tube is evaluated by several observers, the indicated values will not all be the same, but will differ. Even if several detector tubes are used and the evaluation is carried out by one observer, a 'deviation' will be found. To evaluate this 'deviation', detector tube manufacturers determine the standard deviation using statistics.

The following causes of random errors are known in the detector tube method (these errors cannot be avoided, but information as to their magnitude can be given):

- Slight variations in the amount of filling and in the packing density of the detector tube preparation.
- Various observers evaluate the indication differently (practice, eyesight, colour discrimination, influence of lighting).
- Slight variations in temperature and pressure during measurement.

On the basis of the different standard deviations, detector tubes can be subdivided into different groups. This was proposed some years ago (8).

Table 1. Relative Standard Deviation in Detector Tubes

	Relative standard deviation	
Group	At the start of the range of the measurement	At the end of the range of the measurement
1	10%	5%
2	15%	10%
3	20%	15%
4	30%	20%
5	40%	30%
6	Coefficient of variation above 40%, hence only estimation of the order of magnitude of the concentration possible	

In the numerical values given in Table 1, account is taken of the fact that the relative standard deviation at the start of the range of measurement of the tube is higher than at the end of the range. It is not possible to achieve a standard deviation less than 5%, mainly for production reasons (e.g. slight variations in the filling level, in the packing density and in the inside diameter of the glass tubes used). The detector tubes allocated to Group 1, with a standard deviation of 10% to 5%, therefore represent the optimum level. The high standard deviation of 40% to 30% shown by the detector tubes in Group 5 is attributable largely to difficulties in the practical evaluation of the tube indication (e.g. with diffuse discoloration).

Systematic Errors (bias)

In contrast to the random errors, systematic errors cannot be calculated using statistical methods, but can in most cases be avoided. There are a number of causes of systematic error:

- The tubes are not correctly calibrated by the manufacturer.
- The properties of the filling preparation in the detector tube change during storage.
- The pump used to suck in the air sample does not supply the theoretical volume (pump is leaking).

- The pump does not have its full suction capacity.
- The wrong pump is used.
- Interfering influence caused by other gases and vapours, i.e. the selectivity of the reagent system (also called cross-sensitivity).
- The temperature of the air sample when it enters the detector tube does not correspond to the admissible measuring temperature.

4. MANUFACTURING AND CALIBRATION OF DETECTOR TUBES

Detector tubes are manufactured under defined climate conditions in so-called production batches. The whole production process and the calibration procedure for detector tubes take place within a quality assurance system according to the international standard ISO 9001 (9). In this standard the following quality system requirements are laid down:

- Management responsibility
- Quality system
- Contract review
- Design control
- Document control
- Purchasing
- Purchaser supplied product
- Product identification and traceability
- Process control
- Inspection and testing
- Inspection, measuring and test equipment
- Inspection and test status
- Control of nonconforming product
- Corrective action
- Handling, storage, packaging and delivery
- Quality records
- Internal quality audits
- Training
- Servicing
- Statistical techniques.

This quality assurance system is being assessed by an external independent certification body in the framework of regular audits. This means, the detector tube manufacturer maintains a certified quality assurance system.

Every single production batch of detector tubes (e.g. approx. 10 000 tubes) is calibrated; the indicating behaviour of the tubes with respect to test gases of defined composition is determined and the calibration curve profile established, for tubes with graduated scale in the printed-on graduations or in the case of colour matching tubes in the composition of the colour standard.

The generation of calibration gases of defined composition is particularly important (10). At least two different methods are employed. Example for establishing the calibration curve:

A representative random sample (200 detector tubes) is taken from the respective production batch (e.g. 10 780 NH_3 detector tubes). The spacing between the following graduations is to be established: 5, 10, 20, 30, 40, 50, 60, 70 ppm. Several tubes are tested at four of these concentrations and the stain length of the indication is determined. All results are entered in a system of coordinates and the scale to be printed on the tubes is laid down. Additional tests are performed at different concentrations; these results are statistically evaluated (calculation of coefficient of variation).

A further step in this quality assurance procedure is the estimation of the shelf life by 'short term ageing' at elevated temperatures. Finally the detector tubes batch is tested and statistically assessed by the quality assurance department. When the tubes do fulfil the quality requirements, they are released for shipping.

5. TO WHAT EXTENT DOES THE USER RELY ON THE DETECTOR TUBE QUALITY SPECIFIED BY THE MANUFACTURER?

The user of detector tubes has various possibilities for accepting this measuring system for analyses.

First: Application as preserved analysis. The user relies on the data provided by the manufacturer in operating manuals, detector tube handbooks, and brochures. This is probably the most frequently practised procedure. If the detector tube manufacturer maintains a certified system of quality assurance, the user can be confident that a reliable measuring system is at disposal.

Second: Application as preserved analysis with a look behind the scenes. The user ascertains by visits to the manufacturer's plant, under which conditions the detector tubes are manufactured and calibrated.

Third: Some of the detector tubes are tested on a trial basis. The user can recalibrate the detector tubes in a laboratory by checking a representative sample of the detector tube shipment by means of test gases (such as two or three detector tubes per package). It should be emphasized specifically at this point that a very well equipped laboratory is needed for the generation of defined gas concentrations. A competent gas analyst is also needed who has the necessary practical experience in this special field. But in view of the high technical input, recalibration of detector tubes by the user will be limited to a few exceptions.

6. CONCLUSIONS

Detector tubes may be used as an analysis method in the field of occupational hygiene. However, this is only allowed - as with any other kind of analysis method - within a well defined measuring strategy. Detector tubes are designed as 'conserved' analysis means. Therefore, as a major advantage, they allow the user to assess their characteristics; applications and application limits can be predicted. Manufacturers of detector tubes give details (in their booklets) about measuring range, indication errors and about interfering substances. Used by trained operators detector tubes constitute a simple, low cost analysis method which allows reliable measurements of contaminants.

REFERENCES

(1) COLLINGS, A.J. (1982). Performance standard for detector tube units to monitor gases and vapours in working areas. Publication of the International Union of Pure and Applied Chemistry. Pure and Appl. Chem., Vol.54, pp. 1763-1767.

(2) DIN 33882, Teil 1 (1990). Measurement by means of detector tubes; Measuring system with length-of-stain detector tubes for short-term measurement (German standard). Beuth Verlag, Berlin.

(3) BS 5343, Part 1 (1986). British Standard for gas detector tubes. British Standards Institution.

(4) BHATIA, S.P. (1988). Laboratory testing of selected hydrogen sulphide gas detector tubes. Journal Pulp and Paper Canada. Vol.89, 183-188.

(5) ZINDLER, G. (1989). Überwachungspflicht der Schadstoffsituation an Gießplätzen in Eisengießereien - Kohlenmonoxid als Leitkomponente. Zeitschrift Gießerei. Vol.76, 726-727.

(6) LEICHNITZ, K. (1967). Versuch einer Deutung der Eichkurven von Prüfröhrchen (Interpretation of calibration curves for detector tubes). Chemiker Zeitung/Chemische Apparatur Vol.91, 141-148.

(7) DETECTOR TUBE HANDBOOK (1989). Drägerwerk AG, Lübeck.

(8) LEICHNITZ, K. (1983). Detector tube measuring techniques. ecomed-Verlagsgesellschaft, Landsberg. ISBN 3-609-66509-2.

(9) ISO 9001. (1987). Quality systems - Model for quality assurance in design/development, production, installation and servicing.

(10) ISO 6145/4. (1986). Gas analysis - Preparation of calibration gas mixtures - Dynamic volumetric methods.

137X

DIRECT-READING INSTRUMENTS

H. SIEKMANN and H. KLEINE

Berufsgenossenschaftliches Institut für Arbeitssicherheit
Alte Heerstrasse 111, Sankt Augustin, Germany

SUMMARY

Direct-reading instruments have the advantage that they offer immediately available results in the case of exposure measurements at workplaces. Dependent on the application, the devices have to meet a series of requirements. Advantages and disadvantages of the instruments available on the market are discussed. A need for further activities can be seen in the field of standardization and technical development of direct-reading instruments.

1. INTRODUCTION

The present paper deals with direct-reading instruments for measuring gaseous and vaporous atmospheric contaminants in workplace air. A common characteristic of these instruments is a combined sampling and analytical operation.

Direct-reading instruments are used in places where the result of pollutant measurements has to be immediately available. It is either indicated in terms of a concentration value or controls an operational function such as the initiation of a warning signal. Continuous measurement is the rule, i.e. that measuring values are available at any point in time. But there are also devices with a quasi-continuous operation.

The basic principle of any direct-reading instrument consists in converting a characteristic property of the substance under consideration into an electrical signal proportional to the concentration determined; this is done by means of a suitable detector whose operating principle is based on:

- flame or photo ionization
- IR and UV light absorption
- electrochemical reactions.

2. APPLICATIONS

The main advantage of direct-reading instruments is their speed of response. They are therefore especially suited for:

- Quick estimation of the exposure situation at the workplace (screening).
- Ease of determination of temporal concentration variations or local dispersion of the pollutant.
- Possibility of warning workers, wearing a gas monitor, of high, immediately intoxicating pollutant concentrations.
- Possibility of efficiency control of protective measures, e.g. installation of a ventilation system.
- Stationary monitoring: possibility of initiating an alarm as soon as there are fault conditions in the manufacturing system.

In addition to these rather specific applications, direct-reading instruments may also be used for 'classic' exposure measurements with a view to verifying compliance with limit values in force. Among these classic applications are:

- concentration measurements to determine shift averages,
- concentration measurements for comparison with short term values,
- control measurements.

3. REQUIREMENTS

Direct-reading instruments have to fulfil the following requirements. Firstly, there are general requirements applying to all methods for measuring pollutant concentrations. Secondly, there are special requirements concerning the time resolution of the instruments. Also, requirements have to be defined that refer to the reliability of direct-reading instruments with warning functions.

General Requirements

Direct-reading instruments have to comply with the requirements laid down in the European draft standard prEN 482 'General requirements for the performance of procedures for workplace measurements' that apply to all kinds of measuring procedures. These requirements refer to the:

- unambiguity of the measuring result
- selectivity
- overall uncertainty
- measuring range
- averaging time
- influence of environmental parameters
- unit of the measuring result.

The problem of selectivity is going to be discussed later. Requirements with respect to overall uncertainty, measuring range and averaging time are defined in the European draft standard prEN 482 taking into account the application. For example, in the case of measurements for comparison with limit values, the overall uncertainty must not exceed 30% for concentrations around the limit value. In the case of low concentrations, the maximum overall uncertainty is 50%.

Details of prEN 482 are dealt with in the paper 'Performance requirements for measuring methods'.

Special Requirements

In this context, the time resolution is of predominant importance. Whenever the temporal variation of a pollutant concentration is to be measured, the instrument has to be designed so as to detect concentration changes as quickly as possible, for example, in the case of discontinuous concentrations of pollutants generated in manufacturing processes.

This applies in particular to the time constant of the instrument, e.g. the signal initiation time t_{90}. In fact, the time constant must be in the range of one minute or even less.

Whenever a measuring device is carried on the man to warn the worker of suddenly occurring, toxic concentrations, the warning signal has to be generated fast enough to permit the person to take shelter or the necessary protective measures. Again the time constant must be in the range of a few minutes or even seconds.

As to the other above mentioned applications, requirements concerning the time resolution depend on the measuring task to be fulfilled.

Whenever a direct-reading instrument is used for warning purposes, it has to be sure that the warning signal is reliably initiated as soon as the preset threshold value is exceeded. This means that, on one hand, the measuring uncertainty within the range of the threshold value must not be too high and that, on the other hand, the device has to show increased reliability. Faults occurring in the measuring system have to be either immediately detectable or without disadvantageous effect on the warning function (redundant design).

Whenever the measuring devices are intended for use in explosive atmospheres, they have to meet the pertinent requirements in terms of explosion protection.

Up to now, no standards or regulations have been established that define the above specified special requirements with respect to exposure measurements. As far as we know, there is only one EC Directive, 78/680/EEC, that deals with the time resolution of measuring devices for continuous monitoring of vinyl chloride monomers. There is a great need for standardization in this field. This is why we would suggest the European Standardization Committee, CEN, should consider the problem of direct-reading instruments for measuring atmospheric contaminants in workplace air and define particular requirements and test methods for these devices. It would be possible, in this context, to rely on experience related to environmental protection and standardization activities in connection with measuring devices for explosive atmospheres. Appropriate IEC or CENELEC regulations could underlie the determination of electric requirements.

A completely new aspect would be the definition of requirements concerning the warning function in the case of toxic substances. Such requirements would go beyond the limits of established practice. The problem of warning employees of immediate hazards due to toxic atmospheres represents, however, a work safety matter of great importance.

4. CHOICE OF SUITABLE INSTRUMENTATION BY THE USER

What are the criteria that should determine the choice of a direct-reading instrument?

A large spectrum of direct-reading instruments is available on the market; this is why the user tends to believe that there is an appropriate device for any kind of application. This impression may however easily turn out to be misleading. This immense offering does not only prove to be very confusing but also unsatisfactory in terms of compliance with the requirements resulting from a large variety of different measuring tasks.

The user has to consider a series of criteria when selecting an appropriate device. First of all, a suitable detection principle has to be chosen on account of the physical/chemical properties of the substance under consideration. As long as a single substance only is to be measured, the user can choose between several instruments. Initial difficulties are liable to arise in connection with normal atmospheric components such as water vapour or carbon dioxide, that may already affect the measuring result.

In practice substance mixtures predominate in the field, and interferences are most liable to be present that may influence the measuring result. It is possible that an interfering component absorbs at the same wavelengths of radiation as the component under consideration. In this case, absorption spectrophotometry cannot be used. The only solution consists in selecting a different measuring technique with a higher selectivity.

Another aspect to be taken into consideration when choosing an appropriate measuring device is the compliance with the requirements resulting from the intended application. This refers in particular to the measuring range and to the overall uncertainty. Devices designed for high concentrations occurring in the case of emission measurements are normally not suitable for monitoring measurements at workplaces.

In connection with exposure measurements, wearability is another significant criterion. Measuring devices should be wearable on the person or ought to be at least easy to move. Stationary devices can only be employed for exposure measurements at stationary workplaces.

5. TYPES OF INSTRUMENTS

In Table 1, some currently used direct-reading instruments are presented and compared in terms of advantages and disadvantages. The areas in which the different methods have been found useful or less suitable are indicated. As type tests have not yet been carried out on these devices, the assessment is mainly based on practical experience.

The table includes a list of selected measuring principles in the first column. The second column indicates examples of substances and groups of substances that can be measured by use of these measuring principles. The following columns contain information in terms of a quantitative assessment of the selectivity, the time resolution and the framework conditions for use. Suitability of the devices with respect to these properties has been identified by the symbol '+', non-suitability by the symbol '–'.

Table 1. Direct-reading Instruments

Measuring principle	Detectable substances	Selectivity	Time resolution	Application		
				s	m	p
Flame ionization detector (FID)	Organic gases and vapours	– –	+ +	+ +	+	– –
Photo ionization detector (PID)	Organic gases and vapours	–	+ +	+ +	+	+/–
Infrared photometry	Inorganic/organic gases and vapours e.g. CO, CO_2 chlorinated hydrocarbons	+	+	+ +	+	– –
Ultraviolet photometry	Specific gases, e.g. ozone	+	+	+ +	+	– –
Gas chromatography	Inorganic/organic gases and vapours	+ +	+ +	+ +	+	– –
Electrochemical sensors	Specific gases, e.g. H_2S, NO_2, CO	+	+	+ +	+	+
Chemiluminescence	Specific gases, e.g. ozone, NO	+ +	+	+ +	+	– –

s = stationary, m = mobile, p = personal

Flame Ionization Detector and Photo Ionization Detector

Flame ionization detectors (FID) are universally usable, frequently employed detectors for measuring combustible organic gases and vapours. The measuring principle consists in generating a measurable ion current in a hydrogen flame. The ion current is directly proportional to the hydrocarbon concentration introduced into the hydrogen flame by the test gas stream. Therefore, FIDs are sensitive for a series of different hydrocarbon containing substances.

Consequently, all sample components detectable by means of FID are indicated at the same time. The selectivity of these devices is rather low; flame ionization detectors should therefore be employed exclusively for the quantitative monitoring of workplaces where the exposure situation is determined by a single pollutant only. Nevertheless, the instruments show excellent time resolution results and - in most cases - a large linear measuring range.

Due to the above mentioned properties, FIDs are preferably used for the semi-quantitative determination of temporal and local concentration variations. By combining flame ionization detectors with other measuring methods of increased selectivity, precious information can be obtained on exposure situations in the field.

Photo ionization detectors are comparable to flame ionization detectors in terms of use and suitability. As they permit the user to vary the ionization energy, these instruments distinguish themselves by an increased selectivity.

Infrared and Ultraviolet Spectrophotometry

Increased selectivity, i.e. better substance specific measuring results are achieved by use of photometric methods. The measuring principle is based on the light absorbing properties most substances show within a certain frequency range. For applications this range is the infrared (IR) or ultraviolet (UV) spectrum, the IR range playing the more important role. The method can be used for measuring a large variety of inorganic and organic gases such as, for example, carbon monoxide, carbon dioxide or a series of different halogenated hydrocarbons.

UV spectrophotometry is used for special gases, e.g. ozone.

Although worse than that of the FID, the time resolution of IR and UV photometers is still satisfactory.

Gas Chromatography

Higher selectivity than in the case of photometric methods can generally be obtained by means of gas chromatography, an analytical method that is mainly used for laboratory analyses. Yet, automatic devices, the so-called process gas chromatographs, are also used for continuous monitoring of industrial workplaces.

Compared to other types of direct-reading instruments, the time resolution of the gas chromatograph is less satisfactory.

Chemiluminescence

To guarantee maximum selectivity, substance-specifically designed measuring systems ought to be used. Among them are devices for nitrogen and ozone measurements whose operating principle is based on chemiluminescent reactions. These instruments are only suited for measurements of these gases and do not serve to measure other substances.

Electrochemical Sensors

A closer look at the last three columns of the table shows that all instruments that have been mentioned so far are suitable both for stationary as well as for personnel exposures. However, with the exception of some photo ionization detectors, these instruments are not capable of being carried on the man thus making personal monitoring impossible. Electrochemical sensors represent a group of measuring devices that are very well suited for personal monitoring due to their reduced size.

The measuring principle of electrochemical sensors relies on the fact that certain gases and vapours, e.g. hydrogen sulphide, nitrogen dioxide, carbon monoxide, chlorine and formaldehyde, can be oxidized and reduced by electrochemical reactions. These reactions take place within a substance-specifically designed system, the so-called electrochemical cell.

Due to the fact that electrochemical cells and the corresponding technical equipment can be designed as lightweight instruments of reduced dimensions, electrochemical gas sensors are often employed as warning instruments when there is a risk of sudden acute gas intoxication. Whenever these direct-reading instruments are used for current exposure measurements, it must be ensured that they meet the requirements defined for this type of application. Warning instruments available on the market do normally not comply with these requirements.

6. FUTURE NEEDS

There is a need for further work in the fields of standardization and technical development of direct-reading instruments.

Future standardization activities should aim at defining requirements for direct-reading instruments and the corresponding test methods. Apart from the general requirements applying to all measuring techniques, particular attention ought to be paid to specific properties of direct-reading instruments, viz. the time resolution and the warning function in the case of a risk of sudden acute intoxication. The technical committee 137 of the European Standardization Committee, CEN, seems to be the suitable body for this work.

The following aspects should be taken into account when promoting the further technical development of direct-reading instruments:

- *Variety of substances:* it would be desirable to complement the spectrum of substances detectable by means of direct-reading instruments.
- *Selectivity:* selectivity should be improved to ease substance specific measurement.
- *Wearability:* it would be useful to design small size instruments capable of being worn on the person, thus permitting measurements within the breathing zone.
- *Handling:* simple handling and calibration should be the principal goals as well as easy data assessment (e.g. averaging over different times).

If adequate advances are made in the field of technical development, direct-reading instruments are most liable to play a role of growing importance for the measurement of atmospheric contaminants in workplace air, in particular because they offer quickly available results.

1375

Sep-Pak® DNPH-SILICA CARTRIDGES FOR THE ANALYSIS OF FORMALDEHYDE (AND OTHER ALDEHYDES) IN AIR

J.P. BARONE and T.H. WALTER

Millipore
Milford, MA 01757, USA

Formaldehyde and other aldehydes are receiving increasing attention both as toxins and as promoters in the photochemical formation of ozone in the atmosphere (1). Sources of aldehydes in residential buildings include plywood and particleboard, insulation, combustion appliances, tobacco smoke, and various consumer products. Aldehydes are released into the atmosphere in the exhaust of motor vehicles and other equipment in which hydrocarbon fuels are incompletely burned. Much of the recent interest in formaldehyde has been spurred by the scheduled introduction of methanol-powered cars in California in 1992 (2). While methanol bums more cleanly than gasoline, releasing lower levels of hydrocarbons and nitrogen oxides, it produces higher concentrations of formaldehyde. The amount of formaldehyde released by methanol-powered cars will be a key factor determining the effectiveness of this strategy in improving air quality (3).

The most sensitive and specific method for analysing aldehydes and ketones is based on their reaction with 2,4-dinitrophenylhydrazine (DNPH) and subsequent analysis of the hydrazone derivatives by HPLC (see Figure 1). The hydrazones may be detected by absorbance in the ultraviolet region, with maximum sensitivity obtained between 350 and 380 nm.

$$O_2N\text{-}C_6H_3(NO_2)\text{-}NHNH_2 + R\text{-}C(=O)\text{-}R' \longrightarrow O_2N\text{-}C_6H_3(NO_2)\text{-}NH\cdot N{=}C(R)R' + H_2O$$

DNPH **Aldehyde or Ketone** **Hydrazone Derivative**

Air-borne aldehydes have traditionally been collected by drawing a sample through an impinger containing a solution of DNPH (4, 5). However, the impinger collector is generally cumbersome to use, and is not well suited for high flow rates or extended collection times because the solvent evaporates.

The limitations of impingers have created considerable interest in solid adsorbent cartridges containing DNPH coated on a solid support (6, 7). A procedure for preparing such devices from Sep-Pak-Silica cartridges is described in Environmental Protection Agency Compendium Method TO-11 (8). This procedure calls for passing an acetonitrile solution of DNPH through a Sep-Pak cartridge, then evaporating the solvent under a stream of nitrogen to leave a coating of DNPH. Unfortunately, this method is time consuming and may lead to variable DNPH loadings and impurity levels.

Precoated DNPH cartridges (Sep-Pak DNPH-Silica cartridges) are now available from Waters. These devices consist of DNPH-coated silica contained in radially-compressed cartridges constructed from high purity polyethylene. To obtain consistent DNPH loadings, silica is coated with DNPH in multi-kilogram quantities, then packed into Sep-Pak

Column:	Nova-Pak® C18 3.9 x 150 mm
Mobile phases:	A: Water/Acetonitrile/Tetrahydrofuran 60/30/10 v/v/v B: Water/Acetonitrile 40/60 v/v
Gradient:	100% A for 1 min, then linear gradient from 100% A to 100% B in 10 min
Flow rate	1.5 ml/min
Sample:	Mixture of DNPH and DNPH derivatives in acetonitrile
Injection:	20 μl
Detection	Absorbance at 360 nm

1. DNPH
2. Formaldehyde-DNPH
3. Acetaldehyde-DNPH
4. Acetone-DNPH
5. Acrolein-DNPH
6. Propionaldehyde-DNPH
7. cls(2-Butanone)-DNPH
8. Crotonaldehyde-DNPH
9. trans(2-Butanone)-DNPH
10. Butyraldehyde-DNPH
11. Benzaldehyde-DNPH
12. Isovaleraldehyde-DNPH
13. Valeraldehyde-DNPH
14. 0-Tolualdehyde-DNPH
15. m-Tolualdehyde-DNPH
16. p-Tolualdehyde-DNPH
17. Hexaldehyde-DNPH
18. 2,5-Dimethylbenzaldehyde-DNPH

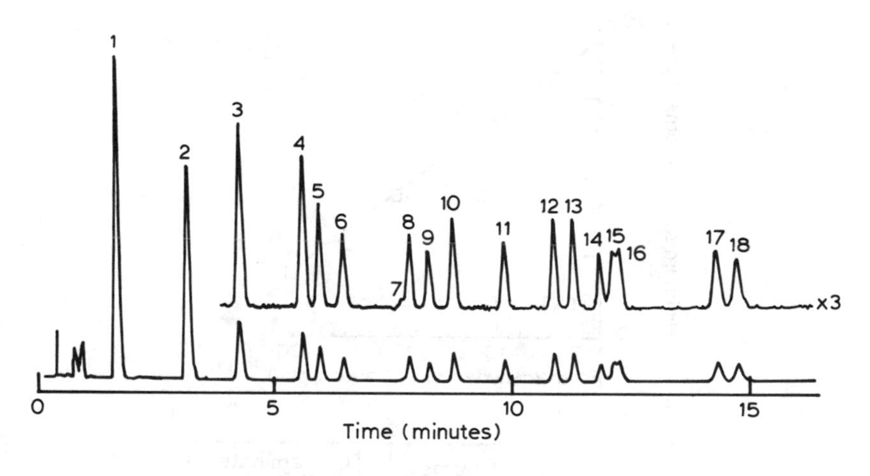

Fig. 1. HPLC separation of DNPH and DNPH derivatives

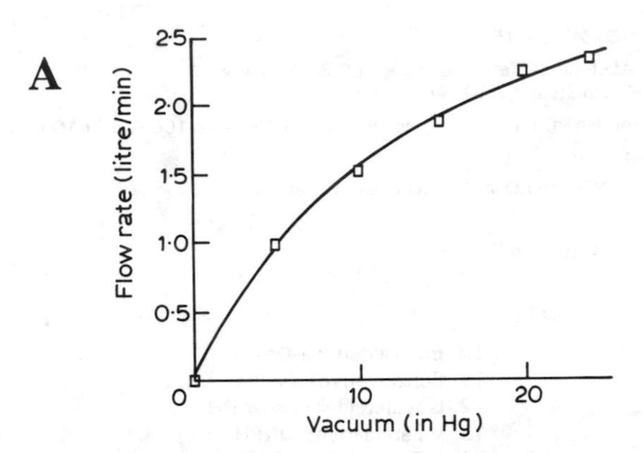

Fig. 2. Air flow rate through DNPH-Silica Sep-Pak cartridges *vs* applied vacuum

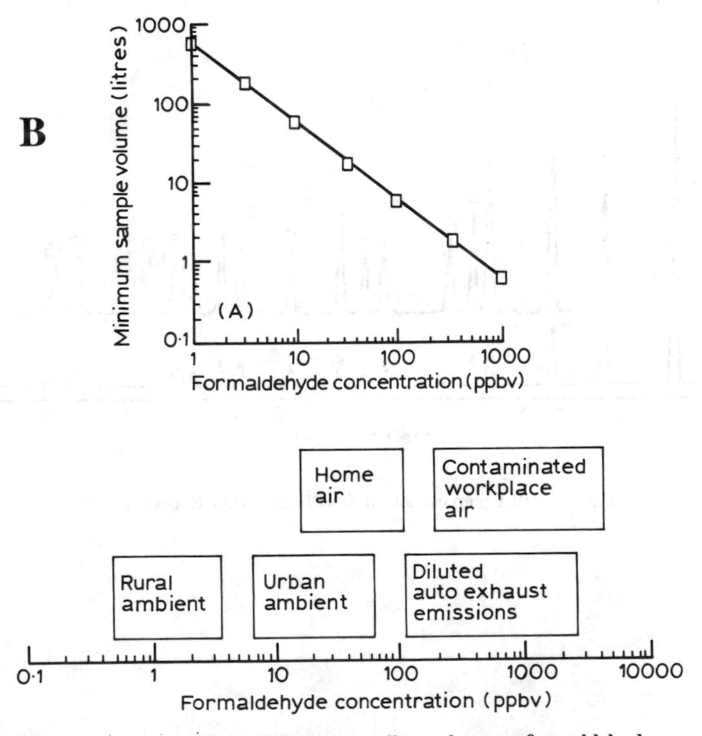

Fig. 3. (A) Recommended minimum sampling volume *vs* formaldehyde concentration for Sep-Pak DNPH-Silica cartridges.
(B) Typical formaldehyde concentrations in different types of air

cartridges. Each cartridge contains 1.0 mg of DNPH coated on 350 mg of silica, sufficient to collect more than 75 μg of formaldehyde. For higher capacity, several cartridges may be connected together. With a single cartridge, sampling rates of 1-2 litres/min may be achieved using moderate vacuum (see Figure 2).

Unused DNPH-Silica cartridges contain traces of hydrazones (see Figure 4B). As a guideline for accurate quantitation, the amount of a particular hydrazone formed from the sample should be at least ten times the amount in the blank. This imposes a minimum sample volume, which depends inversely on the concentration of formaldehyde. The recommended minimum sample volume is shown as a function of formaldehyde concentration in Figure 3A, along with typical concentration ranges for formaldehyde in different types of air (Figure 3B) (1, 7). For analyses of formaldehyde in diluted auto exhaust emissions, typically containing 250–2500 parts per billion v/v (ppbv) formaldehyde (7), the minimum sample volume is in the 0.2–2 litre range. When determining aldehydes in ambient air, which typically contains formaldehyde concentrations in the 0.5–50 ppbv range, 10–1000 L of air must be sampled.

A typical analysis of diluted exhaust emissions from a car fuelled by conventional gasoline is shown in Figure 4A. The sample (provided by Dr S.B. Tejada of the US EPA Atmospheric Research Assessment Laboratory) was collected using a constant volume sampler dilution tunnel, with a vehicle operated on a prescribed driving schedule on a chassis dynamometer. The cartridge was connected to a heated (100°C) sampling manifold using a short piece of teflon tubing. The sample was collected using a metal bellows pump and a mass flow controller. The example shown in Figure 4A represents an 8.4 L sample collected at 1 L/min. Formaldehyde is by far the most abundant carbonyl compound emitted, with smaller amounts of several other aldehydes and ketones also observed. A typical Sep-Pak DNPH-Silica blank, containing only trace levels of DNPH derivatives, is shown in Figure 4B.

In addition to analyses of auto exhaust emissions, Sep-Pak DNPH-Silica cartridges may be used to determine aldehydes and ketones in contaminated workplace air, residential air, and ambient air. For sampling of urban ambient air, an ozone scrubber must be used before the cartridge to avoid decomposition of the hydrazone derivatives (9). Two examples of indoor air analyses are shown in Figure 5. For both analyses, samples were collected at a rate of 0.5 L/min using a portable sampling pump (DuPont Alpha 1). The sample shown in Figure 5A was obtained in a conventional home, and shows typical levels of formaldehyde (47 ppbv) and acetaldehyde (24 ppbv), as well as a significant amount of acetone (20 ppbv). The sample of Figure 5B was obtained in a chemical research laboratory, and shows lower concentrations of formaldehyde (12 ppbv) and acetaldehyde (4 ppbv), but a much higher concentration of acetone (97 ppbv).

It has been demonstrated that cartridges loaded with samples may be stored at room temperature for at least two weeks without compromising sample integrity (7). Thus, with proper packaging, cartridges may be sent from a field sampling site to a central laboratory for analysis. Longer storage times are permissible if cartridges are refrigerated.

Waters Sep-Pak DNPH-Silica cartridges provide a sensitive, selective, and convenient method for the determination of aldehydes and ketones in air samples. These devices provide a powerful tool for analytical chemists involved in the determination of this important class of pollutants.

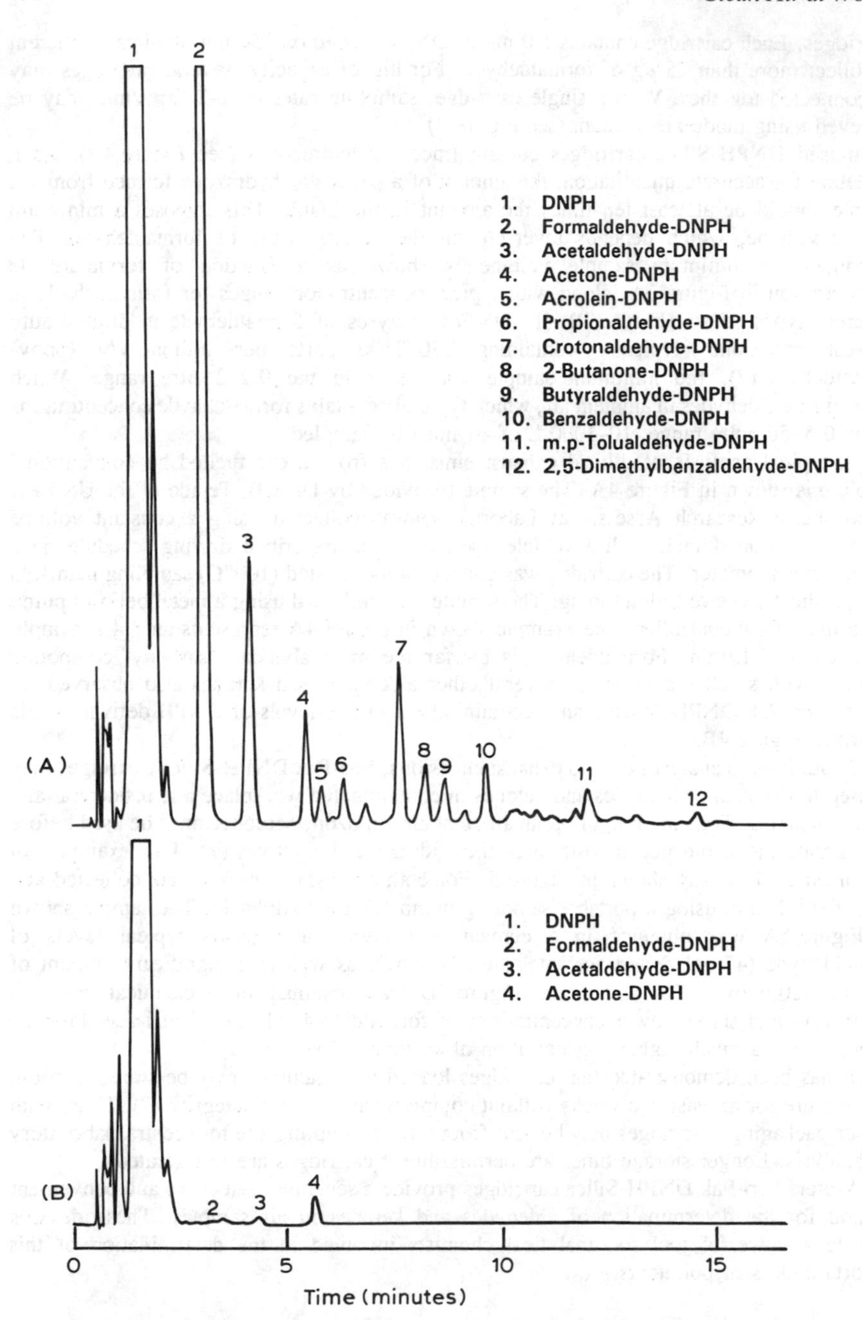

Fig. 4. Chromatograms of (A) Aldehydes and ketones sampled in diluted exhaust emissions from a gasoline-fuelled car, and (B) Sep-Pak DNPH-Silica cartridge blank. See Fig. 1. for chromatographic conditions

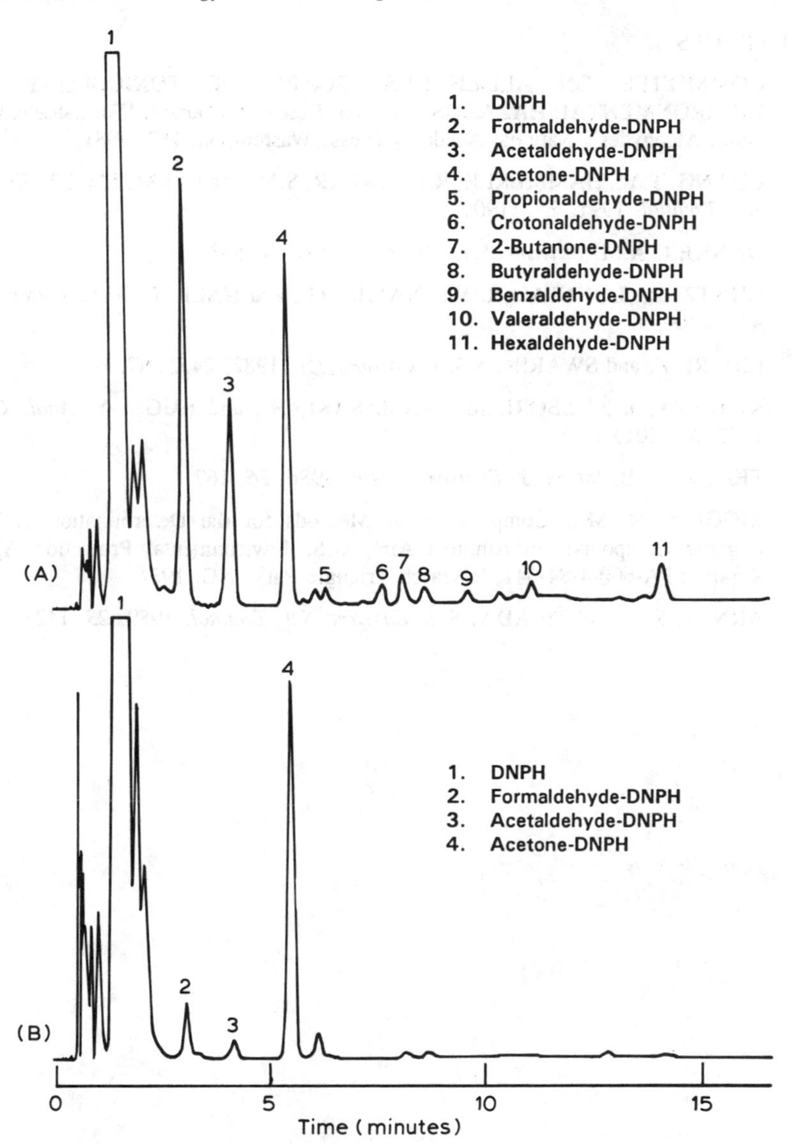

Fig. 5. Chromatogram of aldehydes and ketones sampled from indoor air (A) Home air sample (100 L), (B) Lab air sample (20 L) See Fig. 1. for chromatographic conditions

REFERENCES

(1) COMMITTEE ON ALDEHYDES, BOARD OF TOXICOLOGY AND ENVIRONMENTAL HAZARDS, National Research Council, 'Formaldehyde and Other Aldehydes'; National Academy Press, Washington, DC, 1981.

(2) CHANG, T.Y., HAMMERLE, R.H., JAPAR, S.M. and SALMEEN, I.T. *Environ. Sci. Technol.* 1991, *25*, 1190.

(3) DUNKER, A.M. *Environ. Sci. Technol.* 1990, *24*, 853.

(4) KUNTZ, R,, LONNEMAN, W., NAMIE, G., and HALL, L. *Anal. Letter* 1980. *A16*, 1409.

(5) LIPARI, F. and SWARIN, S.J. *I. Chromatogr.* 1982, *247*, 297.

(6) KUWATA, K., UEBORI, M., YAMASAKI, H., and KUGE, Y. *Anal. Chem.* 1983, *55,* 2013.

(7) TEJADA, S.B. *Intern. J. Environ. Chem.* 1986, *26*, 167.

(8) RIGGINS, R. M., 'Compendium of Methods for the Determination of Toxic Organic Compounds in Ambient Air', U.S. Environmental Protection Agency Report EPA-600/4-84-041, Research Triangle Park, NC, 1984.

(9) ARNTS, R.R. and TEJADA, S.B. *Environ. Sci. Technol.* 1989, *23*, 1428.

1376

MONITORING OF UNDERGROUND AIR IN COAL MINES

A BRAITHWAITE*, M COOPER, and J BOVELL

Nottingham Polytechnic, Nottingham NG11 8NS, UK

SUMMARY

Monitoring underground air in coal mines for trace levels of indicator gases can provide an advanced warning of the spontaneous combustion of coal. The gases produced by desorption and self heating oxidation processes have been studied and potential monitoring systems developed.

1. INTRODUCTION

The atmosphere of a coal mine differs from that of ordinary fresh air due to gases being introduced into the ventilation air by the various activities associated with under ground mining operations. These include use of diesel engines, hydraulic systems and shot firing of explosives which produce a range of pollutants including NO, NO_2, CO, CO_2 and poly aromatic hydrocarbons (PAHs). A second source of pollutants is coal itself. The process of mining coal from the strata releases a range of gases often referred to as fire damp, the primary gas being methane. The aim of this work is to examine trace pollutant gases in order to be able to identify potential fire hazards from the coal heatings and spontaneous combustion of coal (1).

2. MINE ATMOSPHERE MATRIX

Our investigations into the composition of coal mine atmospheres has involved three linked areas of study; an examination of the spontaneous combustion of coals, analysis of coal mine atmospheres and development of instrumentation to monitor coal mine air. This has enabled those gases arising from underground activities and those due to spontaneous heating of coal to be identified (2).

When coal is heated two processes occur that lead to emission of gases, desorption and oxidation. Desorption gases already present in the coal matrix primarily consist of C_1 to C_5 alkanes, with a logarithmic decrease in concentration from methane to the higher alkanes. The second process involves spontaneous combustion which refers to the self heating coal oxidation processes that produce gases, for example, alkenes, aldehydes and ketones, that may be used as indicators of potential spontaneous combustion problems. The stages in the spontaneous combustion process are shown below;

```
              oxdn
Coal + O2 ---------> Heat (ΔHco) ------> Oxidation products
  ^          Rco   T          |          CH3OH, CH3CHO, (CH3)2CO, C2H5OH
  |       self heating        |                      |
  |___________________________|                    Smoke
                                                     |
                                                    Fire
Coal dust explosion <------ CH4 ignition <-----------|
```

where R^{co} is the reaction rate of the coal oxidation process and is directly proportional to the temperature of the coal T; ΔH^{co} is the enthalpy of the exothermic coal oxidation process.

3. EVOLUTION PROFILES OF GASES PRODUCED FROM COAL

An automated coal combustion system was used to oxidise selected coal samples under a controlled heating regime the gases evolved being carried into a sequential set of specific analysers and a gas chromatographic sampling system to provide in situ analyses.

Evolved gases	Monitoring technique
CO, CO_2, CH_4	Continuous specific infra red analysers
O_2	Continuous paramagnetic analyser
Alkanes, alkenes	} Discrete gas chromatographic analyses
aromatics	}
sulphur gases	}
oxygenates	}

Results of the investigation showed that the evolution of the individual gases are related to coal temperature thus providing a useful evolved gas profile. A comparative set of results were obtained using an inert atmosphere of nitrogen to identify those gases evolved without oxidation taking place and which would not serve as useful indicators of spontaneous combustion.

The work has shown that potential indicator gases include CO, CO_2, CH_3CHO, $(CH_3)_2CO$, COS, C_2H_4 and C_3H_6. Other gases which were investigated but were found to be unsuitable indicator gases were C_1 to C_5 alkanes, C_2H_2, H_2S, SO_2, CS_2, CH_3OH, and C_2H_5OH.

4. ANALYSIS OF THE INDICATOR GASES

A gas chromatography system was developed for the analysis and identification of the indicator gases. The system was automated and under the control of a dedicated microprocessor system and hardware interface unit running software written in C to control the system functions, monitor the status of the instrument conditions, collect and process the data and generate chromatographic reports.

The GC system was designed to be portable, running from a 12V supply and included an automated sampling system for introducing samples of air into the chromatographic column. Separation was effected on a wide bore 10 m Poraplot open tubular column incorporated into a specially designed 'oven' unit. A range of detectors were investigated including flame ionisation (FID), thermal conductivity (TCD), and photoionisation detectors (PID). Detectors suitable for a portable instrument include the TCD for monitoring higher concentration components ($>$500 vpm) and the PID for monitoring trace components (sub vpm). Two types of detector were developed, a micro TCD and a PID. The TCD was a commercial detector modified for minimal dead volume and adapted to take the eluent from wide bore capillary columns. The response and sensitivity of PIDs depends on the energy emitted by the source filler gas and the ionisation properties of the compounds to be detected (3).

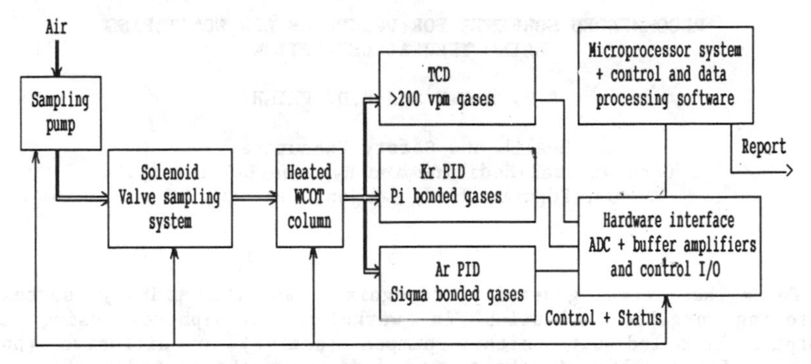

Fig. 1. Schematic diagram of the automated GC system

Thus a krypton source (10.6eV) is suitable for alkenes and oxygenates as shown below, and an argon source (11.7eV) is suitable for alkanes. A micro PID source and ionisation chamber and associated electronics were developed and optimised for energy efficiency and low power requirements. A multidetector GC system as shown above can therefore monitor for methane, ethane, alkenes and oxygenates.

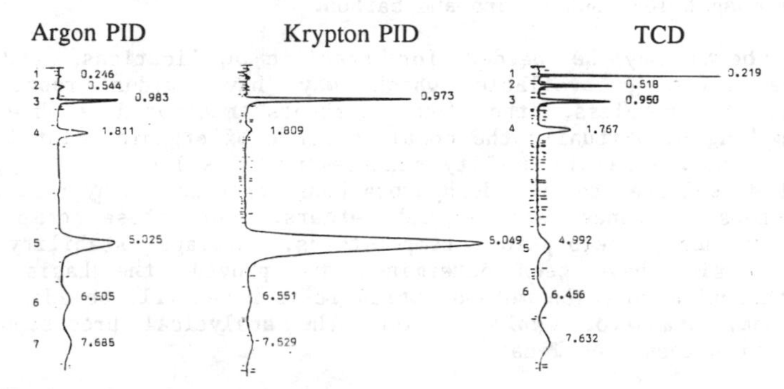

Fig. 2. Chromatograms illustrating the response of GC detectors to mine air contaminants

1. methane; 2. ethane; 3. ethene; 4. propane; 5. propene; 6. methylpropane; 7. n-butane.

5. REFERENCES

(1) CUDMORE, J.F. and SANDERS R.H., 'Spontaneous Combustion of Coal Mine Fires and Interpretation of Mine Gases', Australian Coal Industry Research Laboratories, 1984, Report No. 84-1.
HERTZBURGH, M., 'Mine Fire Detection', US Bureau of Mines IC-8768, 1978.

(2) COOPER, M., PhD Thesis, Nottingham Polytechnic, 1991.
BOVELL, J., Nottingham Polytechnic, unpublished results.

(3) DRISCOL, J.N., J. Chrom. Science, 1985, **23,** 488.

ACKNOWLEDGMENTS

The authors would like to acknowledge the assistance received for this work from British Coal and the European Coal and Steel Community.

RECOMMENDED SORBENTS FOR WORKPLACE AIR MONITORING USING THERMAL DESORPTION

R.H. BROWN and M.D. WRIGHT

Health and Safety Executive,
Occupational Medicine and Hygiene Laboratory,
403, Edgware Road, London, NW2 6LN, UK

Tenax has been generally recognised as the primary sorbent for monitoring organic volatiles in workplace atmospheres using thermal desorption coupled with either pumped (active) or diffusive (passive) sampling. For sampling in the pumped mode, comprehensive data is available on breakthrough volumes and suitable desorption temperatures.

Tenax, however, is not the ideal sorbent for sampling very volatile compounds, where the breakthrough volume may be too low for practical sampling times. For such compounds it is necessary to use stronger sorbents that are nevertheless compatible with the recovery of the compound by thermal desorption prior to analysis. It is proposed that these sorbents should be Chromosorb 106, Spherocarb and carbon.

Other sorbents may be needed for specific applications, and new sorbents are becoming available which may have wider ranges of applicability. Nevertheless, the four sorbents mentioned enable the practical sampling of virtually the complete range of organic volatiles of significance in workplace air quality measurement. A list of 60 organic vapours has been selected to include hydrocarbons, chlorinated hydrocarbons, alcohols, esters, ketones and glycol ethers. For these compounds, breakthrough volumes, desorption temperatures, storage stability and analytical precision have been determined and provide the basis of a generalised thermal desorption method, which it is hoped will be adopted as an international standard. Table 1 lists the analytical precision and storage data for solvents on Tenax.

Table 1

Precision and storage recovery of solvents

Organic compound	Loading ug	%CV* time=0	mean recovery** + %CV*			
			5 months		11 months	
Aliphatic hydrocarbons						
hexane	7.8	10.7	93.6	17.9	100.8	26.1
heptane	8.4	2.4	99.5	2.1	100.0	1.3
octane	8.6	2.4	100.1	1.8	100.0	0.5
nonane	12	0.8	-	-	101.0	0.4
decane	9.2	2.2	100.4	1.5	100.2	0.5
undecane	9.1	2.3	100.7	1.5	100.2	0.2
dodecane	9.9	2.8	101.8	1.5	101.5	0.4

(Table 1, contd.)

Aromatic hydrocarbons						
benzene	11.0	2.5	98.7	2.0	98.6	0.8
toluene	10.9	2.6	(100)	1.8	(100)	0.6
p-xylene	5.3	2.5	99.9	1.7	99.8	0.7
o-xylene	11.0	2.4	100.0	1.7	98.8	0.6
ethylbenzene	10.0	0.5	99.6	0.4	97.9	1.3
propylbenzene	10.5	2.3	99.7	1.5	98.5	0.7
isopropylbenzene	10.9	2.3	98.9	1.8	97.2	1.3
m- + p-ethyltoluene	10.5	2.3	98.8	1.7	96.9	1.2
o-ethyltoluene	5.4	2.2	99.2	1.6	97.6	0.8
1,2,4-trimethylbenzene	10.8	2.2	100.1	1.3	98.9	0.7
1,3,5-trimethylbenzene	10.7	2.2	100.0	1.5	99.1	0.5
trimethylbenzene	10.2	1.7	101.6	0.5	101.3	0.8
Esters and glycol ethers						
ethyl acetate	10.3	0.6	97.6	1.0	100.0	2.5
propyl acetate	10.9	2.4	100.5	1.7	99.1	0.8
isopropyl acetate	9.4	1.0	97.0	0.4	100.0	1.4
butyl acetate	10.8	2.4	100.3	1.6	99.9	0.6
isobutyl acetate	10.7	2.3	100.2	1.4	99.8	0.7
methoxyethanol	8.9	5.4	87.3	5.7	93.1	1.6
ethoxyethanol	10.4	4.2	97.6	2.5	97.2	3.3
butoxyethanol	10.0	2.6	100.6	4.1	100.1	3.0
methoxypropanol	10.4	2.4	95.3	3.6	99.0	1.2
methoxyethyl acetate	12.5	2.1	100.6	1.2	98.9	1.4
ethoxyethyl acetate	11.4	0.9	99.8	2.2	98.7	2.6
butoxyethyl acetate	11.5	2.3	101.3	1.3	99.9	1.1
Aldehydes and ketones						
acetone		not recommended on Tenax				
methyl ethyl ketone	9.2	0.9	97.4	0.8	99.1	0.6
methyl isobutyl ketone	9.3	0.6	100.7	0.6	100.7	0.5
cyclohexanone	10.9	0.8	102.4	1.2	100.7	0.6
2-methylcyclohexanone	10.7	0.7	101.1	0.5	101.1	1.3
3-methylcyclohexanone	10.5	0.8	103.6	1.0	103.0	0.7
4-methylcyclohexanone	10.6	0.9	103.6	1.4	102.7	0.6
3,5,5-trimethylcyclohex-2-enone	10.6	2.3	101.4	0.9	97.7	1.2
Alcohols						
isopropanol		not recommended on Tenax				
butanol	9.0	1.1	94.8	3.0	96.9	1.2
isobutanol	8.9	1.0	93.6	3.5	96.4	1.0

*6 replicates ** normalised to toluene = 100. The stability of toluene has been established in a BCR intercomparison.

SAMPLE TAKING AND DETERMINATION OF ALDEHYDES AND ISOCYANATES BY HPLC

E. EICKELER

Drägerwerk AG, 2400 Lübeck, Germany

SUMMARY

The determination of aldehydes and isocyanates in workplace atmosphere is achieved by a solvent-free sampling procedure and a sensitive analytical method. The measurement method described is suitable for detecting these substances in the ppb range. It makes it possible to monitor the threshold limit concentrations.

1. INTRODUCTION

The older and usual sampling methods are based on washing bottles (impingers) filled with absorbing solutions. This technique has some disadvantages particularly for personal monitoring (reagent solution can run out). An alternative sampling method consists of solid carriers impregnated with reagents which selectively react with air contaminants.

2. DETERMINATION OF ALDEHYDES

Aldehydes are collected on glass fibre filters impregnated with 2.4-dinitrophenylhydrazine according to the procedure of Levin *et al.* (1). Two filters and a support pad are contained in a filter holder which can be stored in a refrigerator for more than six months. A known volume of air is drawn through the filter using a personal air sampler. The resulting hydrazone derivatives are analysed by HPLC with UV detection.

Quality Control

The limits of detection of some important aldehydes are shown in Table 1 (air volume 20 L).

Table 1. Detection Limits of Aldehydes

	μg	μg/m³	ppb
Formaldehyde	0.1	5	4
Acetaldehyde	0.3	15	8
Acrolein	0.3	15	6
Glutaraldehyde	0.6	30	7.5

Impregnated glass fibre filters were treated with various quantities of aldehydes, stored in a refrigerator and analysed at weekly intervals. The recoveries found were about 95–100% after a storage time of four weeks.

3. DETERMINATION OF ISOCYANATES

Isocyanates are sampled on glass fibre filters impregnated with 1-(2-methoxyphenyl) piperazine. Stable urea derivatives are formed, desorbed from the filter and analysed by HPLC equipped with an ultraviolet (UV) and electrochemical (EC) detector according to the MDHS 25 method (2). The same filter holder is used as described above. The sampling procedure is suited for the determination of all known isocyanates.

Quality Control

The detection limit is defined as the amount of analyte which will give a peak whose height is about three times the height of the baseline noise (20 L air volume). For monomers of TDI, HDI and MDI a detection limit of about 1–4 ng or 50–200 ng/m^3 (EC detector) was found.

The precision of the method (sample taking included) was 10–12%, assuming a pump error of 5%.

4. CONCLUSION

The sample taking procedure with impregnated glass fibre filters for determination of aldehydes and isocyanates proved convenient in practice. The use of fragile glassware and hazardous solvents is avoided.

REFERENCES

(1) LEVIN, J.-O., ANDERSSON, K., LINDAHL, R. and NILSSON, C.-A. (1985). Determination of sub-part-per-million levels of formaldehyde in air using active or passive sampling on 2.4–dinitrophenylhydrazine-coated glass fibre filters and high-performance liquid chromatography. Anal. Chem. **57**, 1032-1039.

(2) MDHS 25 (1987). Methods for the determination of hazardous substances. Organic isocyanates in air. Laboratory method using 1-(2-methoxyphenyl)piperazine solution and high performance liquid chromatography. Health and Safety Executive (HSE) London, UK.

DIFFUSIVE SAMPLING OF DIFFERENT POLLUTANTS IN THE ENVIRONMENT

M. HANGARTNER, J. LUSTENBERGER, CH. MONN an B. TRÜSSEL

Institute for Hygiene and Applied Physiology,
ETH Zürich, Switzerland

SUMMARY

Experience with diffusive samplers for the following air pollutants is described: nitrogen dioxide, sulphur dioxide, ozone, formaldehyde and benzene -toluene-xylene. The performance data of these samplers, such as sampling rate, scatter, detection limit etc. are listed. The agreement with independent measurement methods under ambient air conditions is discussed.

1. INTRODUCTION

Diffusive sampling is a widely accepted method in industrial hygiene for monitoring different pollutants. The design of these samplers is adapted to workplace situations, that means in the range of MAK values and exposure times of 8 hours. Due to increasing consciousness of environmental problems, a corresponding interest in area monitoring can be observed. Diffusive samplers were adapted to ambient conditions: long-term sampling of 1 to 2 weeks and low ambient concentrations. Pollutants of interest were nitrogen dioxide, sulfur dioxide, ozone, formaldehyde and hydrocarbons.

2. DIFFUSIVE SAMPLERS

For NO_2 and SO_2 the Palmes tube type of sampler was used, with added glycerole for better adsorption of SO_2. Ozone is trapped by reaction with 1,2-di(4-Pyridylethylene) and determination of the yielded aldehydes is by the MBTH Method (1). A short-term personal version was developed as well (2). Formaldehyde is adsorbed by sodium bisulphite on a glass fibre filter. Hydrocarbons are collected by activated carbon, where they penetrate the sampler through a 0.16 mm mesh steel screen from two sides. The high amount of 780 mg activated carbon is intended to reduce the interference of humidity in outer air to a negligible minimum (3). Shelters were constructed to minimize climatic influences such as rain, wind etc. For more details see (2). Table 1 gives a summary of absorbents and analytical methods and figure 1 the samplers used.

pollutant	Adsorption	Analytical method
Nitrogen dioxide	Triethanolamine	Saltzmann
Sulfur dioxide	TEA/Glycerole	p-Rosaniline
Ozone	Dipyridylethylene	MBTH
Formaldehyde	Sodium disulphite	Chromotropic acid
Benzene/Toluene/Xylene	Charcoal	Gaschromatography

Table 1: Absorbents and analytical methods used

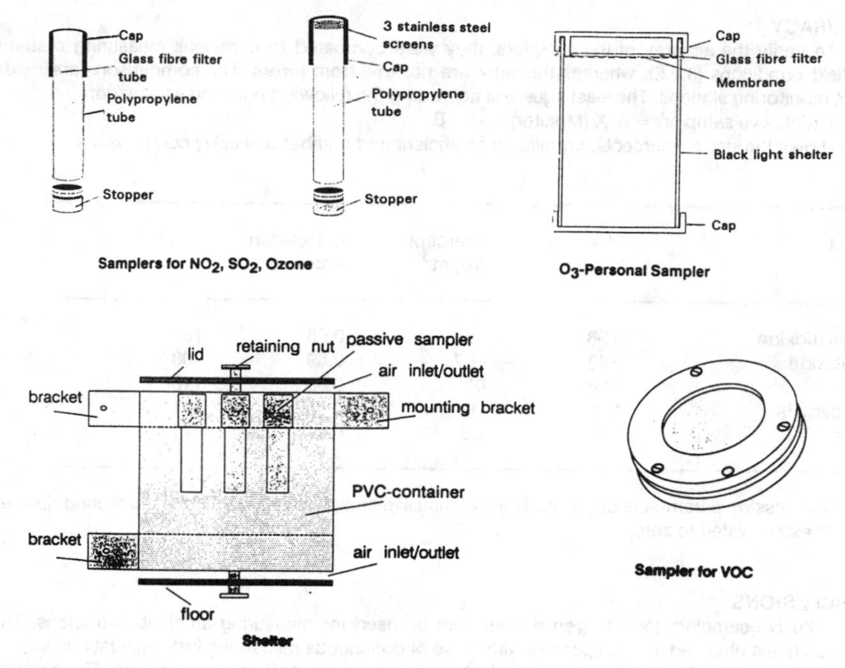

Figure 1: Diffusive Samplers used and shelter for protection

3. PERFORMANCE DATA

The following key data was obtained under laboratory conditions: Sampling rate in ml/min: exposure of the samplers in test atmospheres of constant concentration; variation coefficient from multiple samples; analytical detection limit, as determined by discriminating blanks from measured values on the 99.7 % level, converted into a one week exposure.

Table 2 shows a summary.

Pollutant	sampling rate ml/min	range mg • h/m^3	variation coefficient	detection limit µg/m^3
Nitrogen dioxide	0.89	1 - 33	6 %	3
Sulfur dioxide	0.67	3 - 30	10 %	5
Ozone -ambient	0.85	2 - 27	7 %	3
Ozone-personal	8.5	0.2 - 1.8	9 %	3
Formaldehyde	9.0	6 - 80	25 %	5
Benzene	435	1 - 30	8 %	3.4
Toluene	391	1 - 30	8 %	0.2
Xylene	373	1 - 30	8 %	0.4

Table 2: Sampling rate and range, average relative scatter of multiple samples and detection limit following a one week exposure of the samplers used.

4. ACCURACY

To verify the accuray of the samplers, they were compared to continous measuring systems under field conditions (3,4,5), whereas the latter are not free from errors. The comparisons are made different monitoring stations. The least square fit according the following equation was made:

Y (diffusive sampler) = A X (Monitor) + B

Table 3 shows the slope, intercepts, correlation coefficient and number of weekly comparisons.

Pollutant	Slope	intercept $\mu g/m^3$	correlation coefficient	number
Nitrogen dioxide	0.98	1.9	0.98	184
Sulfurdioxide	0.43	-1.7	0.89	188
Ozone	0.98	0*	0.83	116
Formaldehyde	1.05	-13	0.98	11
Toluene	1.14	2.3	0.87	7

Table 3: Regression parameters of comparisons of diffusive samplers with continous monitoring devices (* extrapolated to zero)

5. CONCLUSIONS

Diffusive samplers for **nitrogen dioxide** can be used for measuring ambient immissions. The measured values obtained are comparable with those of continuous measuring instrumentation.

The diffusive samplers for **sulphur dioxide** which were used form an unstable complex. The sampler can be used for immission loads exceeding 30 $\mu g/m^3$ and is thus suitable for area wide surveillance.

Diffusive samplers for **ozone** can be used for a maximum exposure of one week. Average area ozone concentrations can be determined. It is possible to make an estimate for the probability of ozone concentrations exceeding the hourly limit. This should however not be done without including a reference station equipped with continuous measuring instrumentation.

The sampler for formaldehyde can be used for measuring indoor air as well as outdoor air concentrations. Optimal exposure time is one to two weeks. In case of ambient air measurements, scatter is very likely to be higher than in the case of indoor measurements. Therefore, the number of samplers per station has to be increased correspondingly.

Diffusive samplers for **benzene-toluene-xylene** can be used for measuring ambient immissions. Exposure time is one to two weeks.

All samplers have to be protected from weather influences by a shelter. Wind speed is a critical factor but can be minimized by the special shelter shown in figure 1.

6. REFERENCES

(1) MONN, CH. and HANGARTNER,M.: Passive Sampling for Ozone. J. of Air and Waste Management Association, Vol 40 No 3, (1990).

(2) LUSTENBERGER,J., MONN,CH. AND WANNER, H.U.: Measurements of Ozone Indoor and Outdoor Concentrations with Passive Sampling Devices.Proceedings of Fifth International Conference on Indoor Air and Climate, p. 550 - 556.

(3) TRÜSSEL,B and HANGARTNER,M.: Passive Sampler for Benzene, Toluene and Xylene for Ambient Monitoring. Paper 90-170.13 presented at 83rd Annual Meeting & Exhibition, Pittsburgh, USA, (1990).

(4) HANGARTNER, M.: Einsatz von Passivsammlern für verschiedene Schadstoffe in der Aussenluft. VDI Berichte Nr. 838, S. 515 - 526 (1990).

(5) Hangartner, Meuli, Ch, Isler,R. und Lustenberger, J.: Vergleich von Ozonpassivsammlern mit kontinuierlichen Messgeräten. Umwelttechnik 4/90.

1380

DETERMINATION OF SOME ALIPHATIC AMINES IN AIR BY DIFFUSIVE SAMPLING AND HPLC ANALYSIS OF FLUORESCENT DERIVATIVES.

LARS HANSÉN, BENGT-OLOV HALLBERG
ANN-MARIE NILSSON-HAGELROTH

Division of Analytical Chemistry,
National Institute of Occupational Health
S-171 84 SOLNA
Sweden

SUMMARY

The aim of the project was to find a convenient method for sampling and analysis of some low molecular weight primary aliphatic amines of interest in occupational health.

Diffusive samplers of cylinder shape, usable for absorption solution, were manufactured of polymethyl methacrylate. The uptake of amine from amine/air mixtures was tested in a dynamic gas generation chamber. As comparable techniques were used sampling with pump on either cellulose filters impregnated with phosphoric acid or in gas washing bottles containing hydrochloric acid. The amine coupling reaction was performed with the reagent 7-fluoro-4-nitrobenzo-2-oxa-1,3-diazole (NBD-F) at pH 8.5 and increased temperature. Separations were made with HPLC using a Resolve C-18 column, methanol/water as eluent and a fluorescence detector. One pmole of methylamine could be detected. Eight primary amines in a homologous series were separated. An uptake rate of about 8.5 mL/min for methylamine was calculated for the diffusive samplers. The method was evaluated for methylamine, allylamine and monoethanolamine using ion chromatography and isotachophoresis as reference analytical methods. Methylamine in air can be determined at a concentration of 1/1000 of the Swedish exposure limit value (10 ppm) with a sampling time of 4 hours.

EXPERIMENTAL

Generation of amine/air mixtures

The gas generation set-up was made of glass and consisted of two round flasks each of 4 litres (= mixing camber) and a cylinder (= sampling chamber, 130 cm length, 9.5 cm Ø) (1). Amine gas from a cylinder was led through a mass flow regulator and was mixed with diluting air in the mixing chamber. The flows were regulated and constant. The amine/air gas mixture was passed over six samplers placed in the sampling chamber. They were weighed before and after the experiment. Simultaneously amine/air reference samples were taken out by pump. As comparative sampling methods were used gas washing bottles containing hydrochloric acid (20 mmol/L) and cellulose support pads (Millipore AP10, 13 mm, in a Swinnex 13 mm standard holder) impregnated with phosphoric acid.The concentration level of amine in the chamber was continuously registered by an IR-instrument equipped with a gas cell.

Diffusive sampling

The sampler was made of plexiglass (polymethyl methacrylate). It has the form of a cylinder (Ø 45 mm, height 15 mm) and a weight of 33 gram. The lid has 65 channels for diffusion (5.0 [height] x 1.0 [Ø] mm). A membrane of teflon® (Ø 37 mm), which is penetrable for gases but not liquids, covers a 2 mL cavity containing the absorption solution (hydrochloric acid, 20 mmol/L). A tight lid is pressed over the sampler and is removed only during the sampling period. The sampler is placed near the breathing zone of the worker with a metallic clip.

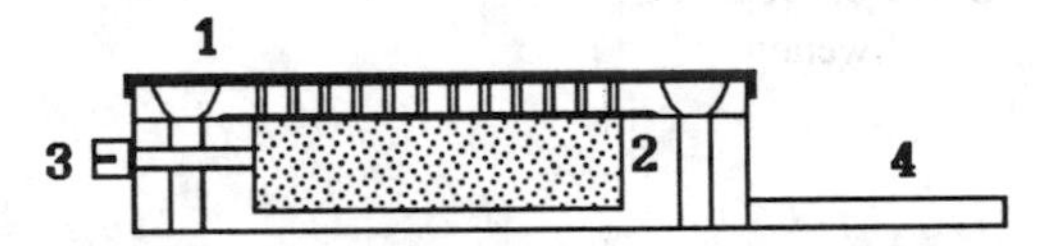

Figure 1. Diffusive sampler made of plexiglass. 1= lid; 2= cavity for absorption solution; 3= channel for filling and taking out sample; 4= clip.

Fluorescence reaction

The method was a modification of one described by Imai and Watanabe (2). In a dark glass tube (4 mL) were put borate/HCl buffer solution (pH 8.8), amine sample and NBD-F solution (at least 10 times in excess). After mixing, the tube was heated to 55°C on a water bath for 5 minutes. The mixture was chilled on ice bath and HCl was added to stop the reaction.

Analytical instrumentation

The separations were performed by HPLC with the following parts: Pump: LKB 2150 (LKB-Produkter AB, Bromma, Sweden); Injector: Rheodyne 7125 (Rheodyne, Cotati, California, USA); Detector: Fluorescence HPLC monitor RF 535 (Shimadzu, Kyoto, Japan); Recording and Data logging: ELDS 900 lab data system for chromatography (Chromatography Systems AB, Kungshög, Sweden).

Analytical conditions: Column: Resolve C-18, 5 µm, steel, 3.9 x 150 mm (Waters, Millipore corporation, Milford, USA); Eluent: Methanol/water (55:45); Flow: 1,0 mL/min; Injection volume: loop 20 µL; Detection: fluorescence, excitation 480 nm, emission 530 nm;

A calibration curve was linear between 1 and 200 pmoles methylamine injected.

RESULTS

Sampling and analysis after generation of methylamine

Methylamine gas from a cylinder was passed through the generation chamber. The calculated generated concentration of methylamine was ca 4.8 ppm. The reference methods confirmed the generated concentration: 3.7 ppm (gas washing bottles, C V 12.2 %) and 4.7 ppm (impregnated filters, C V 15.3 %). At those conditions the uptake rate for the diffusive samplers could be determined to 8.5 mL/min (C V 6.1 %).

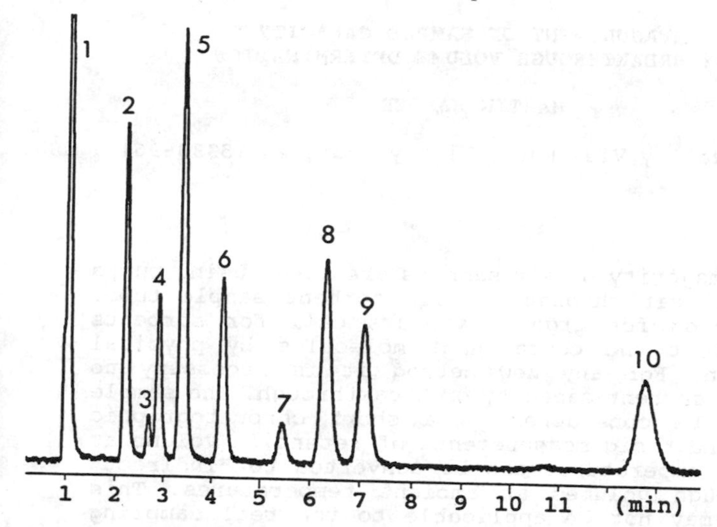

Figure 2. Separation of eight primary amines as their NBD-derivatives. 1= reagent, 2= methyl-, 3= interference from reagent, 4= ethyl-, 5= allyl-, 6= propyl-, 7= sec-butyl-, 8= iso-butyl-, 9= n-butyl-, 10= amylamine, ca 60 pmoles of each.

Comparison of methods

The analytical method was evaluated for methyl- allyl- and monoethanolamine. *Methylamine* was sampled in gas washing bottles (HCl, 20 mmol/L) from a generation chamber. The absorption solutions were analysed by the HPLC-method and by isotachophorseis (ITP; 3) with good agreement. *Allylamine* generated and sampled as above with a calculated concentration of 0.13 and 2.7 mmol/L gave 0.15 and 2.4 mmol/L respectively analysed as NBD-derivative.

Ethanolamine, used as an anti-corrosive agent, was sampled in an engineering industry. Some samples were taken in tubes with aluminium oxide and were analysed as NBD-derivatives with the HPLC-method and as free amine ions by ion chromatography and ITP. Statistical T-test of the mean concentrations showed no difference between the methods at two sampling sites. At the third site it was good correlation only between the HPLC and ITP methods. However, a more careful study at a similar factory is planned to confirm the results.

REFERENCES

1. Rudling J, Hallberg B-O, Hultengren M, Hultman A. Development and evaluation of field methods for determination of ammonia in air. Arbete och Hälsa 20 (1983). Arbetsmiljöinstitutet, Solna. (in Swedish)
2. Imai K, Watanabe Y. Fluoremetric determination of secondary amino acids by 7-Fluoro-4-nitrobenzo-2-oxa-1,3-diazole. Analytica chimica acta 130 (1981) 377-383.
3. Sollenberg J, Hansén L. Isotachophoretic determination of amines from workroom air. J Chromatogr 390 (1987) 133-140.

MEASUREMENT OF SAMPLE CAPACITY BY BREAKTHROUGH VOLUME DETERMINATION

MARTIN HARPER

SKC INC., 334 Valley View Road, Eighty Four, PA 15330-9614, USA.

SUMMARY

The vast majority of air samples are taken using pumps which pull air through a solid sorbent sample tube. Many methods for organic vapours call for sorbents that collect the contaminant molecules by physical adsorption. For any new method it is necessary to determine sorbent capacity or breakthrough. The sample tube can be considered as a short chromatographic column, and rapid measurements of retention volume at elevated temperatures can be converted to "indirect" breakthrough volumes at ambient temperatures. This approach may not be applicable to the real sampling situation. In particular, no account is made of the difference in carrier gas, or the concentration of vapour. An apparatus has been specially designed to measure breakthrough volumes under strict control of air flow-rate, temperature, relative humidity, and contaminant concentration, at levels applicable to the normal sampling situation. Safe sample volumes defined under OSHA/NIOSH/HSE protocols and measured by this equipment differ significantly from those derived from indirect measures of breakthrough. If these indirect estimates of safe sample volume are applied, sample loss might be expected under the worst case conditions of the existing performance standards.

1. INTRODUCTION

Rapid generation of sampling methods is required to support legislation covering the concentrations of organic vapours allowed in the work-place. Many thousands of organic compounds are used inindustry, and the effort to produce fully validated methods for all of them would be time-consuming. The three most important parameters to be considered for any sampler/analyte combination are sampler capacity (normally expressed in terms of safe sample volume), desorption efficiency, and sample storage stability. Safe sample volumes (V_s) are defined by OSHA (1), NIOSH (2), and HSE (3) in terms of the breakthrough volume (V_B). Breakthrough volume is defined as the volume of contaminated air that can be passed through a sampler before the effluent concentration reaches 5% of the applied air concentration. "Worst case" conditions of 25°C and 80% relative humidity are used, and the applied concentration is twice the OSHA Permissible Exposure Level (PEL). The safe sample volume is then defined as two-thirds the breakthrough volume to account for the possible presence of interfering compounds.

2. DIRECT METHOD OF BREAKTHROUGH VOLUME DETERMINATION

The method above requires an apparatus suitable for the generation of standard atmospheres of known contaminant concentration and carefully controlled temperature and relative humidity. Such an apparatus has been designed and built for this study. A schematic is shown in figure 1.

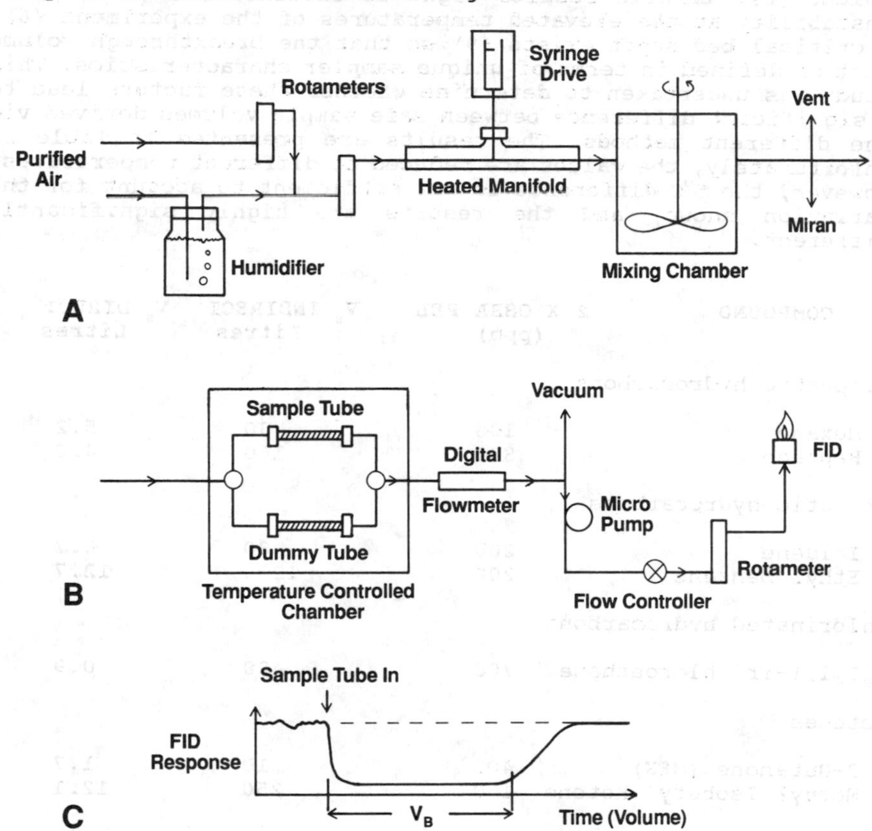

Figure 1: Direct method of breakthrough volume determination. A: Apparatus for generating standard atmosphere. B: Apparatus for breakthrough determination. C: Typical breakthrough curve.

3. INDIRECT METHOD OF BREAKTHROUGH VOLUME DETERMINATION

A sample tube can be conceived as a short chromatographic column. When fitted into the oven of a gas chromatograph, it is possible to obtain rapid measurements of chromatographic retention volume (V_R) at elevated temperatures and extrapolate the results to room temperature. An "indirect" measure of safe sample volume can be obtained by this method (4).

A number of objections have been raised concerning the validity of indirect safe sample volumes. The accuracy and the linearity of the extrapolation have been questioned (5,6). The presence of gases other than nitrogen in real air is not addressed and can be important (7). There is no control over concentration, which is an important determinant of breakthrough volume (8). Erratic results might be obtained due to compound instability at the elevated temperatures of the experiment (6). A critical bed depth exists (9) so that the breakthrough volume must be defined in terms of unique sampler characteristics. This study was undertaken to determine whether these factors lead to a significant difference between safe sample volumes derived via the different methods. The results are presented in Table 1. Unfortunately, the values are reduced to different temperatures. However, the 5°C difference is not sufficient to account for the variation shown, and the results are highly significantly different.

COMPOUND	2 X OSHA PEL (ppm)	V_s INDIRECT[1] Litres	V_s DIRECT[2] Litres
Aliphatic hydrocarbons			
Hexane	100	30	5.2
Heptane	800	160	4.0
Aromatic hydrocarbons			
Toluene	200	80	8.7
Ethyl benzene	200	1200	12.7
Chlorinated hydrocarbons			
1,1,1-Trichloroethane	700	28	0.9
Ketones			
2-Butanone (MEK)	400	10	1.7
Methyl Isobutyl Ketone	100	250	12.1
Esters			
Ethyl acetate	800	20	2.2
n-Butyl acetate	300	730	9.4
Alcohols			
1-Butanol	50*	50	9.1
2-Butanol	100	30	4.0
2-Ethoxyethanol	400	75	9.4

[1] Results at 20°C; [2] Results at 25°C; * OSHA Ceiling Limit

Table 1: Comparison of safe sample volumes by direct and indirect methods.

4. CONCLUSIONS

It should be stressed that, at present, these results are only applicable to pumped sample tubes. The relationship between breakthrough under these conditions, and saturation of a diffusive sampler, is not known, and would likely be very complex. In addition, all investigations to date have employed constant contaminant concentrations, which is not a normal situation that would be encountered during sampling. Further work is underway to investigate these situations.

"Indirect" (GC) measurements of safe sample volumes refer to very low contaminant concentrations, and are not comparable to safe sample volumes as defined by OSHA, NIOSH, and HSE.

"Direct" measurements of breakthrough are relatively quick and simple to obtain if the right equipment is available. There is no advantage to the use of indirect measurements.

The equipment described above can be used to assist in the developing the optimum sampler for any given situation, since it is possible to vary sorbent weight, mesh-size, tube diameter, flow-rate, temperature, relative humidity, contaminant concentrations and interferences.

REFERENCES

1. U.S. DEPT. OF LABOR. OSHA Analytical Methods Manual. OSHA Analytical Laboratory, Salt Lake City, Utah, 1985.
2. U.S. DEPT. OF HEALTH EDUCATION AND WELFARE (NIOSH). NIOSH Manual of Analytical Methods, 2nd ed., Cincinnati, Ohio, 1977.
3. HEALTH AND SAFETY EXECUTIVE (U.K.). Methods for the Determination of Hazardous Substances. Protocol for assessing the performance of a pumped sampler for gases and vapours. MDHS 54. HSE, London, 1986.
4. BROWN, R.H. and PURNELL, C.J. (1979) J. Chromatogr. 178:79-90.
5. POSNER, J.D. Evaluation of sorbents for the collection and analysis of trace levels of airborne vapors: Bis (2-chloroethyl) sulfide (mustard), a case study. (in press)
6. TANAKA, T. (1978) J. Chromatogr. 153:7-13.
7. PIECEWICZ, J.F.; HARRIS, J.C. and LEVINS, P.L. Further Characterization of Sorbents for Environmental Sampling. Environmental Protection Agency Report: EPA 600/7-79-216, Sept. 1979.
8. BERTONI, G.; BRUNER, F.; LIBERTI, A. and PERRINO, C. (1981) J. Chromatogr. 203:263-270.
9. WHEELER, A. and ROBELL, A.J. (1969) J. Catal. 13, 299-305.
10. HEALTH AND SAFETY EXECUTIVE (U.K.). Methods for the Determination of Hazardous Substances. Volatile Compounds in Air. MDHS Draft Document.

DESORPTION EFFICIENCY - A CRITICAL FACTOR AFFECTING THE ACCURACY OF AIR SAMPLING METHODS

MARTIN HARPER AND LLOYD V. GUILD

SKC Inc., 334 Valley View Road, Eighty Four, PA 15330-9614, USA.

SUMMARY

The commonest method of taking air samples involves concentration by adsorption on a porous solid sorbent. A desorption step is necessary prior to analysis and the desorption efficiency contributes to the overall analytical recovery of the method. Solvent desorption is the commonest method, particularly in the U.S.A. Thermal desorption has achieved prominence in the European Community, and is used also in the U.S.A. for the analysis of environmental samples. Solvent desorption in particular has many disadvantages, and thermal desorption is not applicable to every sampling situation. When these methods are critically assessed and compared to a recent innovation (supercritical fluid extraction, or SFE) SFE appears to represent an alternative approach of considerable value.

1. INTRODUCTION

Sampling workplace and environmental air provides measurements of the concentration of organic gases and vapours. The commonest method involves concentration by adsorption on a porous solid sorbent. The advantage of this method is that more than one species may be collected using a single sample. The sorbent must be able to hold molecules tightly during sampling, but release them easily on desorption. Desorption efficiency is the ratio of recovered to collected sample.

The most common sorbent for organic vapours is active charcoal (1). It has a high surface area comprised of numerous micropores of molecular dimensions. Active charcoal will collect both polar and non-polar compounds over a very wide boiling range. Heat of adsorption is high, and difficult to overcome by thermal desorption except at temperatures high enough to cause unwanted reactions. The normal desorption method is to use a solvent of small molecular size and high heat of adsorption, and which does not interfere with the analysis. Carbon disulphide is especially useful because of low background response on flame ionization detectors. However, it is a fairly poor solvent, especially for polar compounds, and rather toxic.

2. FACTORS AFFECTING SOLVENT DESORPTION

SORBENT SURFACE. Pores that are fine enough to hold small molecules sufficiently well to allow collection without sample loss by breakthrough also hold larger molecules very tightly. In addition these pores may be shaped with a constricted neck ("ink-bottle" pores) which sterically hinders desorption (2). Diffusion coefficients are low in liquids and long periods of agitation may be required for complete desorption. Reactive sites may chemisorb the adsorbate or catalyze reactions, including breakdown or polymerization. These factors cause a proportionally greater loss at lower loadings. The effect can be seen in figure 1.

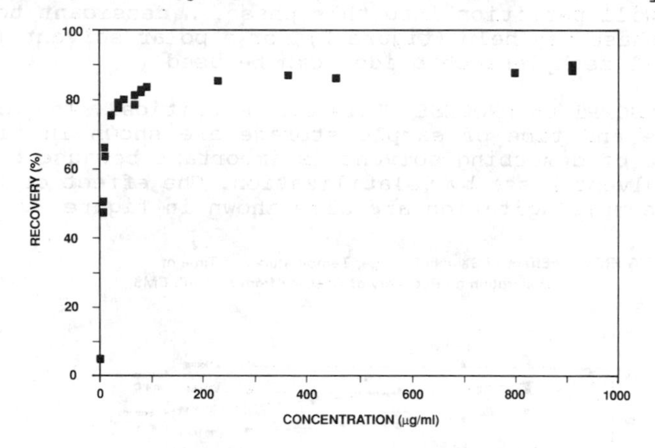

FIGURE 1: Desorption Efficiency of Styrene (Data Supplied by Dr. M. Van Leeuwen, BCO)

SORBENT/SOLVENT RATIOS. A partition equilibrium is set-up between the adsorbate in solution and adsorbed surface. Increasing the amount of solvent, or decreasing the amount of sorbent increases the desorption efficiency (3,4), but may compromise other sampling or analytical requirements such as sensitivity or capacity.

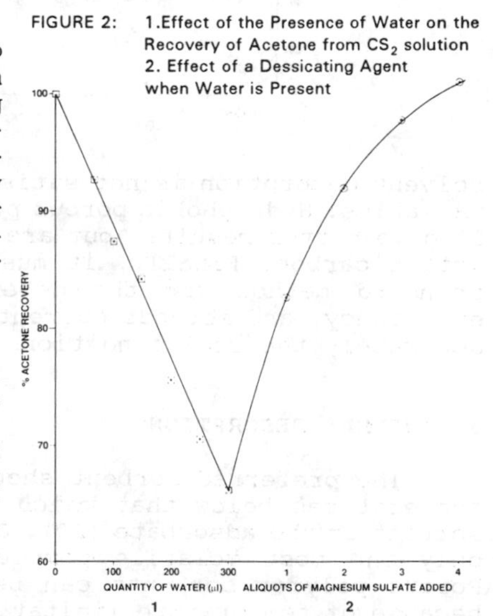

FIGURE 2: 1.Effect of the Presence of Water on the Recovery of Acetone from CS_2 solution 2. Effect of a Dessicating Agent when Water is Present

PRESENCE OF MULTIPLE SORBATES. Non-polar adsorbates on charcoal appear to be unaffected by the presence of co-adsorbed species, but the desorption efficiency of polar compounds can be altered considerable by the presence of other compounds (3,5). This may be due in part to competition for a finite number of reactive polar sites on the sorbent surface, and in part to a co-solvent effect, modifying the polarity of the carbon disulphide. This effect can be made use of by deliberately adding polar modifiers to the desorbing solvent (6,7).

PRESENCE OF ADSORBED WATER VAPOUR. While non-polar compounds are relatively unaffected by the presence of water molecules, polar compounds may be strongly affected (8). Hydrolysis reactions may take place, especially during extended periods of sample storage. In addition, the amount of water adsorbed onto charcoal can be very large. It is also displaced by carbon disulphide and can form a discrete hydrophilic phase. Polar molecules will partition into this phase. A dessicant to remove the water phase may help (figure 2), or a polar solvent (e.g. 5% methanol/95% methylene chloride) can be used (9).

THE DESORPTION PROCESS. This can be critical. The effect of temperature and time of sample storage are shown in figure 3. Temperature of desorbing solvent is important because of sample loss (or solvent loss) by volatilization. The effect of time and degree of sample agitation are also shown in figure 3.

FIGURE 3: Effect of Sample Storage, Temperature and Time of Desorption on Recovery of Acetone from Anasorb CMS

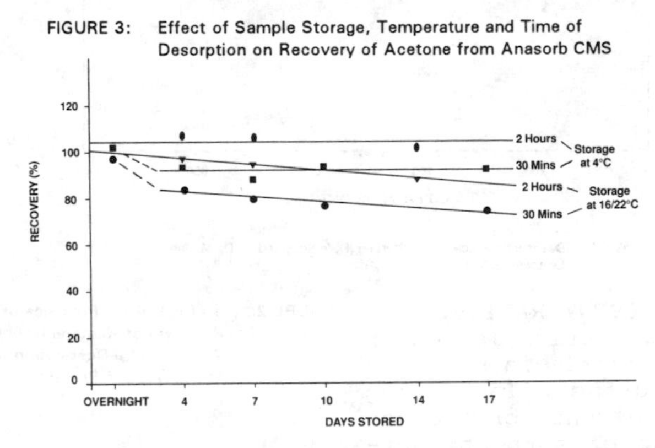

Solvent desorption is not satisfactory when dependent on so many variables. Hydrophobic porous polymers (e.g. Chromosorb 106, XAD-2) give better results, but are more restricted in their use than active carbon. Finally, it must be recognized that there is no standard method for the determination of solvent desorption efficiency, and methods currently in use probably do not reflect accurately the real conditions of sampling and desorption.

3. THERMAL DESORPTION

The preferred sorbent should release the sample readily at temperatures below that which would lead to degradation of the sorbent or the adsorbate (10). Carbonaceous sorbents will release only the most volatile compounds at reasonable temperatures. Porous polymer sorbents can be used for the middle range, but because of temperature limitations they are less useful for high boiling compounds. Tenax, a relatively non-porous polymer has a high maximum temperature, but a rather low capacity. Clearly, a number of samplers, or a multiple sorbent bed, would be required to sample the kind of mixtures (e.g. styrene/acetone) often found in industry if thermal desorption is to be employed.

4. SUPERCRITICAL FLUID EXTRACTION

A recent development is supercritical fluid extraction (SFE). The technique has has been used to extract organic compounds from soil samples (11), which is often laborious and difficult by normal methods such as soxhlet extraction. It is obvious that the same factors which make SFE so attractive in soil analysis, would be useful in the desorption of sorbents used in air sampling (12). Diffusion coefficients in supercritical fluids are orders of magnitude higher than in liquids, so that pore penetration and mass transfer are faster. The agent normally used (carbon dioxide) is a smaller molecule than carbon disulphide, with greater penetrating ability. It is also relatively non-toxic. Polar modifiers can be added easily. Desorption temperatures can be kept reasonably low, and the whole process can be automated and directly interfaced to a supercritical fluid chromatograph (SFC).

5. CONCLUSIONS

Not all of the factors affecting solvent desorption efficiency are under the control of the analyst, and the complexity of the process can lead to the application of an erroneous desorption efficiency correction under certain circumstances. With thermal desorption it is necessary that the sorbent be closely matched to the analyte for optimum desorption efficiency. Supercritical fluid extraction is faster and more effective than solvent desorption and there is less possibility of sample breakdown than with thermal desorption. SFE represents an exciting new development in the field of occupational hygiene sampling for airborne gases and vapours.

REFERENCES

1. WHITE, L.D.; TAYLOR, D.G.; MAUER, P.A. and KUPEL, R.E. (1970) Am. Ind. Hyg. Assoc. J. 31:225-232.
2. GREGG, S.J. and SING, K.S.W. Adsorption, Surface Area, and Porosity. Academic P.
3. POSNER, J.C. and OKENFUSS, J.R. (1981) Am. Ind. Hyg. Assoc. J. 42:643-646.
4. DOMMER, R.A. and MELCHER, R.G. (1978) Am. Ind. Hyg. Assoc. J. 39:240-246.
5. RUDLING, J. (1988) Am. Ind. Hyg. Assoc. J. 49:95-100.
6. FRACCHIA, M.; PIERCE, L.; GRAUL, R. and STANLEY, R. (1977) Am. Ind. Hyg. Assoc. J. 38:144-146.
7. KENNY, J. and STRATTON, G. (1989) Am. Ind. Hyg. Assoc. J. 50:A431-434.
8. RUDLING, J. and BJORKHOLM, E. (1986) Am. Ind. Hyg. Assoc. J. 47:615-620.
9. POSNER, J.C. (1981) Am. Ind. Hyg. Assoc. J. 42:647-652.
10. POSNER, J.C. (in press).
11. McHUGH, M.A. and KRUKONIS, V.J. (eds.) Supercritical Fluid Extraction. Butterworths, NY, 1986.
12. HAWTHORNE, S.B. and MILLER, D.J. (1986) J. Chromatogr. Sci. 24:258-264.

COMPARISON OF MULTI-BED ADSORBENT TUBES TO TRAP TOXIC ORGANIC COMPOUNDS

S. HASTENTEUFEL[1] and W.R. BETZ[2]

[1]Supelco D Gmbh, 6380 Bad Homburg, Germany
[2]Supelco Inc., Bellefonte, PA 16823-0048, USA

SUMMARY

The use of adsorbent tubes for monitoring airborne contaminants has gained widespread acceptance over the past several decades. The airborne contaminant(s) which are adsorbed and subsequently desorbed from the adsorbent surface typically dictate which adsorbent is chosen. Recent work focusing on the use of Type I, non-specific (electronically neutral) adsorbents has allowed the sampling professional to create a single adsorbent tube with one or several Type I adsorbents.

1. INTRODUCTION

According to the classification scheme developed by A.V. Kiselev (1), adsorbents can be categorized according to the chemistry of the adsorbent surface (Table 1). Adsorbates also can be categorized. Only Type I adsorbents interact non-specifically, i.e. London forces) with the four groups of adsorbates (Table 1).

Table. 1. Classification of Adsorbents and Adsorbates

Molecules	Adsorbents		
	Type I without ions or active groups	Type II localized positive charges	Type III localized negative charges
Group A: • Spherically symmetrical shells • σ-bonds	Nonspecific interactions/ dispersion forces		
Group B: • Electron density concentrated on bonds/links • π-bonds			
Group C: • (+) Charge on peripheral links	Nonspecific interactions	Nonspecific + specific interactions	
Group D: • Concentrated electron densities • (+) Charges on adjacent links			

The focus of this study was to compare several types of adsorbents for use in sampling a wide range of airborne contaminants. Type I (i.e. graphitized carbon blacks) and weak Type III (i.e. typical porous polymers) adsorbents were investigated for trapping semi-volatile compounds). Several carbon molecular sieves and activated charcoals, which are stronger Type III adsorbents, were evaluated for trapping volatile compounds. Adsorbents investigated in this study are listed in Table 2.

Table 2. Classification of Adsorbents

Adsorbent	Surface	Classification (Kiselev)
Graphitized Carbon Blacks	Graphitic Carbon	Class I
Carbon Molecular Sieves	Amorphous Carbon	Weak Class III (can approach Class I)
Activated Silica Gel	Oxides of Silica Gel	Class II
Activated Charcoal	Oxides of Amorphous Carbon	Class III
Porous Polymers	Organic "Plastics"	Weak → Strong Class III

2. EXPERIMENTAL

The adsorbents were evaluated using gas-solid chromatographic techniques described in US EPA Document No.500/7-78-084 (2). This analytical approach involves injecting into an adsorbent tube, connected to the injector and detector ports of a gas chromatograph. The data obtained included the specific retention volume, adsorption coefficient, and equilibrium sorption capacity. The specific retention volume, or breakthrough volume, is the volume of gas required to allow migration of the introduced adsorbate from the front of the adsorbent bed to the back (i.e. elicits a detector response). The adsorption coefficient and capacity values are subsequently extracted from this breakthrough volume data.

3. RESULTS AND DISCUSSION

Breakthrough volume, adsorption coefficient, and capacity (3) data for several of the studied adsorbents are presented in Table 3.

Data from the study of weak to strong Type III adsorbents for monitoring volatile compounds are presented in Tables 4 and 5.

Table 3. Breakthrough Volumes (V^t_g), Adsorption Coefficients (K_a), and Equilibrium Sorption Capacity (Q_g) Values for Carbotrap, Tenax GC and Amberlite XAD-2 Adsorbents

	V^t_g 20°C■			K_a			G_g		
Adsorbate	Carbotrap	Tenax GC	XAD-2	Carbotrap	Tenax GC	XAD-2	Carbotrap	Tenax GC	XAD-2
n-Decane	4.79×10^{9}	1.56×10^{7}	3.63×10^{7}	2.60	3.61×10^{-2}	5.40×10^{-3}	28.0	9.14×10^{-2}	2.12×10^{-1}
Benzylamine	2.23×10^{7}	3.57×10^{6}	1.63×10^{7}	1.21×10^{-2}	8.23×10^{-3}	2.42×10^{-3}	9.81×10^{-2}	1.57×10^{-2}	7.18×10^{-2}
Chlorobenzene	1.58×10^{6}	1.51×10^{5}	4.84×10^{5}	8.55×10^{-4}	3.49×10^{-4}	7.20×10^{-5}	7.30×10^{-3}	8.85×10^{-4}	2.24×10^{-3}
p-Xylene	4.24×10^{7}	3.88×10^{5}	7.95×10^{6}	2.30×10^{-2}	8.39×10^{-4}	1.18×10^{-3}	1.82×10^{-1}	1.66×10^{-3}	3.40×10^{-2}
p-Cresol	2.06×10^{7}	1.50×10^{7}	4.96×10^{6}	1.11×10^{-2}	3.45×10^{-2}	7.39×10^{-4}	9.14×10^{-2}	3.70×10^{-3}	2.21×10^{-7}
n-Pentanoic acid	4.31×10^{5}	9.78×10^{5}	1.01×10^{5}	2.34×10^{-4}	2.26×10^{-3}	1.51×10^{-5}	1.76×10^{-3}	3.99×10^{-3}	4.31×10^{-4}
Cyclohexanone	2.04×10^{6}	1.06×10^{6}	6.27×10^{5}	1.10×10^{-3}	2.45×10^{-3}	5.45×10^{-5}	8.19×10^{-3}	4.28×10^{-3}	1.47×10^{-3}
2-Methyl-2-propanol	6.52×10^{3}	6.86×10^{2}	5.42×10^{3}	3.53×10^{-6}	1.58×10^{-6}	8.06×10^{-7}	1.99×10^{-5}	2.09×10^{-6}	1.65×10^{-5}

■ For the three adsorbents, correlation coefficients for breakthrough volume and temperature ranged from 0.97190 to 1.00000.

Table 4. Breakthrough Volumes for Dichloromethane

Adsorbent	Breakthrough Volume (liters)
Carbosieve™ S-III	66.2
Carboxen™-569	43.2
Activated Coconut Charcoal	39.2
Carboxen-564	31.5
Carbosieve S-II	31.5
Ambersorb® XE-347	26.0
Purasieve®	5.05
Carboxen-563	1.56
Ambersorb XE-340	1.49
Spherocarb™	1.05

Table 5. Breakthrough Volumes for Water

Adsorbent	Breakthrough Volume (liters)	Equilibrium Sorption Capacity (grams adsorbate/ gram adsorbent)
Carboxen™-569	0.06	1.66×10^{-7}
Carboxen-564	0.10	3.27×10^{-7}
Spherocarb™	0.22	6.79×10^{-7}
Ambersorb® XE-347	0.23	6.95×10^{-7}
Purasieve®	0.24	7.45×10^{-7}
Ambersorb XE-340	0.24	7.32×10^{-7}
Carbosieve™ S-III	0.32	9.72×10^{-7}
Carboxen-563	0.80	2.43×10^{-6}
Carbosieve S-II	1.02	3.10×10^{-6}
Activated Coconut Charcoal	2.44	7.42×10^{-6}

These results indicate that the Type I adsorbent Carbotrap B (a graphitized carbon black) and the weak Type III adsorbent Carbosieve S-III, possess the greatest adsorption strengths. A similar study showed that a second graphitized carbon black, Carbotrap C, provided effective adsorption and desorption of the C_{12}–C_{22} adsorbates. The subsequent construction of an adsorbent tube containing three hydrophobic adsorbents provided an adsorbent tube with a wide sampling range. This multi-bed tube was then tested for monitoring a wide range of adsorbates cited in US EPA TO (Toxic Organic Compounds) Methods 1–3, using thermal desorption techniques. The results of these studies indicated that near complete recovery is achievable for this wide range of compounds. These adsorbent tubes were also tested for sampling in high humidity conditions, with no measured loss of recovery or sampling volume. Chromatograms 1 and 2 illustrate the analytical results obtained from these studies. Desorption and chromatographic conditions have been previously discussed (4).

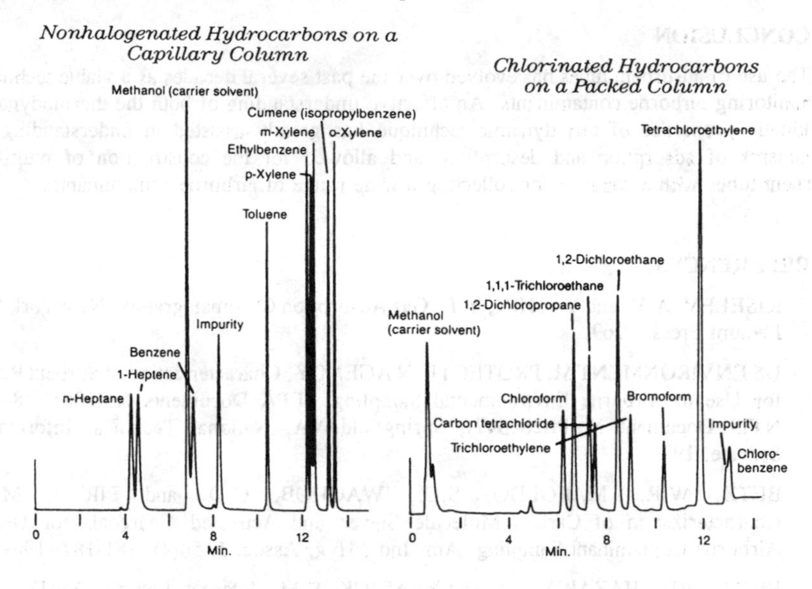

Chromatogram 1. Thermally desorbed TO-1 hydrocarbons

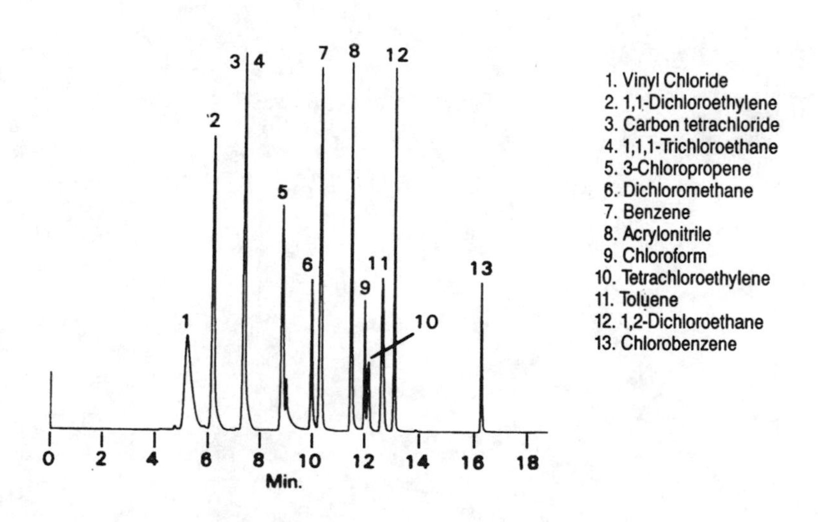

Chromatogram 2. Thermally desorbed TO-2 and TO-3 compounds

4. CONCLUSION

The use of adsorbent tubes has evolved over the past several decades as a viable technique for monitoring airborne contaminants. An effective understanding of both the thermodynamic and kinetic properties of this dynamic technique has greatly assisted in understanding the mechanisms of adsorption and desorption, and allowed for the construction of multi-bed adsorbent tubes with a capacity for collecting a wide range of airborne contaminants.

5. REFERENCES

(1) KISELEV, A.V. and YASHIN, Y.I., Gas Adsorption Chromatography, New York NY, Plenum Press, 1969.

(2) US ENVIRONMENTAL PROTECTION AGENCY, Characterization of Sorbent Resins for Use in Airborne Environmental Sampling. (EPA Documents No.500/7-78-084/ NTIS Document No.PB284347), Springfield VA, National Technical Information Service, 1978.

(3) BETZ, W.R., MAROLDO, S.G., WACHOB, G.D. and FIRTH, M.C., Characterization of Carbon Molecule Sieves and Activated Charcoals for Use in Airborne Contaminant Sampling. Am. Ind., Hyg. Assoc. J. 50(4), 181-187 (1989).

(4) BETZ, W.R., HAZARD, S.A. and YEARICK, E.M., Internat. Labmate Xv(1), 41-44 (1990).

1384

MONITORING OF ANAESTHETIC GAS- AND VAPOUR-CONCENTRATIONS IN THE AIR OF OPERATING THEATRES BY DIFFUSIVE SAMPLING SYSTEMS

KARL-HEINZ PANNWITZ

Drägerwerk AG, Lübeck, Germany F.R.

SUMMARY

Diffusive sampling systems such as ORSA and the Nitrous-oxide diffusion sampler are suited for the sample taking of volatile anaesthetics from the ambient air. Both systems have been verified for this purpose. The influence of temperature, air humidity and interferent compounds has been investigated by sample takings in test atmospheres of known composition. The results of comparison measurements with active sample taking systems, which have been carried out in operating theatres show good correlations.

INTRODUCTION

The volatile anaesthetics enflurane, halothane, isoflurane and nitrous oxide are frequently used particulary as well as in mixtures to anaesthetize patients during operations. Depending on the type of anaesthesia technique selected various amounts of anaesthetic vapours and gases can evaporate into the ambient air and cause impacts of the health of the operating team. A study performed in Switzerland [1] revealed that nitrous-oxide concentrations of 5100 mL/m³ may be encountered in the immediate vicinity of the anaesthetist`s inhalatory organs under unfavourable conditions. To protect the employers working in operating theatres from health risks the air has to be monitored by qualified analysis methods. As personal monitoring methods give the best informations about the exposure, the measuring resp. sampling systems which are applicable for this purpose must be small, light weight and easy to handle. Diffusive samplers may favorably be useful to meet these requirements.

ANALYSIS OF NITROUS OXIDE

Personal air samples of nitrous oxide can be taken with the Dräger Nitrous-oxide diffusion sampler. The technical data and usage conditions are summerized in table 1.

dimensions of unopened sampling tube	115 x 7 mm
dimensions when ready for use	125 x 14 x 12 mm
weight wenn ready for use	approx. 12 g
sampling layer	molecular sieve
adsorption capacity	120 µg N_2O
diffusion barrier	impregnated filter layer which retains moisture and CO_2
measuring range (8 hours sampling)	5 to 500 mL/m³ N_2O
standard uptake rate	0.03 µg /(ppm x h)
minimum velocity of ambient air	1 cm/s
sampling duration (25 to 500 mL/m³)	1 to 8 hours
temperature range	5 to 35°C
air humidity	< 90 % rel. at 25°C

Table 1: Technical data and usage conditions

The sampling system consits of a glass tube for opening on one end, which contains a layer of molecular sieve to collect nitrous oxide from the ambient air. After the sample taking the collected nitrous oxide is thermally desorbed and analysed by infrared spectrometry.

Many tests have been carried out (most in accordance with MDSH 27) to analyse the factors affecting the uptake rate of the Nitrous-oxide diffusion sampler. The validation data and the results of comparison measurements with a portable photoacustic infrared spectrometer type Brüel & Kjaer Multigas-monitor 1302, which have been carried out in operating theatres under field conditions, are summerized in table 2.

description of the experiment	test conditions	results
determination of the standard uptake rate U	25...1600 ppm x h N_2O	U = 0.0296 µg/(ppm x h) s = 0.0023 µg/(ppm x h) v = 7.9 %
influence of temperature and humidity on the uptake rate	temperature 5°C rel. humidity 20 - 90 % exposure 400 ppm x h	U = 0.0296 µg/(ppm x h) s = 0.0016 µg/(ppm x h) v = 5.3 %
	temperature 30°C rel. humidity 20 - 90 % exposure 400 ppm x h	U = 0.0282 µg/(ppm x h) s = 0.0013 µg/(ppm x h) v = 4.7 %
effect of storage	storage duration 3 days 11 days 25 days 35 days	U = 0.032 µg/(ppm x h) U = 0.032 µg/(ppm x h) U = 0.032 µg/(ppm x h) U = 0.031 µg/(ppm x h)
comparison with portable infrared spectrometer	measuring duration 4 hours max. N_2O conc. 119 mg/m³ min. N_2O conc. 2.7 mg/m³	mean conc. 22 mg/m³ diffus. tube 20 mg/m³
	max. N_2O conc. 475 mg/m³ min. N_2O conc. 14.3 mg/m³	mean conc. 195 mg/m³ diffus. tube 181 mg/m³
	max. N_2O conc. 382 mg/m³ min. N_2O conc. 185 mg/m³	mean conc. 298 mg/m³ diffus. tube 313 mg/m³

Table 2: Validation data of the Nitrous-oxide diffusion sampler

ANALYSIS OF ENFLURANE, HALOTHANE AND ISOFLURANE

The diffusive sampler ORSA allows the sample taking of enflurane, halothane and isoflurane vapours as well over periods of several hours by adsorbtion on activated charcoal. The analysis of the enriched anaesthetics ensures after the desorption with toluene by capillary gaschromatography and electron capture detection.

Comparison measurements with active sampling systems (charcoal tubes and pump) have been carried out in operating theatres under field conditions. The results are summerized in table 3.

compound	sampling duration	results of active sampling	results of diffusive sampling
enflurane	5.0 hours 2.5 hours 8.0 hours 8.0 hours	8.00 mg/m^3 0.47 mg/m^3 7.40 mg/m^3 2.40 mg/m^3	8.00 mg/m^3 0.50 mg/m^3 9.50 mg/m^3 2.60 mg/m^3
halothane	8.0 hours	0.60 mg/m^3	0.50 mg/m^3
isoflurane	6.0 hours 2.5 hours	1.80 mg/m^3 0.37 mg/m^3	1.80 mg/m^3 0.38 mg/m^3

Table 3: Comparison between active and diffusive sampling

CONCLUSIONS

The results of the validation tests show, that the diffusive sampler ORSA and the Nitrous-oxide diffusion sampler can be recommended for personal monitoring of volatile anaesthetics in operating theatres.

REFERENCES

[1] J. Buchberger, W. Greuter, S. Kündig. Berufliche Narkosegasexposition des Spitalpersonals in der Schweiz, Arbeitsärztlicher Dienst des Bundesamtes für Industrie, Gewerbe und Arbeit, Bern 1985.

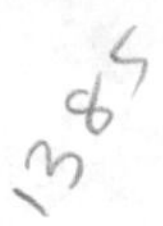

A DIFFUSIVE SAMPLER FOR THE DETERMINATION OF FORMALDEHYDE IN AIR

PFÄFFLI, P., VIRTANEN, H., RIUTTA, O., KAARTINEN, T. and HÄYRI, L.
Institute of Occupational Health
Helsinki, Finland

SUMMARY

A diffusive sampler based on formaldehyde collection onto a 2,4-dinitrophenylhydrazine coated fibre glass filter with a sampling rate of 44.5 ml/min is described. The buffered impregnation showed that the sampling rate is dependent on the pH of the covered filters and on formaldehyde concentration in air in low concentrations at short sampling times.

1. INTRODUCTION

Many of the newer methods for the determination of formaldehyde in air are based on the acid-catalyzed condensation of formaldehyde with 2,4-dinitrophenylhydrazine (2,4-DNPH). In these methods, the collection of formaldehyde is performed either with liquid absorption into 2,4-DNPH solution or with solid adsorbents coated with this reagent. Fibre glass filters coated with the reagent have been used in passive samplers (3).

Most of the published 2,4-DNPH derivations with carbonyls have been performed at a pH range of 0 to 2. The formaldehyde reaction, however, has been found to show the maximal rate (4) and yield (1) at pH 4. Furthermore, the derivatisation of environmental samples should be performed at mild acidic conditions, since many organic com-pounds (e.g. polyols) may liberate formaldehyde in strong acids, thus giving rise to false positive results.

The aim of our study was to examine 2,4-DNPH coated passive samplers further in order to find, whether the method is suitable for routine formaldehyde measurements in the field, and whether the change in pH affects the sampling.

2. EXPERIMENTAL

Coated filters

Chemicals of analytical grade were purchased from E. Merck AG (Darmstadt, Germany), except acetonitrile (AcCN) of HPLC grade, which was from Rathburn Ltd. (Dublin, Ireland). 2,4-DNPH was first recrystallized from methanol for the removal of formaldehyde-2,4-DNPH contamination. **Solution U** (pH 2) : 300 mg of 2,4-DNPH (recrystallised from 4M HCl) was dissolved in 30 ml of AcCN and 0.56 ml of conc. H_3PO_4 and 1.65 ml of glycerin (20% in methanol) were added. **Solution B** (pH 4): 300 mg of 2,4-DNPH and 1.5 ml of diethyleneglycol monomethylether (DEGMME) were added to AcCN (tot.vol. 25 ml). 0.2 ml of saturated KH_2PO_4 solution was added to 5 ml of distilled water and pH regulated with H_3PO_4 to 4.0. These two latter solutions were mixed to get solution B. **Fibre glass filters** (binder-free, Type AE, pore size 0.3 um, SKC, Inc., Eighty Four, PA, USA) were immersed in these solutions for 2 min, and then allowed to dry on a watch-glass at 35 °C for 30 min. After that the filters were covered with aluminium foil and stored in exsiccator over dry silica gel at room temperature until the assembling of the sampler.

Sampler assembling

The sampling cassette designed by us (fig. 1) was made of polypropylene and polyethylene. The filter coated with 2,4-DNPH was installed on the bottom of the cassette dish (f). A polycarbonate membrane (c) (pore size 8 um, Nuclepore[R] 110814, diam. 37 mm, $3.4x10^5$

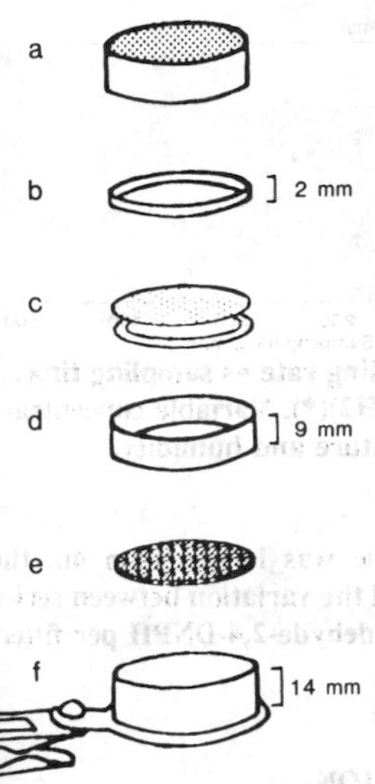

Fig. 1: Diffusive sampler

pores/cm², the pores covered 17% of the filter area, Nuclepore Corp., Pleasanton, CA, USA) glued (with Pelifix, Pelikan Vertriebs GmbH, Hannover, Germany) on a brass ring (37x26x0.3 mm, Tiiviste-valmiste Oy, Helsinki, Finland) was set on an interval ring (d), which in turn lay on the coated filter (e) to provide a diffusion gap (height 9.3 mm, diam. 26 mm) for the sampler. The parts were fixed tightly in the dish with an attachment ring (b) and the sampler was covered with a tight protective cover (a) (to be kept open during sampling). The plastics used for the sampler were declared to be resistant to formaldehyde and not to emit formaldehyde at normal temperatures. The plastic parts were worked up by Muoviyhtymä Ltd. (Espoo, Finland).The cassette set was packed into a laminate bag of polyethylene-aluminium foil-polyester (A. Ahlström Oy, Kauttua, Finland) prepared by hot-seaming. The assembling and packing of the cassette were performed in a closed plastic box (75x50x27 cm, with fixed rubber gloves for handling) into which a dynamic helium flow was conducted. The bags were also sealed in the box by ironing. The unexposed samplers were stored in these closed bags at room temperature. The exposed samplers were stored at -20 °C in the same bags which were folded at the mouth and the folding covered with tape.

Analytical experiments

A standard formaldehyde atmosphere was generated in a dynamic system (5) by heating paraformaldehyde in a warm bath. The gas was conducted with an air stream through a cubic test chamber (1.5 m³) and the concentration was diluted with a side stream of clean air. A metal net for hanging the test equipment (parallelly and perpendicularly) was situated in the middle of the chamber. The formaldehyde concentration varied from 0.1 to 6 cm³/m³. Air samples were simultaneously collected by the diffusion samplers and by pumping (1 l/min) through sodium bisulfite solutions for the determination of formaldehyde by the chromotropic acid method. The sampling time varied from 30 to 630 min.The wind velocity in the chamber varied from 0.05 to 2 m/s, the temperature from 17 to 30 °C and the humidity from 20 to 73 RH%. The analysis was performed by HPLC (3).

3. RESULTS AND DISCUSSION

For pH study, at pH 4, buffered filters containing a small amount of DEGMME to retain some water to dissolve the ionic states of the reaction were used. DEGMME dissolves also formaldehyde well. It was found that the reaction occurred quickly at the beginning, but thereafter (after about 4h) reached a steady sampling rate. The lower the formaldehyde concentration was in air, the faster the sampling rate was at the beginning of the sampling (fig. 2). The same phenomenon was noted with unbuffered samplers (pH 2), but not so strongly (fig. 3). Chemosorption on both types of filters probably caused an affinity to formaldehyde, consequently, affecting the sampling rate, which was much faster than the "theoretical" sampling rate (9.3 ml/min) calculated with the aid of the diffusion coefficient and the cassette dimensions (2).

A linear regression with correlation r=0.995, slope 0.97 and intercept 0.05 was nevertheless found when formaldehyde concentrations were measured in the chamber by the diffusive samplers, and the pumped chromotropic acid method (n=29). On average, the sampling

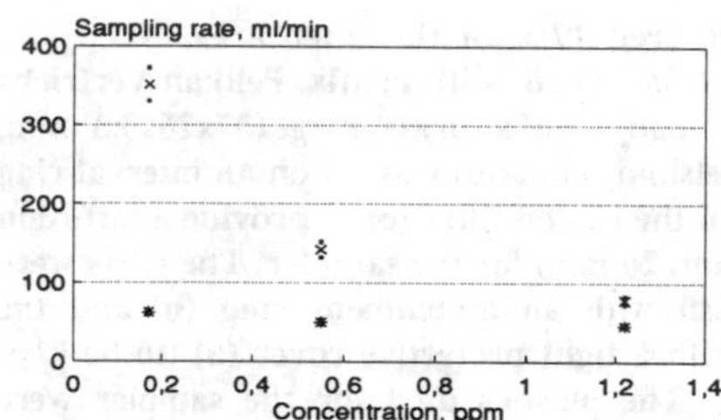

Fig. 2: Sampling rate vs concentration. Sampler U(pH2)(*), sampler B(pH4)(x). Sampl.time 4h, temp. 23 °C, 40 RH%.

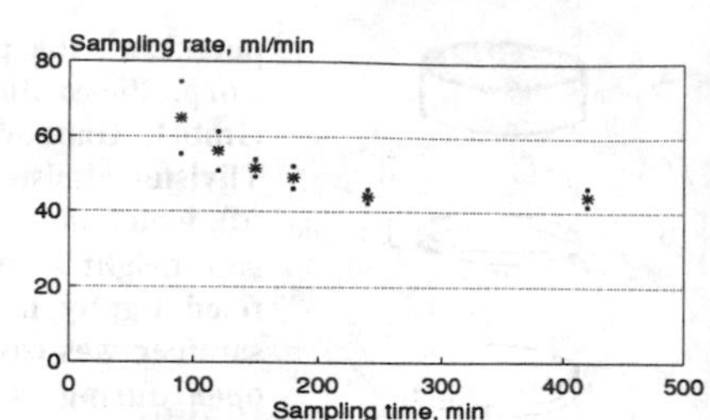

Fig. 3: Sampling rate vs sampling time. Sampler U(pH2)(*). Variable concentration, temperature and humidity.

rate was 44.5 ml/min. For the samples, when the sampling time was longer than 4h, the coefficient of variation was better than ± 4 % (n = 20 samplers) and the variation between series ± 2 (SE)% (N = 24). The reagent blank was at 0.015 ug of formaldehyde-2,4-DNPH per filter. The practical detection limits are shown in table 1.

In air, cm³/m³, (sampling time)	Formaldehyde-2,4-DNPH
0.01 (4h)	5 ng/ml of solvent
0.006 (8h)	50 pg/10ul injection
0.002 (24h)	15 ng/blank filter
	30 ng/sample

Table 1: Detection limits

4. CONCLUSION

Because the sampling rate and yield depend on the pH of the covered sampler filter, the preparation of the samplers should be standardized carefully. Furthermore, because the sampling rate depends on the concentration of formaldehyde in air at short sampling times, short-term samplings are not recommended.

On average, however, rather good results were obtained with the well-prepared monitors, although the scatter was high (≤10%) at short sampling times. The detection limit of the method was well below the current TLVs, but also below the new reference values recommended in many countries. The sampler is mainly suitable for the determination of work-day long average exposures at concentrations found nowadays in work environments. In fact, the detection limit was so low that the sampler could even be used to determine formaldehyde concentrations in the indoor air of offices and homes.

REFERENCES

(1) BICKING, MKL. and COOKE, WM. (1988). J. Chromatogr. 455 310-315.
(2) KRING, EV., THORNLEY, GD., DESSENBERGER, C., LAUTENBERGER, WJ. and ANSUL, GD. (1982). Am. Ind. Hyg. Assoc. J. 43 786-795.
(3) LEVIN, J.-O., LINDAHL, R. and ANDERSSON, K. (1989). JAPCA 39 44-47.
(4) LOWRY, TH. and SCHUELLER RICHARDSON, K. (1981). Mechanism and Theory in Organic Chemistry. 2nd Ed. Harper & Row, Publishers, New York, 635-642.
(5) PERKINS, J. (1981). Ann. Am. Conf. Ind. Hyg. 1 125-169.

1386

DIFFUSIVE SAMPLING ON TENAX: EFFECTIVE BED LENGTH AND CONCENTRATION LIMITS

K. SCHOENE, J. STEINHANSES and A. KÖNIG

Fraunhofer-Institut für Umweltchemie und Ökotoxikologie
D 5948 Schmallenberg

1. EXPERIMENTAL

Test atmospheres (1, 2) containing either 2-xylene ('X', 14.1 ppm), tetrachloroethane ('T', 13.1 ppm) or dimethoxymethylphosphonate ('DMMP' 44.1 and 11.3 ppb) in dry air were passed (linear flow 0.14 cm/s (3)) through a cylindrical channel (stainless steel, i.D. 3.8, length 26 cm), housed in a thermostatically controlled (20°C) cabinet. Along and around the wall of the channel 7 poles served to hold test tubes for diffusive sampling (stainless steel, 0.48 x 8.9 cm, 'ATD-tubes' from Perkin Elmer). The tubes were packed with 100 mg Tenax TA (20-35 mesh, freshly sieved, average diameter 0.67 mm), by use of the filling device from Perkin Elmer, yielding 0.227 mm bed length per 1 mg Tenax. The vapour concentration c_o, was monitored by gas chromatography (GC) daily by active sampling on Tenax followed by thermal desorption/GC (ATD/GC) (4, 5). During the sampling periods of up to 7500 min, the maximum deviation in c_o was 12% of the above mean values.

The diffusive sampling tubes were removed from the channel after different times t of exposure and analysed for their analyte content by ATD/GC. Sampling of X and T was done with the silicone membrane (Perkin Elmer), whereas DMMP was sampled without membrane (because otherwise U_{exp} was $< U_{th}$).

The mass m found on the test tubes was divided by the exposure time t to give the uptake rate U_{exp} (g/min); this was related to the theoretical uptake rate (6). $U_{th} = (A/l) \cdot D \cdot c_o$, with A = cross section area (0.181 cm^2) and l = diffusion path in the tube (1.52 cm), D = diffusion coefficient (X: 4.362; T: 4.782; DMMP: 4.123 cm^2/min).

2. RESULTS

From the time course of U_{exp}/U_{th} (Figure 1) it is shown that, after a certain time t_d, U_{exp} deflects to values lower than U_{th}. In Table 1, estimates of t_d (line 2) are given, together with the amount m_d (line 3) accumulated up to that time on the Tenax: $m_d = A/l \cdot D \cdot c_o \cdot t_d$.

It is assumed that there is some correspondence between m_d and the sorptive capacity of the Tenax. The respective sorption isotherms follow Freundlich's equation

$\log \alpha = p \cdot \log c + q$ with (at 20°C, c in ppm) (7):

	X	T	DMMP
p	0.336	0.428	0.421
q	-2.659	-2.721	-2.426

Setting $c = c_o$, the Freundlich equations yield the amounts α of analyte, which can be maximally adsorbed at the individual exposure concentrations c_o (Table 1, line 4). The other way round, if m_d represented that saturation load α, this would correspond to an amount m_d/α of saturated Tenax (given in Table 1, line 5).

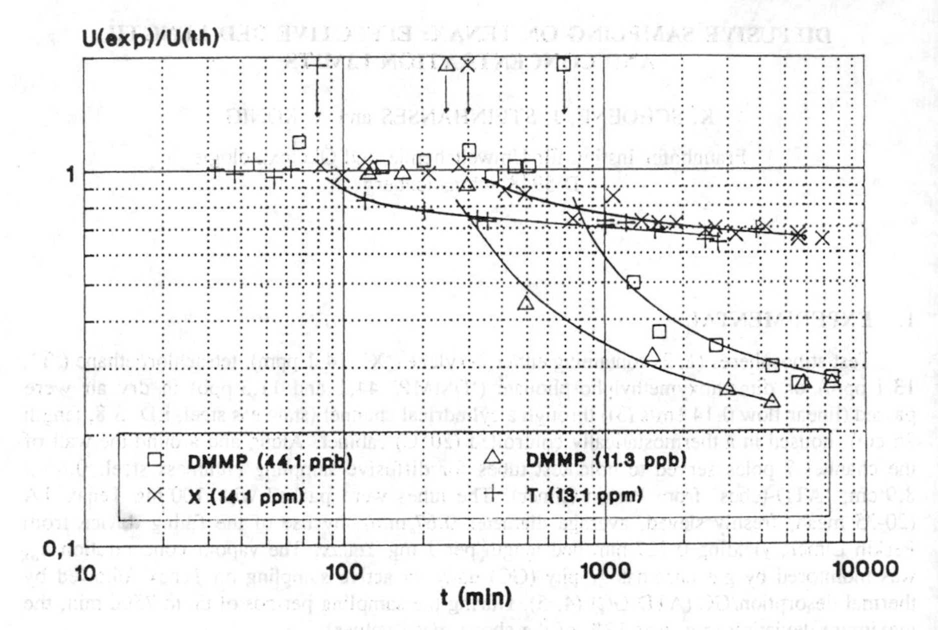

DMMP = Dimethoxymethylphosphonate,
X = 2–Xylene, T = Tetrachloroethane

Fig. 1. Results from diffusive sampling

Table 1. Results and Calculated Data

	X	T	DMMP	DMMP
1. c_o (ppm)	14.1	13.1	0.0441	0.0113
2. t_d (min)	300	80	700	250
3. m_d (ng)	9450	4008	76	7
4. α (mg/g Tenax)	5.3	5.6	1.0	0.57
5. $m_d/1\alpha$ (mg Tenax)	1.8	0.71	0.076	0.012
L (mm)	0.40	0.16	0.02	0.003
6. $m_d/0.1\alpha$ (mg Tenax)	18	7.1	0.76	0.12
L (mm)	4.0	1.6	0.17	0.03
7. $m_d/0.01\alpha$ (mg Tenax)	178	71	7.6	1.2
L (mm)	40	16	1.7	0.27
8. $m_d/0.001\alpha$ (mg Tenax)	1783	711	76	12
L (mm)	400	160	17	2.7

In fact, however, the decline of U_{exp} must have occurred already at loads $< \alpha$, maybe at 0.1α, 0.01α or even lower; the respective amounts of Tenax are given in the Table, lines 6–8, together with the corresponding bed lengths L (1 mg Tenax = 0.227 mm).

On the other hand, the migration of the gas molecules is governed predominantly by the diffusion coefficients of the analytes, which are roughly of the same magnitude. Hence it is reasonable to assume that the depths of intrusion into the sorbent bed will be identical, in other words: the active bed lengths should be the same in all four cases. In the Table, lines 6–8, the only sensible combination fulfilling this postulate, is indicated by underlined figures.

3. CONCLUSIONS

1. The active bed length is shown to be approximately 2 mm, i.e. three times the average diameter of a Tenax ball.

2. The lower c_o, the smaller the fraction of α (and α itself!), up to which theoretical uptake rates can be achieved. DMMP had been chosen as a simulant for highly toxic organophosphonates like sarin. The TLV for 1 hr exposure to sarin is 1 ng/l = 0.2 ppb (8); the detection limit in ATD/GC(FPD) is 2 ng sarin (or DMMP) per tube (5); accumulating 2 ng sarin or DMMP from the TLV atmosphere would require a sampling time of 4200 min, which, for example, is far beyond the t_d found in sampling even 11 ppb. Hence, at such low concentrations there is no choice but active sampling as the only reliable method (1).

4. REFERENCES

(1) SCHOENE, K. and STEINHANSES, J. (1989), Fresen. Z. Anal. Chem. 335, 557.

(2) SCHOENE, K., STEINHANSES, J. and KÖNIG, A. (1990), Fresen. J. Anal. Chem. 336, 114.

(3) BLOME, H. (1988), Staub-Reinh. Luft 48, 177.

(4) STEINHANSES, J., SCHWARZER, N., STEINHANSES, W. and SCHOENE, K. (1989), Fresen. Z. Anal. Chem. 334, 431.

(5) STEINHANSES, J. and SCHOENE, K. (1990), J. Chromatogr. 514, 273.

(6) BERLIN, A., BROWN, R.H. and SAUNDERS, K.J. (1987), 'Diffusive Sampling', The Royal Society of Chemistry, London.

(7) SCHOENE, K., STEINHANSES, J. and KÖNIG, A. (1990), J. Chromatogr. 514, 279.

(8) MCNAMARA, B.P. and LEITNACKER, F. (1978): 'Toxicological Basis for Controlling Emissions of Sarin into the Environment', Edgewood Special Publication EASP 100-98.

1387

DETERMINATION OF ACROLEIN IN AIR WITH 2,4-DINITROPHENYLHYDRAZINE IMPREGNATED ADSORBENT TUBES IN THE PRESENCE OF WATER

S. VAINIOTALO and K. MATVEINEN

Institute of Occupational Health
Dept. of Industrial Hygiene and Toxicology
Topeliuksenkatu 41 a A
SF-00250 Helsinki, Finland

SUMMARY

The determination of acrolein in food processing fumes, which usually contain high amounts of water vapor, is described. The samples were collected in 2,4-dinitrophenylhydrazine (2,4-DNPH) impregnated adsorbent tubes fitted with $CaCl_2$ drying tubes in front. The analysis was carried out by HPLC and UV detection. The sampling tests showed about 20 % decrease in the recovery of acrolein from air of high humidity (RH 85 %) compared with conditions of low humidity (RH <40 %) or with the use of a drying tube. The sampling tube contained 200 mg of impregnated XAD-2 resin (2 %, calculated in terms of 2,4-DNPH) and the drying tube 1 g of $CaCl_2$ (particle size 2-8 mm). Concentrations of acrolein ranging from 0.01 to 0.59 mg/m^3 were measured in the food industry and the catering trade.

1. INTRODUCTION

Acrolein is formed in the incomplete combustion of organic materials, such as fuels, synthetic polymers, food and tobacco. Acrolein is also a typical heat degradation product of fats and oils, in which it is formed from glycerol via dehydration. Acrolein is a strong irritant of the eyes, skin, mucous membranes and respiratory tract.

The concentration of airborne acrolein can be measured with the 2,4-dinitrophenylhydrazine (2,4-DNPH) sorbent tube method (1). However, the water vapor normally present in cooking fumes formed during frying or broiling may interfere with the determination. In the present study, a sampling system fitted with a drying tube was used to remove the interfering water, which otherwise condenses in the sorbent tube.

2. MATERIALS AND METHODS

Tube preparation

The sorbent tubes were prepared in the laboratory. XAD-2 resin (particle size 0.3 - 1.0 mm) was washed first with water and then with acetonitrile and dried at 60 °C. 100 mg of 2,4-DNPH was dissolved in a solution containing 1 ml of concentrated hydrochloric acid and 20 ml of acetonitrile (grade S). This solution was kept in an ultrasonic bath until it was clear, and 5 g of washed and dried XAD-2 resin was added. The mixture was kept at room temperature for two hours after which the solvent was evaporated. 200 mg of the impregnated resin was packed in glass tubes (4 mm i.d.). Both ends of the tubes were closed with glass fibre stoppers.

Sampling and analysis

The air samples were collected in the sorbent tubes at a flow rate of 0.2 l/min. The acrolein was desorbed from each tube with 2 ml of acetonitrile, the solutions were allowed to stand for 5 min and then passed through a cation exchange sorbent (Bond Elut SCX) to remove the reagent. 20 ul of the sample was injected into a liquid chromatograph equipped with a reversed phase column (Spherisorb 5 ODS-2, 250 x 4.6 mm i.d.) and a UV detector. The eluent (acetonitrile:water, 65:35) flow rate was 1.2 ml/min and the detection wavelength 360 nm.

Recovery tests

The effect of humidity was tested using a system which produced air of a known humidity. Air was humidified by bubbling it through a heated water bath and then diluted in a mixing chamber (10 l) containing a device measuring relative humidity (RH). The recovery tests were performed - with the air flow on - by injecting a standard solution of acrolein (dissolved in acetonitrile) into a glass wool plug inserted in the sorbent tube. The humidified air was passed through the tube at a flow rate of 0.2 l/min. The standard solution was injected to the tube in four portions (5 ul of each) to avoid high peak concentrations of acrolein.

The effect of condensed water was investigated using two glass wool plugs inserted in the sorbent tube: the first one for injection of acrolein and the second one containing a known amount of water. The effect of the drying agent was tested by connecting the drying tube immediately in front of the sorbent tube and proceeding as above.

3. RESULTS AND DISCUSSION

The recovery of acrolein measured at low atmospheric humidity was 95 - 97 % (for acrolein concentrations of 8.1 - 1.6 ug/sample). Breakthrough was investigated by passing 6 l of air through two sorbent tubes connected in series at an acrolein concentration of 8,1 ug/sample. No breakthrough occurred. The overall precision of the method (sampling and analysis) was tested at a concentration of 0.8 ug/sample, yielding a coefficient of variation of 0.06 (n=14). The detection limit was 0.3 ug/sample corresponding to 0.01 mg/m^3 for an air sample of 24 l.

The results of the recovery tests concerning the effect of moisture are shown in Fig. 1. Recovery was reduced by about 20 % at high relative humidity (RH 85 %) compared with conditions of low humidity (RH < 40 %) at which the best recovery was achieved. Condensed water in the sampling system also resulted in a 20 % decrease in recovery. In the presence of moisture - originating either from condensed water vapor or high humidity of air - the highest recovery was achieved by using a drying tube in front of the sorbent tube. The drying tube (7 mm i.d.) contained 1 g of $CaCl_2$ (particle size 2 - 8 mm). No adsorption of acrolein on the calcium chloride was observed.

The reason for the decreased recovery and the role of water in the phenomenon is unknown. The acrolein derivative may be converted into another isomeric form (syn/anti) or decomposed. According to previous investigations, the acrolein derivative decomposes in the presence of excess 2,4-DNPH and acid (1 N perchloric acid) at room temperature so that only 30 % is left after one week (2). According to our studies, a decrease of peak height of the acrolein derivative can be seen soon after its desorption in acetonitrile. The decrease is more rapid in samples exposed to water compared with those which are not. Treatment with the cation exchange resin removes most of the reagent, and the acrolein peak remains unchanged. In samples exposed to moisture, water may dissolve some of the reagent and acid in the sampling tube, and thus the reaction resulting in decreased peak height of the acrolein derivative may start already there.

Concentrations of acrolein ranging from 0.01 to 0.59 mg/m^3 were found in the food industry and the catering trade (five restaurant kitchens, one food factory and one bakery). The current Finnish hygienic standard for acrolein in the work atmosphere is 0.25 mg/m^3 (15 min).

REFERENCES

(1) OTSON, R., and FELLIN, P. (1988). A review of techniques for measurement of airborne aldehydes. Sci. Total Environ. Vol 77: 95-131.

(2) LIPARI, F., and SWARIN, S. (1982). Determination of formaldehyde and other aldehydes in automobile exhaust with an improved 2,4-dinitrophenylhydrazine method. J. Chromatogr. Vol 247: 297-306.

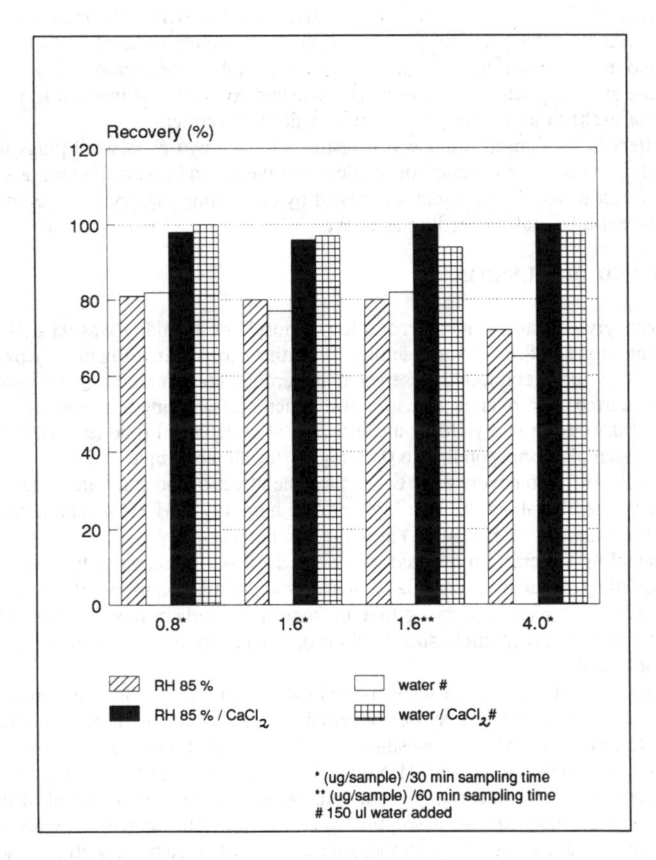

Fig. 1: Effect of moisture on the recovery of acrolein.

MINIMISING BACKGROUND INTERFERENCE BY USING A NOVEL DESIGN OF TWO-STAGE THERMAL DESORBER FOR THE ANALYSIS OF TRACE LEVELS OF VOLATILE ORGANIC POLLUTANTS.

ELIZABETH WOOLFENDEN and GRAHAM BROADWAY

Perkin-Elmer Ltd, Maxwell Road, Beaconsfield, Buckinghamshire, England

SUMMARY

Two stage thermal desorption systems incorporating a packed secondary cold trap offer significant advantages over capillary cryofocusing systems for most environmental applications. However background contamination from the trap packing must be minimised in order to attain the detection limits demanded by much air monitoring work. This can be achieved by extending the time for which the trap is held at elevated temperatures and by diverting the carrier gas away from the trap during idling condition.

INTRODUCTION

Thermal desorption (TD), a technique in which heat and a flow of inert gas are used to extract volatiles from solid or liquid matrices, is now widely used in environmental monitoring. A major advantage of thermal desorption in this field is that by doing away with the inevitable dilution step of conventional solvent extraction procedures, up to 100% of the volatile pollutants retained on porous polymer or charcoal sorbents may be transferred to the gas chromatograph (GC). This leads to improvements in detection limits in the order of 10^3.

However, single stage desorption systems, as illustrated in Figure 1, are of limited application because of the significant band broadening which occurs as components are transferred to the analytical column.

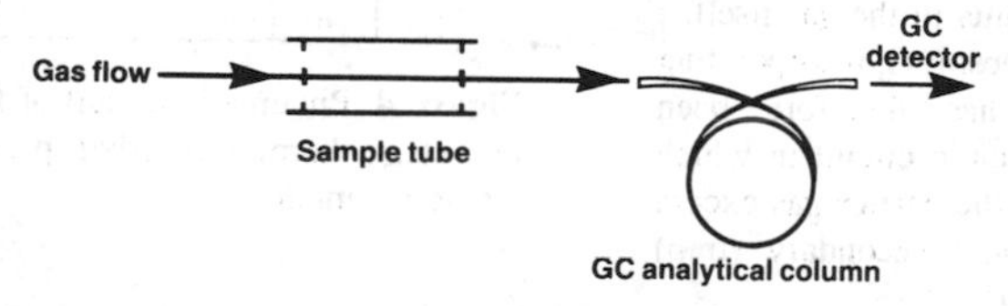

Figure 1. Single stage desorption - transferring the volatiles directly into the GC.

Such peak broadening compromises packed column chromatography and precludes the use of narrow-bore capillary columns.

Two-stage desorption, using some form of component refocusing device, is therefore more common and systems use one of two available options:

1. Capillary cryofocusing

2. Packed cold-trapping

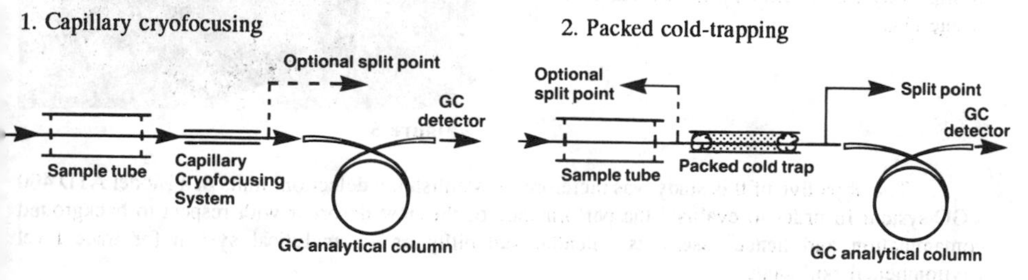

Figures 2.

Figure 3.

For environmental samples, which typically contain significant masses (milligrammes or tens of milligrammes) of water, packed cold traps offer the following advantages over capillary cryofocusing systems:

* Such a system is **compatible with PACKED and CAPILLARY** chromatography.
* Peak widths obtained, when the system is connected to narrow-bore capillary columns, are **comparable to those produced by conventional capillary injectors**.
* **Ice formation is less likely to block the wider bore tubing** of a packed cold trap when samples containing water are desorbed.
* **A liquid cryogen is NOT required** for normal operation as selection of the correct trap adsorbent means that electrical cooling is sufficient for even very volatile species such as nitrous oxide, ethane and vinyl-chloride.
* A packed cold trap minimizes **higher boiling species being lost** through aerosol formation.
* Packed cold traps **facilitate multiple splitting options** as shown in Figure 3.

Two-stage thermal desorbers incorporating a packed cold trap are therefore considered the preferred alternative.

However, systems in which the secondary refocusing unit forms a permanent part of the carrier gas flowpath can suffer from considerable background contamination which interferes with trace level analysis. This is particularly true of systems in which the refocusing device is kept cold at all times during the analysis, except during the secondary desorption stage. The major sources of contamination include the carrier gas lines, filters, etc., or contaminants in the gas itself.

A novel thermal desorber, incorporating a packed cold trap, has therefore been developed using a pneumatic circuit in which the trap is isolated from the carrier gas except during primary (tube) and secondary (trap) desorption (see Figure 4).

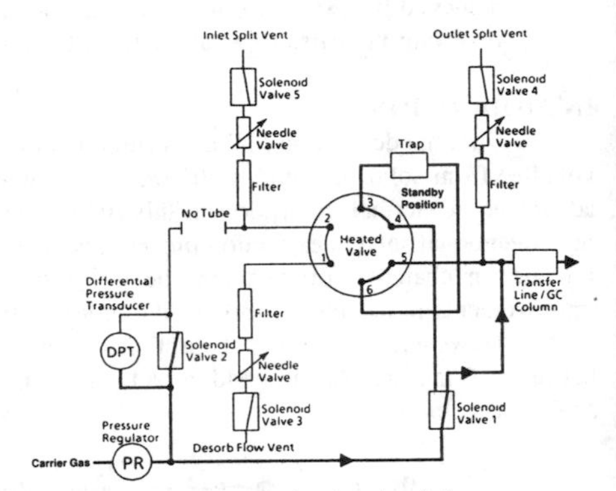

Figure 4. Pneumatic circuit of the new two-stage thermal desorber, presented in 'standby' mode.

The ability to keep the secondary trap at elevated temperatures during the chromatographic analysis stage has also been included in the system. The new instrument, the Model ATD 400 (see Figure 5) was designed by Perkin-Elmer Ltd at Beaconsfield in the UK.

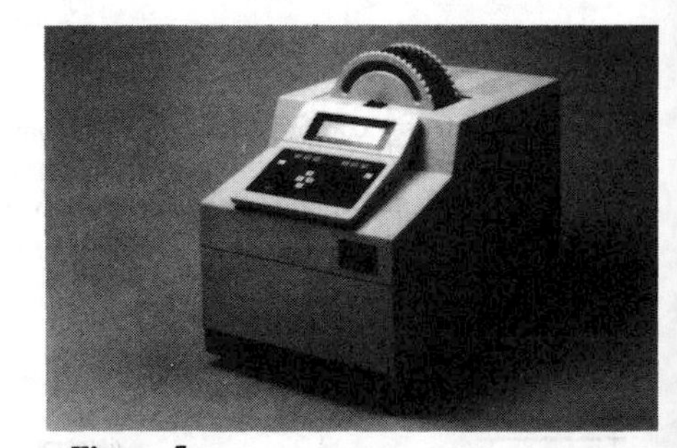

Figure 5.

The objective of this study was therefore to establish the detection limits of a Model ATD 400 - GC system in order to evaluate the performance of the new desorber with respect to background contamination and hence assess its potential suitability as an analytical system for trace level environmental pollutants.

EXPERIMENTAL

Sample tubes packed with 130 mg Tenax TA were spiked with solutions containing 1 ng of benzene and toluene and 1 pg of tetrachloroethylene (see Figure 6).

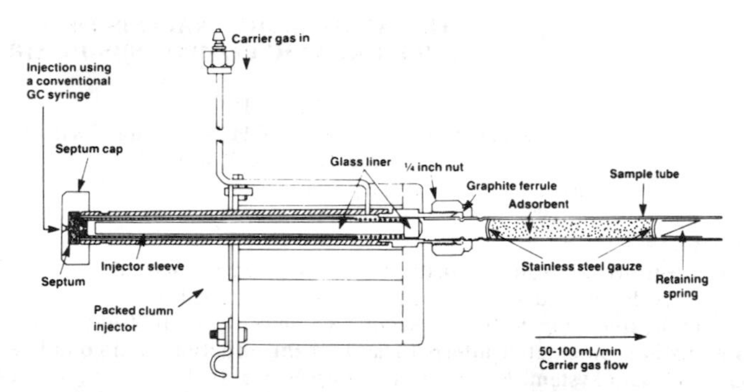

Figure 6. Introduction of Standards onto a Sample Tube in the Vapor Phase via a Packed Column GC Injector

RESULTS AND DISCUSSION

The chromatograms obtained are presented in Figures 7 and 8, and demonstrate the narrow peak widths (~2 sec. wide) obtained from the Model ATD 400.

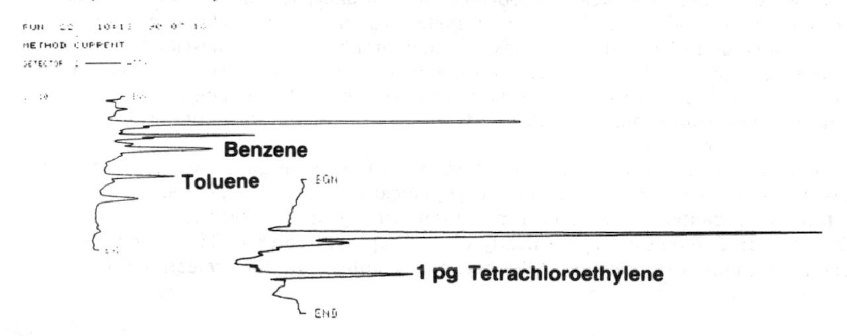

Figure 7. Analysis of a tube spiked with 1 ng each of benzene and toluene.

Figure 8. Analysis of 1 pg of tetrachloroethylene demonstrating the excellent detection limits of the Model ATD 400/GC system.

The levels detected represent extremely low concentrations of these components in ambient air and water:

1. Using the flame ionization detector for benzene:
   - minimum detectable mass: 1 ng
   - equivalent atmospheric concentration in 10 L of air: 30 ppt
   - equivalent concentration in 25 mL of water: 40 ppt
2. Using the electron capture detector for tetrachloroethylene:
   - minimum detectable mass: 1 pg
   - equivalent atmospheric concentration in 10 L of air: <1 ppt
   - equivalent concentration in 25 mL of water: <1 ppt

CONCLUSION

By isolating the packed cold trap from the carrier gas flowpath and by keeping it heated throughout the GC analysis stage, the novel two-stage desorber has been shown to be capable of handling trace levels (ngs/pgs) of components on the sample tubes.

The Model ATD 400 should therefore be compatible with environmental monitoring of volatiles at ppt concentrations.

THE AUTOMATIC ANALYSIS OF VOLATILE ORGANIC COMPOUNDS IN AIR

H. KERN
Varian International AG, CH-6303 Zug, Switzerland
and N. A. KIRSHEN
Varian Chromatography Systems, Walnut Creek CA 94598 USA

SUMMARY

An integrated air/soil gas analysis system based on Varian GC and GC/MS Saturn has been developed and is described here. This system has a built-in cryogenic (or adsorptive) trap to handle sample volumes up to 1000 mL. It also contains sample automation, automated internal standard introduction, and complete control from the GC and data system. Different air samples are analyzed using cryogenic trap and Tenax/charcoal trap. The applications are discussed in respect of accuracy and reliability of measurement at low levels.

1. INTRODUCTION

The concentration of Volatile Organic Compounds (VOC's) in ambient air is increasing with the growing number of mobile and stationary emission sources. Since many of these compounds behave as precursors to ozone formation and/or have adverse health effects, the monitoring of their presence is important. Several analytical methodologies exist for the analysis of VOCs in ambient air. A system for the analysis of halogenated organic compounds in landfill gas by gas chromatography has been desribed (1). Solid phase adsorption followed by either solvent or thermal desorption has been used for many years in industrial hygiene applications and in stack gas monitoring. This technique suffers from problems such as adsorbent contamination, imprecision, and low recoveries for low boiling VOCs.

The systems described use whole air sampling. With this technique, a whole air sample is drawn through an adsorptive trap or a cryogenically cooled trap to freeze out and concentrate VOC contaminants. The trap is then quickly heated and the VOCs are transferred to an ambient or cryogenically cooled capillary column. The column is temperature programmed and the VOCs chromatographed to either selective GC detectors or a mass spectrometer.

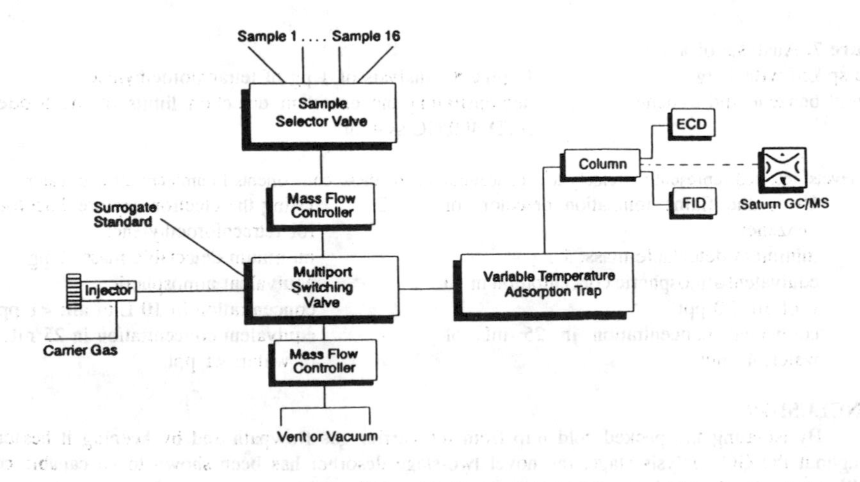

Fig. 1: Schematic of the system for air/soil gas analysis (TO-14)

The integrated air/soil gas analysis systems based on Varian gas chromatographs and Varian Saturn mass spectrometer have been developed and are described here. The systems have built-in adsorptive or cryogenic traps to handle fixed or variable sample volumes up to 1000 mL. They contain sample automation, automated internal standard introduction and complete control from the GC and data system.
The results can be transferred to network or other large computer systems.

2. EXPERIMENTAL

The automated Air/VOCs analysis system is shown schematically in figure 1 and contains the following:

1. GC 3400 (Varian)
2. Variable Temperature Adsorption Trap (VTAT) (Varian) 1/8" ss tube, 2" glass beads bed (60/80 mesh) or Tenax/Chacoal (21 cm bed)
3. Sampling Valve (Valco-W series) with or without surrogate standard introduction (internal standard)
4. Analytical column:
   DB-624, 30 m x 0.32 mm (J & W Scientific) or
   DB-5, 30 m x 0.25 mm (J & W Scientific) or
   Rt_x-502.2, 105 m x 0.52 mm (Resteck)
5. Detectors: ECD and FID or Saturn GC/MS (Varian)
6. Electronic Mass Flow Controller 0-100 mL/min (Sierra)
7. Vacuum pump (Nor-Cal Controls)
8. Optional Stream Selector Valve (Valco)

The built-in trapping or preconcentrating device, the Variable Temperature Adsorption Trap (VTAT) is capable of trapping and preconcentrating VOCs from air on glass beads at -150°C or on an adsorbent, such as Tenax/charcoal or Carbosieve/Carbotrap at ambient temperatures. Subsequently, the VOCs are adsorbed at the required temperature. All temperatures are controlled from the GC or MS data system.
Initially, the internal standard- and sample lines are flushed with the standard and air sample, respectively. The sample is introduced from a stainless steel canister, Tedlar bag or by a small vacuum pump. Upon injection, the internal standard and the sample are deposited onto the trap. The duration of this trapping time can be varied (programmable through the GC method) and depends on the volume of the sample. Sample volumes from approximately 20 to 100 mL are satisfactory. The sample flow during trapping is held constant by the mass flow controller and is usually 30 to 40 mL/min. After the selected volume of sample is deposited on the trap, the latter is heated to the set temperature (120°C for glass beads, 220°C for Tenax/Charcoal) simultaneously the trapped VOCs are backflushed by the carrier gas to the analytical column for separation and subsequent detection.

3. RESULTS AND DISCUSSION

A chromatogram of a standard mixture of VOCs with concentrations ranging from 0.2 - 5 ppb is shown in figure 2. Volumes of 50 to 1000 mL of this standard were introduced to the system to check recovery linearity. The calibration factors were constant (% RSD) in the studied volume range. The precision of these factors demonstrates the linearity of recovery and solute response over the volumes sampled. Therefore wide ranges of VOC concentrations may be detected by simply adjusting the volume passing through the VTAT. Repetitive analysis of 400 mL standard resulted in 0.9 - 6 % RSD for the peak areas.
60 mL of a 21 component VOC standard with concentrations varying from 0.05 to 5.1 ppb were concentrated on a glass beads cryotrap at - 150°C. Excellent sensitivity is exhibit by the target analytes.
Similar results were obtained for all the components.
Library search results for the VOCs on the NIST library gave spectral fits of approximately 900 or greater.

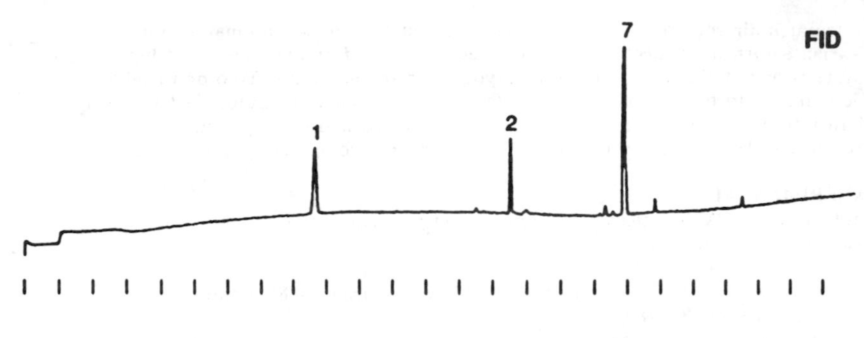

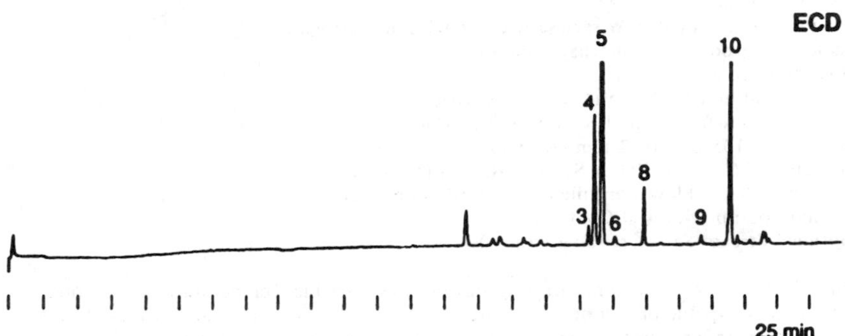

Fig. 2: 400 mL VOCs standard conc. in ppb - 1. Vinylchloride 5.0, 2. Methylene chloride 6.0, 3. Chloroform 0.2, 4. 1,1,1-Trichloroethane 0.4, 5. Carbon tetrachloride 0.5, 6. 1,2-Dichloroethane 2.0, 7. Benzene 5.0, 8. Trichloroethylene 0.5, 9. 1,2-Dibromoethane 0.2, 10. Perchloroethylene 0.5.

Water can be a problem when cryoconcentrating VOCs from large air sample taken under humid conditions. It can plug the trap or the capillary column. The full scan mass spectral sensitivity of the Saturn GC/MS allows one to use much smaller sample volumes so that this problem is eliminated. In the discussed applications sample volumes from 0.25 - 500 mL are taken from various indoor and outdoor air sources.

4.CONCLUSION

The systems described are integrated, compact units with a built-in cryogenic concentrator. Wide ranges of VOC concentrations may be detected, by simply adjusting the volume sampled. Sub-ppb levels of VOCs in indoor and outdoor air or in soil gas can be identified with full scan mass spectra and then quantitated with preconcentration of the sample. Cryogenic trapping at $-150^{0}C$ with liquid nitrogen cooled glass beads or adsorption trapping with adsorbents, such as Tenax or Charcoal based adsorbents, is possible. Finally, the flexibility of these systems is paramount with applications ranging from soil gases in hazardous waste sites to indoor atmospheres to ambient air and source emissions.

REFERENCES

(1) KERN, H. and KIRSHEN N. (1989). Bestimmung flüchtiger halogenierter organischer Verbindungen in Deponiegas mittes Gaschromatographie. VDI Berichte No. 745, 335-344.

1390

DETERMINATION OF VOLATILE AMINES IN AIR WITH DIFFUSIVE SAMPLING ON A REAGENT-COATED FILTER

JAN-OLOF LEVIN, ROGER LINDAHL AND KURT ANDERSSON

National Institute of Occupational Health
P.O.B. 6104, S-900 06 Umeå, Sweden

SUMMARY

A diffusive sampler has been validated for determination of primary and secondary amines in air. The sampler consists of a 20 x 45 mm filter impregnated with naphthylisothiocyanate and mounted in a polypropylene housing. During sampling, a substituted thiourea is formed, which is subsequently desorbed and determined by high-performance liquid chromatography. The sampler has been validated for methylamine, allylamine, isopropylamine, n-butylamine, dimethylamine and diethylamine. The uptake rates were independent of amine concentration, sampling time and relative humidity. The sensitivity of the method is 10 ppb for an 8-h sample.

1. INTRODUCTION

We have previously reported the use of a 1-naphthylisothiocyanate (NIT)-coated XAD-2 sorbent for the determination of ethylenediamine and gaseous polyamines in air (1, 2). A stable thiourea derivative is formed from primary and secondary amines. The reaction is rapid and quantitative on the sorbent at room temperature. The thiourea derivatives have excellent properties for reversed-phase HPLC determination with ultraviolet detection, giving the method high sensitivity.

In recent years, diffusive (passive) sampling has been recognized as an efficient alternative to pumped sampling in occupational hygiene (3). We have previously reported the development of a diffusive sampler specially designed to contain a reagent-coated filter for the sampling of reactive organics (4). The sampler has been validated for formaldehyde, using a 2,4-dinitrophenylhydrazine-coated filter (4). In the present work we report a validation of this sampler for several primary and secondary amines, using a NIT-coated filter. The validation of the sampler for diethylamine has been reported previously (5).

2. EXPERIMENTAL

Diffusive sampler

The sampler housing, measuring 60 x 30 x 5 mm, is made of polypropylene. The impregnated filter, 20 x 45 mm, is placed beneath a 2.9-mm-thick screen of the same size. Within an area of 20 x 20 mm, the screen has 112 holes with a diameter of 1.0 mm. The filter part beneath the holes is used for sampling (sampling filter), and the other half is used to quantitate the filter blank (control filter). The tape is marked into the two sections by a small ridge on the back of the screen plate. A sliding cover is used to seal the holes when the sampler is not in use. The sampler is available from GMD Systems, Inc., Hendersonville, PA, U.S.A.

Laboratory validation of diffusive sampler

For the validation of the diffusive sampler, standard atmospheres of amines were generated in a dynamic system by dilution of certified amine gas. The generation system and exposure chamber were essentially the same as the ones used for generation of formaldehyde (6). The samplers were exposed in the chamber, six at a time, to amine levels from 1 to 10 ppm, with sampling periods between 15 min and 8 hours. Relative humidity was varied between 10% and 85%. Samplers were oriented parallel to the air stream. Six pumped samples using NIT-coated XAD-2 (1) were taken simultaneously from the exposure chamber with a flow of 20 or 50 ml · min^{-1}.

Sample analysis

Samples were analyzed by transferring coated filter or XAD-2 to a 4-mL glass vial and shaking for 30 min with 3.0 mL of acetonitrile. Samples were analysed with HPLC and UV detection as described elsewhere (5).

3. RESULTS AND DISCUSSION

Validation of pumped method for amines

The pumped method with NIT-coated XAD-2 tubes was used to confirm the concentration of dimethylamine, diethylamine and n-butylamine in the validation experiments with the diffusive sampler. In order to validate the pumped method, the recovery of amines from NIT-coated XAD-2 as substituted thioureas was determined. Amounts corresponding to 0.1, 1 and 2 times the TLV of 5 or 10 ppm were added at 20% and 85% relative humidity (1). Some results at the TLV-level are shown in Table 1. As the table shows, the recovery ranges between 90 and 100%.

Table 1. Recovery of amines as thioureans from NIT-coated XAD-2 at different levels and relative humidities. Six parallel determinations. RSD = Relative standard deviation.

Amount added (µg)	Relative humidity (%)	Recovery (%)	RSD (%)
90[a] dimethyl	20	90	2
90	85	93	2
150[a] diethyl	20	94	1
150	85	100	2
75[b] n-butyl	20	94	3
75	85	95	3

[a]Corresponds to 10 ppm in a 5 L air sample.
[b]Corresponds to 5 ppm in a 5 L air sample.

Validation of diffusive sampler for amines

The diffusive sampler was validated according to the protocol published by CEN (7). The effect of concentration, sampling time and relative humidity was investigated. The effects of wind velocity and sampler orientation were studied in connection with validation of the sampler for formaldehyde. In that study the wind velocity at the sampler face was varied between 0.05 and 1.0 m·s^{-1}. The uptake rate was constant within the wind-velocity range studied (4). A wind velocity of approximately 0.1 m·s^{-1} is maintained under most personal sampling conditions (3). The sampler performed well at wind velocities as low as below 0.02 m·s^{-1} (4). Since wind velocity and sampler orientation effects are parameters associated with the sampler and not the analyte, these effects were not further studied.

In the validation experiments with amines, concentrations were varied between 1 and 10 ppm, relative humidity between 10% and 85%, and sampling time between 15 min and 8 hours. Good agreement was generally obtained between the calculated concentrations from dilution of the certified amine gas and the concentrations given by the reference method. The uptake rates were determined from the calculated concentrations of the certified gas. Table 2 shows the uptake rate of the diffusive sampler for various amines. The uptake rates were generally independent of concentration, sampling time and relative humidity.

Table 2. Uptake rates of diffusive sampler for various amines.

Amine	Uptake rate ($ml \cdot min^{-1}$)	RSD (%)	n
Methylamine	17.4	14	54
Isopropylamine	10.2	9	60
n-Butylamine	11.1	11	45
Allylamine	14.9	8	53
Dimethylamine	15.8	6	54
Diethylamine	12.0	6	54

4. CONCLUSIONS

The present diffusive sampler has been specially designed for use with reagent-coated filter tape. The uptake rate for methylamine, allylamine, isopropylamine, n-butylamine, dimethylamine and diethylamine using naphthylisothiocyanate-impregnated filter tape has been determined in validation experiments. The sampling rates are generally independent of amine concentration, sampling time and relative humidity. Since the chromatographic determination of the thiourea derivatives is highly sensitive and specific, the sampler can be used for short-time (15 min) sampling in the ppm range. In an 8-h sample the sensitivity is 10 ppb. It has been previously shown that the sampler performs well at extremely low wind velocities. Thus, the sampler can be used for static as well as personal monitoring of low levels of primary and secondary amines in air.

REFERENCES

(1) ANDERSSON, K., HALLGREN, C., LEVIN, J-O., NILSSON, C-A. (1985). Determination of ethylenediamine in air using reagent-coated adsorbent tubes and high performance liquid chromatography on the 1-naphthylisothiourea derivatives. **Am. Ind. Hyg. Assoc. J., 46, 225-229.**

(2) LEVIN, J-O., ANDERSSON, K., FÄNGMARK, I., HALLGREN, C. (1989). Determination of gaseous and particulate polyamines in air using sorbent or filter coated with naphthylisothiocyanate. **Appl. Ind. Hyg. 4,** 98-100.

(3) BERLIN, A., BROWN, R.H., SAUNDERS, K.J., Eds. (1987). Diffusive Sampling - an Alternative Approach to Workplace Air Monitoring, Royal Society of Chemistry, London.

(4) LEVIN, J.-O., LINDAHL, R., ANDERSSON, K. (1988). High peformance liquid chromatographic determination of formaldehyde in air in the ppb to ppm range using diffusive sampling and hydrazone formation. **Environ. Technol. Lett. 9,** 1423-1430.

(5) LEVIN J.-O., LINDAHL, R., ANDERSSON, K., HALLGREN, C. (1989). High-performance liquid-chromatographic determination of diethylamine in air using diffusive sampling and thiourea formation. **Chemosphere 18,** 2121-2129.

(6) LEVIN, J.-O., LINDAHL, R., ANDERSSON, K. (1986). A passive sampler for formaldehyde in air using 2,4-dinitrophenylhydrzine-coated glass fiber filters. **Environ. Sci. Technol. 20,** 1273-1276.

(7) EUROPEAN COMMITTEE FOR STANDARDIZATION. (1991). Workplace atmospheres - Requirements and test methods for diffusive samplers for the determination of gases and vapours. **CEN prEN.**

1391

A COMPUTER PROGRAM FOR PREDICTION OF UPTAKE RATES FOR DIFFUSIVE SAMPLERS

J. KRISTENSSON and E. NORDSTRAND

Analytical Chemistry Department
University of Stockholm, S-106 91 Stockholm, Sweden

A computer program for prediction of uptake rates, based on the work of Nico van den Hoed and Otto van Asselen, has been further developed and rewritten in Basic. The program can run on any PC computer with a mathematic processor. The program can be used for any type of diffusive sampler.

'In the first step, the program calculates the diffusion coefficient based on the molecular formula of the compound and the ideal uptake rate for the compound.

In the second step, the effective uptake rate for the compound at given sampling conditions is calculated. The calculations are based on adsorption according to the Freundlich adsorption isotherm. A correction factor, tau, is used to correct for the mass transfer inside the adsorption bed. Tau is estimated empirically.

With the program both ideal and effective uptake rates can be calculated if the A and B constants for the Freundlich adsorption isotherm are known. If the difference between the calculated ideal uptake rate and the calculated effective uptake rate is acceptable at the given sampling conditions, the choice of adsorbent for the compound of interest was right. If the difference is unacceptable, another adsorbent must be chosen.

1392

REAL TIME MONITORING INSTRUMENTS FOR EXPOSURE STUDIES: THE PIMEX-METHOD

ROSÉN GUNNAR
Division of Industrial Hygiene
National Institute of Occupational Health
S-171 84 Solna
Sweden

SUMMARY

Control of exposure to air pollutants in the work place is often necessary. In order to devise the most effective technical solution the source and the reason(s) for the high exposure must be determined. The PIMEX-method is a tool that can facilitate this work. Simultaneous measurement of a pollutant in the breathing zone with a real time monitoring instrument is combined with video filming of the work place. The signal from the instrument is superimposed over the picture from the video camera. The results are recorded on a video tape recorder for later analysis. This method has been used to study and reduce exposure to many different air pollutants. PIMEX may be used for control of exposure in a single work place as well as for production of a video film for training purposes or as a research tool.

INTRODUCTION

Steps taken at work for the purpose of reducing exposure to injurious environmental factors, are most effective if the results can be determined immediately. This can be accomplished with the aid of real time monitoring instruments. Problems in analysis and evaluation of the results often arise because the industrial hygienist does not have accurate notes or knowledge of all factors concerning the data collected. A very important factor in this context is the exposed person and his/her behaviour. It is therefore desirable to be able to collate, as completely as possible, information on all important circumstances at a given work place and on the exposure.

In June 1985, experiments were carried out in Sweden with a system consisting of a real time monitoring instrument combined with video filming(1). The first experiments were conducted in a furniture factory where exposure to organic solvents was studied among spray painters. The first results clearly demonstrated that this new use of air sampling instruments might be an excellent tool for industrial hygiene survey(s).

METHOD

Picture Mix Exposure (PIMEX), developed over the period of six years, is the simultaneous measurement of pollutants in the breathing zone with a real time monitoring instrument combined with video recording(2). The signal from the instrument is superimposed in the picture from the video camera showing the work place. A video mixer AVM 2000 (IBC AB, Box 438, S-123 04 Farsta, Sweden) is used for that purpose. The signals are recorded on a video tape recorder for later analysis. Figure 1. shows a sketch of the system. All types of real time monitoring instruments producing a continuous signal with a short lag time (less than 5 s) is usable for the method. The most suitable instrument is small, battery operated and with a fast response time; for example a photoionization instrument for organic solvents and other gaseous substances and a light scattering instrument for dust and smoke.

The technical development of the method has resulted in compact equipment, designed for use by industrial hygienists, safety engineers, and other persons involved in controlling health hazards. The unit includes the necessary video equipment, telemetry, and the video mixer. Combined with a suitable sampling instrument for a pollutant it is ready to use.

Fig. 1: The PIMEX-method for exposure studies. The subject carries a backpack containing a real time monitoring instrument and a telemetry transmitter. The signal is transmitted to the receiver. The work is filmed with a video camera. The output signal from the camera and the signal from the sampling instrument are combined in an AVM 2000 video mixer. The composite signal is recorded by a tape recorder and displayed as a bar on the video monitor.

RESULTS

PIMEX has been used, first of all, as a tool for evaluation of exposure levels of pollutants in work places aimed to reduce exposure (3). It has also been used to determine the reason for high exposure and the results have been utilized in control procedures. The recorded material has also been used for training of workers under operation conditions(4).

Another important use of PIMEX has been production of general video films for specific industrial branches. A film produced for the GRP industry (glass fibre reinforced polyester plastic) demonstrate the PIMEX-method for evaluating different types of ventilation and production systems and different work practices (5). The material has been edited to a video film showing the most important factors resulting in a good work environment. Also films about solvents used in furniture industries, in screen printing shops, wood dust in furniture industries (6), and gasoline vapour when servicing cars have been produced.

The method can also serve as a research tool, especially, for studies of the behaviour and development of technical control procedures. For that purpose, PIMEX has been used to study different ventilation and exhaust systems (7, 8, 9). The method has also been used for modelling of exposure to airborne contaminants (10). The recorded material contains data needed for research.

In addition to the development of a strategy for the use of the method, a lot of work has been conducted to evaluate PIMEX in different kinds of work places with different technical problems. For example, exposures to ammonia in farming work, welding fumes, and chloroform in laboratories have been studied. The results show that the PIMEX-method, used with a photoionization and a light scattering instrument, makes it possible to facilitate the control of exposure to many different air pollutants in many different work places.

CONCLUSION

PIMEX is now a technically and pedagogically developed method used in many countries. The method has also been evaluated with aspect to its ability to serve as a tool for control measurements in work place. The PIMEX method will play an important role in industrial hygiene in the future.

REFERENCES

(1) ROSÉN, G. and LUNDSTRÖM, S. (1987). Concurrent Video Filming and Measuring for Visualization of Exposure. Amer Industr Hyg Ass J 48 (8), 688-692.

(2) ROSÉN, G. and ANDERSSON, I-M. (1989). Video filming and pollution measurement as a teaching aid in reducing exposure to airborne pollutants. Ann Occ Hyg 1, 137-144.

(3) ROSÉN, G, ANDERSSON, I-M. and JURINGE, L. (1990). Reduction of exposure to solvents and formaldehyde in surface-coating operations in the woodworking industry. Ann occup Hyg 3, 293-303.

(4) ANDERSSON, I-M. and ROSÉN, G. (1990). How to educate welders to use local exhaust ventilation. Amer Indust Hyg Conf, Orlando, 13-18 May.

(5) ANDERSSON, I-M., JURINGE, L. and ROSÉN, G. (1990). Reinforced polyester plastic manufacturing in a sound work environment. Exposure to styrene can be kept under control. PIMEX-video. Div. of Ind. Hyg., Nat. Inst. of Occup. Hyg., Sweden.

(6) ROSÉN, G, ANDERSSON, I-M. and JURINGE, L.(1989). Exposure measurements as a pedagogic tool. Proceedings from 2nd International Symposium on Exposure Monitoring in Industry, Antwerpen, 16-17 Nov.

(7) ANDERSSON, I-M., ROSÉN, G. and KRISTENSSON, J. (1991). Evaluation of a ceiling mounted low impulse inlet air unit for local control of air pollutants. Proceedings of the 3rd international symposium on ventilation for contaminant control, Cincinnati. 16-20 Sept.

(8) ROSÉN, G., ANDERSSON, I-M., JANSSON, G. and WEMMERT, B. (1991). Emission and exposure control in soldering operations and laboratories. Proceedings of the 3rd international symposium on ventilation for contaminant control, Cincinnati. 16-20 Sept .

(9) ANDERSSON, I-M. and ROSÉN, G. (1989). Displacement ventilation in the GRP-Industri. Evaluation of styrene exposure. Proceedings from 2nd International Symposium on Exposure Monitoring in Industry, Antwerpen, 16-17 Nov.

(10) ROSÉN, G. and ANDERSSON, I-M. (1990). The PIMEX-method as a research tool for modelling of exposure to airborne contaminants. Amer Indust Hyg Conf, Orlando, 13-18 May.

1393

VISUALISATION OF PERSONAL EXPOSURE TO GASES AND DUST USING FAST RESPONSE MONITORS AND VIDEO FILMING

J. Unwin, P.T. Walsh and N. Worsell

Research and Laboratory Services Division,
Health and Safety Executive,
Broad Lane, Sheffield S3 7HQ, U.K.

SUMMARY

A method is described which combines video filming of the work activity and simultaneous personal exposure measurement to gases and dust using fast-response monitors. The real-time exposure level is dubbed onto the video picture of the work activity. Two visualisation systems are described: a mixer based system producing only visual information and a PC-computer based version allowing data processing of files. Several applications of the technique are described where it is used to evaluate work processes so that specific information on the exposure level for a given activity can be recorded. The video can also be used to demonstrate good working practices and improve awareness of the hazard.

1. INTRODUCTION

It is desirable when surveying the workplace for airborne contaminants to correlate fully the exposure data with the work process in order to implement effective measures to reduce exposure. However, using conventional sampling and analysis techniques, information such as the cause of fluctuations in exposure is not recorded. The visualisation technique described here can help to resolve this problem.

The method involves video filming of the work activity and simultaneous measurement of personal exposure using fast-response gas or dust monitors (1,2). The real-time exposure level is displayed at the edge of the video image. This information is easily interpreted; the correlation between exposure and work activity is clearly and visibly demonstrated, and factors which affect the observed exposure are readily identified. Effective control or hygiene measures (e.g. sites for local ventilation, changes in work practice, machine design or use of personal protection) can then be applied to reduce the overall time weighted average (TWA) exposure.

2. VISUALISATION SYSTEMS

Two Channel Mixer Based Visualisation System

The signal from a video camera, filming the work activity, and the analogue signal from a fast-response personal monitor (carried in a back pack by the worker) are combined in a video mixer (IBC AB, Stockholm, Sweden, AVM2000). The signal from the personal monitor is transmitted by radio telemetry at a frequency of 174 MHz to the receiver unit (MIE Medical Research, Leeds, UK). The mixer displays the exposure measurement as a bar graph at the edge of the video image; the height of the bar is proportional to exposure. The mixed picture is recorded on a VHS video recorder and viewed on a VDU screen. Signals from two personal monitors or two measurement ranges from the same monitor can be displayed on the screen. A schematic diagram of the battery powered system is shown in Fig.1.

Computer based visualisation system

This system is able to use the power and flexibility of the microcomputer for more advanced data processing and analysis. Here the video mixer is replaced by a

386-PC. A schematic diagram of the system is shown in Fig.2. A C-programme controls two specialised pieces of computer hardware: a data acquisition card and a video graphics adapter supporting genlocked multiple video standards (Truevision, UK, Targa card). The programme samples and digitises up to four monitor signals at 10 Hz under interrupt while the graphics adapter mixes the computer graphics with the video image, displaying each data sample as a bar graph and text value on the TV monitor.

An interactive menu has options such as changes in colour, number and position of bars displayed on the screen, numerical bar scale display in appropriate units, time, date and filename display. The maximum, minimum and quasi-TWA exposure values can be determined by manipulating the data file entries. On play-back, using suitable software, any frame of interest (e.g. one where exposure is high) can be captured to a graphics file and printed on a colour printer.

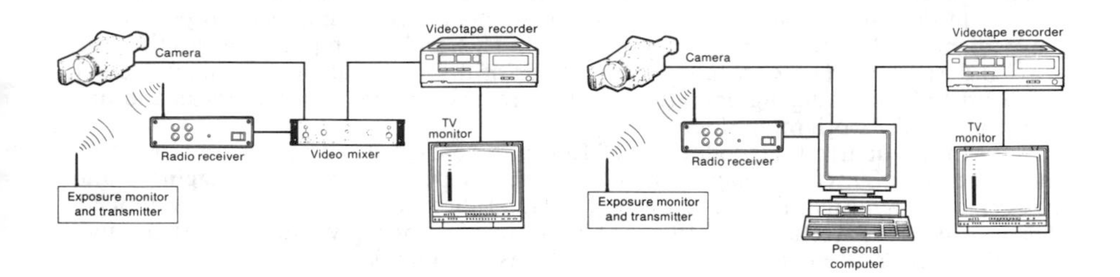

Fig. 1. Mixer-based system. Fig. 2. PC-based system

3. PERSONAL MONITORS

Monitors need a fast response time (< 2 s), in order to follow any fluctuations in concentration, and be compact so that the subject's movement is not hindered. Selectivity is not essential since determining general trends in exposure is of greatest benefit. For organic vapours a modified hand-held photoionisation detector (PID; Photovac Microtip) was used. Its sample probe is located on the lapel near the breathing zone of the subject, and samples at 500 ml/min. The PID can detect many common organic vapours such as benzene and chlorinated hydrocarbons. For a one component system the monitor can be calibrated and concentration changes indicated directly on the screen in units of ppm.

The dust monitor (MIE Inc., Miniram PDM3) operates on an infrared light scattering principle. A flow adapter allows air to be drawn through the sample chamber at 2 L/min using a pump (Casella, UK; AFC123) via a short sample tube (300 mm x 5 mm ID). No particle size selection is achieved although the monitor is most sensitive to respirable dust, the response is thus proportional to the total dust concentration.

4. APPLICATIONS

Organic Vapours

Glass-reinforced plastics are frequently used in the boat construction industry. An initial gel coat is applied to the boat mould by hand brush and then subsequent glass fibre and gel layers are applied using hand rollers. Styrene liberated from the gel resulted in excessive exposure in enclosed sites such as the hull, superstructure and deck sections. The monitor was calibrated with styrene and the concentration indicated on the screen rarely fell below 100 ppm. These findings were confirmed by diffusive organic vapour badges with analysis by gas chromatography where concentrations >250 ppm (20min TWA) were observed for some operations.

During furniture restoration a dichloromethane (DCM) based paint stripper, applied by hand brush, is used to remove old paint and varnish. The monitor was calibrated with DCM. The local exhaust ventilation (LEV) was found to be ineffective when located too far from the work site; the application of the stripper was carried out too close to the breathing zone in some instances; thus the exposure was reduced significantly by better location of LEV and modification of working position.

Dust

The visualisation technique has been applied to identify dust control problems in many industries: wood dust in belt and hand sanding operations; silica dust in potteries and quarries; flour in bakeries; grain dust in agricultural premises.

In the pottery the clay was produced by manually loading several powdered ingredients and water into the mixing vessel. This and other practices resulted in raised dust levels. These included the transfer of dry materials to the mixing vessel by hand and housekeeping operations such as removal of empty paper sacks and dry sweeping of the work room floor.

In the quarry the production of fine and coarser stone aggregates for road construction resulted in many dusty operations. A poorly controlled bagging plant was clearly identified. Exposure was high when filling and closing the sacks; a need for personal protection for the operator in the primary stone crusher and maintenance crew entering the shaker plant was also identified.

5. CONCLUSIONS

The visualisation technique has been used to study a wide variety of processes involving exposure of personnel to hazardous vapours and dust. The critical activities influencing the overall TWA exposure were identified and the information presented in a clear and concise manner. Ideally the system should be used before and after implementation of control measures to assess their effectiveness. The technique can also be used as an educational tool to illustrate sound working practices and improve awareness of hazards.

The information from the visualisation method can be supplemented by personal monitoring using diffusive or gravimetric samplers in order to obtain TWA levels e.g. for specific analysis of a mixture of organic vapours or quantitative dust concentration measurements.

The computer-based system provides more sophisticated data analysis, e.g. TWA calculation, maximum exposure concentration; and generates a hard copy of important visual evidence.

REFERENCES

(1) ROSEN G. and LUNDSTROM S. (1987). Concurrent Video Filming and Measurement for Visualisation of Exposure. Am. Ind. Hyg. Assoc. J. Vol. 48, 688-692.

(2) UNWIN J. and WALSH P.T.(1990). Getting It Taped. Occupational Safety and Health. Vol. 20. 16-18.

Session IV

MEASUREMENT METHODOLOGY
Aerosols

KEYNOTE PAPERS

POSTERS

1394

BASIC ASPECTS WITH RESPECT TO DUST AND AEROSOL MEASUREMENT

W. COENEN

Berufsgenossenschaftliches Institut für Arbeitssicherheit
Alte Heerstrasse 111, Sankt Augustin, Germany

1. INTRODUCTION

This paper introduces dust and aerosols in general, the technical properties of dust in particular, and discusses the state of European standardization in these fields.

2. IMPORTANCE OF DUSTS AND AEROSOLS FOR WORK HYGIENE - DIFFICULTIES AND STATE OF EUROPEAN STANDARDIZATION WORK

Inhalable particles play an important role in work hygiene. They occur at almost all industrial workplaces. Among the latter count, of course, traditional dust producing branches such as the mining industry, the ceramic industry, metal working and metal manufacturing industries, as well as former asbestos working industries. Some more examples could be given by referring to parts of the food industry, the woodworking industry, as well as to the building trade. In the Federal Republic of Germany not less than 33% of all officially acknowledged industrial diseases are related to inhalable particles. The majority of actually diagnosed cases of dust induced occupational diseases may, however, be imputed to exposure situations that existed decades ago. Approximately one quarter of the total number of 500 limit values for hazardous substances in force in the Federal Republic of Germany deal with dusts and aerosols.

Until 1980, the problem of workplace monitoring with respect to exposure to inhalable particles was given priority in our country. It is only since that time that the monitoring of gases and vapours has begun to play a part of comparable significance. These examples reflect the work hygienic situation in terms of inhalable particles which – as I think – may be considered typical for comparable industrial structures in the other EC Member States and in the western world. Thus, it is not surprising that in all these countries standards, measuring techniques and limit values can be found that have been developed over years with a view to the same goal: the protection of man at work. As a matter of fact, there are specific traditional differences concerning technical criteria and nationally introduced monitoring devices as well as a large variety of different formerly collected exposure data which are of invaluable importance for epidemiological research. Consequently, the situation is not such as to ease the task of European harmonization. It is important to develop a future common basis without giving up precious experience, information and methods.

The amendment 88/642/EWG of the Directive 80/1107/EWG for the protection of workers against risks due to chemical, physical and biological agents at the workplace included a method of reference (appendix IIa) for sampling, measuring and assessing hazardous chemicals. This method of reference focusing particular attention on the problem of suspended matter, represents the implementation of the recommendations formulated by the Pneumoconiosis Conference at Johannesburg in 1959 (Johannesburg Convention) and contained in the standards of ISO/TC 146 since 1983; it also refers to the 'General requirements for the performance of procedures for workplace measurement' as well as to the 'Specification for conventions for measurement of suspended matter in workplace atmospheres' established by the European Standardization Committee (CEN).

In the meantime, a lot of energy has been spent on accelerating the process of making decisions; so far, two draft standards could be passed which are of great significance in terms of a harmonized determination and monitoring of workplace exposure to aerosols. The draft standards deal with the following themes:

1. General minimum requirements and test criteria for all measuring procedures in connection with the determination of exposure to hazardous substances.
2. Technical definitions for respirable, thoracic and inhalable dust.

The contents of these draft standards will be presented and discussed in detail within the framework of this session, so I do not have to go into details now. Let me, however, try to give you a brief survey of their contents as shown in Figure 1.

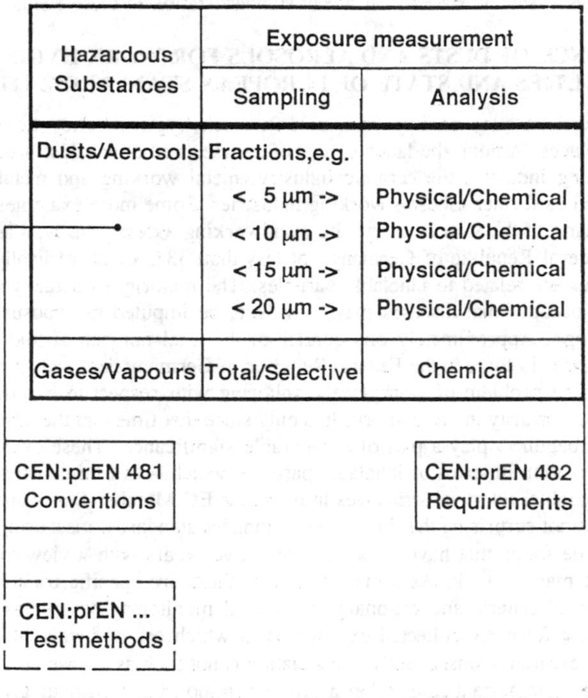

Hazardous Substances	Exposure measurement	
	Sampling	Analysis
Dusts/Aerosols	Fractions,e.g.	
	< 5 µm ->	Physical/Chemical
	< 10 µm ->	Physical/Chemical
	< 15 µm ->	Physical/Chemical
	< 20 µm ->	Physical/Chemical
Gases/Vapours	Total/Selective	Chemical

CEN:prEN 481
Conventions

CEN:prEN 482
Requirements

CEN:prEN ...
Test methods

Fig. 1. Exposure determination - actual state of CEN standardization

If an effect-oriented measurement of the exposure to concentrations of hazardous substances in workplace atmospheres is our goal, we first have to sample the relevant components from or together with the air. Sampling is followed by a specific analysis. This procedure also applies to measurements within a closed device.

While gases and vapours permit total or selective sampling of identical molecules, dust particles must first be fractionated according to their size (which is in the range of 1000 mm). The different particle fractions are then subjected to physical (morphological, crystallographical) or chemical analysis.

This means that for a defined gas there is only **one** possible result. In the case of dusts, however, we may obtain as many results as there are particle fractions. Of course, this is not realistic. The general requirements as specified in prEN 482 apply to the entire procedure of exposure measurement, i.e. for dusts and gases. As a matter of fact, they are sufficiently precise, thus permitting measuring techniques for gases and vapours to be tested and assessed.

In the case of dusts and aerosols, an additional definition of groups of particle sizes is necessary, each group being characterized by an identical or almost identical size specific effect of the particles on human health. The different groups can then be separately subjected to chemical/physical analysis. Definitions of conventions for such effect oriented particle groups were elaborated by CEN on the basis of the latest scientific research results; they are specified in prEN 481 representing the first uniform approach to this difficult problem within CEN and ISO. Unfortunately, elaboration of harmonized conventions is not enough.

To guarantee comparable results, it is also necessary to determine concrete test and assessment methods for particle-selective sampling and measuring devices. These test criteria are in preparation.

In the European Member States, conventions are, of course, already in existence and varying devices have been used to determine exposure to dusts and aerosols in accordance with pertinent work safety requirements. Up to this point in time, two particle groups have been distinguished:

- the totally inhalable particle fraction, often called total dust, and the much finer
- respirable particle fraction, also called fine dust.

This rough two category classification has always formed the basis of practical health protection at the workplace. National measuring methods and measuring devices for dusts and aerosols and, consequently, all formerly collected exposure data as well as - and this is of particular importance - all national dust limit values, do actually refer to this classification. As dust technological conventions differ slightly from state to state, measuring results and particle limit values are not comparable in detail either. This fact is to be taken into account when striving for European harmonization.

A more precise classification of particle fractions including more than two categories only is required by scientists investigating particle deposition in the different areas of the human respiratory tract. Although scientifically founded, this suggestion is difficult to put into practice; in some cases of workplace measurements either further differentiation of particle fractions is impossible or the toxicological accuracy is exhausted.

Again, a compromise is to be found and defined in the form of conventions to which toxicologists, epidemiologists, as well as designers of measuring devices and occupational hygienists can refer.

3. DEPOSITION OF DUSTS AND AEROSOLS IN DIFFERENT RESPIRATORY AREAS

To protect workers against exposure to dusts and aerosols, the question must be answered as to which particles are retained in the human respiratory tract, and which particles are inhalable. It must then be investigated in which respiratory regions these particles are deposited. The transport and deposition within the respiratory system is largely determined by the behaviour of particles in flowing gases, the aerodynamic diameter of a particle being the decisive parameter. The aerodynamic diameter of a particle is not the size determinable by means of a microscope/electronic microscope; it is defined as the diameter of a sphere with a density of 1 which shows the same aerodynamic behaviour as the particle under consideration. In other words: a particle is theoretically substituted by an aerodynamically comparable sphere whose diameter is used to describe the particle. Thus, it is possible to describe the behaviour of any particle, for example, compact, bizarre shape, flocky or even needle-shaped.

Figure 2 gives a schematic overview of the different respiratory regions as well as of particle intake and retention characteristics of these respiratory areas. Only a limited portion of all airborne particles in workplace air are captured by inhalation. This total inhalable fraction is often called 'total dust'. Dusts that do not pass through the nasal-pharyngeal- laryngeal area are deposited as nose-pharynx-larynx-dust. The dust fraction passing through that area and reaching the lungs is the so-called thoracic dust fraction. The tracheobronchial tree has dust selective properties, too, retaining a portion of the incoming dust as tracheobronchial dust; the rest passes through this area and gains access to the bronchioles and alveoli, i.e. to the lowest respiratory regions. This fraction is also called 'fine dust'.

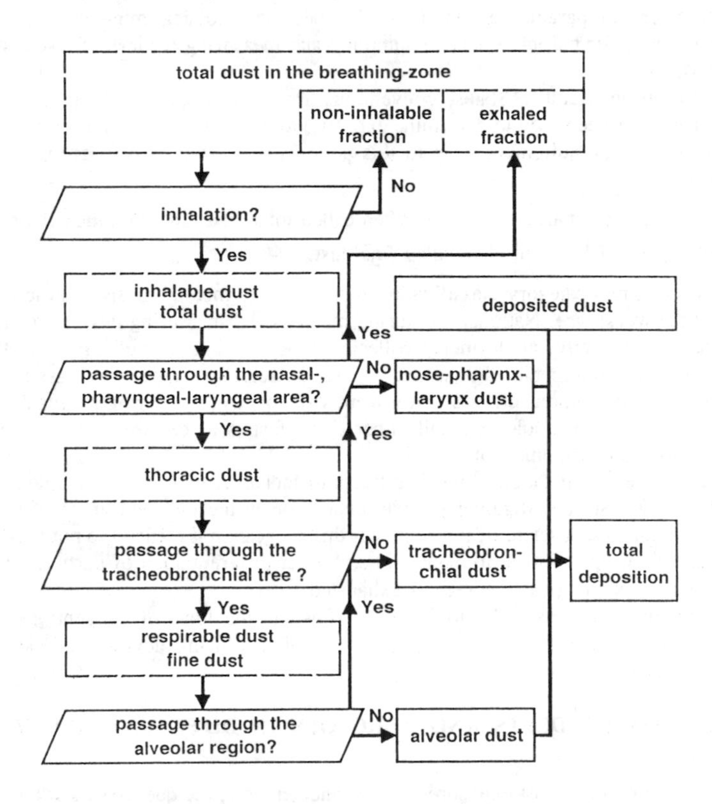

Fig. 2. Particle deposition during respiration

Alveolar dust is deposited in the alveolar region and a part of the finest dust fraction is exhaled on condition that it is not deposited in the upper respiratory areas during exhalation. The sum of all deposited dust fractions is called total deposition.

The entire respiratory cycle is illustrated in Figure 3. The air particle flow is picked up by inhalation (right side), larger particles forming the non-inhalable fraction. While the air particle flow passes through the different respiratory areas, larger particles are deposited, the particle spectrum becoming continuously finer. Eventually, a certain percentage of the finest particles is exhaled. The human respiratory system can be divided roughly into four respiratory borderline regions:

- passage through the mouth or the nose functioning as intake orifice of the human body (E)
- passage through the pharyngeal-laryngeal area forming the borderline between the pharynx and the bronchial area (1)
- passage through the borderline between the bronchioles and the alveolar region (2)
- passage through the alveolar region into the very fine alveoli (3) (exhalation).

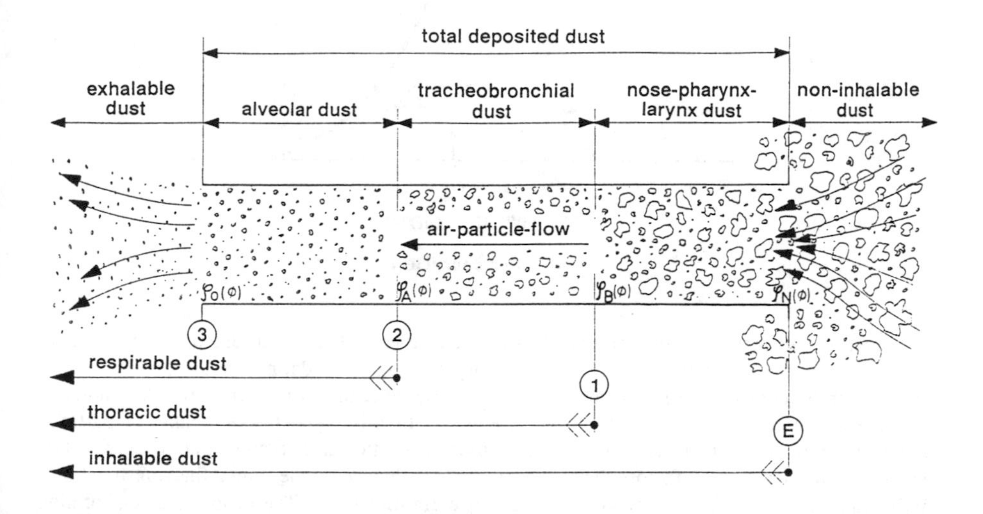

Fig. 3. Filter model for different respiratory areas

The above mentioned borderlines reflect the probability of particle admission to the lower respiratory regions. The probabilities are indicated as functions of the particle size, i.e. of the aerodynamic diameter. On account of the fact that particles are deposited in the different respiratory regions – which means that they are filtered from the inhaled air – it is possible to characterize the respiratory regions by their filtering capacity. Consequently, particle permeability or retention can be described in terms of particle size dependent functions. On the basis of these functions, aerodynamic separation systems can be developed that reflect the deposition pattern of the particles in the respiratory cycle.

Figure 4 is a qualitative representation of such a function. While the permeability of a respiratory filter is rather high with respect to small particles, particles of larger aerodynamic diameters are increasingly retained, which shows that the retention behaviour of the filter is in inverse proportion to particle admission. In terms of quantity, a respiratory filter admits the fraction f of all incoming particles of the aerodynamic diameter $\emptyset$ to the lower respiratory regions. The filter characteristic is determined by the 50% value of the permeability and by the filtering capacity as a function of the particle diameter.

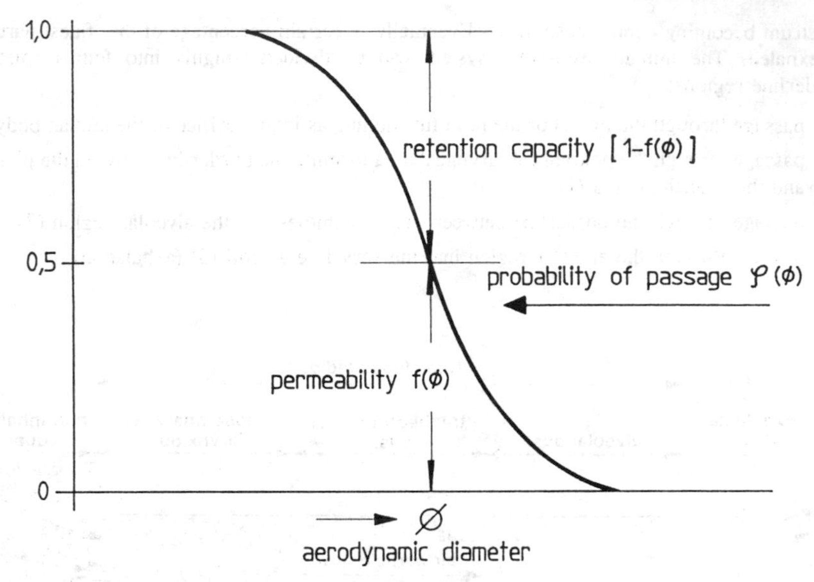

Fig. 4. Filter characteristic

Figure 5 illustrates several filters of different characteristics which are arranged so as to represent the conditions in the human respiratory system. The definition of conventions for these filter functions F_E to F_3 can be compared to the determination of framework conditions for effect-oriented dust and aerosol measurements. In this context, the intake of particles through the mouth or the nose represents a problem of particular difficulty, due to the fact that there are two complicated factors that have a marked influence on the 'outer filtering function': individual disposition and inflowing air from different directions. The problem of elaborating conventions for the different respiratory functions will be broached in the following papers.

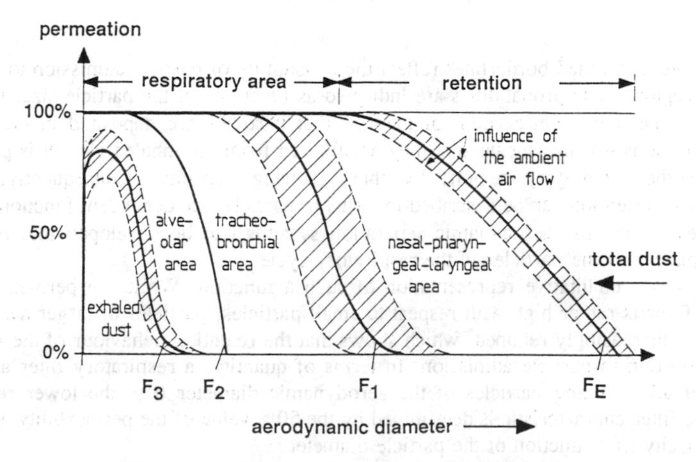

Fig. 5. Permeation of particles during respiration

The same model generally applies to fibres. Attention must however be paid to the fibre length which may be considerably greater than the aerodynamic diameter, thus entailing a modified deposition behaviour of the particles. On account of this fact, as well as on account of the specific effect that bio-resistant fibres may have on human health, different risk potentials have been attributed to certain fibres of identical dimensions.

The EC Directive 83/477/EWG contains detailed information on the problem. Interlaboratory trials as carried out within the framework of AFRICA (Asbestos Fibre Regular Informal Counting Arrangement) are very important for the harmonization of exposure measurements with respect to fibrogenic dusts.

4. CONCLUSION

Let me finally summarize some ideas which ought to underlie all further developments in view of a Europe-wide reorganisation and harmonization of effect-oriented dust measuring techniques:

1. To avoid the epidemiological reliability of the large number of measuring data accumulated over the years being called into question, it is necessary to revise the underlying dust technological definitions.
2. Corrections of existing dust technological definitions must imply decisive modifications to be significant for the field. Due to the measuring errors which are, in any case, unavoidable, small changes would not even affect the results of field measurements. Besides, it is not sure whether epidemiological or toxicological findings would gain in precision if dust technological definitions were subjected to petty modifications. On account of the same arguments, it must be ensured that test requirements are sufficiently flexible to confirm the largest spectrum of nationally proved systems to be suitable for further use or, at least, to permit calculative adaptation of formerly collected data.
3. It will also be difficult to introduce supplementary categories to the existing classification (inhalable dust, respirable dust) because they would have to be 'detectable' by means of current field measuring techniques.
4. The lack of corresponding limit values makes the elaboration of additional dust technological definitions impossible. The very first implementation of new definitions must, of course, happen in the field of toxicological limit values. Consequently, before these definitions can be used for approaching concrete work safety problems, a certain time of transition has to pass during which limit values have to be determined and corresponding sampling, measuring and assessment techniques can be developed and made available.

1395

HEALTH-RELATED MEASUREMENT OF PARTICULATE FRACTIONS - RESPIRABLE AND THORACIC DUST

J.F. FABRIES

Institut National de Recherche et de Sécurité (INRS)
Av. de Bourgogne, BP27, 54501 Vandoeuvre, France

SUMMARY

Samplers able to collect airborne particles susceptible of penetrating the upper respiratory airways or the gas exchange region of the lungs are presented. Some problems concerning their performances are then discussed.

1. INTRODUCTION

Many chemical substances may be present in workplace atmosphere, generated by industrial processes and human activity. When they are contained in dispersed particulate matter, then forming aerosols of liquid or solid particles, they penetrate the respiratory airways during inhalation, and deposit at different levels according to particle size and other parameters. The possible health effects due to inhalation of these chemical substances depend on their physical properties (crystalline form, surface properties, dissolution rate, hygroscopicity,..), chemical nature, characteristics of exposure (concentration profile, duration) and individual factors. They also depend on the deposition sites within the respiratory airways of the particles by which the chemical compounds are transported. For example it is now well known that particles containing crystalline silica (quartz, cristobalite) reaching the gas exchange region can induce a pulmonary fibrosis in workers of the mining industry, or kaolin workers and firebrick makers (1). Other fibrogenic dusts are known, such as those encountered in the hard metal industry, where workers are usually exposed to metallic carbides (tungsten, titanium, niobium, molybdenum,..) and cobalt (used as a binder), which is regarded as the main causative factor for hard metal pneumoconiosis (2, 3, 4). The hard metal production process is mainly aimed at making cutting or polishing tools, drill tips, and other special instruments.

However the respiratory system is not necessarily the only target for the toxicity of chemical compounds, even when their entry to the body is by inhalation. For example beryllium is a potent systemic poison that is retained by various body organs including lung, liver and spleen (5). Beryllium exposure is an occupational risk factor in many sectors of high technological industries such as aircraft, aerospace, electronics, where beryllium is used in alloys and ceramics. In order to assess the exposure to aerosols of insoluble forms of beryllium, it was suggested (6) to measure the concentration of beryllium able to deposit in the thoracic region of the respiratory airways, including the trachea, the bronchi and the deep lung.

In order to get better indicators of exposure at the workplace, taking into account the gradual penetration of airborne particles through the respiratory airways and the specific hazards related to global and regional deposition, several particle size conventions have been defined (7, 8). These conventions give average probabilities of particle inhalation or penetration in some compartments of the respiratory airways as functions of particle size (generally expressed in terms of particle aerodynamic diameter). They are based upon particle deposition data obtained with humans and controlled physiological and physical parameters. However they are essentially sampling criteria applicable to exposure assessment and not really to dose calculation; they are based upon the rough assumption that all particles that are inhaled are deposited. These limitations in the use of sampling criteria in epidemiological studies have been recently pointed out (9).

More recent sampling criteria defining inhalable, thoracic and respirable conventions are now in discussion within ISO and CEN (10, 11). They largely integrate proposals made by Soderholm (12), and it is hoped that these new conventions, which are fully compatible, will be used by all countries in an homogeneous way.

The measurement of the particle size fractions of inhaled aerosols requires specific samplers with a particle size selectivity as close as possible to the conventions. The samplers used for respirable and thoracic dusts are examined in the next paragraphs.

2. RESPIRABLE DUST SAMPLERS

Many respirable dust samplers are now available and well tested. They are generally intended to be worn by the individual workers, in the neighbourhood of their nose and mouth, in order to assess their personal exposure to toxic substances which are biologically active in the non-ciliated airways (alveoli, respirable bronchioles), like silica or hard metals. One of the most known devices is the Dorr-Oliver 10 mm nylon cyclone, which was developed in the United States (13). The aerosol is sampled tangentially through a small rectangular slot entry, and the coarser particles are separated from the air under the effect of centrifugal forces caused by the spiralling motion of air in the cylindrical inner cavity. The finer particles are trapped by a filter. The cyclone is used in connection with a small pump, and the flow rate is usually set at the value 1.7 l/min. With these conditions, the particle collection efficiency was measured, and compared with the standard respirable convention of an aerosol (see Figure 1). In Great Britain a similar model was developed, that operates at a flow rate of 1.9 l/min.

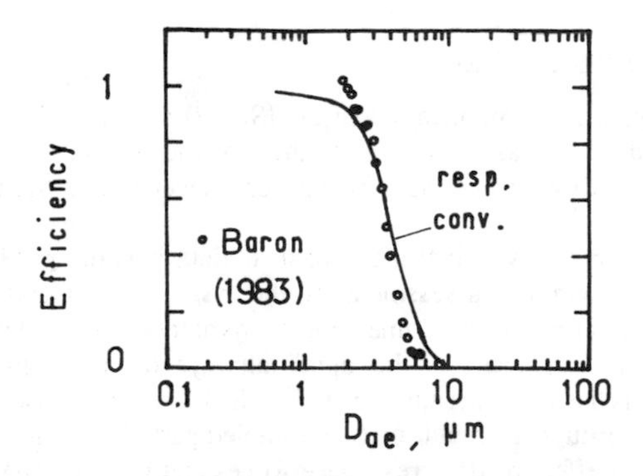

Fig. 1. Collection efficiency of the 10 mm cyclone at 1.7 l/min and respirable convention (Soderholm, 1989) as functions of particle aerodynamic diameter

In France another type of very compact device was elaborated, named CIP10 (14). The aerosol is sucked in through an annular slit; the aspiration effect is obtained by the high speed rotation of a small cup inside a cavity. The coarser particles are trapped by the combination of an impactor stage and a foam filter. The remaining finer particles are then trapped by the rotating cup which contains a fine graded polyurethane foam. Particles smaller than 2 μm are partially rejected.

The measured collection efficiency is graphically reported in Figure 2; it can be seen that it exhibits a maximum, hence resembling the true alveolar deposition curve more realistically than the respirable conventional curve. The main advantages of this instrument are its relatively high flow rate (10 l/min), a large duration of operation with batteries (more than 20 hours), a very low weight (300 g) and a compact shape that makes it generally much more easily accepted by workers than any equipment associating a pump and a sampler, particularly in the case of repeated measurements.

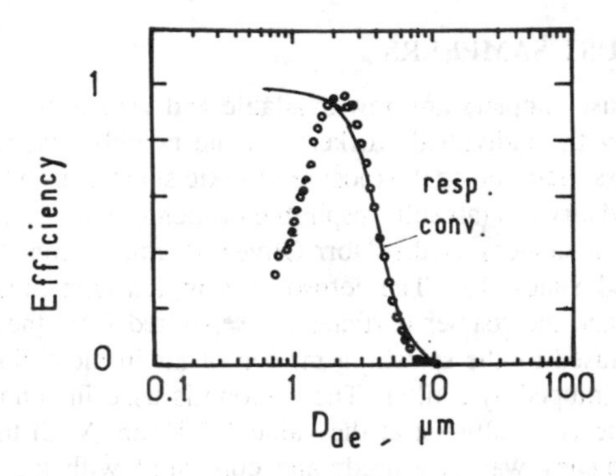

Fig. 2. Collection efficiency of the CIP10 instrument at 10 l/min and respirable convention (Soderholm, 1989) as functions of particle aerodynamic diameter

3. THORACIC DUST SAMPLERS

It is only recently, with the publication of the ISO (8) and ACGIH (7) recommendations, that widespread consideration has been given to the thoracic fraction of an aerosol. Therefore a very limited number of instruments have so far been developed in occupational hygiene for this purpose.

The first thoracic sampler was in fact the vertical elutriator, designed in the United States for the cotton industry and the assessment of the risk of byssinosis (15). The thoracic convention was not yet defined at this time, but the hypothesis of a relationship between the quantity of inhaled dust deposited in the upper airways of the respiratory tract and the prevalence of byssinosis has been already made. In this device the aerosol is drawn by a pump into a vertical cylinder through a circular orifice; sampled particles are sorted according to their size under the combined effect of flow resistance forces and gravity. The vertical elutriator is a static sampling instrument; its nominal flow rate is 7.4 l/min.

The respirable dust sampler CIP10 was recently modified in order to collect thoracic particles. The particle size selector was replaced by a new one, of complex geometry, incorporating eight small circular orifices through which the sampled aerosol is forced to flow (16). Particle velocity change, both in direction and magnitude in the vicinity of the orifices, is responsible for selective deposition on the walls of the sampler. The finer thoracic particles are finally collected in the same rotating cup as in the original CIP10 instrument. This new personal sampler, named CIPT, is now being refined and its performances tested. It is hoped that it will be able to replace the vertical elutriator for air monitoring in the textile industry.

Other prototypes of thoracic dust samplers have been designed, and the adoption of exposure limit values in terms of conventional fractions for chemical substances would certainly enhance the development of new interesting devices.

4. PERFORMANCES OF SAMPLERS

The availability of particle size conventions and many types of samplers gives rise to several questions. For example, how a sampler can be judged suitable for sampling a given aerosol fraction, for what particle size range can it be used, or what accuracy can be expected for the measurements? The problem of the test of performance of instruments for health-related sampling of airborne particles is now a matter of research and discussion within many institutes involved in occupational hygiene. It is also discussed within CEN/TC137/WG3.

One of the major difficulties is the accurate experimental assessment of particle size selectivity (or collection efficiency) of a sampler as a function of particle aerodynamic diameter. This size parameter is essential when dealing with particle dynamics and deposition near air samplers or through the respiratory airways. Several methods are possible, using test aerosols and specific facilities, which are accessible to a limited number of research teams. There is a need in reference materials for generating test aerosols, in order to circumvent some problems that are, for example, related to particle shape or density.

Once sampler efficiency is measured, it is generally observed that the data obtained do not meet perfectly the corresponding conventional curve (as can be seen in Figures 1 and 2). However, as the usual parameter used for expressing the quantity of airborne contaminant is mass concentration, it is very important to study the sampler behaviour in terms of bias between the measured concentration C and the concentration C* that could be measured by a 'perfect' instrument with a particle size selectivity identical to the conventional curve. Such bias can be calculated for real samplers placed in an aerosol with a known particle size distribution by numerical simulation.

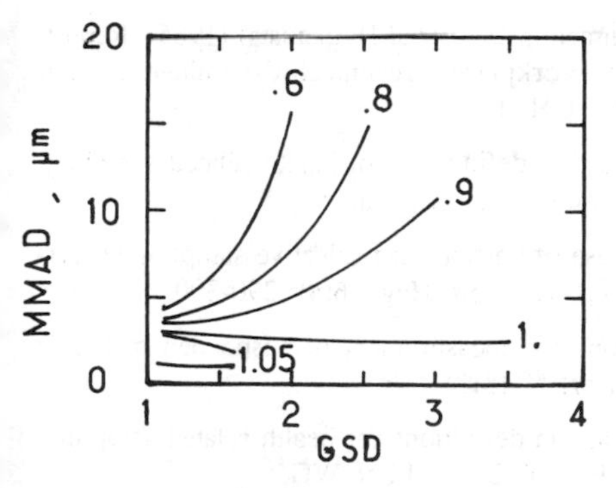

Fig. 3. **Bias map giving the contours β = C/C* in the plane (GSD, MMAD) of aerosol parameters for the 10 mm cyclone/respirable convention**

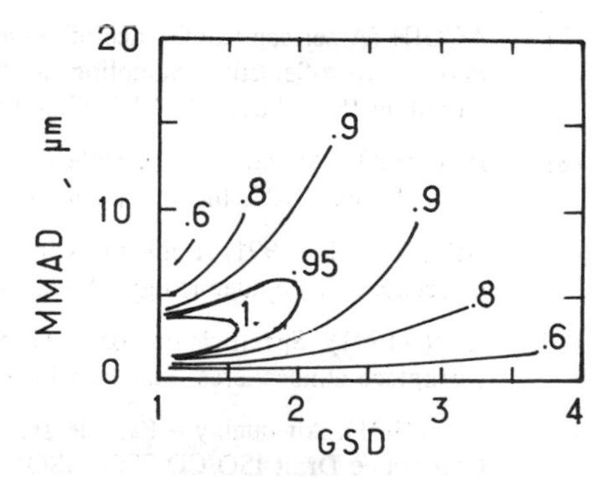

Fig. 4. **Bias map giving the contours β = C/C* in the plane (GSD, MMAD) of aerosol parameters for the CIP10 instrument/respirable convention**

For example, Figure 3 shows the variation of the ratio $\beta = C/C^*$ with particle size distribution, calculated for the 10 mm cyclone placed in a log-normally distributed aerosol with a mass median aerodynamic diameter MMAD and a geometric standard deviation GSD. It can be seen that despite a relatively good agreement between the efficiency data points and the conventional curve for the respirable fraction, an underestimation of concentration close to 20 % can be expected for many industrial aerosols. Figure 4 shows equivalent results for the CIP10 sampler, which are a little bit better within the same particle size range, but with an increased bias for very fine aerosols.

The influence of several factors such as external wind speed, variations in the geometric parameters of sampler inlets or particle size selector needs also to be examined.

The test of performances of samplers is a very sharp task, requiring much time and many specialized equipments; however it remains a necessary step in order to improve the knowledge of exposure to chemical substances, and hence health protection.

REFERENCES

(1) LESSER, M., ZIA, M. and KILBURN, K.H. (1978). Silicosis in kaolin workers and firebrick makers, South. Med. J., 71: 1242-1246.

(2) COATES, E.O. and WATSON, J.H.L. (1971). Diffuse interstitial lung disease in tungsten carbide workers, Ann. Intern. Med., 75: 709-716.

(3) HARTUNG, M., SCHALLER, K.H. and BRAND, E. (1982). On the question of the pathogenetic importance of cobalt for hard metal fibrosis of the lung, Int. Archiv. Environ. Health, 50: 53–57.

(4) BALMES, J.R. (1987). Respiratory effects of hard metal dust exposure, State Art Rev. Occup. Med., 2: 327-344.

(5) ILO (International Labour Office) (1983). Encyclopaedia of Occupational Health and Safety, ILO Publ., 3d ed., Geneva.

(6) RAABE, O.G. (1988). In Advances in Air Sampling, ACGIH, Lewis Publ., 39-51.

(7) ACGIH (American Conference of Governmental Industrial Hygienists) (1985). Report: Particle Size-Selective Sampling in the Workplace. Technical Committee on Air Sampling Procedures, ACGIH, Cincinnati, OH, USA.

(8) ISO (1983). Air quality - particle size fraction definitions for health related sampling, Tech. Report 7708, Int. Organization for Standards, Geneva.

(9) HEWETT, P. (1991). Limitations in the use of particle size-selective sampling criteria in occupational epidemiology, Appl. Occup. Environ. Hyg., 6(4): 290-300.

(10) CEN (1991). Specification for conventions for measurement of suspended matter in workplace atmospheres, Standard Project prEN 481.

(11) ISO (1991). Air quality - Particle size fraction definitions for health-related sampling, Committee Draft ISO/CD 7708, ISO/TC146/SC 2+3 ad hoc WG.

(12) SODERHOLM, S.C. (1989). Proposed international conventions for particle size-selective sampling, Ann. occup. Hyg., 33: 301-320.

(13) VINCENT, J.H. (1989). Aerosol Sampling - Science and Practice, J. Wiley, New York.

(14) COURBON, P., WROBEL, R. and FABRIES, J.F. (1988). A new individual respirable dust sampler: the CIP10, Ann. occup. Hyg., 32(1):129-143.

(15) NEEFUS, J.D., LUMSDEN, J.C. and JONES, M.T. (1977). Cotton dust sampling II - vertical elutriation, Am. Ind. Hyg. Ass. J., 38: 394-400.

(16) FABRIES, J.F., GÖRNER, P. and WROBEL, R. (1989). A new air sampling instrument for the assessment of the thoracic fraction of an aerosol, J. Aerosol Sci., 20(8): 1589-1592.

HEALTH-RELATED MEASUREMENT OF PARTICULATE FRACTIONS - INHALABLE DUST

L. ARMBRUSTER[1] and H.D. BAUER[2]

[1]DMT-Gesellschaft für Forschung und Prufung mbH
Institut für Staubbekämpfung, Gefahrstoffe und Ergonomie, Essen

[2]Bergbau-Berufsgenossenschaft
Institut für Gefahrstoff-Forschung, Bochum, Germany

SUMMARY

The definition of inhalable dust on the basis of experimental results has provided an unambiguous selection curve which can serve as a guide to future instrument development. Suitable static and personal instruments already exist but standards are also required to ensure better comparability of present day test methods.
When exposure is to be assessed by measurement at the workplace the question of whether static or personal instruments are to be preferred must be carefully considered.

DEFINITION OF INHALABLE DUST

In addition to respirable dust as defined at the Pneumoconiosis Conference in Johannesburg in 1959, total inhalable dust has also attracted the attention of toxicologists and practitioners of industrial medicine. It has become clear that soluble, resorbable particles, or particles of substances which react with the body fluids and are potentially injurious to health, must be regarded as risk factors if inhaled. Many metal dusts, soluble toxic salts, wood dust etc. are today so regarded.

In Germany, the MAC Committee established a convention for sampling total inhalable dust. This was to be measured with instruments having a sampling velocity of 1.25m/s ± 10%. Experimental studies, however, have demonstrated that this convention does not yield unambiguous results (1, 2).

Initial findings on the inhalability of dust were published in 1975 by Ogden, who fitted filters behind the oronasal opening in a full-scale model head which he connected to a pump acting as a breathing simulator. He then carried out wind tunnel experiments to measure the probability of inhalation through the mouth and nose for various wind speeds and for various positions of the head in relation to the wind direction (3).

Similar studies were carried out by Armbruster and Breuer and by Vincent and Mark, though with different wind speeds, particulate substances and particle size distributions. Today, sets of individual inhalation probability data are available for particles with diameters ranging from $d_{ae} \leq 1\mu m$ to $d_{ae} = 120\mu m$ at wind speeds of 0.5–8m/s and for representative yaw angles. The studies were carried out with breathing parameters corresponding to respiration at rest and when performing light and arduous work (4, 5).

Analysis of the data yielded mean inhalation probabilities for the various particle sizes and wind speeds. Such a mean inhalation probability is averaged over all yaw angles for breathing through the mouth and nose and over the indicated range of wind speeds.

CEN TC 137, WG 3 based its definition of inhalable dust at the workplace on the mean inhalation probability for particles with aerodynamic diameters of 1–100μm and wind speeds of 1-4m/s. The selection curve so defined shows that the inhalation probability is close to 100%

for small particles (≤ 5μm) falling to 50% for particles of 30μm diameter and then remaining constant. The curve is thus 'open-ended', the upper diameter limit being determined by the particle size distribution of the test dust. However, the inhalation probability is not defined for particles in excess of 100μm (Figure 1) (6).

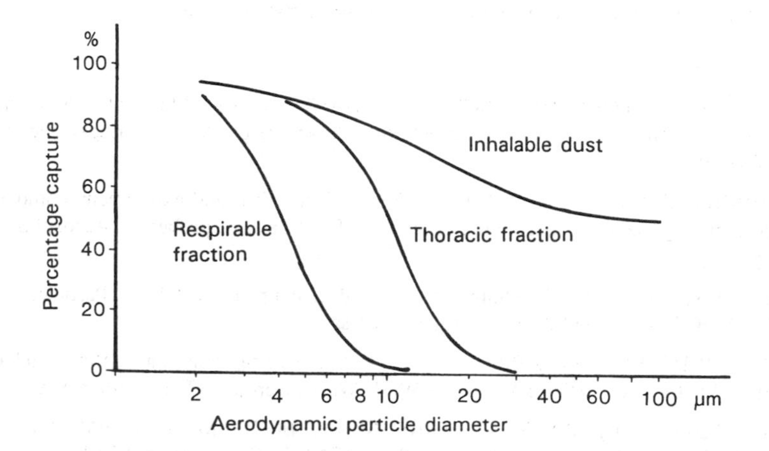

Fig. 1. Selection curves for inhalable dust and for the thoracic and respirable fractions according to CEN TC 137

DEVELOPMENT OF MEASURING AND SAMPLING INSTRUMENTS

As a result of the definition by the MAC Committee, annular slot probes are preferred as the inlet for static instruments, since these are unaffected by the wind direction. A disadvantage, however, has proved to be that their efficiency in capturing coarse particles (> 20μm) declines rapidly with increasing wind speed, the effect being very dependent on the design characteristics.

Ogden, and later Vincent, showed that 'blunt probes' are more suitable for sampling inhalable dust. The first such instrument was the ORB-sampler, which conforms very well to the present definition for wind speeds of less than 2m/s. Vincent constructed a rotating probe in the form of an upright cylinder with a directional inlet in the middle of the cylinder body. This design was based on the human head and within the specified range of wind speeds provides a selection curve for the aspirated particles which comes very close to the definition curve.

A probe developed by Armbruster is based on the same principle but has no rotating parts. This probe, which causes considerable disturbance of the airflow, provides selection curves within acceptable tolerances about the definition curve for wind speeds of up to 8m/s.

The VC 25 with an annular slot probe can also be regarded as a suitable instrument for wind speeds of up to 4m/s. As personal samplers, small instruments with open filters were initially preferred, but did not yield satisfactory results. Such instruments are being modified, for example at the IOM in Edinburgh, and their performance in relation to the definition is being improved. The available data indicate that today the IOM Personal Dust Sampler, the PGP (BIA, St Augustin) and the PERSPEC (Prodi *et al.*, Bologna), can be regarded as suitable for sampling inhalable dust.

CORRECT ASSESSMENT OF EXPOSURE

Use of a suitable instrument does not in itself ensure a result which is representative of an employee's exposure. A choice between fixed-position and personal measurements must be made in the light of local conditions and the airflows at the workplace. In unfavourable circumstances, static measurements may greatly underestimate exposure.

REFERENCES

(1) DEUTSCHE FORSCHUNGSGEMEINSCHAFT (1990). Maximale Arbeitsplatzkonzentration und biologische Arbeitsstofftoleranzwerte. VCH Verlagsgemeinschaft mbH, Weinheim.

(2) ARMBRUSTER, L. and ZEBEL, G (1965). Theoretical and experimental studies for determining the aerosol sampling efficiency of annular slot probes. J. Aerosol Sci. 16, pp 335-341.

(3) OGDEN, T.L. (1977). The human head as a dust sampler. In: Inhaled Particles IV (ed. by W.H. Walton) pp 93-105, Pergamon Press.

(4) ARMBRUSTER, L. and BREUER, H. (1982). Investigations into defining inhalable dust. In: Inhaled Particles V (ed. by W.H. Walton) pp 21-32, Pergamon Press.

(5) VINCENT, J.H., MARK, D., GIBSON, H., BOTHAM, R.A. EMMET, P.C., WITHERSPOON, W.A., AITKEN, R.J., JONES, C.O. and MILLER, B. (1983). Measurement of inhalable dust in wind conditions pertaining to mines. Final Report on CEC Contract 7256-21/029/08. Institute of Occupational Medicine.

(6) CEN (1991). Festlegung von Konventionen von Partikelfraktionen zur Messung von Schwebstoffen am Arbeitsplatz. Draft Standard DIN EN 481, July.

(7) VINCENT, J.H. (1989). Aerosol Sampling, Science and Practice. John Wiley and Sons, Chichester.

1397

HEALTH-RELATED MEASUREMENT OF PARTICULATE FRACTION-FIBRES

G. RIEDIGER

Berufsgenossenschaftliches Institut für Arbeitssicherheit
Alte Heerstrasse 111, Sankt Augustin, Germany

SUMMARY

Health-related measurement of fibres mainly means the measurement of airborne fibres. The technique for sampling has to ensure that fibrous particles meeting predetermined size criteria are sampled quantitatively. The most adequate and common technique for analysing fibres still is microscopy - light microscopy and electron microscopy. Though the instrumental conditions and the counting criteria are well defined, the results are influenced by the individual abilities and by subjective decisions of the counter. This disadvantage could be overcome by using a sophisticated image analysing system instead of a human counter or - to a certain degree - by a proficiency testing scheme for laboratories using microscopic methods.

1. INTRODUCTION

Health-related measurement of fibres means, in the first place, measurement and assessment of airborne fibrous particles in ambient workplace air. In this context, the term fibrous particles applies to long, thin particles meeting certain pre-set criteria. In most cases, these criteria refer to particle dimensions, defining fibrous particles to be longer than 5 μm and thinner than 3 μm with an aspect ratio greater than 3:1. Whether or not the fibres are straight is of no importance at all.

Independent of the analytical method used for fibre measurement, the sampling technique has to ensure that all particles complying with the fibre definition are sampled quantitatively. This criterion is considered to be fulfilled at a suction velocity of at least 4 cm/s. In addition, the sampling method has to be adapted to the analytical method; this aspect will be discussed in detail together with the different analytical procedures described hereinafter.

Due to the requirement for fibrous particles to be assessed according to their shape and dimensions, microscopic analysis is the most suitable analytical method, viz. light and electron microscopic procedures.

2. LIGHT MICROSCOPY

The EC Directive on the protection of workers from the risks related to exposure to asbestos at work (1) requires phase contrast microscopy to be used for determining fibre concentrations. This EC reference method is identical to the Technical Method RTM1, recommended by the AIA - Asbestos International Association (2).

According to this method a membrane filter of cellulose ester is used with a maximum pore size of 1.2 μm. An open filter holder, facing downward, is used with a face velocity of approximately 4 cm/s. Filter preparation preceding the analysis is done by using acetone vapour and triacetin to render the filter transparent. Fibre counting performed in accordance with pre-set counting rules at a total magnification of 500 x and - in most cases - at positive phase contrast. The counting field is determined by the Walton-Becket graticule. Its diameter corresponds to 100 ± 2 μm. A total of 100 fibres are to be counted in different counting fields, evenly spread over the whole filter. For statistical reasons, however, the minimum number of counting fields must be 20. To limit the analytical expenditure, counting can be stopped after a total of 100 counting fields has been analyzed; this corresponds to an evaluated area of approximately 1/1000 of the effective filter area.

Counts based on a large number of counted fibres are more reliable than those obtained on the basis of a low number of fibres. According to counting statistics, the theoretical relative standard deviation is 0.1 for a number of 100 counted fibres, while the same value is increased to 0.32 in the case of only 10 counted fibres. Further to this statistical variation, the results are also influenced by the sampling procedure and by subjective factors as well as by the particular properties of the sample.

This method is relatively easy to use and requires little instrumental expenditure. Unfortunately, there are two serious disadvantages: it is a subjective method and it does only consider the shape of particles without giving any information on the material of which the fibres are made. Besides, it cannot be used to detect fibres that are thinner than approximately 0.2–0.3 μm.

Although device specific parameters and counting criteria are clearly defined, the result is influenced by subjective factors such as visual acuity, concentration, time pressure and fatigue of the counter as well as the interpretation of the often ambiguous counting rules. A split fibre, for example, should be considered as a single fibre, while crossed or overlapping fibres have to be considered separately. In this case the decision which counting rule should be applied often is not clear. Fibre bundles should be resolved and the single fibres in the bundle should be counted, if possible. If not, the whole bundle is to be considered as a unit. Different persons may easily obtain different results when applying this rule.

Another counting rule difficult to use concerns particles attached to fibres. Whenever the particle size exceeds 3 μm, the fibre is to be ignored. Experience has shown that this rule often leads to different interpretations: while one person is convinced of fibre and particle forming a unit, another person cannot see any link between them – no matter whether this is due to a better visual acuity or to a halo around the large particle. In addition, there is no way to decide whether the particle-fibre agglomerate existed already in the air or whether it had been formed on the filter by the sampling process. The latter phenomenon is often the result of a too high air volume sucked through the filter to lower the detection limit at industrial workplaces with high dust exposure; this is done by over-extending the sampling period which leads to an increased particle density on the filter or by increasing the suction velocity which results in an enrichment of coarse particles on the filter making fibre counting more and more difficult. (For a face velocity of *ca* 4 cm/s the theoretical upper limit for the sampled particle size is approximately 35 μm; for a face velocity of 20 cm/s this limit comes up to approximately 80 μm).

For this reason and due to the (low) fibre background of the filter, the above described procedure gives a detection limit at industrial workplaces of *ca* 40 000 fibres/m^3. In the case of workplaces with low dust exposure levels, without coarse particles, the detection limit may be reduced to 20 000 fibres/m^3 or even less.

There is no doubt that the sometimes rather ambiguous counting rules have a decisive influence on the results so that even results obtained by two experienced laboratories may differ by a factor up to 2 in particular cases. More simple and clearer counting rules could improve the situation. Crawford showed that by ignoring attached particles and counting only free fibre ends at an increased aspect ratio of > 5:1 the variation of the counting results could be improved (3).

Because of the subjective influences in fibre counting methods, regular inter-laboratory quality control has been introduced in the form of national and international circulation trials (examples are AFRICA - Asbestos Fibre Regular Informal Counting Arrangement, the WHO/EURO MMMF reference scheme or the RICE scheme in Great Britain). By this means the comparability of results can be achieved as far as this is possible. Since for such circulation trials the 'true' result is unknown, a reference counting or the average of the results of all participating laboratories is considered the best estimate of the 'true' result for a certain sample.

An objective assessment of the method could be obtained by using electronic image analysis. The essential difficulty is the implementing of the ambiguous counting rules into a logical, consistent computer program and in proving that the program yields results comparable to those obtained by the visual fibre counting. The subjective system thus serving to calibrate the objective one, the subjective errors are finally introduced into the objective system.

The fibre counting program for the MAGISCAN system developed in Great Britain permits a semi-automatic analysis - sample change, field change and focusing is still to be done manually, however. The permanent adjusting of the focus which is of decisive importance in visual analysis to consider also fibres beyond the main focus plane, is however impossible - assessment is limited to the main focus plane only.

At BIA we are actually testing our fibre counting program called 'FABIAN', that is run by the IBAS system and which permits automatic sample change, field change and focusing extended by a multi-plane image grabbing enhancing the depth of focus to 5 μm and which enables fibre counting according to the desired counting rules.

In fact, it is possible to improve, within certain limits, the reliability of fibre counting methods using phase contrast microscopy by:

- personnel training
- circulation trials
- clear and simple counting rules
- objective assessment by means of counting machines.

Nevertheless, none of these measures is such as to allow the user to distinguish fibres according to fibre nature and fibre material. Such differentiation is necessary whenever there are unknown fibres or fibre cocktails and the sample cannot be assessed on the basis of a single limit value but, dependent on the fibre materials, is to be assessed on account of different limit values which might even be based on different counting criteria.

In these cases, material analysis of each fibre found is necessary.

Again, phase contrast microscopy would represent a relatively simple solution to the problem; the sample would have to be embedded in a liquid medium of an appropriate refraction index. Ethylcinnamate can be used to identify chrysotile asbestos fibres, which, thus embedded, appear to be bluish when exposed to white light; if, in addition, the fibre turns out to be double refracting in polarized light and morphology is not against it, we can be almost sure to have found chrysotile asbestos. The disadvantage is the reduced optical contrast due to the almost identical refraction index of the fibre and the embedding medium; this is why the method is only suitable for fibres thicker than approximately 1 μm.

3. ELECTRON MICROSCOPY

Over the years, electron microscopy has thus become a current method for fibre analysis. In European laboratories we often find a combination of a scanning electron microscope and an energy dispersive X-ray microanalytical system for routine analysis of workplace samples. The method ZHI/120.46 published by the German Berufsgenossenschaften (4), for example, takes into account the peculiarities of industrial workplaces, whereas VDI 3492 (5) in Germany is to be obeyed to check the efficiency of asbestos clearance measures by controlling measurements and for environmental measurements.

The method (4) is referred to in the German MAK value list as an alternative to the above mentioned European reference method for the assessment of asbestos fibre concentration; this alternative procedure is recommended whenever the sample contains a mixture of asbestos fibres and other mineral fibres that need to be distinguished. Therefore the counting rules defined in the European reference method are to be obeyed also in this case. Since, however, the filter used for phase contrast microscopy is not suitable for electron microscopy, polycarbonate filters with a smooth surface, coated with a thin gold layer, are used.

The nominal magnification of the electron microscope is 2000 x; it is necessary to ensure that even fibres with a diameter of 0.2 μm can be detected; at this magnification, modern instruments give detection limits of $D < 0.1$ μm without problems, assuming that the microscope has been carefully maintained and adjusted. However, such fine fibres no longer give an acceptable signal-to-noise ratio for the energy dispersive X-ray analysis; therefore a detection limit of $D = 0.2$ μm is sufficient. Besides, this value does approximately comply with the detection limit of the European reference method to which this electron microscopic procedure is provided as an alternative.

As for particle density and disturbance due to non-fibrous particles, considerations are in principle the same as for light microscopy. As a result of higher magnification, improved resolution and the negligible fibre background of the filters, low detection limits can be obtained even for industrial workplaces: in low dust level workplaces, detection limits between 15 000 and 10 000 fibres/m^3 can be achieved. A limit of 600 to 300 fibres/m^3 can only be realized in the case of clean rooms such as offices or flats. Organic fibres and particles that are of no analytical interest and are likely to interfere with the counting procedure should be eliminated by low temperature incineration before the evaluation of the sample.

For fibre identification, the elements of the fibre material are excited by the electron beam to emit their characteristic X-ray peaks. Silicon, magnesium, iron, sodium, and calcium peaks are mainly taken into account for asbestos fibre identification, and for other inorganic fibres in addition aluminium, potassium, sulphur and titanium peaks. This requires much higher magnifications than for fibre counting - very often, the magnification comes up to 20 000 x. If possible, it should be ensured that fibre surface contamination that is often found at industrial workplaces due, for example, to mortar, pigments or rubbed-off metal parts, as well as non-fibrous particles close to the fibres, are to be excluded from excitation. In practice, this is not always feasible. Besides, mineral fibres may largely vary in their elementary composition making the unambiguous fibre identification sometimes impossible.

For this very reason and because of the counting rules and their interpretation, electron microscopy has to be considered as a subjective method, too. A regular quality control such as by circular trials is thus definitely desirable.

Again, subjective influences could be largely eliminated by using electronic image analysis. At BIA we have planned complete automation of the whole procedure including field and sample change; the computer will do the image analysis (fibre detection, fibre count and measurement) as well as the controlling of the electron beam in the scanning microscope and the evaluation of the spectra. There is another reason why such an automation is of particular importance for us: in fibre analysis for workplace samples in the Federal Republic of Germany, phase contrast microscopy tends to be increasingly substituted by electron microscopy; while we have a number ratio of 40:60 between electron optical analyses and light optical ones; the same ratio in terms of time expenditure is 80 to 20. Therefore we anticipate a further improvement in our analytical capacities by automation of the procedure.

Whenever dimensions of respirable fibres are to be measured, electron microscopy is to be used. A suitable scanning electron microscopic procedure to determine the distribution of fibre length and fibre diameter has been published as WHO/EURO MMMF reference method (6). Analysis is carried out at a magnification of 5000 x, at which fibres of $D \geq 0.05$ μm and

L ≥ 0.5 μm can be detected without problems. Fibre measurement on the screen is relatively inaccurate and exhausting. At an earlier time, we made such measurements using photomicrographs; for several years now we have made use of electron image analysis in an interactive way for this purpose.

Transmission electron microscopy is almost exclusively employed for special investigations at the workplace in our country. The great advantage of this method lies in its high resolution enabling even the finest fibres down to the chrysotile elementary fibrils (*ca* 0.02 μm) to be detected; in addition to X-ray microscopy, it also permits electron diffraction analysis to be employed to identify thin, crystalline particles on account of their lattice constants. The most important disadvantage of this method compared to scanning electron microscopy is the still increased time expenditure - which is mainly due to lengthy sample preparation - and the expensive technical equipment.

4. OTHER METHODS

Since particle analysis by mass spectroscopy, such as, for example, LAMMA analysis (Laser-microprobe mass analysis) is less common, it will not be discussed in detail within the framework of the present paper.

Direct-reading instruments, in particular light scattering photometers, can be considered to play a certain role for workplace monitoring.

The advantage of such methods consists in the fact that information on particle concentration as a function of time and space are available immediately at the workplace. Generally, these instruments do not yield results comparable to those obtained by any of the above mentioned reference methods. Comparability of results, if possible at all, may only be achieved by calibration at the workplace, i.e. comparison of the method to one of the validated reference methods at the typical workplace conditions. In this case, the fibreunspecific signal given by the light scattering instrument is assumed to correlate with the fibre concentration. Nevertheless, measurements carried out by use of direct-reading instruments do only provide an estimation of the exposure situation; they cannot be used to substitute measurements in accordance with the acknowledged reference methods for exposure assessment.

Direct-reading instruments which are sensitive to fibre shape - for instance because the scattered light pattern is evaluated for fibres aligned for example by an electric field - are also to be calibrated by comparison with a reference method in principle. But of course even such direct-reading instruments cannot give any information on the fibre material.

5. CONCLUSION

From the analyst's point of view, it would be desirable to have identical limit values and assessment criteria for the entire spectrum of different fibres. It would then be possible to do the evaluation without expensive and time consuming fibre material analyses and there would be less room for interpretation. At least, however, the counting criteria should be the same for all kinds of fibres and all possibilities of light microscopy should be utilized before making use of expensive and lengthy electron microscopic analysis. To improve comparability of results, counting rules should be simplified and regular circulation trials should be mandatory. To eliminate subjective influences, advantage should be taken of electronic image analysis.

REFERENCES

(1) Council Directive on the protection of workers from the risks related to exposure to asbestos at work. European Communities Council (1983) [83/447 (EEC)]. Official Journal of the European Communities; (L 263): 25-32.

(2) ASBESTOS INTERNATIONAL ASSOCIATION (1979). Reference method for the determination of airborne asbestos fibre concentrations at workplaces by light microscopy (membrane filter method). London: AIA (Recommended Technical Method No. 1).

(3) CRAWFORD, N.P. *et al.* A comparison of the effects of different counting rules and aspect ratios on the level and reproducibility of asbestos fibre counts. Part I: Report No. TM/82/23; Part II: Report No. TM/82/24. Institute of Occupational Medicine, Edinburgh, December 1982.

(4) Verfahren zur getrennten Bestimmung von lungengängigen Asbestfasern und anderen anorganischen Fasern - rasterelektronenmikroskopisches Verfahren. ZH1/120.46, January 1991. Carl Heymanns Verlag, Köln.

(5) Measurement of inorganic fibrous particles in ambient air. Scanning electron microscopy method. VDI 3492, August 1991. Beuth Verlag, Berlin.

(6) Reference methods for measuring airborne man-made mineral fibres (MMMF). WHO/EURO MMMF Reference Scheme. WHO/EURO Technical Committee, Copenhagen, 1985.

ESTIMATION OF FIBRE CONTAMINATION

T. SCHNEIDER
National Institute of Occupational Health
Lersø Parkalle 105
DK-2100 Copenhagen
Denmark

SUMMARY

Mineral fibre contamination is present if fibres are airborne or have potential for becoming airborne from surfaces or bulk materials. Sampling methods for airborne fibres should be improved. Methods for assessing exposure to thick fibres need to be developed. Deposition on and resuspension from surfaces is an important factor, stressing the need for surface sampling. Methods for determining bivariate log-normal size distributions are given. The limit of detection is in part determined by presence of organic fibres. If most concentrations are below the detection limit, special methods must be used for determining confidence limits. Microscopical analysis of fibres in bulk samples is suitable for detecting very low fibre concentrations, but it is shown that several measurement principles give results that are systematically biased, or have a large variance which only can be reduced by sizing unrealistic many fibres.

1. INTRODUCTION

Mineral fibre contamination is present if fibres are airborne or have potential for becoming airborne from surfaces or bulk materials. To be of health concern the fibres must furthermore be of certain sizes and be durable in the lung fluids [1].

An understanding of the fibre dynamics is a prerequisite for choosing the proper measurement strategy and instrumentation. Spatio-temporal variation is caused by variation in source output rate, by the random nature of dispersion into the room air, and by the removal process. Airborne fibres are removed from the room air by air infiltration, mechanical ventilation and by deposition, mainly by sedimentation onto horizontal surfaces. A useful time-scale is the average residence time Θ of a particle. For a well-stirred room of volume W, room surface A, air exchange rate N, and area averaged fibre deposition velocity v [2]:

$$\Theta(v) = \frac{\int_0^\infty t\, C(v,t)dt}{\int_0^\infty C(v,t)dt} = \frac{1}{N+\frac{A}{W}v} \tag{1}$$

The decay in concentration C resulting from a single burst of fibres into a well-stirred room is

$$C(v,t)=C(v,t=0)\ \exp(-\frac{t}{\Theta(v)}) \tag{2}$$

Stirring can thus not prevent the decay of fibre concentrations. Actually, strong stirring will increase surface deposition. Asbestos fibres can carry high electric charges and their deposition onto charged surfaces, such as plastic sheets used for confinement of asbestos removal areas, can be enhanced manyfold. If $v > 0.1$ NH, where H is ceiling height, more than 10 % of the fibres will escape the general ventilation and settle onto horizontal surfaces. For H=3 m and N=1 h^{-1} this occurs for fibre geometric diameters larger than about 1 µm.

Upon deposition, fibres will be available for resuspension at a rate, which depends on type of surface and transfer of impulse, such as type and level of human activity. For a constant source of S fibres sec^{-1} into the room, the steady state fibre concentration in the air will be:

$$C_\infty = \frac{\frac{S}{W}}{\alpha \frac{v}{H} + N} \quad where \quad \alpha = \frac{F}{F+R} \quad ; \quad Since \quad 0 \leq \alpha \leq 1 \quad : \quad \frac{\frac{S}{W}}{\frac{v}{H}+N} \leq C_\infty \leq \frac{\frac{S}{W}}{N} \qquad (3)$$

F is the fixation (including removal by cleaning) and R the resuspension rate. Expression 2 limits the possible range in concentration for all possible combinations of resuspension and fixation.

In the indoor environment, fibre release is often episodic. An analysis of EPA data on asbestos dust in 49 buildings [3] determined the probability of events with high releases of dust. In the total of 298 two-day samples no concentrations above 0.2 fibres cm^{-3} with length > 5 µm were found. This limited the probability of occurrence and thus set a maximum for the contribution of such episodes to the long term average to less than 0.0015 fibres cm^{-3}. To illustrate further the role of episodes, consider the following example. Suppose that M fibres are dispersed into the room P_e times (P_e episodes) during a reference period T. The time weighted average over the reference period T is:

$$C_{TWA} = \frac{M}{W} P_e \Theta \qquad (4)$$

This shows that one should take as much effort in determining the total number of events and Θ as in determining M. Since surfaces accumulate material from episodical releases, resuspension should be minimized by regular cleaning of floors and other surfaces. Information on cleaning programs and their quality is thus important.

The EEC directive 83/477 allows for calculation of 8 hour TWA or of cumulative doses over 3 month. A general model for estimating concentrations at periods not measured has been suggested [4]. The model uses using exposure modifiers such as time spent at given distance from source, distance from source, and time source is active.

2. SAMPLING AIRBORNE FIBRES

Sampling with an open filterholder is dictated by the analytical method, requiring an even dust distribution on the filter. The filter has to be protected by a cowl extension, but even a conductive cowl acts as a fibre sink due to deposition. To reduce the effect [5], the cowl should be kept as short as possible, the filter diameter as big as possible (the change from 37 to 25 mm diameter was a step in the wrong direction) and the airflow should be as high as possible. Furthermore, fibres should be counted in the central part of the filter [5], because the electrostatic forces only affect the particles that are close to the cowl wall. Only filterholder supports should be used that provide an evenly distributed pressure drop across the entire filter surface. A cowl-rinsing procedure has been suggested [6], but this method does not reclaim fibres that should have been sampled onto the filter, but that deposited on the edge and on the outside of the cowl [7]. Assembly of filter cassettes is a critical step and measurement of pressure drop as a leakage test [8] must be included in the quality control procedure. Leakage may be one of the causes of cowl deposition [9].

The most direct effect of man-made mineral fibres (MMMF) exposure is skin eye, and upper airway irritation from thick fibres. It is therefore somewhat surprising, that methods for assessing exposure to thick fibres has received little attention. The reference method does not conform to the definition of inspirability. Human data on the diameter and length dependence of the irritation in the upper respiratory tract and in the eyes is not available. It is likely that the best measure of this potential is the fibre number, but it is not known whether air samples will be suitable to quantify the irritation risk.

3. SIZE CHARACTERIZATION

The number and size distribution is affected by preparation techniques and instrument parameters. Also counting rules affect the result as any method overestimating the long fibres (e.g. sizing all fibres appearing wholly or partly within the counting area) will obviously overestimate the median length, but, as diameter and length are correlated, also the median diameter [10]. There is an apparent inconsistency in the results obtained by different European laboratories using electron microscopy regarding the size distribution of MMMF [11]. It has recently been shown [12] that analysis on the CRT screen misses thin fibres compared with photographing, even for a modern SEM with frame-store facility.

Size distributions are often reported as the marginal distributions of the diameter and of the length. This does not sufficiently characterize the size distribution since length and diameter are correlated. Size distributions of MMMF [13] and of asbestos [14] are well described by the bivariate log-normal distribution. This distribution is characterized completely by the five parameters GM(D) = $\exp(\mu_D)$, GSD(D)= $\exp(\beta_D)$, GM(L) = $\exp(\mu_L)$, - GSD(L)= $\exp(\beta_L)$, and τ_{DL} = correlation (Pearson) between ln(D) and ln(L). The distribution can be formally written as

$$(D,L) \in LN(\mu_D,\mu_L,\beta_D^2,\beta_L^2,\tau_{DL}) \tag{5}$$

The marginal distributions of diameter and length are log-normal with parameters

$$D \in LN(\mu_D,\beta_D^2) \quad ; \quad L \in LN(\mu_L,\beta_L^2) \tag{6}$$

Given the 5 parameters, the distribution of aspect ratio, aerodynamic diameter, mass and several other measures can then be calculated [15].

Before using the properties of the bivariate log-normal distribution one must test for bivariate log-normality. A test for sample means and variances unknown has been given [16] based on the generalized distances from the data points to the mean vector.

The maximum likelihood estimators μ_D and μ_L, respectively β_D^2 and β_L^2 are:

$$\mu_{MLE}=\frac{1}{n}\sum_{i=1}^{n}\ln(x_i) \quad ; \quad \beta^2_{MLE}=\frac{1}{n}\sum_{i=1}^{n}(\ln(x_i)-\mu_{MLE})^2 \tag{7}$$

The minimum variance, unbiased estimators are

$$\hat{\mu}=\mu_{MLE} \quad ; \quad \beta^2=\frac{n}{n-1}\beta^2_{MLE} \tag{8}$$

used on the marginal distributions. The Pearson correlation between ln(D) and ln(L) is a maximum likelihood estimator of τ_{DL} [17]. The joint distribution of the estimates of μ_D and μ_L is:

$$(\hat{\mu}_D,\hat{\mu}_L) \in N(\mu_D,\ \mu_L,\ \frac{1}{n}\beta^2_D,\ \frac{1}{n}\beta^2_L,\ \tau_{DL}) \tag{9}$$

Notice, that the correlation is not changed. The distributions of the parameter estimators for μ_D, μ_L, $ß_D$, $ß_L$ and τ_{DL} unknown are rather complex and the reader is referred to the literature [17].

4. SURFACE CONTAMINATION

Contaminated indoor surfaces play an important role. They act as reservoirs of settled, airborne fibres and they contribute directly to skin and eye exposure [18]. It has also been demonstrated, that surface sampling is more sensitive than air sampling in detecting presence of MMMF in indoor environments [19].

Many different surface sampling methods are in use [20]. Priority should be given for validation and standardization. Gelatinous foils, specially developed for forensic purposes (foot-print lifters) well suited for sampling MMMF from non-textile surfaces [18]. They have excellent sticking properties. The optical quality is excellent also under phase contrast, since the surface is very smooth and under polarization since the gelatine in isotropic. Unfortunately, autofluorescence interferes with fluorescence microscopy. The sticky foil technique has been used by others for asbestos clearance sampling from surfaces [20]. However, the sampling efficiency for very thin fibres and the lower limit of visibility in an optical microscope needs to be established.

The variation in concentration across a surface can be very large and this may require a large number of samples. A comprehensive method for designing a surface sampling strategy, including determination of optimum spot sample area and number of samples per object has been developed [21]. The sticky foil technique has been developed into a comprehensive surface dust detector system [22].

5. DETECTION LIMIT

The NIOSH lower limit of detection, 7 fibres mm^{-2}, has been determined by taking the mean of blank counts obtained by laboratories participating in the PAT program, plus 3 standard deviations [23]. The PAT samples are synthetic, and do not contain large amounts of interfering particles. In dusty surroundings, the limit is around 0.1 f cm^{-3} [24]. There is an inevitable background of textile fibres from the workers garment, cellulose fibres from wood and paper and many other organic fibres, which cannot be distinguished from asbestos by the reference method. They occur in concentrations up to 0.1 f cm^{-3} [19]. A detection limit below this is an illusion unless these fibre sources can be controlled. Organic fibres cannot be removed by plasma etching of the cleared filter surface will leave a residue looking much like asbestos [25]. They leave a mineral residue, looking like a fibre.

If a TLV is set equal to the detection limit, LOD, conventional decision rules for stating compliance if the upper confidence interval is below, or non-compliance if the lower confidence interval is above the TLV cannot be used. If one only considers compliance for the time interval during which the given sample has been taken and assumes that the only variance is due to the Poisson nature of fibre distribution on the filter, VDI 3492 specifies a lower limit of detection as the upper 95 % Poisson confidence limit for no fibres detected. However, there will be other sources of variance, in particular day to day variations. If most results are below the limit of detection, there is no way by which the distribution type can be estimated. To assume that the concentrations are log-normal at the extreme upper part of the distribution tail can lead to erroneous results [26]. If the distribution is indeed log-normal, censored at LOD, one should use [27] LOD/√2 for all non-detectable values, but LOD/2 if the geometric standard deviation GSD≥3 If a TWA is calculated from measurements of episodic releases and from periods without exposure that are not sampled, much lower concentrations can be obtained reliably. The risk is however, that the assumption of no exposure does not hold.

6. BULK MATERIAL

Microscopy is widely used to determine volume % of a given phase in a matrix, such as low concentrations of asbestos on a filter [28]. Optical microscopy can detect the same fibre sizes as can the reference method for the fibres, once they become airborne, and optical microscopy is used extensively for analysis of bulk samples [29]. Three types of quantitative measures of asbestos in bulk samples are used: fibre number pr unit of weight, projected area and volume percentage. If fibre number is used as a measure of fibre content, the result will depend critically on the preparation procedure. If fibre area percentage as determined by point counting of dust on slides is used [29], the result will depend on fibre dimension and grain size of the matrix and not only fibre volume. Determination of fibre volume percentage in bulk samples or of volume concentration in air samples is usually based on measurement of length and diameter of individual fibres using optical or electron microscopy and postulating an average fibre cross section [28]. If one uses the simple sum of individual fibre volumes one gets a negatively biased result. This is seen from Fig.1, which shows the result of

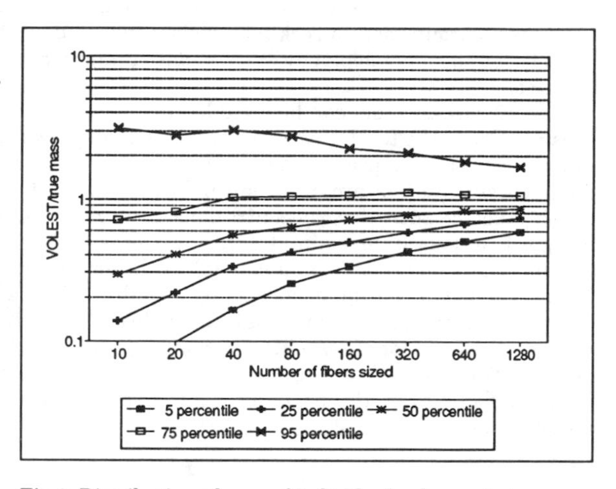

Fig.1: Distribution of sum of individual volumes to true volume for 1000 simulations on the same theoretical sample.

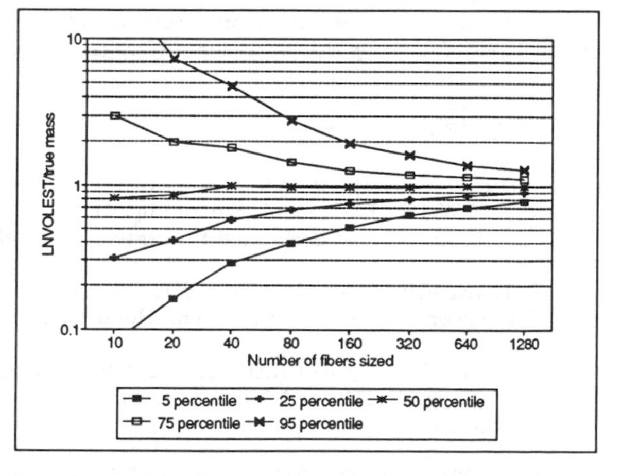

Fig.2:As Fig.1, but log-normal distribution of individual volumes assumed.

1000 computer simulations for a population of fibres with the relevant parameters GSD(D) = 2.5, GSD(L) = 3 and τ_{DL} = 0.5. If the individual volumes are log-normally distributed, an unbiased estimator based on log-transformed individual fibre volumes can be designed (Fig.2). Details of these calculations will be described elsewhere [30]. The figures show that the large, inherent variance can only be reduced by increasing the number of fibres sized. This is unobtainable for very low concentrations and it shows that even though very low fibre concentrations (below 0.001 % v/v) can be detected by optical microscopy, statistics alone gives a large variance. Other uncertainties are added on top of this, such as non-representative bulk samples and biased laboratory preparation procedures.

A ban on import and use of tremolite asbestos containing rock materials was enacted in Denmark in 1990. This created an instant need for analysis of bulk samples and consequently a stop-gap analytical method [31]. Action should be taken to harmonize the definition of hazardous tremolite and other amphibole fibres in bulk samples, and reference samples should be produced. A pilot interlaboratory comparison of tremolite asbestos determination in rock materials conducted by the authors laboratory showed a need for method development and validation and for a more extensive intercomparison exercise.

Dustiness tests can be included to assess the potential for release of airborne fibres relative to the total dust release. It has been shown that as little as 0.001 % asbestos in loose clay soil can produce around 0.1 fibre cm^{-1} in air while at the same time the total dust concentration is around 5 mg m^{-3} [32]. The need for analytical methods, which can reliably detect such low concentrations is obvious. Many laboratory tests have been designed and used to estimate the dustiness of MMMF insulation material during handling. The final test should be a full scale simulation by standardized insulation jobs. This is done on a routine basis, e.g. by Scandinavian manufacturers.

Test of building materials for release of particles during use is much more involved than test for off-gassing. A laboratory and full scale test for measuring the release of fibres from MMMF ceiling boards after installation has been described [33]. This test principle could form the basis for testing release of any type of particles from indoor building materials.

7. CONCLUSION

There is a need for improving present, and developing new methods for sampling and analysis of fibres, particularly for surface contamination and for bulk material.

REFERENCES

(1) POTT, F., BLOME, H., BRUCH, J., FRIEDBERG, K.D, RÖDELSPERGER, K. and WOITOWITZ, H.-J. (1990). Einstufungsvorschlag für anorganische und organische Fasern. Arbeitsmed Sozialmed Praventivmed 25:463-466.

(2) SCHNEIDER, T. and LUNDQVIST, G.R. (1986). Man-made mineral fibres in the indoor, non-industrial environment. Build Environ 21:129-133.

(3) CRUMP, K.S. and FARRAR. D.B. (1989). Statistical analysis of data on airborne asbestos levels collected in an EPA survey of public buildings. Regulat. toxicol. pharmacol. 10:51-62.

(4) SCHNEIDER,T., OLSEN,I., JØRGENSEN, O. and LAURSEN, B. (1991). Evaluation of exposure information. Appl. Occup. Environ. Hyg. 6:475-481.

(5) BARON, P.A., and DEYE. G.J. (1990). Electrostatic effects in asbestos sampling I: Experimental measurements. Am Ind Hyg Assoc J 51:51-62.

(6) BREYSSE, P.N., RICE, C., AUBOURG, P., KOMOROSKI, M.J., KALINOWSKI, M. VERSEN, R,. WOODSON, J., CARLTON, R. and LEES, P.S.J. (1990). Cowl rinsing procedure for airborne fiber sampling. Appl Occup Environ Hyg 5:619-622.

(7) BARON, P.A. (1989). Asbestos measurement and quality control. 6. international colloquium on dust measurement technique and strategy, Jersey, 28-30 Nov. Asbestos International Association pp. 142-151.

(8) FRAZEE, P.R. and TIRONI, G. (1987). A filter cassette assembly method for preventing bypass leakage. Am Ind Hyg Assoc J 48:176-180.

(9) WANG, C.C., FLETCHER, R.A., STEEL, E.B. and GENTRY, J.W. (1990). Measurement of the uniformity of particle deposition of filter cassette sampling in a low velocity wind tunnel. J Aeosol Sci 21 Suppl.1:S621-S624.

(10) SCHNEIDER, T. (1978). The influence of counting rules on the number and on the size distribution of fibers. Ann Occup Hyg 21:341-350.

(11) International Agency for Research on Cancer (1988). IARC monographs on the evaluation of carcinogenic risks to humans. Man-made mineral fibre and radon, International Agency for Research on Cancer, Lyon.

(12) KAUFFER, E., SCHNEIDER, T. and VIGNERON, J.C. (1991). Assessment of the fibre size distributions in the man made mineral fibre industry. In preparation.

(13) SCHNEIDER, T., and HOLST, E. (1983). Man-made mineral fibre size distributions utilizing unbiased and fibre length biased counting methods and the bivariate log-normal distribution. J Aerosol Sci 14:139-146.

(14) CHENG, Y.S. (1986). Bivariate lognormal distribution for characterizing asbestos fiber aerosols. Aerosol Sci Technol 5:359-368.

(15) HOLST, E., and SCHNEIDER, T. (1985). Fibre size characterization and size analysis using general and bivariate log-normal distributions. *J Aerosol Sci* 16:407-414.

(16) KOZIOL, J.A. (1982). A class of invariant procedures for assessing multivariate normality. Biometrika 69:423-427.

(17) ANDERSON, T.W. (1958). An introduction to multivariate statistical analysis, John Wiley and Sons, New York.

(18) SCHNEIDER, T. (1986). Man-made mineral fibers and other fibers in the air and in settled dust. Environ Int 12:61-65.

(19) SCHNEIDER, T., NIELSEN, O., BREDSDORFF, P. and LINDE, P. (1990). Dust in buildings with man-made mineral fiber ceiling boards. Scand J Work Environ Health 16:434-439.

(20) BURDETT, G.J. (1988). Annex 1. Asbestos risk in buildings and building maintenance. In IPCS. International programme on chemical safety. Report of an IPCS

working group meeting on the reduction of asbestos in the environment. 12-16 December 1988, Rome, Italy. World Health Organization, Geneva.

(21) SCHNEIDER, T., PETERSEN, O.H., AASBJERG NIELSEN, A., and WINDFELD, K. (1990). A geostatistical approach to indoor surface sampling strategies. J Aerosol Sci 21:555-567.

(22) SCHNEIDER, T., PETERSEN, O.H., NIELSEN, T.B. and LØBNER, T. (1991). A simple and comprehensive surface dust detector system for assessing the standard of cleaning. This conference.

(23) BARON, P.A. (1991). Phase contrast microscope asbestos fiber counting. Appl Occup Environ Hyg 6:182-182.

(24) ASTM (1983). D 4240-83. Standard test method for airborne asbestos concentration in workplace atmosphere. In Annual book of ASTM standards, section 11, volume 11.03:439-453.

(25) BARON, P.A. and PLATEK, S.F. (1990). NIOSH method 7402-asbestos fibers-low temperature ashing of filter samples. Am Ind Hyg Assoc J 51:A-730-A-731.

(26) BERRY, G. and DAY, N.E. (1973). The statistical analysis of the results of sampling an environment for a contaminant when most samples contain an undetectable limit. Amer J Epidem 97:160-166.

(27) HORNUNG,R.W., REED,L.D.(1990). Estimation of average concentration in the presence of non-detectable values. Appl Occup Environ Hyg 5:46-51.

(28) POOLEY, F.D. and CLARK, N.J. (1979). Quantitative assessment of inorganic fibrous particulates in dust samples with an analytical transmission electron microscope. Ann Occup Hyg 22:253-271.

(29) WEBBER, J.S., JANULIS, R.J., CARHART, L.J. and GILLESPIE, M.B. (1990). Quantitating asbestos content in friable bulk samples: Development of a stratified point-counting method. Am Ind Hyg Assoc J 51:447-452.

(30) SCHNEIDER, T. Estimation of fiber size distributions and of fibre volume in air and in bulk samples by microscopy using fiber size, point counting and test lines. To be submitted for publication.

(31) JØRGENSEN, O. (1991). Asbestos in rock materials. Staub reinhalt Luft (In Press)

(32) ADDISON, J. and DAVIES, L.S.T. (1990). Analysis of amphibole asbestos in chrysotile and other minerals. Ann Occup Hyg 34:159-175.

(33) CHRISTENSEN, G., KNUDSEN, F.E., NIELSEN, P.A., LUNDQVIST, G.A. and SCHNEIDER. T. (1988). Måling af mineralfiberafgivelse fra loftplader. Byggeindustrien 4:3-7.

1399

OCCUPATIONAL HYGIENE INTERPRETATION OF INTERMITTENT DUST EXPOSURES

H. HERRERA, P.O.DROZ and M.P. GUILLEMIN

Institute of Occupational Health Science, University of Lausanne
1005 Lausanne - Switzerland

SUMMARY

Exposure to airborne contaminants varies widely in workplaces, both during the workshift and from one day to the next. Portable direct reading instrumentation is used to study these fluctuations in various workplaces. Interpretation of the gathered data is based on several aspects :
(1) Exposure profiles can be used in actual ventilation estimations
(2) Statistical procedures are used to describe the observed variability with lognormal or combinations of Poisson and lognormal distributions
(3) The sources of variability are studied using a video mixing procedure

In conclusion the combination of today's technology with appropriate statistical technics gives a new understanding of workplace factors, leading to a better assessment of related health risks.

1. OBJECTIVES

Exposure of workers to airborne contaminants in industry varies widely across time. Although these fluctuations were suspected and roughly estimated for a long time, recent advances in direct reading instrumentation have made industrial hygienists more aware of them.

Today's technology (both detectors and dataloggers) allows continuous recording of personal exposures to aerosols. The large amount of data thus collected is often not used to its full extent, and only basic statistical considerations are used.

The objectives of this paper are :
(1) to describe variability in a few typical occupational situations
(2) to show how profiles can be used in ventilation estimations
(3) to discuss the statistical estimation of peaks of exposure
(4) to comment on the validity of the lognormal model
(5) to identify the sources of exposure variability

Points (1) to (4) will be presented in this paper, point (5) refers to the use of a video-signal mixing system, the results of which cannot be shown in this abstract.

2. METHODOLOGY

The instrumentation used in this study included detectors (measurements), dataloggers (data storing), video systems (sources identification) and personal computers (data processing).

Dust monitor

A MiniRAM personal monitor model PDM-3 (MIE Corporation USA) was used in the active sampling mode at 21/min, without a cyclone preseparator. A filter was placed in line after the measuring chamber for gravimetric calibration checks.

Datalogger

A Metrosonics DL-332 (Metrosonic Inc. USA) datalogger was connected to the analog output of the MiniRAM instrument. The datalogger was operated with a measuring frequency of 1 s^{-1}, and data were averaged over periods of 1 to 5 minutes.

Video recording

A home made mixing system was developed to allow simultaneous visualisation of the MiniRAM signal and the video image of the worker. This system includes :

(1) a home made radio transmitter for the MiniRAM analog signal
(2) a home made receiver and an RS-232 signal converter
(3) a Sony video image interface (HB-G900P) and computer (HBI-G900P)
(4) a Panasonic video camera, model WVP-F-102

A complete program was developed to visualized on the side of the video screen the signal from the direct reading instrument in the form of a bargraph.

3. RESULTS AND DISCUSSION

Figure 1 gives 4 examples obtained with the monitoring system described above : a truck driver exposed to diesel particulates, a baker exposed to flour dust, a cabinet maker exposed to wood dust and an assistant exposed to dust and fumes in an indoor firing range. Are also shown on Figure 1 the frequency distributions, with the corresponding lognormal plots, and the autocorrelation functions (ACF) (Mc Dowall et al., 1980). The frequency distribution plots indicate that the data do not perfectly fit the lognormal distribution, although the general pattern looks similar. Also the ACF shows that data are autocorrelated showing that exposure at a given time depends on the previous situation.

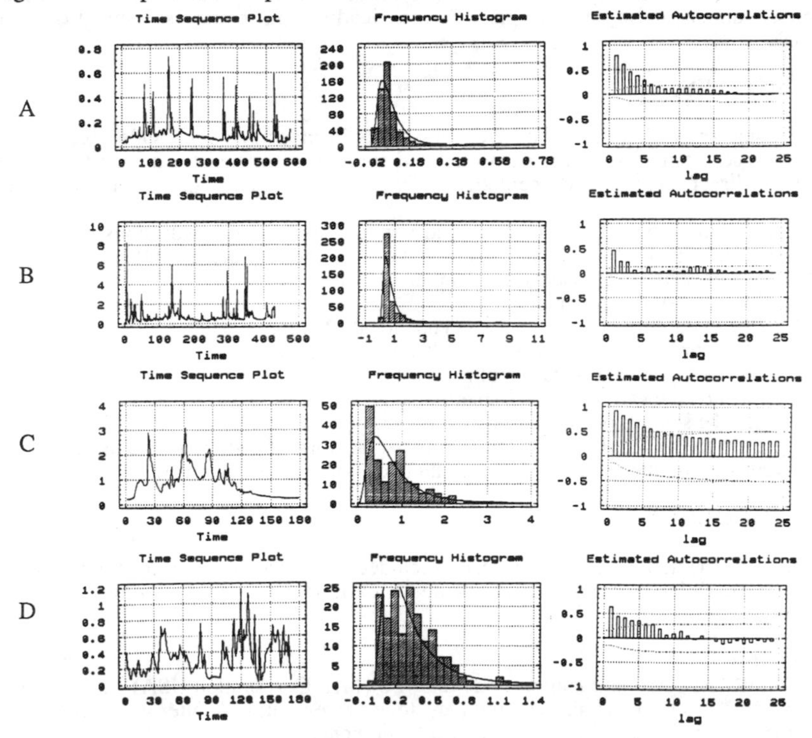

Fig. 1. Selected examples of exposure profiles, frequency distributions, and autocorrelation plots for a truck driver (A), a baker (B), a cabinet maker (C) and a indoor firing range (D).

Estimation of ventilations

If exposure is stationary, i.e. the mean does not change with time, a first order autoregressive model (ARIMA) (Mc Dowal et al., 1980) can be used to described the time-series of exposure data :

$$C_t = \mu\,(1-b_1) + b_1\,C_{t-1} + a \qquad (1)$$

with $C_{t,t-1}$ = concentrations at time t or t-1

μ = exposure concentration

b_1 = first lag autoregression coefficent

a = term describing random fluctuations

On the other hand, a simple physical model for the workplace can also be developed : the workplace exposure model (WEM). With perfect mixing in the room, or at least in a certain volume including the measuring locations, the decay of concentration, without any source of contaminant, can be described by :

$$C_t = C_o\,e^{-Ft/V} \qquad (2)$$

with Ct,o = concentration at time t or o

F = ventilation rate of the workplace

V = volume of the workplace

A comparison of both formula indicates that air exchange rate F/V (h^{-1}) can be estimated by the formula :

$$F / V = -\ln(b_1) / t \qquad (3)$$

It has to be mentioned that this also includes clearance by sedimentation or other processes for aerosols. Table I gives results obtained for the above situations.

Table I. Estimation of ventilation rates form exposure profiles

Industry/operation	t [min]	b_1	F/V [h^{-1}]	Comments
Truck driver 1	1	0.78 ± 0.03	15	smoker
Truck driver 2	1	0.99 ± 0.01	0.6	non smoker
Baker	1	0.44 ± 0.04	49	sedimentation
Cabinet maker	5	0.93 ± 0.03	0.9	natural ventilation
Indoor firing range	5	0.62 ± 0.06	5.7	measured 9.0 h^{-1}

Description of exposure

Continuous exposure recording can be used to decribe the frequency of peaks of exposure. This is often done by simple observation or frequency distribution plots. Lognormal distributions are also used to summarize the data (Jaha, 1987). Table II presents the observed frequency based on the lognormal assumption.

Table II. Observed frequencies for the lognormal top 10, 5 and 1 percentiles

Industry/operation	10%	5%	1%
Truck driver	9	7	3
Backer	11	6	3
Cabinet maker	11	3	0
Indoor firing range	6	2	0

In the situations show in Table II, the lognormal assumption seems to be relatively satisfying although discrepancies may be rater important in some situations. Nevertheless, in situations where emission is very intermitttent, one can consider exposure as being the result of occurence of small incidents of variable sizes. To describe this situation, a combination of Poisson (frequency of the incidents) and lognormal distributions (magnitude of the incidents) could be used. Figure 2 presents typical simulated situations. If the autocorrelation structure is determined, more information may be derived from the distribution parameters (geometric, mean, standard geometric deviation) and the exposure pattern may be better undertood and interpreted (Spear et al., 1986).

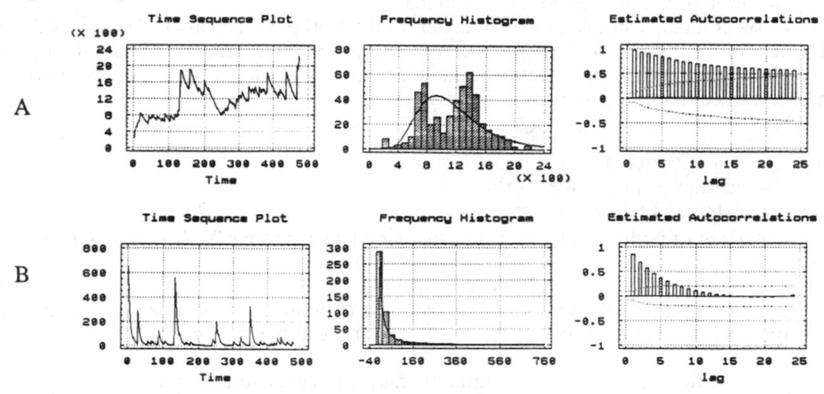

Fig. 2. Simulated exposure profiles based on Poisson and lognormal distributions. (A) Frequency of the incidents has been set to 1/5 min. and the variability of incidents to a GSD of 3.0, together with a ventilation rate of 1 hr^{-1}. (B) In this case, frequency of the incidents has been set to 1/15 min and the variability of incidents to a GSD of 3.0, together with a ventilation rate of 10 hr.

4. CONCLUSION

Recent advances in monitoring technology allows industrial hygienists to gather better information on workplace exposures. Statistical examination of exposure profiles can provide the occupational health specialist with more information than simple descriptive statistics. Time series analysis of exposure data has been shown to be important for industrial hygenists (Francis et al., 1989). More specifically, some technics exist to allow estimation of actual ventilation rates during the measurements and to study the frequency of peaks of exposure. Systematic recording of exposure in various occupational situations will also reveal that in some cases the lognormal model of exposure fluctuations is no more valid, and alternate models have to be identified. When the toxicological properties of the considered agent are taken into consideration, it becomes possible, on the basis of pharmacokinetic and pharmacodynamic considerations, to better evaluate the health risk associated with such exposures.

REFERENCES

Jahr I. (1987) Calculations for workplace measurements and other data with a lognormal distribution Staub Reinhalt, Luft **47.**

Francis M., S. Selvin, R. Spear and S. Rappaport (1989) The Effect of Autocorrelation on the Estimation of Workers' Daily Exposure Am. Ind. Hyg. Assoc. I. **50**, 37-43.

Mc Dowall D., R. Mc Cleary E.M. Meidinger E.M. and R.A. Hay Jr. Eds (1980) Interrupted Times Series Analysis. Sage Public Series/Number 07-021 London.

Spear R.C., s. Selvin and M. Francis (1986). The Influence of Averaging Time on the Distribution of Exposures Am. Ind. Hyg. Assoc. I. **47**, 365-368

1400

EVALUATION OF A DENUDER FOR PERSONAL SAMPLING OF INORGANIC ACIDS

A. HULTMAN and B-O. HALLBERG

National Institute of Occupational Health
171 84 Solna, Sweden

SUMMARY

The aim of this work was to develop a convenient personal sampler for inorganic acids. For this purpose we have used a denuder consisting of a 10 cm long glass tube coated with KOH and an impregnated hydrophilic filter. Laboratory tests with HCl (gas), HNO_3 (gas+aerosol) and H_2S0_4 (aerosol) showed that the sampler effectively separates gas and aerosol. All samples were analysed by ion chromatography. It is our belief that the denuder technique provides a useful method for personal sampling of mixtures of aerosol and gas.

INTRODUCTION

Inorganic acids appear depending on their vapour pressure as vapour and/or aerosol. Since the aerosol fraction often represents a greater health hazard than that of the vapour fraction, it is important to collect them separately. Other gaseous substances may also be absorbed by an aerosol, which will transport these to the respiratory system. Animal tests have shown that the respiratory resistance effect from sulphur dioxide increases with higher humidity. Gases are thus often less harmful alone than in the presence of an in itself harmless aerosol (1, 2).

With the denuder technique the gas is first separated from the aerosol in a coated tube. The aerosol is then collected on a filter. The gas molecules will diffuse into the wall coating, while the aerosol, due to its much lower diffusion rate, will pass through the glass tube. For the denuders to separate the aerosol from the test air it is important that the flow is laminar. The gas collection efficiency of the tube is influenced by tube length and diameter and the test gas flow rate.

Tests with nitric acid gas and aerosol, hydrogen chloride gas and sulphuric acid mist have been performed in the laboratory. The particle size distribution of the generated aerosol was 0.25-7 μm (3), and the denuder satisfactorily separated gas and aerosol.

EXPERIMENTAL

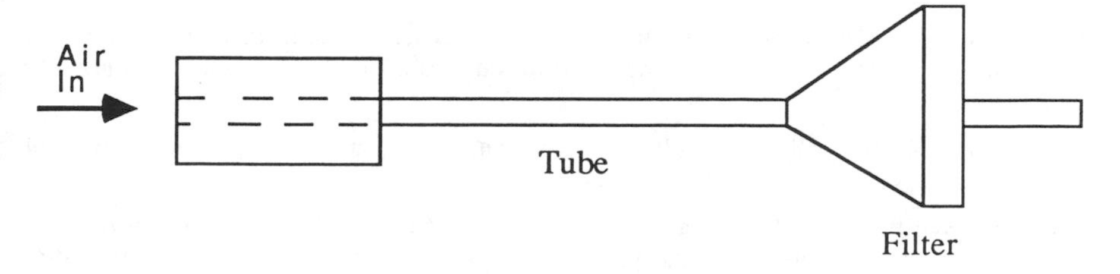

Fig. 1. Denuder

The denuder (see Figure 1) used in this work consisted of a glass tube (length 100 mm, id 6 mm) followed by a filter (Millipore GVWP 0.22 μm dia. 25 mm) impregnated with potassium hydroxide 50 g/l (4). The glass tube inner wall was also coated with potassium hydroxide 80 g/l methanol. Air was pumped through the sampler at a flow rate of 0.2 l/min. This low flow rate excludes particles with a diameter larger than 10 μm. Sampling time was between 30 and 120 min. All tests were performed in a test chamber (400 l) made of polycarbonate. Aerosol was generated by nebulizing acid. In all tests, fritters coupled in series were used as a reference sampling method. All samples were analysed by ion chromatography.

RESULTS

To establish collection efficiency for the glass tube, vapour of hydrogen chloride in dry air was pumped through the sampler. No acid was found on the filters while 100% was recovered from the walls of the tube. A similar test with only filters showed that their ability to collect vapour was 99%. To evaluate the filter collection efficiency for particles, tests were performed with sulphuric acid mist. In this case no acid was recovered from the tube walls and 99% from the filters.

The denuder was further tested for sampling of hydrogen chloride and nitric acid at different humidities. At very high relative humidity, >70% RH, condensed water wetted the wall coating, but this did not influence the recovery of the test gas. Table 1 shows the aerosol/gas ratio recovered from the denuder at different humidities for nitric acid.

Table 1. Distribution Aerosol/Gas for HNO_3, Mean Values from 6 Denuders Sampling time 60 min

RH %	Conc[1] ppm	Gas %	Aerosol %	CV[2] %
80	1.10	70.9	26.9	13.8
45	1.65	102	[3]	3.8

1) Value from two fritters coupled in series
2) Coefficient of variation for the denuder
3) Less than 0.8 μg (= 2 times the background.value of the filter)

In our opinion the developed denuder offers a convenient method for sampling and separating aerosols and gases of inorganic acids effectively.

REFERENCES

(1) AMDUR, MARY, When one plus zero is more than one. Am. Ind. Hyg. Assoc. J. 46(9): 467-475 (1983).

(2) AMDUR, MARY, McCarthy John, and Gill Michael, Effect of mixing conditions on irritant potency of zinc oxide and sulfur dioxide. Am. Ind. Hyg. Assoc. J. 44 (1):7-13 (1983).

(3) MAY K, The collision nebulizer: description, performance and application. Aerosol Science,Vol.4, 235-243 (1973).

(4) GUNDERSON, ELLEN and ANDERSSON, CLARINE. Collection device for separating airborne vapor and particulates. Am. Ind. Hyg. Assoc. J. 48(7):634-638 (1987).

1401

MULTIFRACTION AEROSOL SAMPLING: THE POTENTIAL OF THE PERSPEC SAMPLER

L.C KENNY, C. VOLTA, V. PRODI AND F. BELOSI*
Physics Dept., Bologna University,
Via Irnerio 46, 40126 Bologna, Italy

*Lavoro e Ambiente SCrL,
Via Mazzini 75, 40137 Bologna, Italy

SUMMARY

The results of an experimental study of the PERSPEC multifraction aerosol sampler are used to devise an optimised scheme for subviding the sampled aerosol into fractions according to CEN conventions. The residual bias in sampled mass is comparable to that of existing sampling instruments.

1. INTRODUCTION

European (CEN) and International standards have recently been agreed for the sampling of workplace aerosols with a view to assessing their potential health effects. Three sampling conventions are defined, for the inhalable, thoracic and respirable aerosol fractions. The majority of current sampling instruments collect either the inhalable or the respirable fraction, not both. However the PERSPEC personal sampler is designed to collect all three aerosol fractions simultaneously. Multifraction sampling, apart from providing low-resolution estimates of the size distribution of the sampled aerosol, may allow the sources of workplace aerosol to be identified and better controlled, and the health effects arising from exposure to be better understood. This paper considers how the performance of the PERSPEC can be optimised to the new CEN sampling conventions.

2. EXPERIMENTAL STUDY OF THE PERSPEC

The PERSPEC works by drawing aerosol into an annular nozzle, the centre of which is filled by a clean air core. The airstream is impacted perpendicularly onto a filter situated below the nozzle, thus forcing the aerosol-laden air to turn through a tight bend, during which the particles are separated according to their aerodynamic diameter; small particles follow the air streamlines and deposit towards the outside of the filter, whilst large particles, due to their inertia, diverge from the streamlines and deposit closer to the centre of the filter. The positions in which particles would land on the filter have been predicted using a simple physical model of the instrument(1).

An experimental study was carried out in which a group of five PERSPEC specimens, operating at an aerosol flow rate of 1.75 LPM, was exposed to test aerosols consisting of atomised suspensions of monodisperse polystyrene latex particles.

Particle sizes between 1 μm and 18.6 μm were used. The PERSPECS were commercial specimens and were used with their associated pumps and flow splitters. However, the flow circuit was modified to include an expansion volume in the clean air supply line, as pulsations from the pump were found to have a serious effect on the size separation characteristics of the sampler.

The PERSPECS were loaded with neutralised cellulose ester filters, which were mounted after exposure for microscopic examination using the DMF method(2). The position of the particle deposit along the filter radius was measured using an optical microscope.

For each monodisperse particle size, the particles were seen to deposit in a band whose location and width depended on the particle size, as predicted by the physical model. The width of the band, particularly for smaller particles, was larger than the model predicitions. By counting the number of particles per field in a selection of slides, it was established that the radial number distribution was approximately gaussian, and that the deposit edge positions identified subjectively enclosed roughly 95% of the deposited particles. Future work will include the counting of the slides with an automated image analysis system, in order to remove any subjective influence from the results obtained.

For particles sizes above 10 μm, the particles were seen to be spread over a wide area of the filter and not to have deposited in a narrow band as predicted physically. This result was attributed to the poor adhesion of the particles to the smooth cellulose ester filters, an effect which would probably not occur with the glass fibre filters used with the sampler in the field. A further experimental study to examine the particle bounce problem in more detail is planned.

3. CALCULATION OF PERSPEC SAMPLING EFFICIENCY AND BIAS

The aerosol which is collected on the PERSPC filter can be subdivided into the various fractions by physically cutting it into suitably-shaped parts. Calculations were made to assess the effective sampling efficiencies for the PERSPEC using a range of rectangular and circular cutters of various sizes, shapes and positions. The gaussian distribution of the deposit was taken into account in the calculations. For particle sizes above 10 μm, the model predictions for the particle deposit were used in place of the experimental data, due to the unresolved particle bounce problem described above.

Changes in the size, shape and position of the filter cutter allow one to obtain considerable flexibility in the effective PERSPEC sampling efficiencies, in a way which is not possible with conventional size-separating samplers. For the cutter shapes giving apparently good agreement with the sampling conventions, the bias in aerosol mass sampled was calculated for a range of unimodal lognoral aerosols. The best cutter shapes were chosen by comparing the bias maps obtained.

4. RESULTS

For the respirable convention, the best match to the sampling convention was obtained using a circular cutter of radius 14.5 mm, followed by doubling the aerosol mass on the on the relevant portion of the filter, i.e. the part outside the cutter edge. For the thoracic convention, the best result was obtained with a circular cutter of radius 7.5 mm, offset by 5 mm from the centre of the filter. Note that in each case, the calculations assumed that the process of selecting particles from the airborne state onto the PERSPEC filter matches the CEN inhalable convention exactly; preliminary wind tunnel tests suggest that this assumption may be valid for low-velocity winds(3), but a more thorough experimental study will be necessry to quantify the PERSPEC inlet efficiency in more detail.

The bias calculated for the respirable fraction is shown in Figure 1. The bias is within acceptable limits (i.e. bias<10%) over a wide range of size distributions, and the performance compares very favourably with samplers currently in regular use such as cyclones(4). The thoracic bias is higher, as shown in Figure 2, particularly for size distributions in which the thoracic fraction is less than 30% of the inhalable aerosol. However, in some cases it may be possible to reduce the thoracic bias by the use of correction factor derived from the ratios of the three fractions sampled.

5. CONCLUSIONS

The results obtained show that the PERSPEC has the potential to sample the respirable fraction of workplace aerosols with very low bias with respect to the new CEN sampling convention. The thoracic aerosol fraction may be sampled reasonably well where it forms more than 30% of the inhalable aerosol, but is otherwise significantly oversampled. The practical realisation of these results requires the manufacture of a suitable filter cutting tool, and the incorporation of a pulsation damper into the PERSPEC flow circuit. Although more work is needed for a thorough validation of the sampler, in particular to evaluate its inlet efficiency and precision, it appears capable of meeting both the performance requirements and the practical requirements of a multifraction aerosol sampler.

REFERENCES

(1) PRODI, V., BELOSI, F., MULARONI, A. and LUCIALLI, P. (1988). PERSPEC: A personal sampler with size characteristisation capabilities. Am. Ind. Hyg. Assoc. J., 49(2), 75-80.

(2) LE GUEN, J.M. and GALVIN, S. (1981). Clearing and mounting techniques for the evaluation of asbestos fibres by the membrane filter method. Ann. Occup. Hyg., 24(3), 273.

(3) VINCENT, J.V. (1989). Aerosol sampling - Science and practice. John Wiley UK, ISBN 0 471 92175 0.

(4) LIDEN, G. and KENNY, L.C. (1991). The performance of respirable dust samplers: sampler bias, precision and inaccuracy. Ann. Occup. Hyg. (in press).

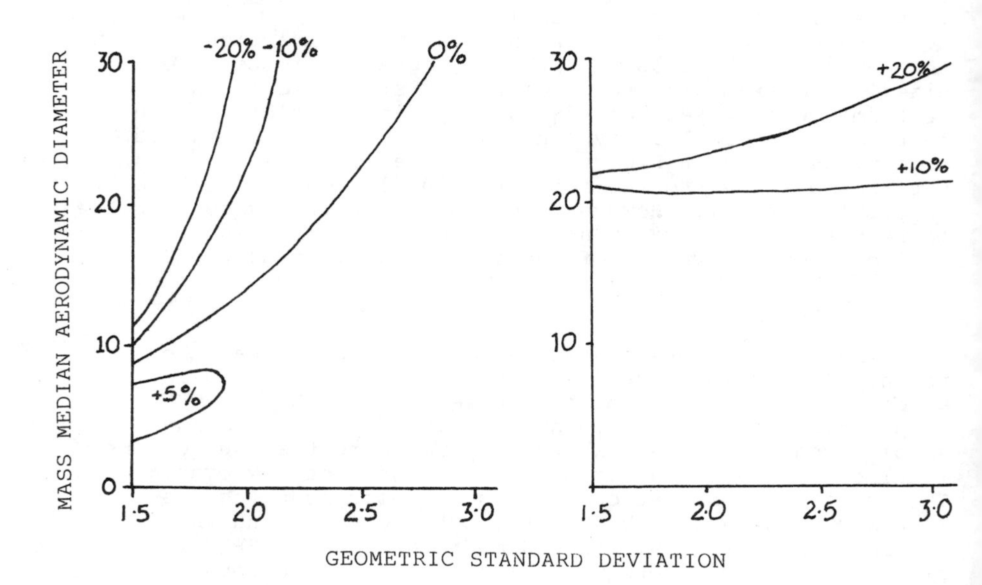

Fig. 1: Bias in the respirable mass sampled with PERSPEC, relative to the CEN sampling convention, for a range of lognormal aerosols.

Figure 2: Bias inthe thoracic aerosol mass sampled with the PERSPEC, relative to the CEN sampling convention, for a range of lognormal aerosols.

CONSTRUCTION OF A PORTABLE FIBRE MONITOR MEASURING THE DIFFERENTIAL LIGHT SCATTERING FROM ALIGNED FIBRES

A.P. ROOD E.J. WALKER & D. MOORE

Health and Safety Executive, Occupational Medicine and Hygiene Laboratories, 403 Edgware Road, London NW2 6LN, UK.

SUMMARY

A cheap portable instrument to measure airborne fibres has been constructed. Fibres are sucked into the instrument and deposited on a standard microscope slide aligned with the flow. The light scattering signal from this deposit is used to indicate the number of fibres present. The fibres can be identified by polarised light microscopy on the slide, and the instrument has been calibrated against the European Reference Method.

DESIGN AND CONSTRUCTION DETAILS

A portable instrument needs to be battery operated, simple in operation and robust in use. We have stripped down a standard 4l/min sampling pump (Rotheroe and Mitchell L2SF) and fitted it with associated electronics for fibre precipitation and sensing into a hand-held unit running from rechargeable nickel-cadmium cells. The instrument top block (see Fig. 1) is cast from epoxy resin in which the corona needles, the precipitator electrode, and the detectors are set. The baseplate on which the block seals is made from aluminium.

Fibre alignment in the instrument is based on the aerosol spectrometer of Prodi et al (1982), but a simplified design of inlet was used to accelerate the incoming air; two razor blades set with their edges parallel and 1mm apart are angled at 20° to the incoming air (Fig. 1), and cause the fibres to align in the direction of flow. The sheath air used by Prodi to surround the aerosol was not found necessary in our design, but mechanical flatness and accurate positioning of the blades was found important. The air containing the aligned fibres then passes through a corona discharge generated by three needles operating at 3.5KV with respect to the baseplate (Whitby, 1961). Positive ions collide with the incoming dust and fibres giving them an overall charge. A precipitator plate downstream from the charging region precipitates the dust and fibres (Liu et al 1967) onto a removable glass slide under the sensors. Both corona and precipitator voltages are generated by battery driven generators (Brandenberg 512AA).

The fibres precipitated onto the glass slide retain their alignment and the difference in light scattering parallel and at right angles to the flow (Timbrell & Gale, 1980) is used to determine the number of fibres. A calibration curve can be derived by counting, in the PCOM fibres sampled from a range of concentrations. Four sensors (RS BPW21) are mounted below the slide to detect the scattered light emanating from a green light emitting diode collimated to a spot in the middle of the slide, the optimum angle of these detectors has been determined from a separate experiment looking at the angular dependence of the scattered light. The collimation produces a 3mm spot of uniform intensity within the area of dust deposit. A matt black surface under the microscope slide serves to absorb light transmitted throught the glass. Signal differences can be observed by phase-locked amplification of the output from detectors at right angles to each other, holding the signals electronically, and displaying the difference on a voltmeter. The signal sum from all four detectors can also be acquired and displayed.

Identification of the fibres on the slide is possible using polarised light microscopy. A drop of a suitable refractive index liquid is introduced between the slide and a cover-slip by means of a pipette, and the dispersion observed in the normal way. A similar observation under phase-contrast conditions allows the fibres to be counted using the European Reference Method a cross-check on the instruments performance.

RESULTS

The performance of the instrument has been checked on two groups of fibres (asbestos and MMMF), and on a range of tests dusts including talc. These tests have been carried out in a laboratory dust box and during field trials of the instrument. The figure below shows the sum and difference signals from the instrument for a range of aerosols, two distinct forms of response can be seen, the fibres show a strong positive slope, and the compact dusts (including talc) a flat response. The slope of the response reduces with fibre diameter, and the curly nature of chrysotile has a similar effect.

Exposure standards can be measured in just a few minutes even in the presence of non-fibrous background dusts. The UK clearance level for asbestos in buildings of 0.01 f/ml, can be measured in about 300 minutes using this instrument.

REFERENCES

Liu, B Y H; Whitby K T; and Yu, H S (1967) Rev. Sci. Ins. 38.1, 100-2

Prodi, V; De Zaiacomo, T; Hochrainer, D; and Spurny, K; (1982) J. Aerosol Sci. 13.1, 49-58

Timbrell, V; and Gale, R W; (1980) "Biological Effects of Mineral Fibres I" IARC Scientific Pub. No. 30, 53-60

Whitby, K T; (1961) REv. Sci. Inst. 32.12, 13251-55

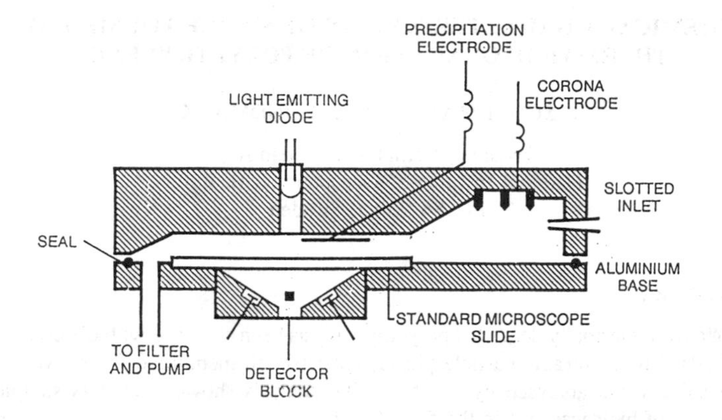
PRECIPITATION ELECTRODE
CORONA ELECTRODE
LIGHT EMITTING DIODE
SLOTTED INLET
SEAL
ALUMINIUM BASE
STANDARD MICROSCOPE SLIDE
TO FILTER AND PUMP
DETECTOR BLOCK

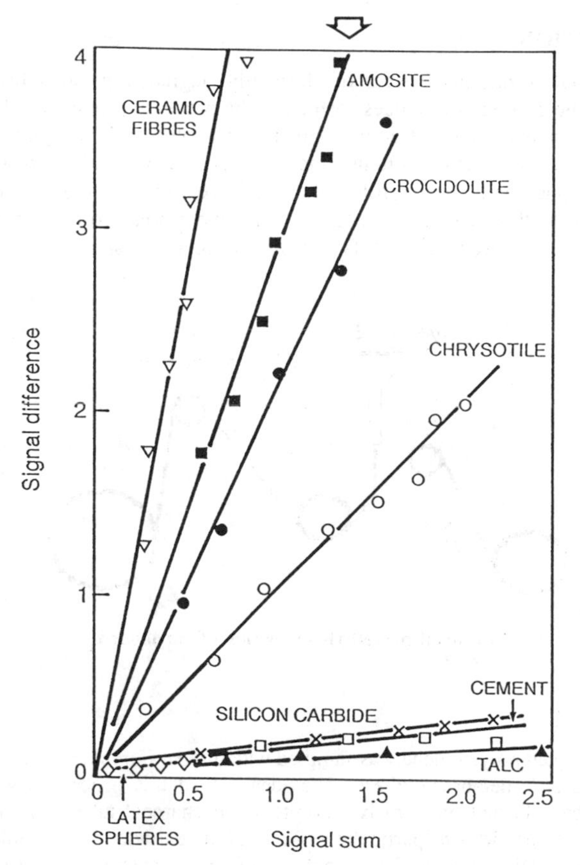
5 f/ml for 5 min
CERAMIC FIBRES
AMOSITE
CROCIDOLITE
CHRYSOTILE
CEMENT
SILICON CARBIDE
TALC
LATEX SPHERES
Signal difference
Signal sum
0
0.5
1.0
1.5
2.0
2.5
1
2
3
4

CHEMICAL CHARACTERIZATION OF SMOKE FORMED BY THERMAL DEGRADATION OF POLYETHYLENE

J. SCULLMAN and B-O. LUNDMARK

Division of Analytical Chemistry
National Institute of Occupational Health
171 84 Solna, Sweden

SUMMARY

Smoke from thermally degraded polyethylene has been analysed with different analytical methods. The respirable particle phase, collected on membrane filters, was extracted with solvent and analysed by GC/MS. The analysis showed that this sample mainly consisted of hydrocarbons in the field C_{20}–C_{30}.

1. INTRODUCTION

Research about hazardous effects of thermally degraded plastics has mostly been directed towards the gaseous substances formed in the processing industry (1). In our study we have focused on the aerosol fraction, inspired by another investigation about diesel fumes where it was shown that no lung function decrease was indicated when the diesel vehicles were equipped with aerosol filters that effectively trapped the exhaust particles (2, 3). Thus the aim of this work was to investigate the smoke from thermally degraded polyethylene polymer in order to identify hazardous compounds.

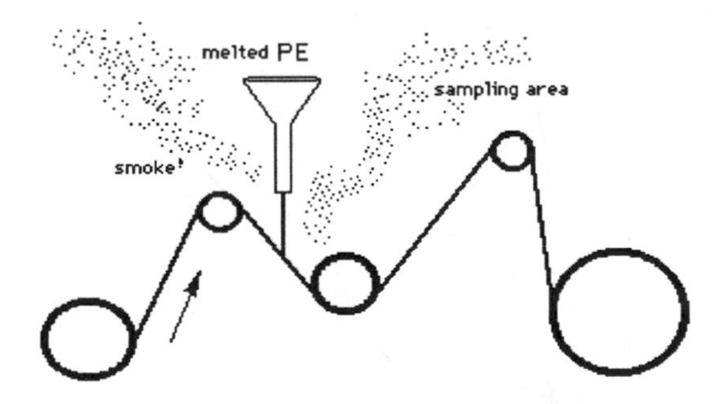

Industrial polyethylene coating of cardboard

2. EXPERIMENT

Smoke form heated polyethylene was sampled with three different methods: charcoal tubes, diffusive monitors and membrane filters. The idea behind this procedure was that organic vapours were to be collected on diffusive monitors and charcoal tubes. The latter were also supposed to collect some airborne particulates. Most of the particle phase should be trapped on the membrane filters. Diffusive monitors, charcoal tubes and membrane filters were desorbed and analysed by GC/MS, ion chromatography and HPLC. Some filters were also examined by electron microscopy. The particle size distribution was determined with a particle counter. Particles with a diameter less than 0.1 μm could not be detected (Figure 1).

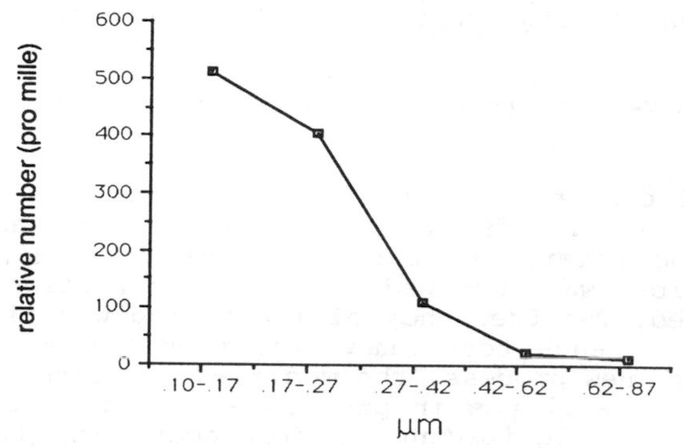

Fig. 1. Particle size distribution

3. RESULTS

The diffusive monitor analysis showed no traces of organic vapours at all. The charcoal tubes contained small amounts of more volatile compounds. The membrane filters contained at least one homologous series of hydrocarbons, C_{20}–C_{30}. These hydrocarbons were the main constituents. Laboratory generated smoke from the polyethylene gave a very complex mixture. After separation on a silica column the unpolar fraction when analysed also gave one or more homologous series. The dominant series seems to consist of n-alkalenes. The electron microscopic examination showed that the particles were well rounded and with a diameter less than 1 μm. The size distribution measurement indicated that the particles had diameters from about 0.8 μm down to 0.1 μm. The distribution was greatly displaced towards smaller diameters.

4. CONCLUSION

With the experimental conditions used it was found that smoke from thermally degraded polyethylene mainly consists of a respirable mixture of homologous series of high boiling hydrocarbons principally in the field C_{20}–C_{30}.

REFERENCES

(1) HOFF, A. *et al.* (1982). Degradation products of plastics. Scand. j. work environ. and health 8.

(2) ULFVARSON, U. *et al.* (1985). Hälsoeffekter vid exponering för motoravgaser. Arbete och Hälsa 5.

(3) ULFVARSON, U. *et al.* (1988). Minskad lungpåverkan hos stuveriarbetare när dieseltruckarna försågs med filter. Läkartidningen 85 (15).

1404

AIRBORNE MICROORGANISMS IN THE PHARMACEUTICAL INDUSTRY

YOUSEFI V
NCOH P O Box 4788,
Johannesburg, Republic of South Africa (RSA)
and
Dias L
Squibb/B-Meyers Ltd P O Box 48, Isando Johannesbug RSA

SUMMARY
An environmental monitoring programme was designed and adhered to in order to establish an acceptable base-line level. Whenever, the base-line was violated, process and production was stopped, and a correction action was requested. The frequency of monitoring was dictated by the history of each test site. The need for modified testing were for: new process, the modified existing processes, new component, a change in procedure, the installation of new equipment, significant differences in test results, facilities revision, start-up operation and personnel changes.

INTRODUCTION

It is known that the quality of the environment ultimately affects the quality of a product. In the pharmaceutical industry, contamination of the environment is attributable to various sources including manufacturing activities, facility design, unhygienic operation, equipment, materials and type of process.

Contaminated products have led to outbreaks of illness, and in some instances human disability/death. It may lead to allergies, ineffective preservatives, loss of drug potency, problems such as formation of toxic substances in cosmetics and pharmaceuticals, and pyrogens in medical devices. These changes in turn, may lead to the recall, reprocessing or destruction of the product.

MATERIALS

1. SAS (Surface-Air-System) air sampler(1)
2. Biotest RCS (Reuter-Centrifugal-System) air sampler(1)
3. Agar petri-dishes
4. Medium coated plastic strips

METHODS

The collection of microorganisms is complicated by the viability of the sample. Special handling, processing and analytical techniques are needed for the enumeration and identification(1-4,6).

RESULTS

Figure 1 is one example of weekly trends over eight month periods of sampling in Raw Material Warehouse. A similar results were obtained for each environment. The genera of the resultant isolates were determined by colony characteristics. Table 3 is the result of a parallel sampling with two different samplers and media.

DISCUSSION

Cladosporium in soil and aerial inhabitants is associated with chromomycosis in humans; a more or less localized and chronic infection of the skin.

Bacillus species are strictly aerobic gram positive rods. These endospore formers can be of economic loss to man, since they are capable of spoiling paper, food, drugs and wood. Their elimination in the manufacturing environments is essential.

Other aerial contaminants which were isolated in a pharmaceutical environmental were common soil inhabitants such as Acinetobacter, Alternaria, Aspergillus, Phoma, Rhodoturula, Pseudomonas and Mycelia Sterilia (Table 3). Phoma and Mycelia Sterilia are plant pathogens and hence their presence has no serious implications in a pharmaceutical plant. Acinetobacter calcoaccticus on the other hand can be a nuisance in a penicillin production area since it is highly resistant penicillin. Several species of Alternaria are human pathogens and like Aspergillus terreus should be absent in community services such as hospitals and clinics. Aspergillus in particular causes various types of aspergillosis of which the most serious is pulmonary aspergillosis, similar to tuberculosis.

Baselines are the mean CFUs/m3 obtained over 14 consecutive days for each site, which serve as guidelines for comparing routine results. Once routine results exceed grade limits corrective action is required (Table 1), to re-establish acceptable microbial levels. The proceeding action taken depends upon the extent by which a grade is exceeded. Trend analyses, (Figure 1) illustrate the frequency at which a grade is exceeded and reveal seasonal changes.

An action report (Table 2) serves to communicate the action required to the person responsible for ensuring that hygiene levels are improved, within the recommended time. This in turn leads to a safe product being manufactured and/or may potential pathogens which could either degrade a product or infects its users.

CONCLUSIONS

An Environmental Control Programme, be it in a pharmaceutical company or in a community service, creates confidence in the hygiene practices within an operation and in the safety of the drugs/cosmetics produced, stored, transported/distributed and prescribed. Most importantly it guarantees the health of the community utilizing pharmaceutical products/drugs.

REFERENCES

1. Mark A Chatigny "Sampling airborne Microorganisms" Air Sampling Instruments ACGIH 1980
2. H. Seyfarth (1981) "Determination of Microbiological Quality of Air in the Production of Sterile preparations" Paper presented at the Sterilization Debate. Mayorca, March 27-29, 1981.
3. Recommended Standard concerning Bacterial Levels in Industrial plants. Federal Republic of Germany.
4. Bergey's Manual of Determinative Bacteriology. 8th Edition. R.E. Buchanan; N.E. Gibbons.

5. Fundamentals of Microbiology. 9th Edition. M. Frobischer; R.D. Hinsdill; K.T. Crabtree; C.R. Goodheart.
6. Introductory Mycology. Third Edition. C.J. Alexopoulos; C.W. Mims.
7. Leonard A Dell (1979) "Aspects of Microbiological monitoring for Nonsterile and Sterile Manufacturing Environments" Pharmaceutical Technology August 1979.

Table 1: Corrective Action Required for Weekly Trend Results

Grade	Acceptable limit CFU/m3	Allowable* limit CFU/m3	Action Limits	
1	100	<250	>250	<500
2	250	<500	>500	<750
3	500	<750	>750	<1000
HEPA filters/ airlines	100	<100	>100	<250
Flow Bench	10	< 10	> 10	

* Report only CFUs/m3 exceeding more than 1 grade above their acceptable limit.

Table 2: Reporting of Corrective Action

ENVIRONMENTAL CONTROL - WEEKLY ACTION REPORT

Date of Monitoring: 9 June 1988
Date of Report: 11 June 1988
From: Chief Microbiologist
To: Factory Manager

Sampling	**Grade**	**Acceptable** Limit	**Measured** CFU/M3	**Action** Clean before Next monitoring	Clean Immediately
Raw Material Warehouse	1	100	372		X

Action Recommended by: Chief Microbiologist

Actioned by: Pharmacist

Table 3: Parallel Sampling

Location	Colony Count per m^3 *SAS(CRB)	*RCS (Blood)
Store 1	83	1100 Bacteria
	55	100 Mould
Store 2	33	150 Bacteria
	76	75 Mould
Weighing	11	50 Bacteria
	33	25 Mould
Powder Blending	256	175 Bacteria
	56	- Mould
Canteen	NG	825 Bacteria
	-	125 Mould

Types of Micro-organisms:

Acinebacter Caloaceticus; Alternarium; Aspergillus Terreus; Bacillus Species; Cladosporium; Diphococci; Epicoccum; Gram positive coccus; Micrococcus species; Mycelia Sterila; Penicillium; Phoma; Pseudomonas species.

*Two different samplers and Media

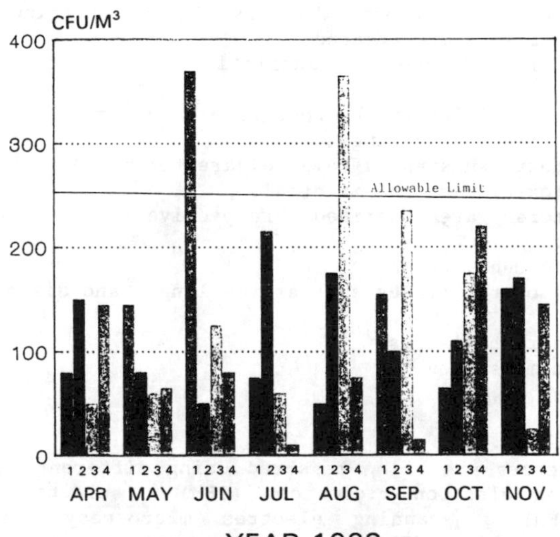

Fig. 1. Weekly trend – airborne microorganisms in warehouse

RISK ASSESSMENT OF FIBROUS MATERIALS USED FOR INJECTION MOULDING OF CERAMIC COMPOSITES

A. VANDE GUCHT, J. PAUWELS, J. SLEURS

S.C.K./V.I.T.O.
Boeretang 200,
B - 2400 Mol

SUMMARY

The health hazard by inhalation of ceramic whiskers is evaluated primarily by characterization of the whiskers. Different techniques such as aerodynamic particle size laser spectrometry, active laser aerosol spectrometry, cascade impact analysis, scanning electron microscopy and transmission electron microscopy are in use. The results indicate that laser optical methods might be used as monitoring technique to reduce the amount of fiber counts in workplaces.

1. INTRODUCTION

A technology assessment study of the injection moulding technique for whiskers reinforced ceramics includes the risk assessment of fine mineral fibres. The purpose of this study is to develop an economically feasable monitoring technique for airborne whiskers. In this particular case study silicon carbide whiskers are treated.

The health hazard due to inhalation of airborne whiskers is determined by :

a) the region of deposition in the respiratory system
b) the residence time in the lungs
c) the adsorbed toxic substances (e.g. cigarette smoke)
d) the chemical toxicity of the material.

These factors are defined respectively by the following characteristics :

a) the aerodynamic behaviour
b) the morphology of the fibres such as the length and diameter distribution
c) the surface characteristics
d) the chemical composition.

2. STRATEGY

The risk factors are being measured using different techniques such as : transmission electron microscopy (TEM) ; selected area electron diffraction (SAED) ; scanning electron microscopy (SEM) ; energy dispersive X-ray analysis (EDXA) ; active scattering laser spectrometry and electron scattering analysis (ESCA).

The aerodynamic behaviour is measured by impaction techniques and aerodynamic particle size laser spectrometry.

The surface charge will be determined using an electrical aerosol size analyzer. Laser spectrometry might fulfill the need for a continuous monitoring system with an alarm function during the processing of the whiskers.

The results of this characterization and measuring strategy are also intended to support eventual biological experiments in vitro as well as in vivo.

3. GENERATION OF AEROSOL

A known amount of SiC whiskers is suspended in very pure water. The suspension is sprayed with a Collison atomizer. The droplets are dried in a diffusion drier filled with silicagel.

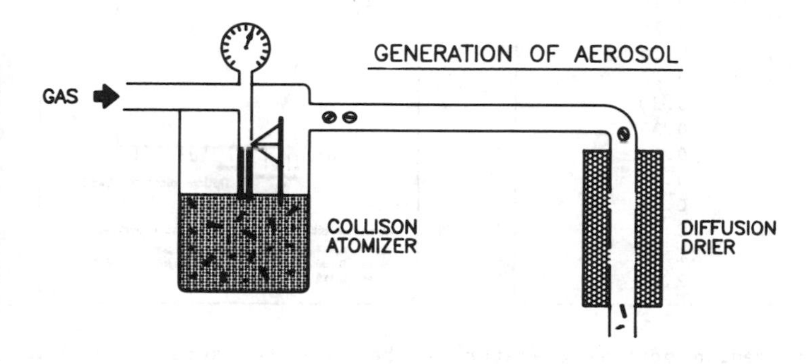

4. CHARACTERIZATION OF THE WHISKERS

4.1. Aerodynamic particle sizer

The particle size is calibrated with standard latex particles.

Indicated diameter (μm)	Measured diameter with APS (μm)
0.527	0.54
0.620	0.58
0.726	0.58
	0.72
1.088	1.0
2.020	1.7
	2.0

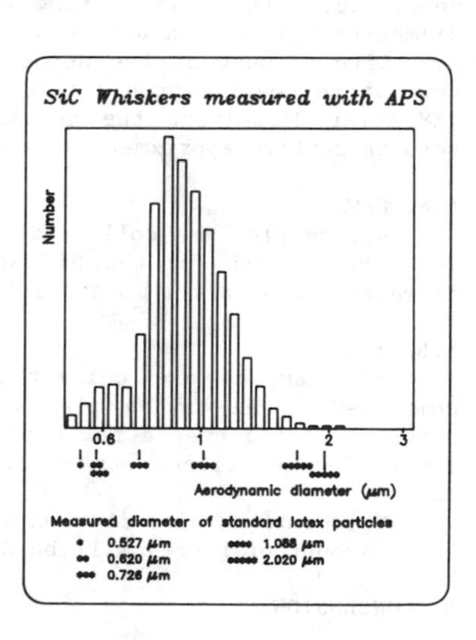

The count median aerodynamic diameter of the generated aerosol is 0.87 μm ± 0.04 μm (95 % probability). The distribution is approximately log normal with a geometric standard deviation of 1.25 ± 0.06 (95 % probability).

4.2. Active laser aerosol spectrometer

The ASAS is calibrated with standard latex particles.

Indicated diameter (μm)	Measured diameter with ASAS (μm)
0.109	0.10 ?
0.173	0.14
	0.18
0.176	0.14
0.180	0.18
0.190	0.16
0.232	0.20
0.312	0.30
0.360	0.34
0.460	0.46
0.481	0.45
0.527	0.48
0.726	0.70
1.09	1.1
2.02	2.0

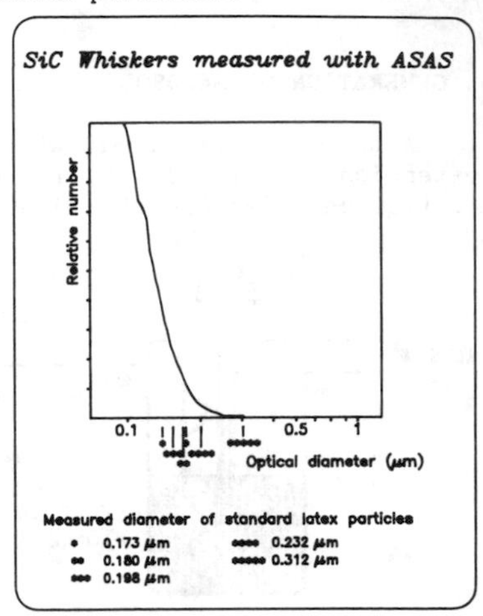

The count median optical diameter of the generated aerosol is 0.12 μm. The distribution is approximately log normal with a geometric standard deviation of 1.2.

4.3. Cascade impactor

A cascade impactor with 5 classical and 3 reduced pressure stages was developed. With an inlet flow of 0.5 m^3/h the stage cut-off aerodynamic diameters are between 0.1 μm and 3 μm.

Direct count of the deposited fibres in SEM was not possible up to now due to coating problems for the stages. Fibre count is executed with TEM after dissolving the coating with the embedded fibres. Preliminary results confirm apprixmately the measurements with the APS.

4.4. SEM

Air samples are collected on Nuclepore filters and coated with gold. The fiber lengths of the SiC whiskers are between 5 μm and 120 μm, the diameters between 0.1 μm and 1.5 μm.

4.5. TEM

Air samples are collected on membrane filters. The filters are combusted in an oxygen plasma. The ashes are suspended in ultra pure water, filtered over a Nuclepore filter, and coated with a carbon layer. The Nuclepore filter mounted on grids is removed by gently washing with chloroform.

The combined distribution of diameters and lengths for a sufficiently large amount of fibres will be determined in a very near future.

5. CONCLUSION

Different characterization and counting methods for whiskers were tested and calibrated. The results indicate that laser optical methods might be used for control and alarm during the use of whiskers thus reducing considerably the response time and the number of tidy and costly electron microscopic measurements.

1406

DEVELOPMENT OF CRITERIA AND INSTRUMENTATION FOR SAMPLING OF TOTAL INHALABLE DUST.

RJ Aitken

Institute of Occupational Medicine Limited, 8 Roxburgh Place, Edinburgh, EH8 9SU, Scotland

SUMMARY

Important work, relating to the definition of inhalable dust and the development and testing of instrumentation, capable of sampling according to this definition, has been carried out at the Institute for many years. This paper highlights the essential elements of this work as well as recent developments.

THE DEFINITION OF INHALABLE DUST

Initial work considered the definition of inhalable dust, ie what fraction of airborne dust, present in a volume of air was capable of being inhaled by a human who breathed that air in.

This was investigated by exposing a tailors mannequin (comprising a full torso and head) to a series of test environments, in a large working cross-section (1.5m x 2.5m) wind tunnel. The mannequin was aspirated at 20 l/min with a sinusoidal flow, provided by a piston driven breathing machine, so as to simulate the flow pattern of a worker engaged in light duties. This was carried out for a series of relatively monodisperse aloxite test dusts in the range 6 to 100 microns and for a range of wind speeds (from 1 to 4.5 m/s), with the mannequin rotated through 360 degrees with respect to the wind.

The concentration of dust entering the mouth of the mannequin, measured by weighing the mass collected on a filter located behind the mouth was compared with the concentration measured by an isokinetic probe, sampling in the freestream alongside.

The ratio between these measurements was the basis for the first definition of inhalability (8) and subsequently, the basis of the inhalability criteria adopted by the International Standards Organisation (2) and the American Conference of Government Industrial Hygienists (1).

More recent work (6) measured this ratio for windspeeds up to 9 m/s, more relevant to environmental conditions.

DEVELOPMENT OF INSTRUMENTATION FOR THE MEASUREMENT OF INHALABLE DUST.

Subsequent to this work, a programme was carried out to develop a series of instruments, capable of measuring inhalable dust. Instruments developed include a 2 l/min personal sampler (4), a 2 l/min personal spectrometer (3), a 3 l/min static sampler (5), a 70 l/min static sampler (6) and a 10 l/min static spectrometer (7).

Personal samplers for inhalable dust

The 2 l/min personal sampler, which is worn on the chest, incorporates an inhalable entry along with a cassette system which retains the filter. The personal spectrometer is an 8 stage impactor with which a full size distribution of the inhalable fraction can be measured. For these devices, entry efficiency was measured by mounting the instruments on the chest of the mannequin and comparing the mass collected by the instrument with that entering the mouth of the mannequin. The entry efficiency results, as a function of aerodynamic diameter, are shown in Figure 1 below where the efficiency is expressed as a fraction of the airborne concentration.

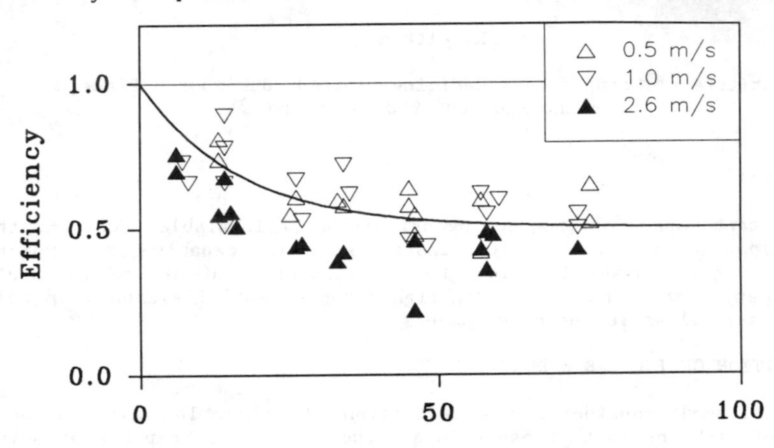

Figure 1.Sampling efficiency of the IOM personal sampling instruments

Over the range of conditions tested, the performance of the samplers is close to the inhalability curve (also shown).

Static samplers for inhalable dust

The inhalable static samplers all incorporate rotating heads and provide for measurement, and in some cases, sub-classification of inhalable dust over a wide range of sampling flow rates.

The entry efficiencies of these instruments were, measured by exposing the instruments to the same range of windspeeds and particle sizes as in the earlier experiments. These results are shown in figure 2 below.

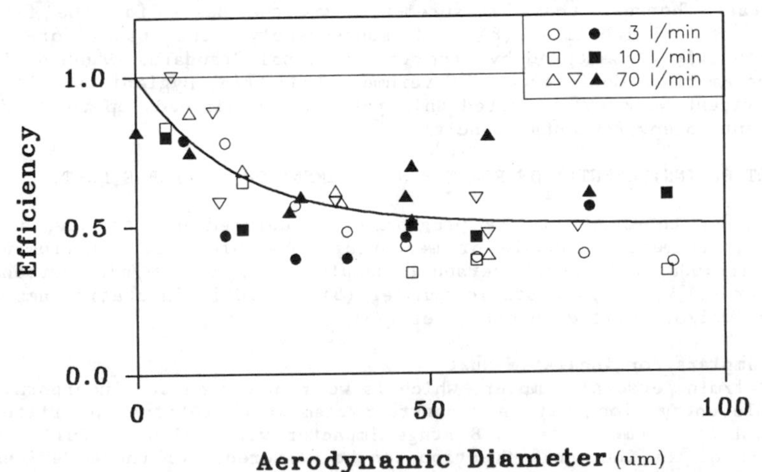

Figure 2. Sampling efficiency of the IOM static sampling instruments (open symbols for windspeeds 1.0 and 2.0 m/s, closed for 3 and 4 m/s)

As in the case of the mannequin, entry efficiencies were calculated by comparing the mass collected by the instrument under test with that collected by an isokinetic probe.

As for the personal instruments, it may be seen that the performance of these devices is in good agreement with the inhalability criteria over the range of windspeeds commonly found in the workplace.

Further developments now being pursued include a personal sampler, capable of separating the inhalable aerosol into the thoracic and respirable subfractions, as defined by the conventions currently being discussed by the various international advisory and regulatory bodies (CEN, ACGIH, ISO). In this instrument, porous foams will act as both size selectors and dust collection substrates. The device will thus offer simultaneous measurement of the three important fractions.

REFERENCES

(1) AMERICAN CONFERENCE OF GOVERNMENT INDUSTRIAL HYGIENISTS (ACGIH) (1985). Particle size selective sampling in the workplace. Report of the ACGIH Technical Committee on air sampling procedures, ACGIH, Cincinnati, Ohio.

(2) INTERNATIONAL STANDARDS ORGANISATION (ISO) (1983) Air quality-particle size fraction definitions for health related sampling. Technical Report ISO/TR/7708-1983 (E), ISO, Geneva

(3) GIBSON, H, VINCENT JH, MARK D, (1987). A personal inspirable aerosol spectrometer for applications in occupational hygiene research. Annals of Occupational Hygiene; 31:463-479.

(4) MARK D, VINCENT JH. (1986). A new personal sampler for airborne total dust in workplaces. Annals of Occupational Hygiene; 30: 89-102.

(5) MARK D, VINCENT JH, GIBSON H, LYNCH G. (1985). A new static sampler for airborne total dust in workplaces. American Industrial Hygiene Association Journal; 46: 127-133.

(6) MARK D, VINCENT JH, LYNCH G, AITKEN RJ, BOTHAM RA. (1990). The development of a static sampler for the measurement of inspirable aerosol in the ambient atmosphere (with special reference to PAHs). Final report of IOM work on CEC Contract 7261-04/424/08. Edinburgh: Institute of Occupational Medicine Ltd. (IOM Report TM/90/06).

(7) MARK D, VINCENT JH, GIBSON H, AITKEN RJ, LYNCH G. (1984). The development of an inhalable dust spectrometer. Annals of Occupational Hygiene; 28: 125-143.

(8) VINCENT JH, ARMBRUSTER L. (1981). On the quantitative definition of the inhalability of airborne dust. Annals of Occupational Hygiene;24: 245-248.

1407

DETERMINATION OF POLYCYCLIC AROMATIC HYDROCARBONS IN FOOD DUST CONTAMINATED AIR

B. ANDREJS and C. SCHUH

Berufsgenossenschaft Nahrungsmittel und Gaststätten
Mannheim, Germany

SUMMARY

When collecting samples of polycyclic aromatic hydrocarbons in the food processing industry, other dusts are also collected. This has an influence on the recovery. An HPLC method has been examined for its suitability in PAH analysis in the presence of food dust.

1. INTRODUCTION

During thermal degradation of food, polycyclic aromatic hydrocarbons can develop. In the following we will analyse how far fibre glass filters influence the recovery of polycyclic aromatic hydrocarbons which are contaminated with product dust during air measurements. Preparation of samples and analysis by HPCL and a Diode-Array Detector (DAD) are described.

2. METHOD

Collection of Samples

The air samples were taken with a sample collecting system PGP (BiA)-Gesamtstaub and a personal air sampler with a flow rate of 3.5 l/min (1). The amount of the collected product dust was determined gravimetrically.

Preparation of Samples and Analysis

Filter, extracted by toluene/ultrasonicated for 5 min

- concentrate to dryness at 40°C (rotary vacuum evaporator)
- solid phase extraction by C_{18} cartridges

Washed with water/methanol

Eluted with methanol

Dissolved in methanol HPLC Diode-Array Detector

The analyses were made by HPLC with a Diode-Array Detector in the wavelength range of 210-400 nm. Using a DAD has the advantage to acquire UV spectra online and to monitor signals of up to 8 wavelengths simultaneously. The separation of PAHs was made by a 150 x 4.0 mm Nucleosil 5 C_{18} column, mobile phase: acetonitril/water, gradient, flow rate: 0.8 ml/min, temperature: 30°C, Inj. Vol.: 20 μl. For the determination of the wavelength with maximum absorption for a mixture of Benzo(k)–, Benzo(b)fluoranthene, Benzo(e)pyrene, Benzo(a)pyrene and Perylene spectra of the whole wavelength range were recorded versus time.

The quantification of each component can occur with the wavelength with the highest signal. The identification of the PAHs by comparing the UV spectra with those stored in a library gives clear results (Figure 1) even when there is no complete base line separation. By comparison of the spectra at different retention times the peak purity can be controlled (Figure 2).

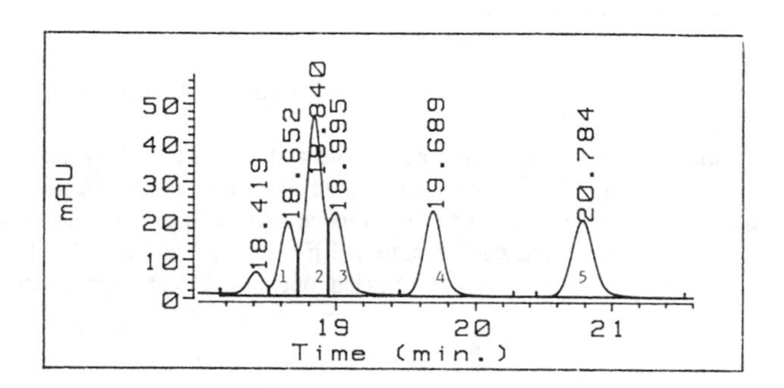

Fig. 1. HPLC chromatogram of a standard mixture of 5 PAH (1 ng/μl)
1 - Benzo(e)pyrene; 2 - Benzo(b)fluoranthene; 3 - Perylene
4 - Benzo(k) fluoranthene; 5 - Benzo(a)pyrene

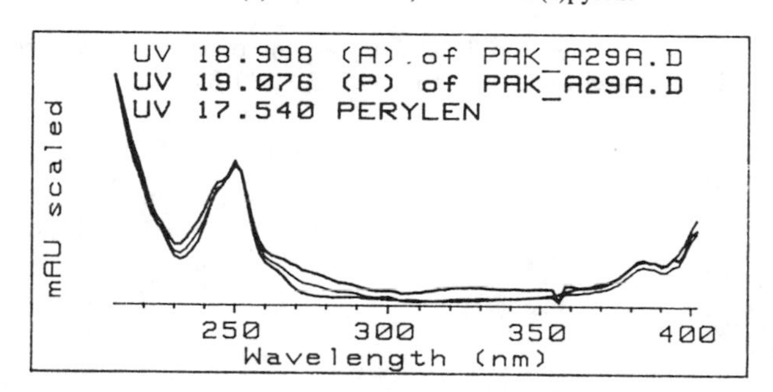

Fig. 2. The UV spectrum of perylene and purity control

3. RESULTS AND DISCUSSION

When collecting higher amounts of air (for example $1m^3$) the filters can actually be contaminated with 5–10 mg of product dust. The PAHs are mixed with the product dust and are adsorbed (for example flour, fat content about 2%) at the matrix of fat containing dust (2). A treatment with organic solvents (toluene) does not lead to the complete dissolution of, for example, Benzo(a)pyrene from glass fibre filters which are contaminated with flour. Only 10 mg of flour reduces the recovery of Benzo(a)pyrene by approx. 20%. A four-fold rise of the amount of product on the filters leads to a 22% drop in the recovery.

Table 1. Recovery of Benzo(a)Pyrene

Number of samples	Flour on filters (mg)	Recovery of B(a)P (%)
15	0	95
8	10	79
8	40	73

The Benzo(a)pyrene concentration was 500 ng in all samples. That means a quarter of the TRK value, when collecting 1 m^3 of air.

A longer sonication (40 min) leads to an improvement of the results. We realized, however, that after a treatment of only 10 min under ultrasonic exposure a classification of the UV spectra was not possible. UV chromatograms of a toluene extract of loaded and unloaded glass fibre filters indicate that better results can be obtained if a saponification with ethanolic KOH and a liquid/liquid extraction with toluene is carried out subsequent to the extraction with toluene.

REFERENCES

(1) Von den Berufsgenossenschaften anerkannte Analyseverfahren, ZH 1/120, Carl Heymanns Verlag KG.

(2) Beitrag zur Bestimmung von Benzo(a)pyren im pflanzlichen Material, K. Speer, Deutsche Lebensmittel-Rundschau 83 (1987) 3 S. 80–83.

1408

A NEW APPROACH TO THE ASSESSMENT AND MEASUREMENT OF CHROMIC ACID MISTS

J. BUTT

British Steel plc, British Steel Technical
Welsh Laboratories, Port Talbot, West Glamorgan, UK

Recent experience in British Steel has cast doubt on the validity of data obtained in studies of chromic acid mist using the form of sampling currently recommended by the UK Health and Safety Executive for the measuring of airborne particulate matter. In addition, epidemiological work published in the past decade has indicated the possible link between exposure to acid mist and laryngeal cancer. Knowledge of the deposition characteristics of particulate matter in the respiratory airways suggests that the size characteristics of emissions of this type are extremely important in the aetiology of associated occupational disease. It is therefore essential that the size collection efficiency of the currently recommended sampling technique is defined with regard to the mists being sampled.

Therefore, a technique capable of simultaneously assessing particle size and mass concentration quickly and reliably would prove invaluable in assessing the risk arising from exposure to acid mist in inhaled air.

Thin layer chromatographic plates are used to measure the fall-out of hexavelant chromium throughout the workroom over a 50 hour period. The plates are subsequently removed from the workroom, sprayed with suitable strength solutions of sulphuric acid and diphenyl carbazide in acetone solution. Finally, once the colour has developed, the plates are colour photographed in order to provide a permanent record despite colour fade on the plate.

This technique provides a quick, simple empirical method for assessing the effectiveness of engineering control measures designed to reduce the risk of exposure by inhalation, ingestion and skin contact. It may also be used to measure the dispersion and deposition of chromic acid over large distances. Further, a quick reliable assessment of the environmental impact of chromic acid can be made at the area of direct entry into the atmosphere e.g. discharges from stacks or ducts.

This new approach forms part of a current combined British Steel/ECSC sponsored project (Study Agreement No.7261/04/465/08) entitled 'A Study of the Sampling of Acid Mists to Assess their Impact on Exposed Personnel and to Study Exposures of Personnel on Different Types of Process'. Photographs of plates obtained from different types of processes will be shown on the poster.

1409

A SIMPLE AND COMPREHENSIVE SURFACE DUST DETECTOR SYSTEM FOR ASSESSING THE STANDARD OF CLEANING

Schneider T[1], Petersen OH[1], Nielsen TB[2] and Løbner T[3].

(1) Danish National Institute of Occupational Health,
Lersø Parkalle 105, DK-2100 Copenhagen.
(2) Bygge- og Miljøteknik a/s,Birkehavevej 3, DK-3460 Birkerød.
(3) Danish Association of Service Trade Employers,
Rosenørns Alle 1, DK-1970 Frederiksberg.

SUMMARY
The BM DUSTDETECTOR system (patented) is a portable, easy to use instrument for on-site determination of dust on surfaces. Dust from hard surfaces is lifted off with specially designed BM DUSTLIFTER gelatin foils. The foils are inserted into a laser based detector for determination of percent area covered by dust. For carpets the BM CARPETTESTER vacuum mouthpiece with pressure gauge is used in combination with a conventional vacuum cleaner for determination of resuspendable dust. The dust is deposited on the same type of foils in a radial impactor mounted after the mouthpiece. The fixed geometry of, and pressure drop in the mouthpiece standardizes the measurement. The surface dust detector system is the first system for objective assessment of the need for, and result of cleaning and is already in use in the Scandinavian countries. The system will have a major impact on quality assurance for cleaning, for objective specification of cleaning programs, and for development and test of new cleaning methods.

1. INTRODUCTION

Particle contamination of the indoor air must be controlled by both ventilation and cleaning. Significant research and development have produced effective ventilation systems and measurement techniques for documenting the ventilation efficiency. In contrast, methods for evaluating cleaning effectiveness has previously not received much scientific input. This is surprising, since cleaning Scandinavia is a billion ECU pr. year industry. The professional companies offer cleaning in buildings specified in cleaning programs. Choice of programs will depend on use of rooms, types of surfaces and furniture, and not least on the customer's budget. The programs have been developed over the years based on practical reasoning, customers reactions etc. The quality of the cleaning has been assessed at different levels. Level 1: no assessment at all. Level 2: walk-through visual inspection. Level 3: quality index, calculated from systematic visual inspection. The question: was it necessary to spend 5 billion ECU or was it not sufficient, and the question: what did we actually get for the 5 billion ECU, could not be answered on the basis of objective measurements.

Systematic research and development was thus initiated which has now resulted in a patented system for simple measurement and evaluation of dust on surfaces. A preliminary version, applicable on hard surfaces only, has been described previously [1].

2. MATERIALS AND METHODS

Dust on non-textile, hard surfaces is sampled with BM DUSTLIFTERS. They are specially designed gelatinous, sticky foils. They were originally developed for forensic purposes (finger and foot print lifters). They have very good sticking properties, are thick and soft so they can reach the bottom of surface roughness elements, and have excellent optical properties [2].

The foil with dust sample is placed in the BM DUSTDETECTOR. This is a diode-laser light extinction meter which measures the projected area of the dust particles [3]. The foils measure 3 x 7 cm. The battery powered light extinction meter measures 8 x 11 x 34 cm. A blank reading has to be made prior to sampling. The result is given as percent of foil area covered by particles. For particle diameters ≥2 µm, the extinction will be independent of shape and refractive index. The extinction is affected by the presence of the foil surface, but this effect has not been evaluated theoretically.

Dust in carpets is evaluated with the BM CARPETTESTER, which consists of two parts [3]. The first part is a special vacuum mouthpiece. The orientation, distance from the floor, and volume flow rate are kept constant. The dust in carpets is defined as the dust resuspended from the carpet by moving the mouth piece 4 m along the carpet with a speed of about 10 cm sec^{-1}. This dust passes through a specially designed radial impactor, mounted behind the mouthpiece. This is the second part. The radial impactor is basically a tube with a foil placed over a longitudinal slit, cut into the tube wall. Particles will deposit on the tube walls and on the foil with a size dependent probability, due to turbulence. Turbulence and thus deposition is increased by a circular constriction inserted immediately upstream of the slit [4]. The same foil as above is used, and the deposited dust can thus be evaluated with BM DUSTDETECTOR. A pressure gauge monitors the pressure drop across the constriction. This serves to maintain the pressure drop and thus the volume flow rate at a predetermined level. A conventional vacuum cleaner can be used as air mover provided there is a by-pass valve in the hose for manually down-regulating the flow rate. The fixed geometry, volume flow rate, speed of moving the mouth-piece, and pressure drop in the mouthpiece standardizes the measurement.

3. RESULTS

The BM DUSTDETECTOR has been calibrated against optical microscopy with image analysis, and the calibration is built into the BM DUST DETECTOR microprocessor so that the result is in area percentage. The BM CARPETTESTER has been calibrated against total weight of dust dispersed onto a test floor [3]. Determination of the size dependent deposition efficiency is ongoing.

The BM DUSTDETECTOR and BM CARPETTESTER have been tested in a Scandinavian field study [3]. 17 rooms in a total of 14 buildings were investigated. A total number of 2931 individual samples were taken on hard surfaces and 263 on carpets.

For samples taken from clean, hard surfaces, the standard deviation is 0.2 % area. There is a sligh positive bias, which tends to increase for rough surfaces. The resolution of the digital display is 0.1 % area.

4. DISCUSSION

The surface dust detector system is the first system for objective assessment of the need for, and result of cleaning and is already in use in all the Scandinavian countries. The system will have an impact on the way cleaning is dealt with, since it provides an objective way of specifying cleaning programs and allows objective quality assurance . The system can also be used to test different mop methods and vacuum cleaners, to assess the effect of polish treatment, and to develop new cleaning methods. The pressure gauge can be used to quality control vacuum cleaner performance, since low air-flow indicates poor vacuum, filter clogging or leaks.

The analytical system is much easier to use and much faster than optical microscopy and image analysis. The lower limit of detection is comparable to the lower limit of perception of black dots on a white surface, which is 0.2 % [5]. The field trial has shown that reproducible results can be obtained.

It is not likely that guidelines for surface dust based on health effects can be determined in the near future. It is even possible that it never will be since the relationship between surface concentration and airborne concentration is influenced by so many parameters [6] that no usable relationship between surface and air concentration can be established. Source control in terms of limiting surface concentration seems most appropriate. For use in such a pragmatic approach, the guidelines in Tab.1 have been suggested, based on present knowledge of which levels are obtainable with "good cleaning practice". The values are obviously specific for the BM DUSTDETECTOR and BM CARPETTESTER.

The sampling strategy is an integral part of the system. A preliminary strategy has been formulated[1,3], based on theoretical considerations [6]. Work continues on development of a comprehensive strategy. It will include detailed decision rules for selecting number of samples per object and surface area. Methods for calculating average surface dust levels and expressions for minimum variance estimators for surface contamination on areas not sampled will be included.

Tab. 1 Preliminary recommandation for surface dust levels[2]

Surface type	After cleaning	Between cleanings
Furniture, low position	≤1%	≤2%
Furniture, high position	≤1%	≤3%
Hard floors	≤5%	≤8%
Carpets *	≤10%	≤20%

* as measured with BM CARPETTESTER

The Danish service trade association is presently preparing norms for cleaning based on the dust detector system.

REFERENCES

(1) SCHNEIDER, T.,PETERSEN, O.H.,ERIKSEN, P.,VINZENTS, P. (1989). A simple method for the measurement of dust on surfaces and the effectiveness of cleaning. Environ Int 15:563-566.

(2) SCHNEIDER, T. (1986). Man-made mineral fibres and other fibres in the air and in settled dust. Environ Int 12:61-65.

(3) PETERSEN, O.H., KLOCH, N.P., ABILDGAARD, A., KONGSTAD, K.S., NIELSEN, T.B., SCHNEIDER, T. (1991). Udvikling af instrument og metoder til måling af støvniveau på overflader. In Danish. English summary report entitled: Measurement of dust levels on surfaces. Bygge- og Miljøteknik, Birkehavevej 3, 3460 Birkerød, Denmark.

(4) KIM, C.S., LEWARS, G.L., ELDRIDGE, M.A., SACKNER, M.A.(1984).Deposition of aerosol particles in a straight tube with an abrupt obstruction. J Aerosol Sci 15:167-176.

(5) HANCOCK, R.P., ESMEN, N.A., FURBER, C.P.(1976). Visual response to dustiness. J Air Pollut Control Assoc 26:54-57.

(6) SANSONE, E.B.(1987). Redispersion of indoor surface contamination and its implications. In:Treatise on clean surface technology, Mittal, K.L., ed., Plenum, vol 1:261-290.

(7) SCHNEIDER, T., PETERSEN, O.H., NIELSEN, A.A., WINDFELD, K.(1990). A geostatistical approach to indoor surface sampling strategies. J Aerosol Sci 21:555-567.

Session V

DATA AND INFORMATION MANAGEMENT

Keynote Papers

Posters

1410

IMPLEMENTING A DATA BASE OF CHEMICAL EXPOSURES: COLCHIC DATA BASE

B. CARTON

INRS Research Centre, Avenue de Bourgogne - BP 27
54501 Vandoeuvre, France

INTRODUCTION

Within the framework of their respective assignments, the research departments of the INRS and the risk prevention departments of the Regional Health Insurance Funds (CRAM) carry out sampling operations at workplaces.

From the beginning of the 1970s the corresponding analyses were carried out either at the INRS, or in one of the four existing interregional chemistry laboratories. The Commission for the Prevention of Work Accidents and Occupational Diseases proposed, in 1983, to create enough laboratories so that each one of them covered, on average, the territory of two Regional Funds; this proposition was made to emphasize the necessity of safeguarding:

> "The centralization of information by the INRS, which must have, vis-à-vis these laboratories, a triple task of coordination, harmonization and training".

This wish of the Commission led to the creation of a new instrument, at the level of the institution, made up of the Regional Insurance Funds (CRAM), the National Health Insurance Fund (CNAM) and INRS: the collection system of data gathered by the chemistry laboratories of the INRS and the Health Insurance Funds (COLCHIC system), operation coordinated by a working group of the future users.

THE INFORMATION COLLECTED

The laboratories specialized in physicochemical analyses at the INRS or the interregional laboratories of the CRAM participate in the policy of prevention of work accidents and occupational diseases. As far as occupational diseases are concerned, the activity of these laboratories brings them, within variable contexts and objectives, to visit certain work stations and firms that come under the Social Security general insurance scheme; air samples at work stations or samples of products used in the manufacturing process are taken, lists of information concerning firms and processes involved are made, analysis results usefully complete the files.

These intervention reports and the data they contain are often only used once, at conclusion time; one can hope that the confrontation of different data is likely to provide additional information. Reusing results acquired at the cost of large financial effort is also worthwhile.

The system, therefore, collects all of the corresponding data:

- those which concern globally the intervention (characteristics of the firm, references, preliminary information),
- those which concern samples of air or product (work station, sampling method, volume, sampling device, representativity, etc),
- the results of the analyses done, the analytical methods used.

This collection is only possible because minimal requirements of quality and homogeneity of the sampling techniques, of analysis and data acquisition were, in general, satisfied.

THE COLLECTION SYSTEM

The application comprises two levels corresponding to two sub-applications whose objectives are different but complementary: the local (laboratories) and the national (data base) level (see Figure 1).

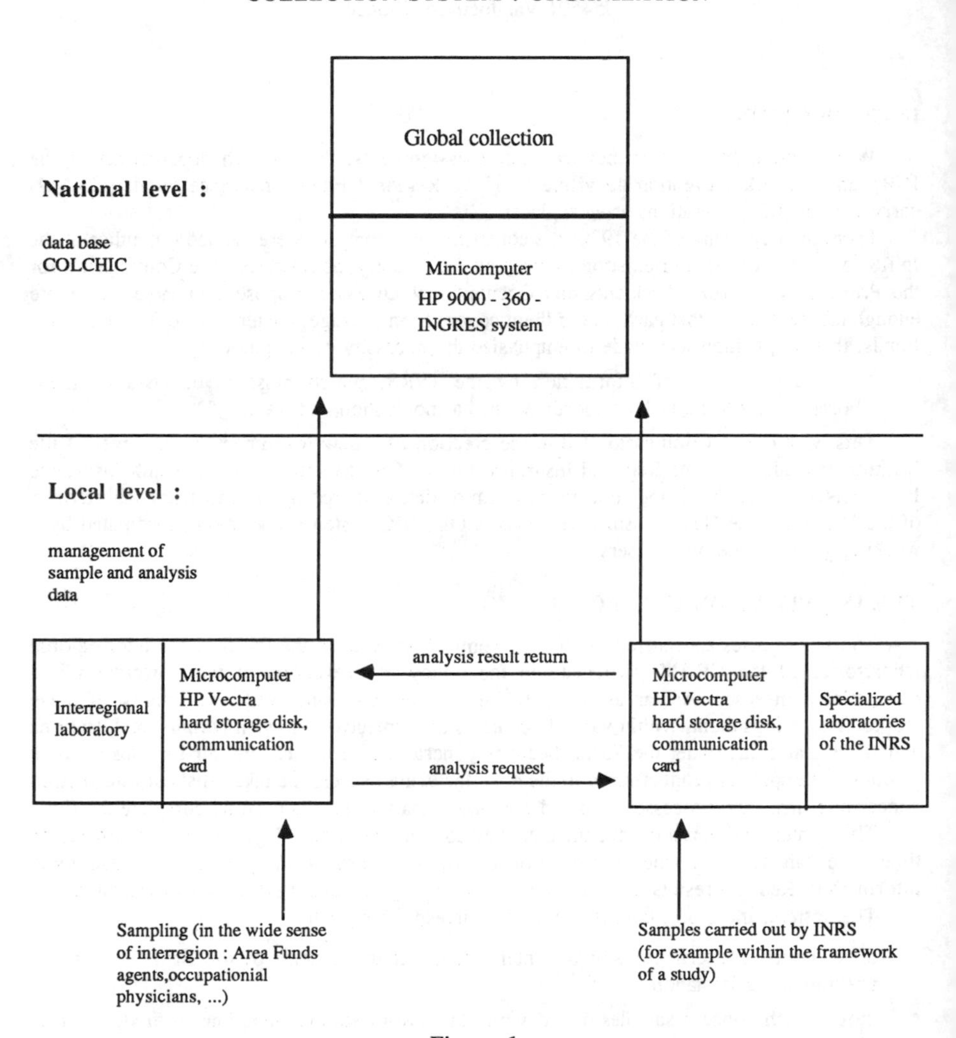

Figure 1

Putting the system into place answers three preoccupations, **three objectives**:

- the desire to help each of the chemistry laboratories in providing it with computer equipment which allows it to process, stock and access all of the data produced by the sampling/analysis activity of the laboratory in its interregion;

- the concern to homogenize the practices of sampling and analysis in use in the different laboratories.

This falls within a larger perspective of dialogue between the laboratories and samplers of the CRAM and INRS since annual specialized meetings (sampling/analysis methods, meeting of laboratories, meeting of specialized engineers) preexisted. Standardization efforts concern the transcription of the activities' characteristics, of the firm and work station; the exchange of procedures for analysis requests and results, which take place in a broader quality control perspective; as well as the creation of the actual data bank, at the INRS, which groups all of the collected information at the interregional level. This national bank must be able to provide useful information as much for the establishing of standards (exposure/pollutant matrices) as for the injection of exposure data in epidemiological studies.

Therefore, the aim is not to set up a centralized system, but to collect all of the information provided by the nine laboratories situated at the same level (the eight interregional laboratories and the INRS analysis laboratory). All of the data are grouped at the level of a particular site, the INRS central computer acting as feeder: receiving information that it does not modify and making it available to all the laboratories participating in the operation.

At the **interregional level**, the laboratories dispose of information processing computer software and are responsible for transfer of analysis requests and results. The software being identical in the eight CRAM laboratories and at the INRS, this ensures a good homogeneity of the services provided and fulfils the requirements of quality management.

In addition to the actual data processing software, a certain number of programs are set up allowing us to follow locally the activity of a laboratory: processing of reports, of analysis requests, of archives consulted, of companies visited by the laboratory, of names of products analyzed, of analysis dates which should allow to plan different operations and the corresponding execution times. Local data base software will be used later to retrieve this local information.

Thanks to a specific program, standardized reports can be generated, describing the activity of the laboratory.

The **national part** answers the third objective described, the setting up and management of the actual data base.

The information gathered by the collection system is transmitted via telecommunication to the computer of the INRS research centre. This information allows setting up and operation of a national data base on chemical exposures.

This data base, grouping all of the sampling and analysis information, can be consulted. The data do not constitute a statistically representative sample of the work stations of French industry, but a collection of measurements made to answer specific and different types of questions, occurring in not necessarily homogeneous regional contexts. This non-representativity must of course be borne in mind at the time of extracting data from the base; the multiplication of data associated with one result allows this to be taken into account.

The constituent tables are established in a way that preserves some access confidentiality, mainly in relationship with the corporate name of the firm. The base functions to the advantage of its users and cannot be questioned directly from outside.

The relational data base management system (SGBDR) can be used for answering three types of questions:

- questioning by menu of all the standard questions which have been automated;
- questioning the tables of the base in using the SGBDR;
- use of the fourth generation language and writing of the procedures by the manager of the base.

In the system's present state, access to the last two questioning types is unusual at the local level.

THE REFERENCE FILES

The reference files make up the application dictionary, thus a high level of coherence between users is reached. Each datum, to the extent that it can be coded, has its reference file; such a datum can be questioned at the data base level and can constitute selection criteria.

The list of reference files is given in Table 1. Each of them was established with the help of the specialists concerned. Of course, these files are not frozen and their evolution is constant with the appearance of new techniques, new sampling aids, new pollutants, etc.

Table 1. Reference Files

1) List of main files:

1	Origin of the request
2	Origins of the sampling equipment
3	Activity: Technical Committee, risk
4	Work station
5	Chemical pollutants
6	Product categories
7	Reasons for intervention
8	Sampling methods
9	Types of analysis requested
10	Objective – representativity of the sample
11	Analysis of methods used
12	Measuring units
13	Sampling equipments code

2) Summary of the file 'Work stations':

A1	Storage and transport
A2	Grinding and sifting
A3	Mixing, compression, melding, compression
A4	Melting, sintering, roasting, drying, casting
A5	Machining, assembling, welding, gluring
A6	Preparation, treatment, protection
A7	Bottling, barrelling, winding, titring
A8	Control, cleaning, preparation
B1	Iron and steel-making and general development of metals
B2	Foundry
B3	Construction and public works
B4	Textile
B5	Printing, serigraphy, document reproduction.

Such a procedure allows conservation of an excellent homogeneity at the same time as authorizing the necessary evolution of the files and the integration of new validated techniques or of new preoccupations. Each file can be questioned and edited at the local level.

A special effort will probably have to be made in the future to homogenize as much as possible the dictionaries used by the different factual data bases as much inter- as intranational. This evolution, which will respect the specific characteristics of national contexts, is likely to facilitate the building of bridges, of passages, between data bases. Pertinence of classical reference files is really questionable; would it not be possible to use a software developing new reference files during the operations?

CONCLUSION

COLCHIC was installed by all its users and has been operational since 1987 (see Table 2). The objectives defined at the outset of the project have been reached, in particular the storage and questioning of information concerning exposure to chemical pollutants at the work station.

Use of this new tool has started. Only the small volume of stored data limits for the moment the interactive interrogation. Two examples illustrate this, one concerning the silicosis risk and another one giving an extract of the job-exposure matrix for cadmium.

The flexibility of the system allows to follow the evolution of sampling and analysis techniques; it could authorize, so long as the rules in use are respected, the parallel collection of data coming from outside partners in a structure so organised.

A number of side packages have been progressively linked to the core collection system: sampling methods, analytical methods, specific documentation, computation programs, labelling, simulation, etc). The system can be now depicted as a management and information system for chemical exposure assessment.

Table 2. COLCHIC

- FIRST COMPLETE YEAR: 1987
- VOLUMES STORED IN 1990 (LAST COMPLETE YEAR)
  - Information concerning 1450 facilities
  - 1567 analyses of 'industrial products'
  - 13 524 air samples
  - 34 737 analytical results
- THE DB MAY BE USED **INSIDE** OUR INSTITUTION DIRECTLY (ON-LINE)
- ANY QUESTION COMING FROM THE OUTSIDE IS TREATED INDIRECTLY

Development and Management of Occupational Hygiene Databases

PL BEAUMONT
HEALTH AND SAFETY EXECUTIVE
TECHNOLOGY AND HEALTH SCIENCES DIVISION
MAGDALEN HOUSE
STANLEY PRECINCT
BOOTLE
MERSYSIDE
L20 3QZ
UNITED KINGDOM

INFORMATION NEEDS

The need for adequate information has been given special emphasis in the U.K with the advent of The Control of Substances Hazardous to Health Regulations 1988 (COSHH). Similar developments are taking place throughout the European Community as a result of E.C Directives. Wherever they are needed assessments of risks to health and of the steps to be taken to maintain control of the risks can only be made if there is adequate information available.

In some cases exposure measurements and possibly routine monitoring are required. Failure to collect the correct information may result in wasted effort and possibly failure to adequately control health hazards.

COMPUTER SYSTEMS

Many organisations have found that the volume and complexity of their occupational hygiene data are such that computer systems provide the most efficient means of storage and selective retrieval.

The number of companies which produce occupational hygiene software is increasing in order to meet this demand. There is a tendency for individual organisations and even for the various units within large organisations to individually tackle the common problems of development and management of occupational hygiene databases. There has been no concensus on the nature of the basic information to be stored and processed or on the objectives to be achieved by the use of the databases.

SHARING EXPERIENCE

Sharing experience helps to avoid wasteful duplication of effort and mis-guided projects. There is clearly a need to define the pre-requisite parameters to be recorded when exposure measurements are taken and to standardise aspects of the storage and retrieval of such data.

In order to address those needs the British Occupational Hygiene Society has established a Special Interest Group on Data and Information Systems in Occupational Hygiene.
The membership includes occupational health professionals from industry, government and consultancies and includes software suppliers and publishers. The Chairman is Paul Beaumont and the secretary/treasurer is Ian Bartlett. The work includes all aspects of information technology in occupational hygiene. The Group has two high priority projects concerning exposure data.

CORE INFORMATION - PRESENTATION OF EXPOSURE MEASUREMENTS

The minimum information requirements for collection and recording of exposure data have been defined. A professional standard for the presentation of occupational exposure data was presented at the BOHS Conference 1991.

The standard applies for both paper and computer systems. The form which goes with the standard is shown in Figure 1. A detailed guide to completion of the form is about to be published in the Annals of Occupational Hygiene.

MANIPULATION OF DATA

This project considers the objectives of occupational hygiene software and will provide guidance to aid software companies in the production of suitable systems and assist purchasers to choose appropriately.

CONCLUSIONS

The British Occupational Hygiene Society strongly urges the use of the standard core set of information for occupational exposure data, as illustrated in the form. This should ensure that good quality exposure data are available for:-

- controlling occupational exposure to hazardous agents
- epidemiological studies
- setting occupational exposure limits

The data set advocated is compatible with the UK Health and Safety Executives National Exposure Database. International adoption of the common base data set would enable powerful studies using large populations of data.

Environmental monitoring Data

BOHS

Author	
Tel. no.	

File reference	Reference to related records	Date of sampling	Total no. of people on site	Agent			
Occupier of premises				CAS No.			
Address of premises/Location/Identity				Units			
				Sampling/Analysis Details			
		Post Code		MSDS Ref			
Department/Area				Males exposed			
Building/Room				Females exposed			

	Reference number	Sample type	Sample description (eg name/task/process/equipment)	Male Female	NI no. Personal no.	Sample period	Duration (Mins)	Result	TWA	Result	TWA	Result	TWA
1													
2													
3													
4													
5													
6													
7													
8													
	Exposure limits			8 Hour									
				10 min									

NEDB / BOHS 1

Industry and SIC Code

Reason for monitoring

Biological monitoring

	Exposure details			Control measures		Related records
	Conditions	Frequency	Metabolic rate	RPE	LEV	
1						
2						
3						
4						
5						
6						
7						
8						

Comments on origin of sampled material e.g. product name

Comments on exposure modifiers, e.g. skin contact, other relevant jobs, confounding factors, biological monitoring

1412

GESTIS - A REGISTRY ON OCCUPATIONAL HEALTH AND WORK SAFETY OF THE STATUTORY ACCIDENT PREVENTION AND INSURANCE INSTITUTIONS IN INDUSTRY

K. MEFFERT and R. STAMM

Berufsgenossenschaftliches Institut für Arbeitssicherheit
Alte Heerstrasse 111, Sankt Augustin, Germany

SUMMARY

The authors present goals, constituent elements, structure, information flow and possible use of GESTIS. Particular attention is paid to the central database on substances and products – ZeSP – representing an essential element of GESTIS. Its structure, contents as well as experience gathered during its generation are discussed in the light of the necessary European harmonization process.

1. INTRODUCTION

Some of the work safety documentations used by the statutory accident prevention and insurance institutions in industry (in the following called professional associations) have already been in existence for about 20 years: Documentations on occupational diseases and exposure data at the workplace, a literature database orientated towards the missions of the statutory accident prevention and insurance institutions as well as a documentations with regard to particularly exposed workers (asbestos and other carcinogenic substances).

Since the Gefahrstoffverordnung (ordinance on hazardous substances) came into effect in 1986, both, the professional associations as well as their member enterprises have empathised an increasing need for safety-relevant data on hazardous substances and their user friendly, field oriented preparation and use. On account of this development a central database on substances and products was created by the central federation of the statutory accident prevention and insurance institutions in industry while decentralized information systems were established to furnish branch and application oriented data.

Apart from the provision of information on products and substances, the exhaustive and integral use of all documentations available appeared to be an urgent necessity. Consequently, GESTIS was called into being in 1988 as a common project of all professional associations in industry interested in creating and exploiting an integrated information system.

The coordination of GESTIS lies with the Berufsgenossenschaftliches Institut für Arbeitssicherheit (BIA) which is assisted by an advisory group including representatives from all professional associations interested.

2. CONSTITUENT ELEMENTS OF GESTIS

GESTIS consists today of four central documentations – i.e. they are supervised by the central federation of the professional associations, and four decentralized information systems – i.e. they are developed and used by single professional associations.

Central Database on Substances and Products – ZeSP

ZeSP was founded in 1987 for online and offline use. It contains information on pure chemical substances and preparations; these data are especially processed for control and consultative tasks in the field of work safety. As ZeSP is developed in coordination with the database on hazardous substances of the Länder, its structure and contents largely parallel those of the federal information system. The information provided concerns substance and product identification, physical and chemical properties, data on handling and use, classification, identification and transport, legal regulations and guidelines as well as hazardous effects on human health.

The information is available both, in its integral form or - in view of an online use in the firms - as selected extracts treating single aspects of interest such as, for example, substance identification, protective measures, physical/chemical properties or legal regulations.

Immediate access to the data is offered to the professional associations only. Data extracts may also be passed to the member enterprises within the framework of advisory and training activities. Commercial use is, however, excluded. Due to the fact that as a result of the information provided by ZeSP, the technical inspectors of the professional associations may be obliged to take legally relevant measures in the member enterprises, particular attention is paid to the validity of the data. As far as this is possible, all information is based on validated literature and the entered data are checked by experts from the professional associations.

Documentation on Measurement Data of Hazardous Substances at the Workplace - DOK-MEGA

DOK-MEGA was created in 1971; at this point in time it contains about 400,000 data on workplace measurements. DOK-MEGA is used, for example, to analyse efficiency of preventive measures taken to avoid or improve exposure situations. DOK-MEGA - as a compound of GESTIS - is mainly developed with a view to concentrating data and creating a register on hazardous substances and workplaces.

Documentation on Occupational Diseases - BK-DOK

Created in 1975, BK-DOK contains information on all cases of occupational diseases that have been reported and acknowledged since that time; at present, its total data pool comes up to 500,000 registered cases. BK-DOK is used offline in particular by the professional associations but also by occupational medicines, works doctors, political bodies as well as by the industry and trade unions. The database may serve to identify substance/disease relationships and to analyse regional distribution or temporal trends of occupational diseases.

GESTIS is meant to improve the substance and product oriented use of these data within the framework of preventive work medicine.

Database on Literature References - ZIGUV-DOK

ZIGUV-DOK provides literature references with respect to hazardous substances, occupational diseases, limit values and working procedures; it also contains details on all literature referred to in the central database on substances and products, thus functioning as background information system for BK-DOK, DOK-MEGA and ZeSP. The decentralized GESTIS databases serve varying purposes. The 'Central register for employees exposed to asbestos-containing dust' (ZAs) as well as the 'Organisation of medical follow-ups of employees exposed to carcinogenic substances' (ODIN) are immediately used for organizing individual work-medical surveillance for all exposed persons registered.

The anonymous data material may as well be employed for the purpose of epidemiological studies.

There are two decentralized information systems: 'Hazardous substances and safety' (GeSi) of the professional association of the ceramics and glass industry and the 'Information system on hazardous substances' of the professional associations for the building trade (GISBAU); both databases serve to process substance and product data with a view to drafting operating instructions and safety data sheets for member enterprises, insured persons and work safety experts.

It is just this domain where a lot of work and commitment is still necessary to close the gap that exists between the complex and differentiated central documentations and their specific branch, workplace, product, user oriented and – above all – easily understandable presentation.

3. LOGIC STRUCTURE OF GESTIS (Figure 1)

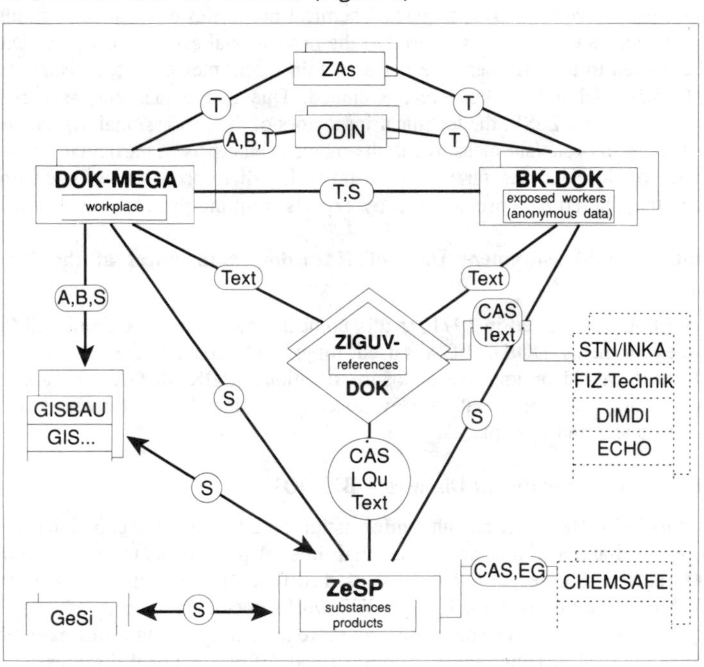

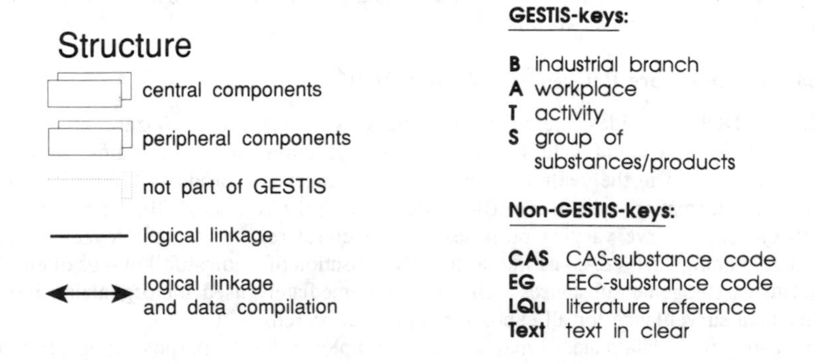

Fig. 1. Logic structure of GESTIS

All databases and registers are logically interconnected by use of commonly coded keys. The key system includes the central compounds of GESTIS: Substances and products, industrial branches, workplaces, exposed persons. Individual keys for these compounds permit information linkup; systematic access and logic information linkup according to characteristic categories are possible by means of systematically structured group keys for the above mentioned compounds (Table 1).

By using the GESTIS key system:

- single information from different databases may be linked,
- data from different databases may be compiled for assessment,
- information from central databases may be transferred to peripheral information systems and the other way round.

Table 1. Uniform GESTIS key system; origin of the keys in brackets
(BG: professional association; BI: Berufsgenossenschaftliches Institut für Arbeitssicherheit (BIA)

Object	Individual key	Group key
SUBSTANCE, PRODUCT	Central registration number (BIA)	Key for substance and product groups
ENTERPRISE	Membership number (BG)	Branch of industrial activity (BIA in accordance with the system of industrial branches established by the German national bureau of statistics
WORKPLACE	Individual industrial workplace	Workplace (BIA)
EXPOSED EMPLOYEE	Annuity insurance number	Professional activity (BIA in accordance with the system of professional activities established by the German national bureau of statistics

4. INFORMATION FLOW AND USE OF GESTIS (Figure 2)

The basic principle of data processing and use underlying GESTIS is reflected by its structure comprising centralized and decentralized compounds: Division of tasks and differentiation between a central information bank on one hand and peripheral, immediate and intermediate users in the field on the other hand.

The central databases contain a large variety of detailed information. The latter is accumulated by central collection and assessment as well as, for example, by decentralized measuring activities of the different professional associations. They provide the necessary data to support the professional associations in their preventive work for their member enterprises and the insured employees, viz. workplace monitoring, administration of occupational diseases, work medical care and surveillance, generation of product safety information.

Data are not only collected centrally but they are also subjected to central analyses with respect to general work safety problems, such as the publication of statistics on occupational diseases, provision of exposure registers, realization of epidemiological studies. These centrally processed data are not only meant to be used by member enterprises and insured persons but also by the public at large.

The principle of task division (centralized/decentralized) and differentiation is a particular characteristic of GESTIS guaranteeing adequate data quality and quantity and permitting user oriented implementation of the supplied information.

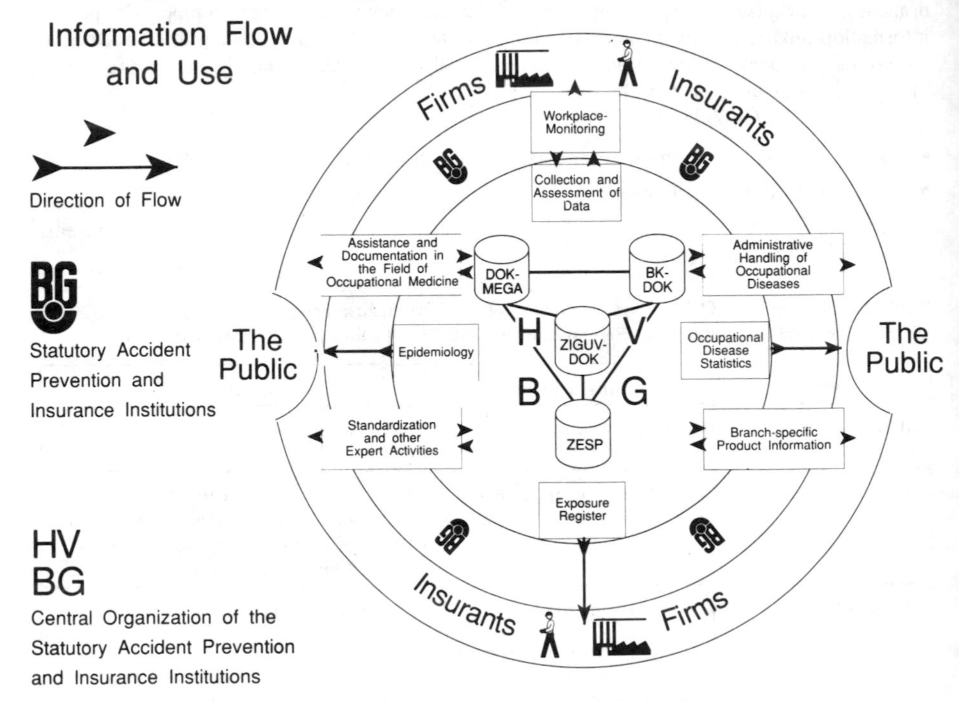

Fig. 2. Information flow and use of GESTIS

5. COOPERATION AT THE EC LEVEL

Immediate action has to be taken to respond to the fast progressing activities of the CEC (e.g. as regards the publication of Directives on hazardous substances or the development of the European database on chemical substances, EUCLID) and to take them into account when further developing our own information systems. This is especially true with respect to the key registers in use (GESTIS key for groups of substances and products), to the harmonization of logical and formal database structures as well as with regard to a common use of data pools and agreements on uniform conditions for use (which should e.g. exclude commercial exploitation).

As to documentation on exposure situations and occupational diseases it is necessary to define the essential contents and to harmonize or adapt the different keys employed.

An example in this context would be harmonized information systems on exposure situations using uniformly coded keys for substances and products as well as for workplaces, professional activities and industrial branches.

As a matter of fact, this is the only way that permits comparable analyses of the actual work safety situations in the different EC Member States and thus lays the basis for any further commonly concerted action.

1413

A NATIONAL EXPOSURE DATABASE

P.E. FJELDSTAD and T.WOLDBÆK

National Institute of Occupational Health
PO Box 8149, 0033 Oslo, Norway

SUMMARY

Since 1985 occupational exposure data from about 45 000 samples are stored in the Norwegian database EXPO. The data can be used to provide evidence of the conditions in a factory or an enterprise over time, to support the setting up of occupational health standards (national TLVs) and in epidemiological research. In addition the database can be exploited to reveal new hazards and to evaluate the use of compounds whenever new toxicological experience becomes available.
Examples of data from the mechanical industry are shown. Lead exposures in connection with glass and battery production are compared. Recorded data are used to evaluate what vapour solvents seem to be of most importance in the work atmosphere, and in which types of companies they are found.

1. INTRODUCTION

The National Institute of Occupational Health in Oslo is a governmental research institute in the fields of occupational medicine, hygiene, physiology and toxicology. Our collaboration with the Labour Inspection is extensive. The institute is used as a regional laboratory, and is receiving work environment samples from Norwegian companies for chemical analysis. In addition to our institute, there are three regional laboratories belonging to the Labour Inspection.

Annually about 30 000 concentration values of compounds in work environment are reported by the regional laboratories. These values are the analytical results of about 10 000 samples and are used to evaluate health hazards to workers. The samples contain solvent vapour, blood, urine, aerosols etc. About 70% of the analytical work is done at the Institute of Occupational Health in Oslo.

The companies are responsible for carrying out the necessary chemical analyses to be sure they are in compliance with Norwegian legislation and rules for work environment. The results from these analyses shall be available for the Labour Inspection on request. The analytical work is made at our institute, by one of the regional laboratories or by some commercial laboratory.

Since 1985 about 45 000 samples have been analyzed at our institute and stored in the database EXPO. In collaboration with the Labour Inspection the database has now been further developed. During 1992 the data analyzed at the other three governmental regional laboratories will be included.

The data can be utilized in several ways, for example: to provide evidence of the conditions in a factory or an enterprise over time, to support the setting up of occupational health standards (later referred to as TLVs), in epidemiological research or as evidence in court. In addition the database can be used to reveal new hazards and to evaluate the use of compounds whenever new toxicological evidence becomes available.

The Labour Inspection contemplated such a database for several years. A pilot project was run in the early eighties with data from glass-fibre reinforced plastics and formed a basis for the development of EXPO.

As our institute needed a system for laboratory administration, the two tasks were combined. The resulting system satisfies the demands of the Labour Inspection as well as the additional demands placed on a system for laboratory administration. At present EXPO covers mostly the eastern part of Norway as far as air samples are concerned, while biological samples to monitor the work environment are arriving from all over the country. The Labour Inspection and the Institute of Occupational Health are aiming at making EXPO cover exposure data from most kinds of work from the whole country.

2. SYSTEM DESCRIPTION

Until now the system has been based on a Norwegian-developed fourth-generation tool for database construction (FICS, Kvam data a.s.). It is running on a ND-100 minicomputer (Norsk Data a.s.). From spring 1992 EXPO will be run under the relational database system ORACLE (Oracle Corporation) on a Nokia Alphascope 30 (NOKIA Data).

The data have been organized at three levels of tables as demonstrated in Table 1.

Table 1. The Types of Data Stored in the Database

Level 1 -Series
Identification No. - Series
Project No.
Receiving date
Reporting date
Employers identification No.
Company name and address
Industry type - code and description
Sampling method - filter, charcoal tube etc.
Sampled by - type of institution
Analytical method - atomic abs., gas chrom. etc.
Analyst in charge
Free text - used as heading in reports
Level 2 - Samples
Identification No. - Series and sample
Sampling date or time
Job description
Sampling code - personal, stationary, biological etc.
Exposure type - long term, peak exposure etc.
Working conditions related to normal
Hours per. week for this operation
Person's name, sex and birth day
Protecting equipment
Level 3 - Compounds
Identification of series, sample and compound
Measured concentration

Level 1 contains data belonging to series, that is one batch of samples from one company. Data concerning a company like the employers id. no., company name and address and industry type, are gathered in an address table which is continually maintained. There is a table with the different industry types, and only descriptions from this table are allowed. For later use of the data, it is important to make sure that the same object is not identified in different ways.

Sampling made by a local industrial hygienist will not always be judged as reliable as the sampling made by a skilled and trained hygienist from the Labour Inspection. Hence, we have recorded the sampling method and the kind of institution that has been responsible for the sampling.

Level 2 contains data belonging to samples in the series. As job description the common notation used in the companies is recorded. This important information is tedious to use because the same kind of work may be recorded in different ways. In the future the job descriptions will be standardized in the same way as the industry types are to day.

Level 3 contains data belonging to compounds in the samples.

Most of the registration work is done by our institute secretary. Part of the analytical results are transferred electronically from the analytical instruments. This decreases the amount of wrongly reported results. The results may be typed in by the analyst himself, who also will decide when an analytical series is finished, and should be marked so that the data can no longer be altered.

The system language is Norwegian. We have sent data from EXPO abroad. Final data reports may easily be translated to help the exchange of information across borders.

3. EXAMPLES

The Mechanical Industry.

EXPO contains data from 935 samples from 80 different mechanical companies as shown in Table 2.

Table 2. Samples from the Mechanical Industry

Type of sample	Number	Comment
Lead in blood	60	
Materials/products	96	Mainly asbestos
Air samples	779	Elements : 260 Solvents : 297 Fibres : 43 Oil mist : 42

This is a complex type of industry that involves several work processes, for instance welding, soldering, moulding, cleaning and painting, and many different work environmental problems could be expected.

Solvent vapours were found in 297 samples. Xylenes, ethylbenzene and toluene are the most frequently detected, but also the aliphatic hydrocarbon mixtures are found quite often. Compared to the TLV the highest concentration is found for methylethylketone (MEK), but also benzene and 2-ethoxyethylacetate are measured in high values compared to their TLVs. However, none of these compounds is found very often. Toluene is among the most frequently detected compounds, but the average value is only about 25% of the TLV (TLV=40 ppm). The data in EXPO do not indicate that solvent vapours represent a great problem to the work environment in the mechanical industry.

About 300 samples containing aerosols from the mechanical industry are recorded. Figure 1 shows that Fe, as would be expected, is frequently measured in high values, but also Al, Mn and Mg are found quite often, but in lower concentrations compared with the TLV. In addition to the elements and dust, oil mist was found in about 40 samples with an average value of about 50% of its TLV.

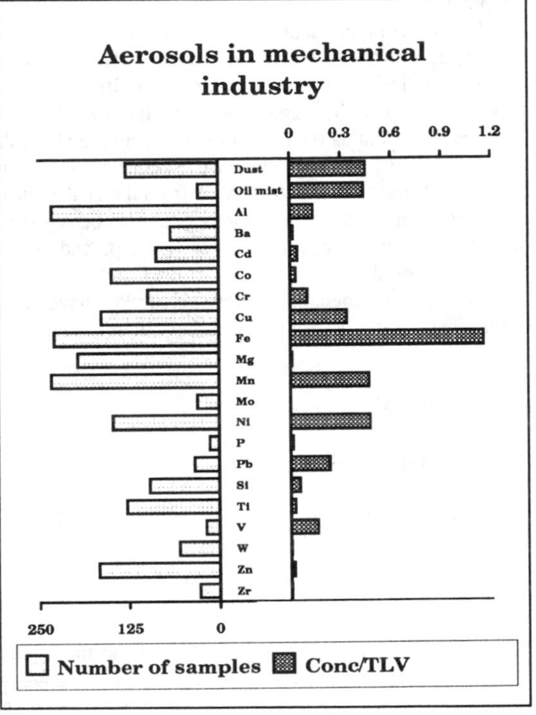

Figure 1

In the 60 blood samples most of the lead concentrations are normal values, but some elevated values have been detected. High values are found in a factory in connection with soldering on radiators. Blood samples from the workers in this factory are analyzed annually.

Samples of different kind of materials have been analyzed. Typical examples are raw materials and dust that has settled on horizontal surfaces. In these samples asbestos fibres are frequently found. 50 samples contain chrysotile asbestos.

Lead in Blood Samples from Battery and Glass Production

In the production of glass and batteries lead represents a health risk. Since 1985 4673 samples have been recorded from the battery production. These samples mainly originate from three different factories. In early 1990 only two of them were left, and during 1990 the largest one was closed. The remaining factory uses a commercial laboratory and the data are no longer recorded in EXPO. Only 73 samples were analyzed in 1990 while in 1985 the number was 1355.

Figure 2 shows the distribution of concentrations of lead that was found in blood samples in 1990 and 1985 from battery production. Regarding to the few samples in 1990 the distribution is estimated to be the same for both years, and a significant number of high values are found. The median and average values for 1985 were 1.7 μmole/litre, while for 1990 the corresponding values were 1.6 μmole/litre. In Norway 2.5 μmole/litre is the highest concentration allowed for a worker before she or he should be taken out of work which might involve risk of lead exposure. The worker must stay away from any work involving lead exposure until the blood has a lead concentration below 2.0 μmole/litre.

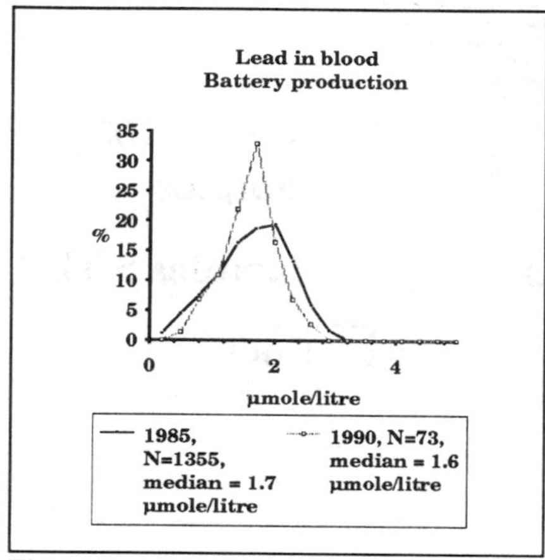

Figure 2

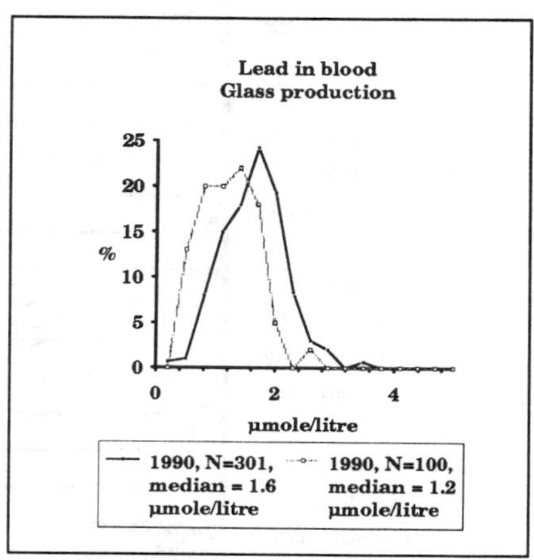

Figure 3

For females who might get pregnant the concentration of lead in the blood must not exceed 1.5 μmole/litre.

In the glass industry there are somewhat better conditions than in the production of batteries. From the glass production there are 1227 samples from 16 different companies. For this industry the concentrations of lead in blood samples were lower in 1990 than in 1985 (Figure 3). The median and the average values were 1.7 μmole/litre in 1985 and 1.4 μmole/litre in 1990. In Figure 4 the two kind of industries are compared.

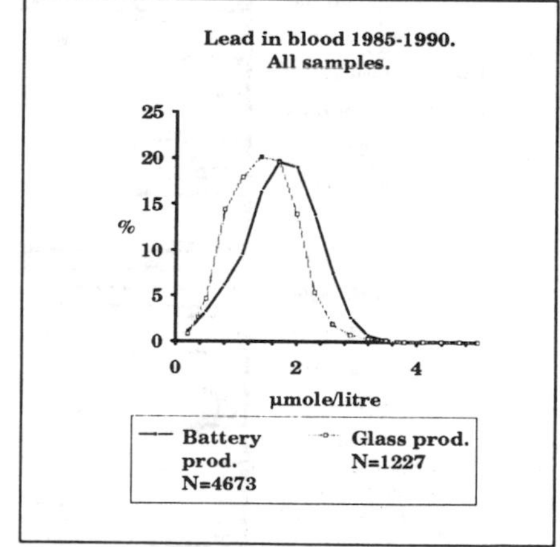

Figure 4

Solvent Vapour

In EXPO are recorded 7705 samples containing solvent vapour from a variety of different industries.

Figure 5 shows what compounds are found. Only compounds detected in more than 100 samples are shown. The white bars give the percentage distribution of each compound in the 7705 samples. Toluene and the xylenes are the most frequently detected. Toluene is found

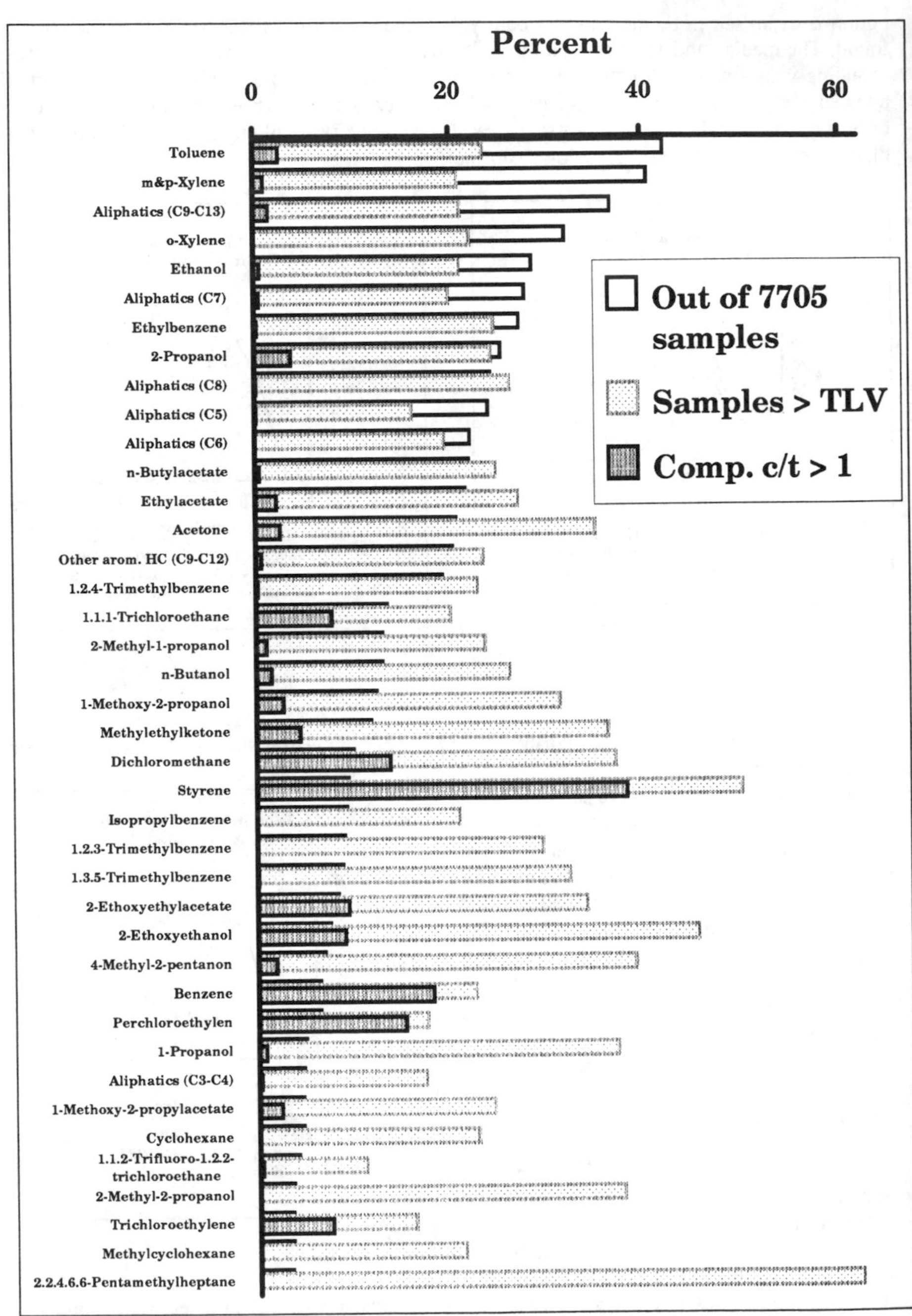

Figure 5. Distribution of solvents.

in more than 40% of the samples. But also the aliphatic hydrocarbons, ethanol, ethylbenzene and 2-propanol are found quite frequently.

1667 (21%) of the samples exceed the TLV for mixtures. The average value for the additive exposure value is 0.74, and 50% of the samples have an additive exposure value less than 0.16. The light shaded bars in Figure 5 show the percentage of the samples containing a given compound, that exceed the TLV for mixtures. Compounds, with bars higher than 21%, are found more often than the average in samples above the TLV. The most frequently detected compounds are close to the 21% average. As could be expected compounds found in fewer samples have the bars that differ most from the average. About 50% of the samples containing styrene exceed the TLV, and about 70% of the samples containing 2,2,4,6,6-pentamethylheptane are above TLV.

The darker shaded bars in Figure 5 show the percentage of the samples where the concentration of a compound alone exceeds the TLV. In total 1178 samples, about 70% of all the samples exceeding the TLV for mixtures, contain at least one compound exceeding the TLV alone. In more than 40% of the samples containing styrene, styrene alone is responsible for the high additive exposure value, while 2,2,4,6,6-pentamethylheptane that was mainly found in samples exceeding the TLV, not even in a single case have been responsible for the high exposure value alone.

Styrene is mainly found in the production of plastics, so also is dichloromethane, while benzene is found in the oil industry and perchloroethylene in laundries. These were the compounds, that were most frequently measured with high values compared to their TLVs.

The highest amount of samples exceeding the TLV is, as expected, found in the production of plastics (41% of 862 samples) since styrene is mainly found in this industry. 632 samples have been taken in connection with printing, 169 exceeding the TLV. Also in the production of paintings there are several high values (34% of 401). Among the industries with fewer samples, 79 of the 160 samples taken in connection with airplane maintenance exceed the TLV.

4. DISCUSSION

A database like EXPO does not contain a randomized population of samples. Data to be recorded in the database have been collected by occupational hygienists to answer specific questions. It is necessary to know the sampling background to extract correct information from the database. Possible causes of data bias to be considered are:

- Samples taken to confirm suspected high exposure levels will be higher than 'true' average.
- A test to evaluate a new ventilation system will most likely yield low results.
- Tests carried out to calm workers upset for some reason, in fact not connected to the exposure level, might result in low values.

In the examples of lead exposure in the battery and glass industries the average blood values will not represent the workers' average values. Samples are taken from those workers who are possibly exposed to lead, and the persons with the highest values will deliver blood samples more often than persons with lower values.

To be able to judge the significance of an exposure, it is necessary to take into account what kind of exposure the measurement is supposed to represent. Was the sample taken at peak exposure or is it an average of an eight hours' exposure? How many hours out of the total working time are spent with this particular type of work, and is the exposure level a representative one?

In the examples above, none of these points have been considered. This would be possible, because all the necessary data are systematically stored in EXPO, as seen from Table 1. But it would be considerably more tedious than just to make simple surveys like those in the examples above.

We are convinced that data collected to serve specific occupational hygiene purposes will be of valuable help solving problems elsewhere, if made available. When they are pooled together, they will form an important source of information, that may accelerate the struggle for better work environment.

1414

CRAM* INTERREGIONAL CHEMICAL LABORATORIES:
A STRUCTURE FOR THE ASSESSMENT OF WORKERS' EXPOSURE TO HAZARDOUS CHEMICALS

Authors : Pierre LARDEUX, CNAMTS**
Jean-François CERTIN, Laboratoire Interrégional de Chimie de l'Ouest (LICO), CRAM des Pays de la Loire

Eight laboratories have been established in France ; they are an integral part of the CRAM' s Occupational risk prevention department and are funded by the National occupational risk prevention fund financed by employers' occupational accident and disease contributions. About a hundred persons are employed for air sampling in factories and analysis. In coordination with the CNAMTS, a close collaboration with the INRS is set up. It concerns analysis and sampling materials and methods.

Their function is : to detect and to estimate chemical hazards at workplaces, to give advice on prevention issues, to inform, to make aware.

Questions arise from various sources : CRAM's themselves ; factories : managers, occupational physicians, Health, safety and working conditions committees ... Actions may take the form of campaigns, studies in special occupational areas - woodworking industries, printing industries, quarries, dental prosthetists, , ...- or on specifics demands - ventilation device control, new processes ... -

A significant amount of data are collected in COLCHIC, a national data bank set up at the INRS. Common working methods and strategies are used by all laboratories.

Every year	-	1000	firms examined	:	in the metals, woodworking, building industries ...
	-	10000	air samplings	:	silica, asbestos dusts, welding fumes, solvent vapors ...
	-	1200	products analysed	:	paints, solvents, adhesives ...

In accordance with the strategy agreed upon, priority is given to personal sampling during several hours to allow comparison with the recommended threshold limit values.

The knowledge of workers' exposure so collected will grow from year to year. For example, in the last three years several hundreds of air samplings have been carried out during silver-cadmium brazing operations.

Concrete actions are suggested to firms : Laboratories' measurements lead, for example, to assess ventilation device efficiency or location and size of pollution sources. Observation of actual working practices during the whole work day is an essential complement for assessing exposure in accordance with the different work phases. Surveys are followed by reports sent to the employer and to the occupational physician, in which the results are presented with practical advice in order to better prevent the chemical hazards (ventilation, selection of chemicals ...).

* Regional sikness insurance fund
** National sickness insurance fund (Salaried employees)

1415

DOCUMENTATION MEGA: MEASUREMENT DATA ON HAZARDOUS SUBSTANCES AT THE WORKPLACE

M. STÜCKRATH

Berufsgenossenschaftliches Institut
für Arbeitssicherheit (BIA)
Alte Heerstraße 111
Sankt Augustin
Federal Republic of Germany

The German statutory accident prevention and insurance institutions in industry are held to support their member-enterprises by means of technical control and advice on all kinds of work safety problems. Among this count the monitoring of workplace exposure to hazardous substances.

In support of these missions, the German statutory accident prevention and insurance institutions in industry (professional associations) created a special measuring system for"Hazardous substances" run by the professional associations and the BIA together.

In this system, the professional associations are responsible for

- immediate contact with the enterprises and provision of technical advice
- decentralized sampling
- decentralized quality assurance

The Berufsgenossenschaftliches Institut für Arbeitssicherheit is responsible for

- organisation of the system
- coordination of quality assurance
- development of measuring methods and software
- central analysis.

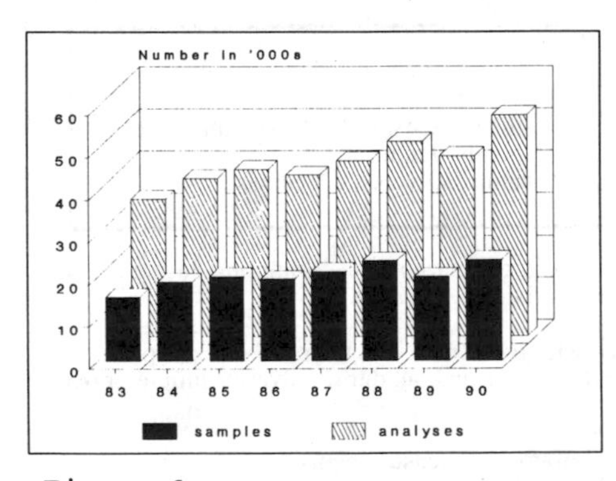

Figure 1:
OMEGA: Samples and analyses between 1983 and 1990

The technical instrumentation of the BIA comprises sampling- and analytical methods for a total of 290 different substances. The institute's annual capacity in terms of pollutant measurements comes up to

- 25,000 samples from 3,900 firms on which 54,000 analyses are carried out

During the measuring activities all necessary data are collected

- for recording the measurement
- for assessing the exposure situation at the workplace
- for evaluating the data pool for comparative studies (identification of typified characteristics)
- for complementing the information system on hazardous substances - GESTIS - created by the professional associations.

The data collected refers to

- enterprise
- branch of industrial activity
- workplace
- manufacturing procedure
- working materials and products
- spatial conditions
- exposure conditions
- measuring method/conditions
- analytical methods
- application of technical rules for assessment.

Primarily, the data are used to establish an individual measuring report completed by hints for assessment based on legal regulations. Further, they are entered into the database MEGA. At present, the database has the following volume:

- since	1972
- information fields per record	150
- number of records	400,000
- number of analysed substances	290
- enterprises	16,000
- types of workplaces	4,000

Validity and completeness of data are essential. This implies high quality requirements with respect to sampling, analysis, data collection and -processing. This means

- regular exchange of experience, training
- uniform application of tested measuring methods
- provision of uniform working means, e.g.
  * formulars
  * information on sampling methods and exposure assessment
  * key registers
- uniform reporting based on the technical rules for assessment
- double input and plausibility check during data entry

At present, the data pool is comprehensive enough to be used for an individual description of all types of industrial workplaces, manufacturing procedures and working materials in terms of exposure to hazardous substances.

Use of data

Due to the strict requirements applying to the protection against misuse of data in data processing in the Federal Republic of Germany, full data analysis is restricted to the professional associations that provided the data.

More complex statistic approval is possible on the basis of anonymous, typified data extractions and with the agreement of the data owner.

The documentation is used for the following purposes:

- Determination of exposure registers dependent on the branch of industrial activity and on the type of workplace.
- Determination of workplace registers dependent on the exposure situation.
- Trend analyses for temporal comparison of exposure situations at certain workplaces.
- Efficiency control to investigate the effectiveness of technical and organisational protective measures.
- Determination of limit values, determination of the state of art and estimation of the impact of lowered limit values in industry.
- Determination of exposure situations at particular or comparable workplaces within the framework of investigations into the occupation-bound development of diseases for which benefit is claimed.

The example of a trend analysis presented in figure 2 was taken from the foundry industry; due to the large number of persons working in this branch of industrial activity in the Federal Republic of Germany, the data collected in foundries represent a large portion of the entire data pool comprised in MEGA.

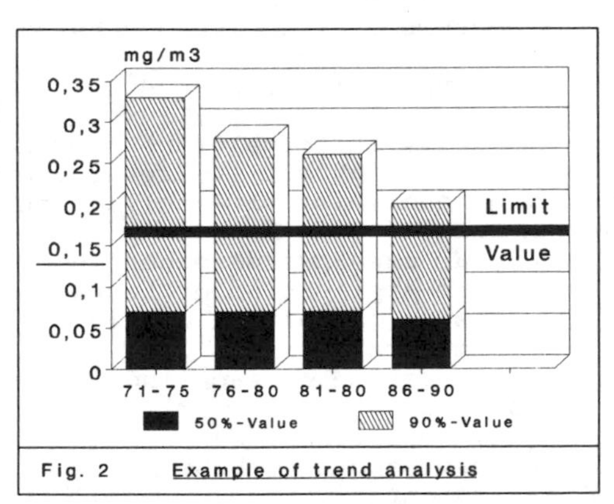

Fig. 2 Example of trend analysis

Number of measurement data per collection period:
1971-75: 1124 1976-80: 7690
1981-85: 7634 1986-90: 2951

Last but not least, MEGA is a long-term database aimed at serving epidemiological purposes.

1416

DEVELOPMENT AND MANAGEMENT OF OCCUPATIONAL HYGIENE DATABASES

P.L. BEAUMONT

Health and Safety Executive, Technology and Health Sciences Division
Magdalen House, Stanley Precinct, Bootle, UK

INFORMATION NEEDS

- UK Control of Substances Hazardous to Health Regulations 1988 (COSHH)
- EC Directives
- Assessment of risks to health
- Control of risks to health
- Exposure measurements

COMPUTER SYSTEMS

- Efficient means of storage and selective retrieval
- Occupational hygiene software is increasing
- No consensus

SHARING EXPERIENCE

- Avoids wasteful duplication of effort
- Need to standardise
- British Occupational Hygiene Society Special Interest Group on Data and Information Systems
- Industry, Government, Consultancies
  Software suppliers and publishers
- Two high priority projects
- Chair - Paul Beaumont, Health and Safety Executive

CORE INFORMATION

PRESENTATION OF EXPOSURE MEASUREMENTS

- Minimum information requirements
- Professional Standard BOHS Conference 1991

MANIPULATION OF DATA

- Guidance to aid software

CONCLUSIONS

The British Occupational Hygiene Society with the UK Health and Safety Executive's National Exposure Database

- International adoption
- Standard core set of information
- Good quality exposure data for:
  - controlling occupational exposure to hazardous agents
  - epidemiological studies
  - setting occupational exposure limits.

PLASTBASE. A database for PCs on chemical emissions and occupational health in the plastics processing industry.

B. Jensen, A. Schmidt and P. Wolkoff
National Institute of Occupational Health, Denmark

Introduction

The consumption of plastics is steadily increasing as well as the number employed within the plastics processing industry. Presently more than 70 different types of plastic polymers are in use in Scandinavia [1], and the number of new types of plastics (copolymers, blends, composites, etc.) and additives is ever increasing. Due to the complex spectra of thermal degradation products from most plastics, work on occupational health in the plastics processing industry is faced with an immense complexity of the chemical environment, leaving the non-specialist in shortage of relevant information for assessing the risk of possible exposures or determining the cause of identified symptoms.

	number of polymer types		
x	number of additives		
x	number of manufacturing processes	=	"∞"

This situation has prompted the Danish National Institute of Occupational Health to construct a database for PCs, named PLASTBASE [2), 3)], giving easy and fast access to key informations about chemical emissions, toxicological evaluations, and occupational health in the plastics processing industry. The database, based on relevant chemical and toxicological informations from the international literature, is intended as a tool for the industrial safety organisations, environmental consulting firms, industrial hygienists, and the occupational health authorities.

Working environment of the plastics processing industry

Thermal degradation of plastics covers a number of phenomena occuring to varying degree simultaneously. There are:

Process:	Emissions:
Evaporation increases with temperature	Monomer- and oligomer-residues, volatile additives
Pyrolysis sets on at given threshold temperature (x time), usually well above recommended values	Products from degradation of the polymer backbone (true "thermal degradation") and from degradation of additives
Thermooxidation on prolonged heating in air, dependent on degree of stabilization from antioxidants	Products from oxidation of the polymer and additives
Condensation, mechanical treatment	Smoke, fume, aerosol, dust

Secondary manufacturing processes take place everywhere in the plastics processing industry: gluing, welding, crimping of foils, cutting, and other mechanical treatments, surface treatments, and painting. Also here, one is faced with a multitude of volatiles and the risk of overheating with resulting thermal degradation of polymers, additives, and evaporation of residual monomers. Occasionally fatal incidents are seen outside the plastics processing industry when plastic coated metals are welded or polymeric materials are otherwise inadvertently heated[4)].

The concept of PLASTBASE

Information is depending on the context. In PLASTBASE mutually relevant data on chemical emissions from plastics and their human health effects are closely coupled. A database should utilize the inherent ability to present related data from both ends. In PLASTBASE this is done through related files, two main routes being possible: from polymer type to toxicological findings (chemical section, left side of diagram), or from symptom to polymer types or single chemicals (toxicological section, right side of diagram):

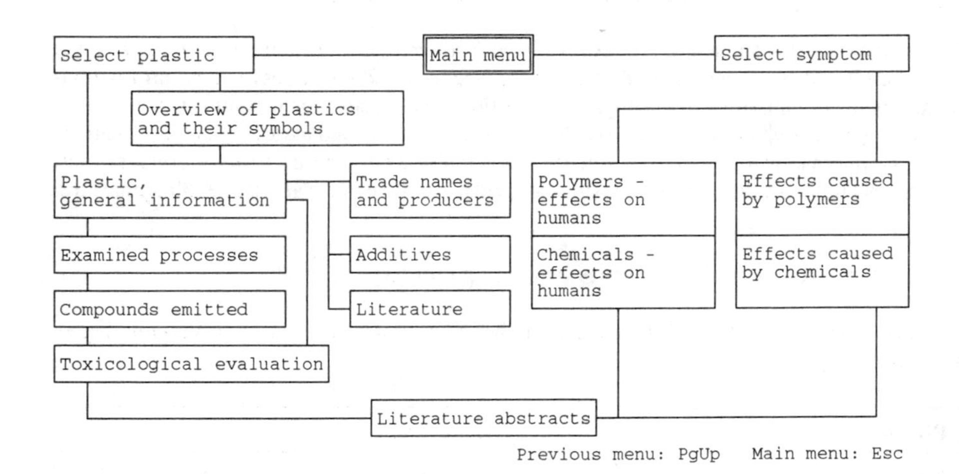

Structure diagram of PLASTBASE

All information is immediately obtainable from interrelated, self-explanatory menus.

Sources of information

PLASTBASE is based on peer reviewed articles obtained from systematic searches in the international databases CISDOC[5)] and TOXLINE[6)]. Search profiles like

plastic or	and	*degradation or
polymer or		thermal degradation or
composite or		thermooxidation or
...		...
polyethylene or		occupational health or
polypropylene or		working environment or
...		toxicology or ...

have been used. Moreover, informations from safety sheets, trade journals, etc. have been included. Toxicology on single chemical compounds is collected from standard handbooks, IARC Monographs, etc. Around 380 references are used. Data regarding fire and combustion are omitted. Adjacent areas like varnishes, adhesives, rubber, and elastomers are only slightly touched.

Conclusions

Some conclusions may be drawn from the contents of PLASTBASE:

Thermoplastics:
Although varying amounts of thermal degradation products will always be present during the processing of thermoplastics, regular processing on modern and maintained machinery does not normally yield massive exposures. However, the industrial environment suffers from problems similar to indoor air problems with symptoms arising from prolonged exposure to low levels of irritants, possibly in the form of particulates or aerosols, and often in conjunction with heat, dry air and static electricity[7)]. Irritation of eyes, nose and mucous membranes are the most frequently recorded symptoms[8)]. Malodours often occur, giving rise to anxiety about possible exposures. Traditional workplace measurements toward threshold limit values of one or two components are often grossly inadequate considering the complexity of the chemical environment. There is a need for more detail in the approach of the industrial hygienist and a better understanding of thermal degradation of plastics. It is well known, that special precautions are needed against overheating of poly(vinyl chloride) (PVC), polytetrafluoroethylene (PTFE) and polyoxymethylene (POM).

Thermosets:
In contrast, the manufacture of glassfibre reinforced polyester, polyurethane foam, phenol-formaldehyde thermosets, and epoxy resins may give risk of high acute levels of reactive chemicals with danger of CNS symptoms and sensitization. Increased use of closed processes and automation, and improvements in ventilation techniques seem to be the only solutions[9)].

Acknowledgement

PLASTBASE has been financially supported by the Danish Work Environment Fund.

References

1) Anon.
Plastforum Scandinavia 1990;12:32-43

2) Jensen B, Schmidt A, Wolkoff P.
PLASTBASE. A database on chemical emissions and occupational health in the plastics processing industry. (in Danish)
Danish Work Environment Fund 1989

3) Jensen B, Schmidt A, Wolkoff P.
PLASTBASE. (in Danish)
Plastorama Scandinavia 1989;(4):4-5

4) Fogh A, Brangstrup JPH, Rasmussen K.
Death after machining of polyurethane plastic with exposure to 1,5-naphthylene-diisocyanate (NDI).
J Toxicol Clin Toxicol 1985;23:437

5) CISDOC
International Occupational Safety and Health Information Center, ILO.
CH-1211 Geneve 22

6) TOXLINE
Deutsches Institut für medizinische Dokumentation und Information.
D-5000 Köln 41

7) Indoor air pollutants: exposure and health effects. EURO Reports and Studies 78.
WHO, Regional Office for Europe, Copenhagen 1983

8) Jensen B, Schmidt A, Wolkoff P
Occupational health problems in the Danish plastics industry - an interview investigation. (in Danish)
Danish Work Environment Fund 1991

9) Astrup Jensen A, Breum NO, Bacher J, Lynge E.
Occupational exposure to styrene in Denmark 1955-88.
Am J Ind Med 1990;17:593-606

PROVISION OF INFORMATION FOR ASSESSING RISKS ARISING FROM THE USE OF HAZARDOUS SUBSTANCES AND PREPARATIONS

David J. Evans
Head of Health and Safety
British Steel plc, London

Introduction

Accurate and reliable measurements of pollutants in the working environment provide data which is generally helpful and sometimes essential for the assessment of health risks and the precautions which are necessary. A systematic assessment, as required in the UK by the new Control of Substances Hazardous to Health (COSHH) Regulations[1], starts by identifying potential hazards associated with the substance or preparation and evaluating the risks arising from the way it is used. The need to measure exposures in order to compare with occupational exposure standards[2] might be indicated by the assessment for which the line manager is responsible.

Assessment of Risk

The purpose of the assessment is to determine whether the work of an operator is likely to give rise to any risk to health which should be eliminated or for which control is necessary. This must be followed by the provision of any necessary precautions and then monitoring their effectiveness. The core of the assessment is therefore a judgement depending on knowledge of the nature of the hazard associated with a substance being handled or generated and the work operations which might result in exposure to them.

The line manager having the responsibility for the work is best suited to make the assessment, provided that he is supplied with adequate information about the substance, trained in the application of that information and that he is supported by professional specialists where necessary. Within British Steel a major management training programme was undertaken when the new Regulations were introduced backed up by the Company's own team of occupational hygienists and its medical service. The problem of how best to provide information was anticipated in the period appeared before the Regulations were made and this lead to the establishment of a health and safety database (HASD).

Hazardous Substances Data

A basic requirement of the UK Legislation[3] is that suppliers of substances for use of work should provide information about risks which could arise and the precautions which might need to be observed. Information and advice provided by suppliers have no common content or format although this will be remedied by a European Directive[4]. Experience has shown that data supplied tend to over-or under-emphasise the hazard according to the possible effect on the market, while data sheets have been inaccurate as well as inadequate. Generally, the data sheets have been incapable of providing readily available, understandable and reliable information which is essential for a manager making an assessment. It was recognised that a piecemeal approach to this problem by individual line managers would be inefficient and costly, particularly where the substance was used at several locations within a works and in a variety of different works.

This same issue had been faced some years earlier in one Division of the Company which had developed a card index system to consolidate information about hazard and precautions related to substances used in its works. This provided valuable experience in recognising the need for a central data service to provide edited and supplemented information for simple presentation to managers in a standard format.

During the two years before the introduction of COSHH, a database was prepared for operation on the Company's mainframe computer with access via a national communications network. An initial feasibility study identified suitable software and enabled the requirements to be specified in terms of use as well as supply.

It was established that the information should be:-

- comprehensive and include all preparations and substances, so that absence of hazard could be positively confirmed rather than merely assumed from the absence of a record. Each substance record was required to provide all the necessary information that a manager might reasonable require including a grading of the hazard to identify those substances most urgently requiring substitution. The data were required to include advice on precautionary measures and extend to matters of fire safety, although not required by COSHH.

- reliable in order to win the ready confidence of those who use the database, so requiring all entries to be validated by an occupational hygienist and occupational physician.

- easy to use by a line manager with general rather than specialist training. This required a consistent format for presentation and a simple means of access.

Shortly before the introduction of COSHH, an extensive training programme was undertaken to provide line managers, not only with a basic understanding of the principles of occupational health protection, but particularly in the use of HasD and making assessments. Within a year of the coming into force of the Regulations, nearly 9,000 substances and preparations were recorded in HasD and virtually all the first assessments completed with a start made in refining and improving assessments especially those relating to the most hazardous categories. In addition to British Steel's use of the database via its communication network, some other organisations are able to gain access to HasD by means of modems.

Conclusions and Further Developments

HASD has provided a cost-effective and reliable database which has been invaluable in the Company's strategy for ensuring the proper protection of the health of its workers and in complying with COSHH. The experience gained has facilitated the current development of a more general compendium of information on all aspects of health and safety[5] which will be of value to the line manager, including summaries of occupational hygiene survey results, guidance and legislation. This health and safety information service (THESIS) will also hold computer based training modules on various aspects of health and safety.

It is planned to incorporate HASD into the more comprehensive THESIS system and to make it available to other users by means of a CD-ROM.

Further information may be obtained from British Steel plc, 9 Albert Embankment, London, SE1 7SN.

REFERENCES

1. Control of Substances Hazardous to Health Regulations 1988, (SI 1988 No. 1657), HMSO.

2. Occupational Exposure Limits: HSE Guidance Note EH40 (1991), HMSO London.

3. Health and Safety at Work etc Act 1974, (1974 c37), HMSO London.

4. EC Directive (91/322/EEC) on establishing indicative limit values by Implementing Council Directive 80/1107/EEC on the Protection of Workers from Risks Related to Exposure to Chemical, Physical and Biological Agents at Work. Official Journal of the European Communities, No. L177, Pages 22-24 of 5.7.91.

5. Developing and Assessing Means of Rapidly Distributing Information, ECSC Ergonomics research project 7239/13/011.

1419

THE SUBTEC SOFTWARE PACKAGE

E. OLSEN

National Institute of Occupational Health
Lersø Parkallé 105, DK 2100 Copenhagen Ø, Denmark

A computer software package for substitution of chemicals and relative exposure assessment in indoor and outdoor environments has been developed at the Danish National Institute of Occupational Health, in co-operation with EnPro ApS, Symbion Science Park, Copenhagen.

The software package consists of:

1. DATABASES, containing 50 technical and environmental properties for more than 800 substances, e.g. vapour pressure (as function of temperature), melting and boiling points, Henry's law constants and octanol-water partition coefficients, occupational and outdoor environment exposure limits etc.
2. A CALCULATION FUNCTION for calculation of physico-chemical and environmental properties of pure components and mixtures, based on the UNIFAC model, e.g. calculation of the SUBFAC-index (relative risk of violating the accepted air quality standards), calculation of emission rates, vapour pressures, flame points etc.
3. A SUBSTITUTION FUNCTION for 'Computer Aided Design' (CAD) of mixtures, based on C M Hansen's solubility parameter system, combined with facilities for searching alternatives using technical and environmental criteria.

The user needs no profound knowledge of molecular thermodynamics or physical chemistry in general, but must be a professional in occupational or outdoor environmental hygiene.

THE UNIFAC MODEL

The UNIFAC model is a group contribution model for calculating properties of mixtures from pure component properties. The only information needed (besides the properties of the pure components) is the chemical structure of the components, the composition of the mixture, and the mixture temperature. For the last 15 years, the UNIFAC model has been used worldwide for the calculations of e.g. rectification or extraction process equipment.

C M Hansen's solubility parameter system

In this system, the solubility properties of pure substances or mixture of substances can be described by their energy content. The energy is divided into energy corresponding to dispersion-, polar- and hydrogen bonding forces. It is convenient to depict substances and mixtures in a three-dimensional space, called the solubility space. According to theory, substances or mixtures situated close in the solubility space are energetically similar and are therefore able to substitute each other.

The development of this software package has been financed by the Danish Environmental Protection Agency and Labour Inspection.

Session VI

INDOOR AIR

KEYNOTE PAPERS

POSTERS

327

1420

WHY MEASURE INDOOR AIR POLLUTANTS?

P. WOLKOFF

National Institute of Occupational Health
Lersø Parkallé 105, DK-2100 Copenhagen Ø, Denmark

SUMMARY

Several sources contribute to the pollution of the indoor environment in non-industrial buildings. In spite of low level pollutant concentrations measured in non-industrial buildings recent field and chamber studies indicate that symptoms in form of odour and mucous irritation may be the causative agents. Further research is needed into how to measure and identify agents responsible for poor indoor air quality.

INTRODUCTION

Imagine a train smoking compartment with six heavy smokers and you enter it from the fresh outdoor air. You will experience a pronounced irritation in the eyes. You decide to do a measurement of the mean concentration of, let us say, aldehydes over an hour. You will not be able to relate your results with the experienced symptoms. On the other hand if you develop or already have at hand a method to measure the short-term peak concentrations in the plumes from the smokers you will find that these concentrations do correlate with your experience of annoyance (1). This is more or less in a nutshell the indoor air situation. You have to understand your indoor environment, you have to know how to perform an investigation of a problem or non-problem building, you have to know how to design a sampling strategy, you have to be able to evaluate the results, and to know the possible causative agents of annoyance. You also have to comprehend the indoor parameters influencing the indoor air quality and how they may interact with each other.

North Europeans including Americans, especially in urban environments, spend more than 90 percent of their time indoors, at home, at work or during transport. This amount is even greater for children, elderly or sick people. The more time spent indoors the greater is the total exposure of indoor pollutants. About half of the working forces in the Western hemisphere work in non-industrial workplaces, i.e. offices, schools, kindergartens, etc. The proportion of people to be employed in office buildings or office-like buildings is likely to increase during the next decades due to modern technology, encapsulation, etc. An increase of people working at home is also anticipated (2).

The energy crisis in the seventies demanded a need for tighter buildings and reduced ventilation requirements (2). At the same time use of synthetic building materials increased together with the introduction of modern office technology like photocopying machines, computers, video display terminals, laser printers, carbonless copy paper, etc. These changes resulted in an increase of new organic indoor air pollutants. Indoor/outdoor concentration ratios of typical indoor pollutants were found to be significantly greater than one. Since the indoor/outdoor concentration ratios of organic pollutants generally are much greater than one it is not surprising that adverse health effects are influenced by the indoor air quality. It is therefore surprising that indoor air pollution prevention has been largely ignored in comparison with the efforts of clean up of the outdoor air.

SICK BUILDING SYNDROME (SBS)

Within this time period a series of symptoms became apparent and they were associated with poor indoor air quality. These are related to irritation of eye, nose and throat, sensation of dry mucous membranes including dry skin; general symptoms like headache, mental fatigue, dizziness, nausea, difficulty of concentration, and odour (3, 5). These symptoms are now internationally recognized by the indoor air quality scientific community and WHO as the *s*ick *b*uilding *s*yndrome (SBS) (6).

For a building to be defined as sick the complaint rate shall exceed 20 percent. The consequence of indoor air quality problems as SBS can lead to disability in form of sensory reactions, discomfort, annoyance reactions, absenteeism, low worker productivity, and even chronic effects.

INTERNATIONAL RESEARCH ACTIVITY

The research activity on indoor air quality problems is reflected in the more than tenfold increase of participants from 120 at the First (triennial) International Conference on Indoor Air Quality and Climate in 1978 in Copenhagen to the Fifth conference in 1990 in Toronto (7).

WHO

WHO concluded that organic compounds may cause odours, mucosal and sensory irritation, and airway effects at levels encountered indoors (6). It has also been concluded that biological contaminants together in indoor air account for a substantial proportion of absenteeism in schools and workplaces (8).

WHO recommended in 1989 that complaints about the indoor air quality should not exceed 20 percent. Complaints related to irritating (organic) pollutants and unwanted odours should not exceed 10 and 50 percent, respectively. Further the emission of volatile organic compounds from building materials and consumer products should be determined and evaluated, and approaches to source control should be developed. Methods for assessing the biological burden and activity should be further developed for organic compounds and their metabolites (6). Further methods for sampling and analysis of aero-allergens and biological irritants should be standardized (8).

INDOOR AIR POLLUTANTS

The indoor air quality is influenced by the universe of indoor air pollutants. These are characterized by the following: Concentration levels are generally sub threshold limit values by a factor $10\text{-}10^3$; they have a dynamic behaviour; and they present a multitude and complex matrix of pollutants. The pollutants may be divided into:

1. Gases and vapours (inorganic and organic)
2. Particulate matter inclusive radioactive particles and environmental tobacco smoke
3. Biological air contaminants.

The organic vapours are by far the largest group of pollutants, several hundred have been identified in the indoor environment.

The pollutants exhibit a dynamic behaviour in their continuous or discontinuous presence. The continuous type may be characterized by a constant or variable emission occurring within months to days, while a constant or variable time pattern characterize the discontinuous type occurring within hours and minutes (9). The complexity of pollutants, their multitude of sources, their different emission patterns, the dynamics of the indoor environment, and different effects on humans are paramount to understand and to be able to plan an investigation, understand, and evaluate results related to sick building problems. Therefore the traditional industrial hygiene approach shall be strongly discouraged.

Sampling can be an effective tool of an investigation, as long as it is sufficiently sensitive to the general low concentrations encountered in non-industrial workplaces and environments, and it is planned according to the sampling objective. Because of the multitude of pollutants and plethora of sources the methods shall be specific in their characterization of chemicals and biological materials present. Recent field investigations in non-problem buildings have discussed the relevance of simultaneous indoor air quality measurements and monitor of symptoms (10, 11).

FIELD AND EXPOSURE STUDIES

Field studies using questionnaires together with field measurements have identified many indoor air quality parameters which correlate with complaints of SBS. Some of the causative agents so far found are particulate matter, (12, 13) volatile organic compounds (b.p. 50-300°C), (10, 13-15) the fleece and shelf factor (m^2/m^3) in a room (16), and macromolecular organic matter in dust (MOD) (17).

Volatile organic compounds are the most investigated type of indoor air pollutants. They are important because of their cause of mucous irritation of eyes and the upper airways or perception of malodours. Several personal exposure studies performed under controlled conditions in climatic chambers have further substantiated that volatile organic compounds are likely to cause SBS (18-22). Recent chamber studies with significantly lower exposure levels of typical indoor pollutants than in previously performed studies have shown that SBS in form of identified eye irritation can be provoked (23).

The exposure pathway of indoor air pollutants are in general through the airways, however, exposure of semi-volatile organics and particulate organic/biological matter may also be through deposition on skin and by particulate matter whereupon they are deposited.

From the point of and need for total exposure assessments of the public it is important to have a battery of methods and sampling strategies ready (24).

CONCLUSION

The sick building syndrome is now internationally recognized as an important research field.

The tools to reduce the indoor air pollution and human exposure thereof is to know all potential sources of pollution, how to measure and identify them in different indoor environments. Further it is relevant to know how to measure their typical emission characteristics and behaviour from building materials and consumer products, and to find out how to minimize their emission. Research is needed to identify the causative agents responsible of SBS and their mechanism of annoyance thus calling for healthy building materials and consumer products.

Present activities within BCR should take into account the future need for accurate measurements of typical indoor pollutants at concentration levels a factor $10\text{-}10^3$ lower than threshold limit values. International research activities about indoor air quality should be encouraged by programmes. Further standardized methods for sampling, analyzing, and interpreting the results should be encouraged.

REFERENCES

(1) AYER, H.E. and YEAGER,D.W. (1982). Irritants in Cigarette Smoke Plumes. *American Journal of Public Health* **72** 1283-1285.

(2) MAGE, D.T. and GAMMAGE, R.B. (1985). Evaluation of Changes in Indoor Air Quality Occurring Over the Past Several Decades. Chapter 2 in: Indoor Air and Human Health. Lewis Publishers.

(3) BURGE, H.A. and HOYER, M.E. (1990). Focus on Indoor Air Quality. *Applied Occupational and Environmental Hygiene*. **5** 84-93.

(4) KREISS, K. (1989). The Epidemiology of Building-Related Complaints and Illness. *Occupational Medicine: State of the Art Reviews* **4** 575-592.

(5) MOLINA, C., *et al.* (1989). Report No. 4. Sick Building Syndrome - A Practical Guide. Luxembourg: Commission of the European Communities.

(6) WORLD HEALTH ORGANIZATION (1989). Indoor air quality: organic pollutants. EURO Reports and Studies 111. WHO. Regional Office for Europe, Copenhagen.

(7) Proceedings of the 5th International Conference on Indoor Air Quality and Climate, Toronto, Canada. 1990. Vol. 1-5.

(8) WORLD HEALTH ORGANIZATION (1990). Indoor air quality: biological contaminants. WHO Regional Publications European Series No. 31. WHO. Regional Office for Europe, Copenhagen.

(9) SEIFERT, B. and ULLRICH, D. (1987). Methodologies for evaluating sources of volatile organic chemicals (VOC) in homes. *Atmospheric Environment* **21** 395-404.

(10) HODGSON, M.J., *et al.* (1991). Symptoms and Microenvironmental Measures in Non-problem Buildings. *Journal of Occupational Medicine* **33** 527-533.

(11) RODES, C.E., KAMENS, R.M. and WIENER, R.W. (1991). The Significance and Characteristics of the Personnel Activity Cloud on Exposure Assessment Measurements for Indoor Contaminants. *Indoor Air* **1** 123-145.

(12) ARMSTRONG, C.W., SHERERTZ, P.C. and LLEWELLYN, G.C. (1989). Sick Building Syndrome Traced to Excessive Total Suspended Particulates (TSP). The human equation: health and comfort. Proceedings IAQ 89 3-7.

(13) NORBÄCK, D., TORGÉN, M. and EDLING, C. (1990). Volatile organic compounds, respirable dust, and personal factors related to prevalence and incidence of sick building syndrome in primary schools. *British Medical Journal* **47** 733-741.

(14) BERGLUND, B. and LINDVALL, T. (1986). Sensory reactions to 'sick buildings'. *Environmental* International **12** 147-159.

(15) NOMA, E., *et al.* (1988). Joint Representation of Physical Locations and Volatile Organic Compounds in Indoor Air from a Healthy and a Sick Building. *Atmospheric Environment* **22** 451-460.

(16) NIELSEN, P.A. (1987). Potential pollutants – Their importance to the sick building syndrome – and their release mechanism. In: Proceedings of the 4th International conference on Indoor air Quality and Climate, Berlin West. Vol. **2** 598-602.

(17) GRAVESEN, S., *et al.* (1990). The role of potential immunogenic components of dust (MOD) in the sick-building syndrome. In: Proceedings of the 5th International Conference on Indoor Air Quality and Climate, Toronto. Vol. **1** 9-14.

(18) MØLHAVE, L., BACH, B. and PEDERSEN, O.F. (1986). Human reactions to low concentrations of volatile organic compounds. *Environmental International* **12** 167-175.

(19) KJÆRGAARD, S.K. *et al.* (1989). Human Reactions to Indoor Air Pollutants: n-Decane. *Environmental International* **15** 473-482.

(20) HUDNELL, H.K. *et al.* (1990). Odour and irritation of a volatile organic compound mixture. In: Proceedings of the 5th International Conference on Indoor air Quality and Climate. Toronto, Vol. 1 263-268.

(21) KJÆRGAARD, S.K. *et al.* (1991). Human reactions to a mixture of indoor air volatile organic compounds. *Atmospheric Environment* **25A** 1417-1426.

(22) WOLKOFF, P. *et al.* (1991). Controlled Human Reactions to Building Materials in Climatic Chambers. Part II: VOC Measurements, Mouse Bioassay, and Decipol Evaluation in the 1-2 mg/m^3 TVOC Range. Indoor Air, submitted.

(23) WOLKOFF, P. *et al.* (1991). Human Reactions to the Emissions of Office Machines in a Climatic Chamber (in Danish with English summary) The Danish Environmental Working Foundation. Copenhagen, 67 pages.

(24) LIOY, P.J. (1990). Assessing total human exposure to contaminants. *Environmental Science and Technology* **24** 938-945.

THE CONCERTED ACTION 'INDOOR AIR QUALITY AND ITS IMPACT ON MAN': GUIDELINES FOR INDOOR RELATED INVESTIGATIONS AND MEASUREMENTS

H. KNÖPPEL

Commission of the European Communities
Indoor Pollution Unit

Joint Research Centre, Environment Institute
21020 Ispra, Italy

SUMMARY

The Concerted Action 'Indoor Air Quality and Its Impact on Man' implements a European collaboration in this rather recent field of environmental research. The first part of this paper briefly describes the objectives of this collaboration, the means to achieve them and both the work performed so far and that currently ongoing. Identification and characterization of indoor air pollutants and their sources is an essential aspect of indoor air quality (IAQ) research. An important part of the Concerted Action's work has been focused on the fact (often overlooked) that sampling and analysis of indoor pollutants are only the last steps in a process which primarily needs to establish why, where, when and under which environmental conditions air should be analyzed. Without a correct answer to these questions however, the results of indoor pollution measurements are at risk of being misinterpreted or even meaningless. Work by the Concerted Action related to this important issue is described in more detail in the second part of this paper.

1. INTRODUCTION

At the beginning of the eighties, the Commission of the European Communities started to take an interest in indoor air quality, when a small research activity was included in the Environmental Protection Programme of the Communities' Joint Research Centre in Ispra, Italy. From the start, this activity was accompanied by an effort to organize a collaboration of European workers in this new field. Such collaboration was and remains important for two reasons in particular:

- There are no scientific structures, as exist for research in the field of outdoor air or water quality, dedicated to indoor air quality research. Therefore, research is fairly scattered and often performed by small groups from a wide range of different institutions.
- Indoor air quality is an environmental issue which more than any other requires interdisciplinary collaboration. In fact, representatives from a wide range of scientific/ technical disciplines are involved in indoor air quality research: heating ventilation engineers, architects, psychologists, chemists, hygienists, biologists, toxicologists, medical doctors, epidemiologists – to name only those most frequently involved.

At the end of 1986, it was decided to start the Community Concerted Action 'Indoor Air Quality and Its Impact on Man' (former COST project 613).

The first part of this paper briefly describes the scope, implementation, activities and results from this Concerted Action. The second part outlines in more detail, work related to the identification and characterization of indoor air pollutants and their sources.

2. THE CONCERTED ACTION 'INDOOR AIR QUALITY AND ITS IMPACT ON MAN'

The Concerted Action 'Indoor Air Quality and Its Impact on Man' is the most important activity in the field of IAQ at Community level. It is included in the Community multiannual research programmes for the protection of the environment 1986-1990 and 1989-1992. Work began in March 1987.

Participants

Ten Community countries (Belgium, Denmark, France, Germany, Greece, Ireland, Italy, The Netherlands, Portugal and the United Kingdom) and the Commission are participating actively. Sweden and Switzerland joined the Concerted Action in 1988. Norway and Finland are going to join soon and representatives of these countries are already participating actively. The Environment Institute of the Joint Research Centre participates on behalf of the Commission in the Concerted Action and provides the scientific secretariat and organizational support.

Goals

From the beginning the ultimate goal of this Concerted Action has been to help understand the consequences to human health and comfort caused by air pollution or, more generally speaking, by inadequate air quality in non-industrial indoor environments (homes, schools, offices, etc.). Therefore the main objective is to promote

- identification and characterization of indoor pollutants and their sources
- assessment of population exposure to indoor pollution
- assessment of health effects of indoor pollution
- investigations into complaints about indoor air quality in office buildings.

Implementation

The Concerted Action 'Indoor Air Quality and Its Impact on Man' is implemented through a CONCERTATION COMMITTEE composed of members from all the participating countries, the Commission and a representative of the World Health Organization (WHO), the SECRETARIAT and WORKING GROUPS (WGs).

The Concertation Committee is responsible for the work performed by the Concerted Action. In view of the fact that the Concerted Action has no proper research funds available, the Concertation Committee has essentially three means to achieve the above-named objectives:

- collation, synthesis and dissemination of knowledge and data
- development and validation of guidelines and harmonized (reference) methods for investigations or measurements
- organization of workshops, symposia, seminars and similar venues.

In particular, the Concertation Committee

- decides on the working programme
- prepares reports summarizing available knowledge of important issues relating to indoor air quality
- identifies ongoing research in the participating countries and major research needs
- establishes working groups for specially defined tasks

- discusses and evaluates the results of the work performed, and
- provides for the exchange of information and collaboration with other international and national organizations active in the field of indoor air quality (e.g. WHO, Nordic Committee on Building Regulations NKB, NATO/CCMS, US EPA)

Typical tasks for Working Groups include

- developing working instruments like guidelines for measurements and investigations in order to promote the efficiency of research and the comparability of results
- organizing inter-comparison exercises for the validation of measurement guidelines
- assessing the status of knowledge in specific areas and making proposals for solutions to IAQ problems.

The secretariat provides for co-ordination and organization of the work, for editing of the results and their distribution to the scientific community interested, national bodies and Community services.

Work Performed So Far

In an attempt to overcome the increasing difficulty of having essential information in a concise form at hand, the Concertation Committee, assisted by the Secretariat and by Working Groups, issues summary reports on single pollutants which are of high priority. Three such reports have been published: 'Radon in Indoor Air' (1), 'Indoor Pollution by NO_2 in European Countries' (2) and 'Indoor Air Pollution by Formaldehyde in European Countries' (3). In all these reports health effects, IAQ standards or guidelines, sources, occurring concentrations, preventive measures and (where applicable), national and Community policies are addressed briefly.

Until now ten WGs have been established. Four of them have already finished their work and issued the following reports (see section 3 for more details):

WG 1: 'Sick Building Syndrome - a practical guide' (4)

WG 2: 'Strategy for sampling chemical substances in indoor air' (5)

WG 3: 'Formaldehyde emission from wood based materials: guideline for the determination of steady state concentrations in test chambers' (6)

WG 8: 'Guideline for the characterization of volatile organic compounds emitted from indoor materials and products using small test chambers' (7).

Six Working Groups are presently preparing state-of-the-art reports or guidelines for various IAQ issues and investigations. The following documents are currently being prepared:

- 'Effects of indoor air pollution on human health'; the final draft is presently being reviewed by the Concertation Committee.
- 'Strategy for sampling biological particles in indoor air'; a draft document has been prepared.
- 'Guidelines for ventilation requirements in buildings'; these guidelines will be aimed not only at avoiding unhealthy indoor air but also to provide a good perceived air quality.
- 'Sick building syndrome: The design of intervention studies'; a preliminary draft document with this tentative title is scheduled for October 1991.
- 'Strategies for VOC measurements in indoor air'; building on the report of WG 2 this report is aimed at giving more in depth guidance for indoor measurements of VOC. A preliminary draft is scheduled for spring 1992.

- Review of the state of knowledge of sensory stimulation by indoor air pollution and resulting sensory, neurological and psychological effects.

An important working tool for the Concerted Action is an inventory of investigations and research projects in the field of indoor air quality ongoing in the participating countries (8). An updated version of this inventory has recently been published (9). Besides other information, it contains short descriptions of 326 investigations and research projects performed within the 14 participating countries. The inventory is of particular use for the identification of research gaps and for planning new research projects or collaborations.

3. GUIDELINES FOR INDOOR RELATED INVESTIGATIONS AND MEASUREMENTS

Over the past few years, much work has been dedicated to developing and further refining sampling and analytical methods for a wide range of air pollutants. This is also true for many indoor air pollutants. Often however, it is overlooked that the sampling and analysis of indoor pollutants are only the last steps of a procedure during which it previously has to be established why, where, when and under which environmental conditions an air sample should be analyzed. Without a correct answer to these questions, results of indoor pollution measurements are at risk of being misinterpreted or even meaningless. Work by the Concerted Action refers to various aspects of this important issue.

Strategy for Sampling Chemical Substances in Indoor Air (5)

Many measurements of chemical substances in indoor air have been published over the past ten years. Although sampling strategy has a strong impact on the results of these measurements, it has only been discussed in a small number of publications. Therefore the Concertation Committee authorized WG 2 to prepare a guideline 'Strategy for sampling chemical substances in indoor air'. The guideline is broken down into two parts.

The first part discusses in general terms the factors which mostly determine the sampling strategy or are their more important components:

- the dynamics of the indoor environment
- the sampling objective
- the point in time when a sample is taken
- duration and frequency of sampling
- the sampling location, and
- quality assurance.

The dynamics of the indoor environment are determined by the variability of the pollution sources, which may emit in a continuous or discontinuous way and at regular or irregular time intervals; by the many different types of indoor spaces; and by different ventilation and climatic conditions; e.g. two samples taken in the same room at the same location will usually yield very different analytical results depending on whether windows were open or closed before sampling.

There are many different sampling objectives which require different sampling strategies. The guideline refers to two groups of objectives requiring the measurement of average and of peak concentrations, respectively. The guideline does not look at the sampling of pollutants for the evaluation of their potential impact on IAQ complaints in large office buildings (see next subchapter). WG 9 – currently developing a more detailed strategy for measuring VOC in indoor air – defines six objectives:

- determination of the distribution of concentrations and personal exposures

- determination of the potential impact of VOC on complaints
- evaluation of compliance with IAQ guidelines
- evaluation of mitigation measures
- identification of a source, and
- validation of models.

The point in time at which a sample is taken is of extreme importance because of the often observed strong variations of indoor pollutant concentrations with time. Besides the ventilation conditions, the outcome of an analysis will strongly depend on whether or not activities such as smoking or cooking or a cleaning or hobby activity using VOC emitting products were ongoing at the time of sampling.

Duration and frequency of sampling themselves depend on the sampling objective and on the sources contributing to indoor air pollution and influence the choice of the sampling method (e.g. active or passive).

Because of the spatial variation of indoor pollutant concentrations usually observed, the sampling location also has a strong influence on the outcome of an analysis. Once more the objective of a measurement will mainly determine the most appropriate choice. But also the location of sources and differences in ventilation efficiency in a room have to be considered. The guideline indicates the breathing zone of a person (personal sample), the centre of a room at breathing height and a position near a source as the most common positions.

In the second part of the guideline some special cases are addressed regarding the sampling of formaldehyde, nitrogen dioxide, suspended particulate matter, asbestos, radon and volatile organic compounds.

Sick Building Syndrome – A Practical Guide (4)

A major problem of IAQ in non industrial workplace environments such as offices is an elevated number of complaints in the irritation of mucous membranes of eyes, nose and throat and unspecific systemic symptoms such as headache and lethargy. These symptoms, where occurring at levels of above 20% of the occupants of a building, constitute the so-called 'Sick building syndrome' (10).

Often the occupants of such buildings suspect indoor air pollution from chemical compounds, mineral fibres or biological particles as the cause for their complaints. However, measurements of these pollutants have often given negative or inconclusive results.

Based on this experience and on more systematic investigations of the SBS, the guideline recommends a procedure involving four steps when conducting investigations of complaints associated with buildings. The first step involves a technical survey of the building and its systems and a questionnaire inquiry into symptoms and complaints of the occupants. In the second step random measurements of climate indicators and an examination of potential sources of indoor pollutants are performed such as siting of an outside air intake near strong air pollution sources, new building or furnishing materials, location of copying machines and laser printers, tobacco smoking, irregularly cleaned humidifiers, water damage or stains or occurrence of moulds. During this step sources of strong odours are also located. Occasionally orientative pollutant measurements may be required at this step in order to locate a suspected source or estimate its strength.

At this point, correction measures should be taken and evaluated. Further steps should only be undertaken if the correction measures are inefficient. These steps may require pollutant measurements, in particular the measurement of strong irritants like aldehydes if odours or sources are detected and of the dust content in the air and its composition if poor cleaning is suspected. Finally, medical examinations may result in a request for pollutant measurement such as a qualitative study of volatile organic compounds together with a toxicological evaluation or a microbiological study, together with provocation tests.

Many useless and disappointing analyses can be avoided by following this guideline.

Guidelines for the Characterization of Emissions of Chemical Compounds from Indoor Materials and Products

The characterization of sources of chemical indoor air pollution such as building/furnishing materials or household and hobby products is a particularly clear example showing that sampling, separation and identification of air pollutants is only part of the overall analytical effort required. Repetitive and comparable results can only be obtained if the materials are analyzed under well defined environmental conditions. Such conditions can only be established and controlled in test chambers.

The Concerted Action has issued two guidelines for the characterization of (a) the formaldehyde emission from wood based products and (b) the VOC emission from a wide range of indoor materials and products which address all aspects of the entire procedure required for this scope such as chamber design and construction, the environmental conditions for testing, specimen preparation, sample collection, chemical analysis and data analysis. As an example of the many aspects associated with the characterization of source emissions, Table 1 summarizes the content of guideline (b). Sample collection and chemical analysis is only one out of seven chapters.

The paramount importance of the issue of when exactly a sample should be taken is illustrated in Figure 1. It shows the concentration vs time profiles of four individuals and the total VOC emitted by a wood stain after application. Depending on the time of sampling concentrations varying by more than two orders of magnitude may be detected.

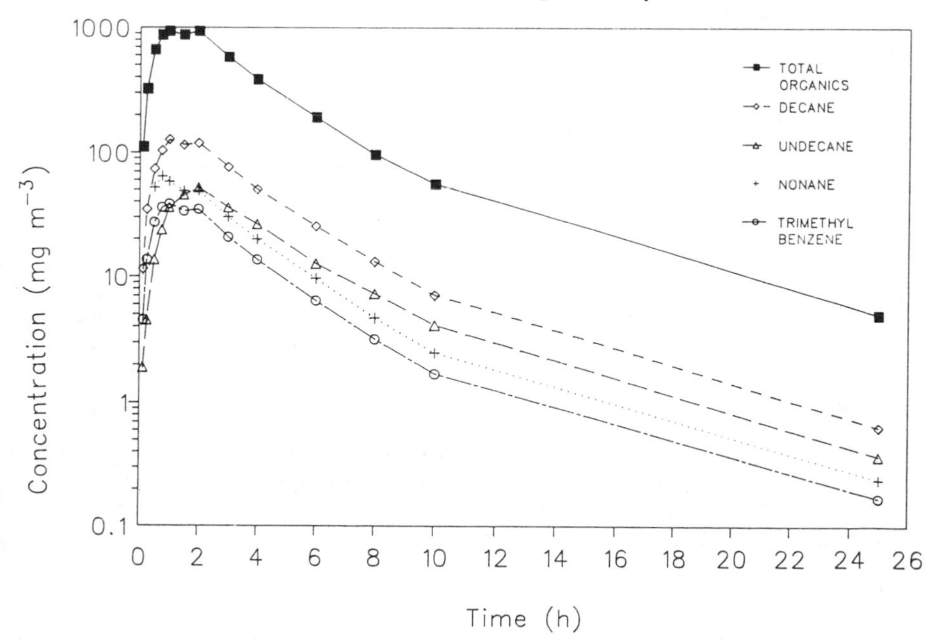

Fig. 1. Chamber concentrations vs time for a wet source (wood stain applied to a support)

Table 1. Subjects discussed in the guideline for the characterization of VOC emissions from indoor materials and products

Preliminary considerations

- testing objectives, mass transfer (i.e. transfer of VOC from a source to the air) considerations including variables affecting mass transfer
- use of the results

Facilities and equipment

- environmental test chambers (chamber construction, internal mixing and leaks, surface air velocity, temperature control system, lights, clean air generation system, humidification and control system)
- environmental measurement and control

Sample collection and analysis

- sampling devices
- sample collection media
- organic compound analysis instrumentation
- standards generation and system calibration

Experimental design

- definition of test objective(s)
- experimental parameters
- product history
- test matrix
- recommended test conditions

Data analysis

- environmental data
- gas chromatography data
- emission description

Quality assurance/quality control

- Reporting test results

The two guidelines described above are both under validation by inter-laboratory comparison exercises. Guideline (a) is validated by a CEN group with the scope to provide a European standard procedure. An inter-laboratory comparison for the validation of guideline (b) has been launched by the JRC in the framework of the Concerted Action and involves 22 European and North American laboratories.

Further work by the Concerted Action to provide guidance for indoor measurements is ongoing. As outlined in section 2, two documents are currently being prepared which will discuss strategies for sampling biological particles and volatile organic compounds respectively, in indoor air. The guideline 'Sick building syndrome: The design of intervention studies' will also address the issue of pollutant measurements.

REFERENCES

(1) 'RADON IN INDOOR AIR'. COST project 613 – Report No. 1. Prepared by James P. McLaughlin on behalf of the Community-COST Concertation Committee.

Commission of the European Communities, Directorate-General for Science, Research and Development, Joint Research Centre – Ispra Establishment. EUR 11917 EN, 1988.

(2) 'INDOOR POLLUTION BY NO_2 IN EUROPEAN COUNTRIES'. COST project 613 – Report No. 3. Prepared by the Community-COST Concertation Committee. Commission of the European Communities, Directorate-General for Science, Research and Development, Joint Research Centre – Ispra Establishment. EUR 12219 EN, 1989.

(3) 'INDOOR AIR POLLUTION BY FORMALDEHYDE IN EUROPEAN COUNTRIES'. COST project 613 – Report No. 7. Prepared by the Community-COST Concertation Committee. Commission of the European Communities, Directorate-General for Science, Research and Development, Joint Research Centre – Environment Institute. EUR 12219 EN, 1990.

(4) 'Sick building syndrome – a practical guide'. COST project 613 – Report No. 4. Prepared by Working Group 1 on behalf of the Community-COST Concertation Committee. Commission of the European Communities, Directorate-General for Science, Research and Development, Joint Research Centre – Ispra Establishment. EUR 12294 EN, 1989.

(5) 'STRATEGY FOR SAMPLING CHEMICAL SUBSTANCES IN INDOOR AIR'. COST project 613 – Report No. 6. Prepared by Working Group 2 on behalf of the Community-COST Concertation Committee. Commission of the European Communities, Directorate-General for Science, Research and Development, Joint Research Centre – Ispra Establishment. EUR 12617 EN, 1989.

(6) 'FORMALDEHYDE EMISSIONS FROM WOOD BASED MATERIALS: GUIDELINE FOR THE ESTABLISHMENT OF STEADY STATE CONCENTRATIONS IN TEST CHAMBERS'. COST project 613 – Report No. 2. Prepared by Working Group 3 on behalf of the Community-COST Concertation Committee. Commission of the European Communities, Directorate-General for Science, Research and Development, Joint Research Centre – Ispra Establishment. EUR 12196 EN, 1989.

(7) 'GUIDELINE FOR THE CHARACTERIZATION OF VOLATILE ORGANIC COMPOUNDS EMITTED FROM INDOOR MATERIALS AND PRODUCTS USING SMALL TEST CHAMBERS'. COST project 613 – Report No. 8. Prepared by Working Group 8 on behalf of the Community-COST Concertation Committee. Commission of the European Communities, Directorate-General for Science, Research and Development, Joint Research Centre – Ispra Establishment. EUR 13593 EN, 1991.

(8) 'PROJECT INVENTORY'. COST project 613 – Report No. 5. Prepared on behalf of the Community-COST Concertation Committee. Commission of the European Communities, Directorate-General for Science, Research and Development, Joint Research Centre – Ispra Establishment. S.P./1.89.33, 1989.

(9) 'PROJECT INVENTORY – 2ND UPDATED EDITION'. COST project 613 – Report No. 9. Prepared on behalf of the Concertation Committee. Commission of the European Communities, Directorate-General for Science, Research and Development, Joint Research Centre – Ispra Establishment. EUR 13838 EN, 1991.

(10) WORLD HEALTH ORGANIZATION (WHO). Indoor air pollutants: exposure and health effects. EURO Reports and Studies No. 78, WHO Regional Office for Europe, Copenhagen 1983.

ACTIVE SAMPLING OF VOC IN NON-INDUSTRIAL BUILDINGS

H. ROTHWEILER

Institute of Toxicology, Swiss Federal Institute of Technology and
University of Zurich, CH-8603 Schwerzenbach, Switzerland

SUMMARY

Active sampling of volatile organic compounds (VOC) in the non-industrial environment on solid sorbents with following thermal desorption, cryotrapping, separation in a high resolution gas chromatograph (HRGC) and detection with a mass spectrometer (MS) or a flame ionisation detector (FID) is widespread in indoor air investigations. All solid sorbents have limitations. The analysis of apolar VOC is satisfactory. Tenax TA is suitable for a wide range of mainly apolar to slightly polar VOC. It shows low blank values and few artefacts (acetophenone, benzaldehyde, phenol). On Carbotrap reaction products of sampled hexanal and α-pinene were found. Sorbents for polar VOC (Porapak S, R and N; Chromosorb 106) show high blank values or cause analytic problems (Carboxen 563, 564, Carbosieve-SIII) due to water absorption. Specific methods for e.g. aldehydes, amines and carboxylic acids might be useful in certain investigations. Simple equipment based on photoacoustic, infrared or photoionisations detectors show at present many disadvantages. Interpretation of obtained data is difficult due to the occurrence of many VOC. The widespread approach to use only total volatile organic compounds (TVOC) to characterize indoor air is very limited because VOC may differ in their toxicity by orders of magnitude.

1. INTRODUCTION

The quality of life of up to 30% of employees in new and refurbished buildings throughout the world is affected. High costs to society must be considered due to absenteeism and reduced working efficiency (1). Moreover, it might be assumed that the situation in private homes is similar.

It is known that these effects are partly caused by volatile organic compounds (VOC). Hundreds of VOC have been detected in indoor air and still the list is increasing. VOC belong to different chemical classes. They are emitted from building products (e.g. paints, floorings, wood, etc.), polluted soil, microorganisms, activities like smoking, cleaning agents, cooking, hobbies, photocopying, etc., and enter the rooms through infiltration from other rooms and outdoor air.

In two case studies (2, 3) it was suspected that hexanal and capronic acid at elevated levels caused negative health effects such as eye, throat and skin irritations. The sources were in both cases reaction products, in one case due to a radiator lacquer and in the other of a wall paint. 2-ethylhexanol together with heptanal caused complaints about eye irritation, dermatological problems and odour (4). Both compounds were liberated when the phthalate plasticisers of the PVC backed carpet were hydrolysed. Annoyance and discomfort are often caused by odorous compounds. Thus aldehydes emitted from a pretreated insulation material caused odour complaints (5). Naphthalene emitted from damp-proof membranes lead to complaints mainly of bad odour and a few about sore throats, headaches and lassitude in a number of houses (6). Complaints about poor air quality and discomfort could be traced to 4-phenylcyclohexene, an extremely odorous compound, off-gassed from carpets with latex backing (7). However, in many investigations about complaints due to bad air quality in buildings the causes of the complaints could not be detected (8). In several investigations, where people complained about unpleasant effects, increased loads of VOC were found in a concentration range of one order of magnitude or even more above the concentration usually encountered in outdoor air or older houses. However, the measured concentrations were well below actual TLV levels, although annoyance and discomfort occurred.

Analyzing and interpreting VOC in non-industrial buildings is exacting. At industrial workplaces people may be exposed to a few compounds at high levels, while in non-industrial buildings persons are exposed to mixtures of VOC at low to moderate concentrations. At industrial workplaces, mainly healthy adults are exposed for eight hours a day and 40 hours a week, whereas in the non-industrial environment the exposure time may be up to 24 hours a day and seven days a week. There you also find persons more at risk like elderly, hypersensitives, allergic persons, children and pregnant women. The longer exposure duration and the persons at higher risks demand lower recommended guideline values.

Several approaches exist to characterize indoor air quality. Some are based on the findings of Mølhave (8) on mixtures and others are based on a compound by compound approach with different fractions of actual TLV values. There, mostly fractions of 1/10 to 1/420 (9) are taken.

In indoor air studies non-selective sampling methods are demanded, because a broad spectra of different chemicals should be analyzed at the same time preferably with the same method. For that purpose VOC are sampled in the non-industrial environment mainly by active or diffusive sampling. Active sampling is mainly used to measure short term average concentrations and diffusive sampling for long term average concentrations (10). Active sampling is generally more expensive than diffusive sampling. It needs sampling pumps and more skilled manpower to obtain the samples. In contrast, diffusive samplers are cheap and may be mailed directly to houses, where they can be exposed for some days. Active sampling is an important tool in indoor air analysis. Most often active sampling of VOC is performed on a solid sorbent with following thermal desorption, cryofocusing and separation in a high resolution gas chromatograph (HRGC) together with, for example, a flame ionisation detector (FID) or a mass spectrometer (MS). With a FID the identification is more complicated and a higher risk of misinterpretation does exist. Either the compounds are identified simply by comparison of the retention indices of known compounds or also together with another detector, e.g. electron capture detector (ECD). For active sampling a broader collection of sorbents is available for a wide range of VOC compared to diffusive sampling.

Sampling and the transfer of VOC to the analytical separation system is only one part of determinations of VOC to be considered. Moreover, one should be aware of the task, that the sampling strategy has to be planned very carefully prior to sampling (10) and that the selection of the separation column in the analytical system is of importance. Capillary columns totally inert and suitable for all kind of VOC (neutral, polar and apolar, basic and acidic compounds) are not known.

In this paper the use of solid sorbents, their reactivity and advantages as well as disadvantages are discussed, and some specific complementary methods are presented. Data interpretation is briefly discussed, conclusions are drawn and a short overview is given.

2. SAMPLING METHODS

Non-industrial environment investigations of VOC are usually performed with active sampling on solid sorbents with thermal desorption (3, 11, 12) or diffusive sampling on activated carbon and liquid desorption (13, 14). The compounds sampled are separated mostly with high resolution gas chromatography (HRGC) with, for example, a flame ionisation detector (FID) or a mass spectrometer (MS).

Other possibilities to sample and analyze VOC exist. Grab sampling by summa-polished stainless steel (15) or aluminium canisters (16) is performed to take samples either within seconds (10-30s) or time integrated up to a week. Then the air samples are analyzed in the laboratory by passing an aliquot of the sample through a cryogenic trap, the main components of air not being trapped while the trace organics are efficiently collected (17). The cryotrap is afterwards thermally desorbed. At high humidity of the sample analytical problems are to be expected. Another possibility is the adsorption of the air from the canister on a solid sorbent, e.g. Tenax, and thermal elution, cryotrapping and analysis with HRGC.

Aluminium canisters showed to be more reactive for polar oxygenated compounds such as acetone and 1,4-dioxane, than the passivated stainless steel canisters (16). Neither stainless steel nor aluminium canisters are well adapted to collect polar compounds. Additional serious problems in the cryogenic trap, such as clogging, may occur due to the co-collection of water.

Canisters are suitable for many VVOC and VOC. However, they do not seem suitable for reactive VOC and VVOC (18), nor for VOC in the upper range of boiling points, due to losses (19).

The above mentioned methods with solid sorbents and canisters are rather complex, demand skilled personnel, are time consuming and therefore expensive, too. A demand for simple routine methods exists. However, other methods are lacking in selectivity or are limited to few compounds. A possibility is to use a portable GC together with photoionisation detector (18). This method allows nearly real time measurements. However, the presented method is limited to few compounds.

Equipment for the estimation of total volatile organic compounds (TVOC) is on the market, like the photoionisations detector (TIP) or for selective compounds, e.g. photoacoustic or infrared detector for some VOC. However, these methods show several interferences. High humidity is known to influence the results. It is known that comparative measurements with other methods show bad agreements, so far.

Direct-reading gas detection tubes, e.g Dräger ® and Gastec ®, are often used in industrial settings and some use them for indoor air investigations. The tubes have rather low sensitivities and show many interferences.They are not designed for indoor air studies in the non-industrial environment.

3. SOLID SORBENTS

The selection of a suitable solid sorbent is important. In Table 1 some solid sorbents for active sampling of VOC in indoor air are compared.

A wide range of different VOC has to be determined. Therefore, it is important to have information about reaction products of or on the adsorbent, and it is advisable to use sorbents with low blank values.

Tenax TA is well investigated and some artefacts are known. Small amounts of benzaldehyde, acetophenone (20) and phenol (21) were released from the adsorbent in the presence of reactive compounds. Moreover, recoveries are low of very volatile, polar compounds and some other compounds such as α-pinene (22, 23). The latter occurs often in indoor air. The measured concentration of single VOC may be influenced by the composition of the air sample. Thus, clear losses of a few compounds, e.g. hexanal, were found in some field measurements in new buildings although full recoveries were expected (24). In laboratory experiments lower retention volumes of benzene in the presence of increased xylene concentrations were seen (25). Specific displacement mechanisms are suspected to have caused these to capacity breakthroughs.

Other adsorbents are not as well investigated. Reaction products of α-pinene and hexanal were found on Carbotrap (23, 26) .Other adsorbents (e.g. Porapak S and Chromosorb 106) (27)

Table 1: Some solid sorbents for active sampling of VOC in indoor air

Sorbent	Desorption technique	Compounds Sampled	Starting at [°C]	Main Advantages and Disadvantages
Tenax TA	thermal	-most non-polar VOC -(terpenes)* -slightly polar VOC -aldehydes $>C_5$ -(acids $>C_3$)*	> 60	-low background -well investigated -some decomposition products (benzaldehyde, acetophenone)
Carbotrap	thermal	-most non polar VOC slightly polar VOC	>60	-low background -reactions of some compounds (i.e. aldehydes, terpenes)
Activated carbon	solvent/ (thermal)	-most non-polar VOC -slightly polar VOC	>50	-high capacity -reactions of some compounds
Porapak Q	thermal	-most non-polar VOC -slightly polar VOC	>60	-high background -low thermal stability
Porapak S or R,N	thermal	-VOC incl. moderate polar -terpenes	>40	-high background -low thermal stability
Organic molecular sieves (e.g. Carboxen 563,564, Carbosieve-S-III)	thermal	-polar and non-polar VOC	>-80	-water adsorption
Silicagel	thermal	-polar VOC -phenols -amines		-water adssorption -decreased capacity by high humidity

* can be used (mainly low recoveries)

being more suitable for enriching polar compounds cause more difficulties due to blank values when thermally desorbed. There is a risk of contamination through decomposition products at increased temperatures, therefore extended cleaning procedures are necessary for these adsorbents. They can be operated only at limited temperatures. On the other hand, sorbents based on molecular sieves, like Carbosieve S-III co-adsorb water. This may lead to cooltrap clogging in the thermal desorption unit. There are different approaches to overcome these difficulties. An increasing purging time of the sorbent just before thermal desorption should enable the satisfactory removal of retained water because it is only weakly adsorbed on the surface (28). Another approach is to remove water from the air selectively in front of the sampling trail. Thus, Schmidbauer and Oehme (29) as well as Maljaar and Nielen (30) used a pre-tube filled with potassium carbonate, or the water was cryotrapped at −10-0°C (3) in front of the sampling tube. Semipermeable ionomer membrane (Nafion®) tubes are also used (17). Thereby the air to be analyzed is sucked through this tube and outside the tube dry zero air in counterflow is blown. Several losses of mainly polar compounds have been observed (31).

In indoor air studies activated carbon is rarely used with active sampling. When it is used, the activated carbon is desorbed mainly with a liquid, e.g. carbon disulphide, but it can also be desorbed thermally. Short open capillary tubes with an activated charcoal layer can be desorbed without major technical expense (32, 33). However, the strong adsorbent has many active sites and losses of reactive VOC are to be expected (34).

Several groups also use tubes with combined sorbents to retain a greater fraction of VOC (35). Compared to some advantages, several disadvantages are known. The sorbents in the same tube differ in their optimal desorption temperature and the problem of avoiding water retention, if an adsorbent for polar compounds is included, cannot be overcome with this approach.

Sampling VOC in different tubes with other solid sorbents and separate analysis is another approach to increase the fraction of VOC measured (3). Instead of repetitive measurements of VOC on adsorbent tubes with the same material, this may be done with different adsorbents.

4. SPECIFIC METHODS FOR VOC

With non-selective solid sorbents a wide range of VOC can be analyzed. However, losses and underestimations of several compounds, mostly with importance in view of toxicology, are to be expected. Therefore it might be advisable to use specific methods, which are supplementary to solid sorbents. However, some of the compounds measured with specific methods do not strictly belong to the group of VOC (bp approx. 60-250°C). In the following some specific methods used in indoor air studies are presented. However, experience with some of these is scarce.

Aldehydes may be sampled in an impinger containing a liquid with dinitrophenylhydrazine, or adsorbed on coated filters and analyzed according to a HPLC method of de Bortoli *et. al*, 1985 (11). Amines, known for their unpleasant odour, may be analyzed after sampling in an alkaline liquid and derivatisation with 2,4-dinitrofluorobenzene with HPLC and UV-detection (36) and low carboxylic acids (C_1-C_7) can be sampled in weakly alkaline solutions, derivatised to the corresponding p-bromophenylesters and separated on a HRGC with FID (37). Isocyanates, known as strong irritants and allergens at very low concentrations, may be determined after derivatisation with 1-(2-methoxyphenyl)piperazine (38, 39).

5. DATA INTERPRETATION

Data interpretation is more difficult in the non-industrial environment than at industrial workplaces. Threshold limits in the non-industrial environment are almost lacking with very few exceptions like formaldehyde.

Approaches for data interpretation are either based on single compound by compound risk estimations or directly on mixtures. Additive effects of different toxic compounds do often occur, but synergistic and antagonistic effects cannot be excluded. Besides several other authors Calabrese and Kenyon (1991) (9) propose to use in many cases a fraction of the actual threshold limit at workplaces (40 hours/week) as recommended maximum allowable average concentrations. The fraction is derived from the exposure time of 40 hours a week compared to a maximum of 168 hours in the non-industrial environment (factor 4.2). Additionally, it is assumed that also a safety factor of 10 to 100, depending on the kind of toxicity, should be introduced to prevent the main sensitive part of the population from negative health effects. Verhoeff *et al.* (1988)(40) supposed to consider additive effects to examine mixtures of VOC. Another proposal to evaluate indoor air quality is based on chemical classes of VOC and TVOC (41).

Mølhave (8) found annoyance and irritations at rather low concentrations of a mixture of 22 compounds in a chamber experiment. Thereof, he derived a lowest effect concentration of 2 mg/m^3. Values of TVOC are used in the field as an indicator for indoor air quality. However, it must be considered that these values reflect only a fraction of the true content of VOC in air. Deviations depend on the composition of VOC, the sorbent, sampling volume and reference standards. Single VOC may differ orders of magnitude in their toxicity. Therefore, a high risk of misinterpretation of TVOC values does exist. A serious interpretation demands information about single compounds or at least of the chemical classes present.

6. QUALITY ASSURANCE

Quality assurance is important in indoor air studies, although physical factors like air exchange and temperature may influence the results. Standards and blank values have always to be analyzed. Calibration is not easy. Generating a standard atmosphere of VOC especially for rather reactive compounds is difficult. In practice these problems are overcome by applying a known amount of the VOC of interest directly onto the adsorbent. Inter-comparison studies between laboratories are recommended, and the use of certified standards. However, the latter are only available in a small variety and high concentrations so far.

7. RESEARCH NEEDED

Analysis for polar VOC still needs some improvements. Either other sorbents for polar compounds, which specifically do not retain water, or methods to remove water only in front of the adsorbent should be found. Apolar VOC can be analyzed in general satisfactorily, although some specific displacement mechanisms seem to occur on Tenax, which influence the results. Equipment to determine easily short term VOC patterns needs to developed.

Efforts should be taken for simple approaches to characterize indoor air quality reliably.

8. CONCLUSION

All adsorbents have limitations. Rarely, one adsorbent is most suitable for all VOC of interest. Therefore, in practice only moderately suitable adsorbents for some VOC of interest may be used and still fulfil the requirements. On the other hand, at least for some objectives it is advisable to introduce more than one sampling method or adsorbents to obtain clear results. The tendency to use more polar compounds in building sites instead of aliphatic and aromatic hydrocarbons demands improvements of adsorbents for polar VOC. The interpretation of obtained data is difficult because little is known about effects caused to groups like pregnant women, elderly people, children and hypersensitives by single compounds or mixtures of VOC in the range of μg/m^3 to a few mg/m^3.

9. REFERENCES

(1) MOLINA, C., PICKERING, C.A.C., VALBJØRN, O. and DE BORTOLI, M. (1989). Sick Building Syndrome, COST Project 613 Indoor Air Quality and Its Impact on MAN. Report No.4. Commission of the European Communities EUR 12294 EN.

(2) ULLRICH, D., NAGEL, R. and SEIFERT, B. (1982). Einfluss von Lackanstrichen auf die Innenraumluftqualität am Beispiel von Heizkörperlacken. Aurand K., Seifert B. und Wegner J. (ed) Luftqualität in Innenräumen. Gustav Fischer Verlag, Stuttgart- New York, 283-298.

(3) ROTHWEILER, H., WAGER, P.A., and SCHLATTER, C. (1991). VOC and some VVOC in new and recently renovated buildings in Switzerland (submitted for publication).

(4) MC LAUGHLIN, P. and AIGER, R. (1990). Higher alcohols as indoor air pollutants: source, cause, mitigation. Indoor Air '90, Proceedings of the 5th International Conference on Indoor Air and Climate, Toronto, Canada, 29 July-3 August, 3, 587-591.

(5) VAN DER WAAL, J.F., MOONS, A.M.M., STEENLAGE, R. (1987). Thermal insulation as a source of air pollution. Indoor Air '87. Proceedings of the 4th International Conference on Indoor Air Quality and Climate, Berlin 17-21 August, 1, 79-83.

(6) BROWN, V.M., COCKRAM, A.H., CRUMP, D.R.and GARDINER, D (1990). Investigations of the volatile organic compound content of indoor air in homes with odorous damp proof membrane. Indoor Air '90, Proceedings of the 5th International Conference on Indoor Air and Climate, Toronto, Canada, 29 July-3 August, 3, 575–580.

(7) SINGHVI, R., BURCHETTE, S., TURPIN R. and LIN, Y. (1990). 4-Phenylcyclohexene from carpet and indoor air quality. Indoor Air '90, Proceedings of the 5th International Conference on Indoor Air and Climate, Toronto, Canada, 29 July-3 August, 4, 671-676.

(8) MØLHAVE, L. (1986). Indoor air quality in relation to sensory irritation due to volatile organic compounds. Ashrae Transl. 92, Part IA, 306-316.

(9) CALABRESE, E.J. and KENYON, E.M. (1991). Air toxics and risk assessment. Lewis Publishers, Chelsea, Michigan USA.

(10) COST Project 613 (1989). Report No.6. Strategy for Sampling Chemical Substances in Indoor Air. Commission of the European Communities, Eur 12617 EN, Luxembourg.

(11) DE BORTOLI, M., KNOPPEL, H., PECCHIO, E., PEIL, A., ROGORA, L. SCHAUENBURG, H., SCHLITT H. and VISSERS, H. (1985). Measurements of indoor air quality and comparison with ambient air: A study in 15 homes in Northern Italy. Commission of the European Communities Report EUR 9656, Brussels-Luxembourg.

(12) WALLACE, L., JUNGERS, R., SHELDON, L. and PELLIZARRI, E. (1987). Volatile organic chemicals in 10 public-access buildings. Indoor Air '87 (Seifert, B., Esdorn, H., Fischer, M., Rüden, H. and Wegner, J., eds). Institute for water soil and air hygiene, Berlin 1, 188-192.

(13) KRAUSE, C., MAILHAIN, W., NAGEL, R., SCHULZ, C., SEIFERT B. and ULLRICH, D.(1987). Occurrence of volatile organic compounds in the air of 500 homes in the Federal Republic of Germany. Indoor Air '87. (Seifert, B., Esdorn, H., Fischer, M., Rüden, H. and Wegner, J., eds). Institute for water soil and air hygiene, Berlin 1, 102-106.

(14) LEBRET, E., VAN DE WIEL H.J., BOS, H.P., NOIJ, D. and BOLEIJ, J.S.M. (1986). Volatile hydrocarbons in Dutch homes. Environ Int.12, 323-332.

(15) MICHAEL, L.C., PELLIZZARI, E.D, PERRIT, R.L., HARTWELL, T.D., WESTERDAHL, D. and NELSON, W.C. (1990). Comparison of Indoor, Backyard, and Centralized Air Monitoring Strategies for Assessing Personal Exposure to Volatile Organic Compounds. Environ. Sci. Technol. 24 (7), 996-1003.

(16) GHOLSON, A.R., JAYANTI, R.K.M. and STORM, J. F. (1990). Evaluation of Aluminium Canisters for the Collection and Storage of Air Toxics. Anal. Chem. 62, 1899-1902.

(17) McCLENNY, W.A., PLEIL, J.D., HOLDREN, M.W. and SMITH, R.N. (1984) Automated cryogenic Preconcentration and Gas Chromatographic Determination of Volatile Organic Compounds in Air. Anal. Chem. 56, 2947-2951.

(18) BERKLEY, R.E., VARNS, J.L. and PLEIL, J. (1991). Comparison of portable gas chromatographs and passivated canisters for field sampling airborne toxic organic vapors in the United States and USSR. Environ. Sci. Technol. 25, 1439-1444.

(19) SWEET, C.W. (1988). Sampling and analysis of toxic volatile organic pollutants in ambient air using an automatic canister-based sampler. Proceedings of the EPA/APCA Int. Symposium Measurement of Toxic and Related Air Pollutants. APCA, Pittsburgh, USA., 299-304.

(20) HUTTE, R.S., WILLIAMS, E.J., STAEHLIN, J., HAWTHORNE, S.B., BARKLEY, R.M. and SIEVERS, R.E. (1984). Chromatographic analysis of organic compounds in the atmosphere. J. Chromatogr. 302, 173-179.

(21) PELLIZZARI, E., DEMIAN, B. and KROST, K. (1984). Sampling of organic compounds in the presence of reactive inorganic gases with Tenax GC. Anal. Chem 56, 793-798.

(22) RIBA, M.L., RANDRANALIANA, E., MATHION, J., TORRES; L. and NAMIESNIK, J. (1985). Preconcentration of atmospheric terpenes on solid sorbents. Int. J. Environ. Anal. Chem. 19, 133-143.

(23) ROTHWEILER, H., WAGER, P.A. and SCHLATTER, C. (1991). Comparison of Tenax TA and Carbotrap for sampling and analysis of volatile organic compounds in air. Atmospheric Environment 25B (2), 231-235.

(24) WAGER, P.A., ROTHWEILER, H. and SCHLATTER, C. (1991). Porapak S as complementary adsorbent to Tenax TA for the determination of VOC in non-industrial buildings. Clean Air '91 Luxembourg (this conference).

(25) VEJROSTA, J., ROTH, M. and NOVAK, J. (1983). Interference effects in trapping trace components from gases on chromatographic sorbents. Sorption of benzene in the presence of oxylene. J. Chromatogr. 265, 215-221.

(26) DE BORTOLI, M., KNOPPEL, H., PECCHIO, E. and VISSERS, H. (1989). Performance of a thermally desorbable diffusion sampler for personal and indoor air monitoring. Environ. Int. 12, 343-350.

(27) FIGGE, K., RABEL, W. and WIECK, A. (1987). Adsorptionsmittel zur Anreicherung von organischen Luftinhaltsstoffen. Fres. Z. Anal. Chem. 327, 261-278.

(28) MOSESMAN, N.H., BETZ, W.R. and CORMAN, S.D. (1988). Development of a New Trapping System for Purge and Trap Analysis of Volatile Organic Compounds. LC GC Int. 1 (3), 62-68.

(29) SCHMIDBAUER, N. and OEHME, M. (1988). Comparison of solid adsorbent and stainless steel canisters for very low ppt-concentrations of aromatic compounds ($\geq C_6$) in ambient air remote areas. Fresenius Z Anal Chem 331, 14-19.

(30) MALJAARS, S.E. and NIELEN, M.W.F. (1988). Evaluation of adsorption tubes for air sampling of C_2-C_4 unsaturated hydrocarbons. Intern. J. Environ. Anal. Chem. 34, 333-345.

(31) BAKER, B.B. (1974). Measuring trace impurities in air by infrared spectroscopy at 20 meters path and 10 atmospheres pressure. Am. Ind. Hyg. Assoc. Journal 34, 735-740.

(32) BURGER, B.V. and MUNRO, Z. (1986). Headspace gas analysis. Quantitative trapping and thermal desorption of volatiles using fused-silica open tubular capillary traps. J. Chromatogr. 370, 449-464.

(33) GROB, K. and HABICH, A. (1985). Headspace gas analysis. The role and design of concentration traps specifically suitable for capillary gas chromatography. J. Chromatogr. 321, 45-58.

(34) NUNEZ, A.J., GONZALEZ, L.F. and JANAK, J. (1984). Preconcentration of headspace volatiles for trace organic analysis by gas chromatography. J. Chromatogr. 300, 127-162.

(35) HODGSON, A.T., BINENBOYM, J. and GIRMAN, J.R. (1988). A Mulitisorbent Sampler for Volatile Organic Compounds in Indoor Air. in: Advances in air sampling. American Conference of Governmental Industrial Hygienists. Lewis Publishers, Chelsea, Michigan USA, 143-157.

(36) VEREIN DEUTSCHER INGENIEURE (1989). Messen der Konzentration primärer und sekundärer aliphatischer Amine mit der Hochleistungsflüssigkeitschromatographie (HPLC). VDI Richtlinien VDI 2467 Blatt 2 Entwurf. VDI - Handbuch Reinhaltung der Luft 5, Beuth Verlag GmbH, Berlin BRD.

(37) KAWAMURA, K. and KAPLAN, I.R. (1984). Capillary Gas Chromatography Determination of Volatile Organic Acids in Rain and Fog Samples. Anal. Chem. 56, 1616-1620.

(38) KEHL, S. (1990). Erfassung und Beurteilung der Belastung durch Isocyanate am Arbeitsplatz sowie im Wohnbereich. Diss. ETH, Nr. 9222, Zürich.

(39) SCHMIDTKE, F. and SEIFERT, B. (1990). A highly sensitive high-performance liquid chromatographic procedure for the determination of isocyanates in air. Fresenius J Anal Chem 336, 647-654.

(40) VERHOEFF, A.P., SUK, J. and VAN WIJINEN, J.H. (1988). Residential indoor air contamination by screen printing plants. Int Arch Occup Environ Health 60, 201-209.

(41) SEIFERT, B. (1990). Regulating Indoor Air. Indoor Air '90. The Fifth International Conference on Indoor Air Quality and Climate. Toronto, Canada 5, 35-49.

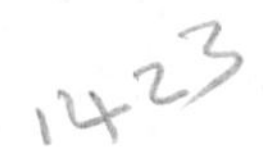

DIFFUSIVE SAMPLING OF VOLATILE ORGANIC COMPOUNDS

D. ULLRICH

Institut für Wasser-, Boden- und Lufthygiene
des Bundesgesundheitsamtes, D-1000 Berlin 33

SUMMARY

Passive sampling is a simple and reliable tool to sample volatile organic compounds (VOC) in indoor air on a routine basis. The accuracy as well as the reproducibility of passive sampling proved to be comparable to active sampling. The decrease of the sampling rate due to very low air velocities can be neglected when assessing the general VOC concentration of indoor air. A sampling time of two weeks is feasible with constant sampling rates to achieve a detection limit of about $1 \mu g/m^3$.

1. INTRODUCTION

During the past ten years the interest and the necessity of evaluating the contamination of indoor air has grown substantially. The first paper of this session gave a general overview of this field of public health. In the early years of indoor air monitoring all analytical methods used were the same as for ambient air measurements. Due to the low concentrations, particularly for volatile organic compounds (VOC), an enrichment step was necessary prior to the analysis. Active sampling by drawing a known volume of air through a tube filled with a sorbing or reacting material fulfils most of the requirements of enrichment. An earlier paper gave a detailed overview of the state-of-the-art for this type of air sampling.

There are two fundamental drawbacks of active sampling. One is the necessity of using bulky, noisy and expensive sampling equipment. The other is that generally the result will represent the situation of a relative short sampling period only. Workplace monitoring was faced with the first problem as well, and to cope with these problems, simpler methods and new samplers were developed. Most of these samplers are specially designed using a well-defined diffusion process requiring no pumps. Therefore, this method is called 'diffusive sampling' or 'passive sampling'. The proceedings of the symposium 'Diffusive Sampling', Luxembourg, 1986, can serve as a detailed and most recent general source of information about this type of sampling (1). Because of its inherent simplicity, we introduced this sampling method on a large scale.

2. PRINCIPLES OF DIFFUSIVE SAMPLING

Assuming a difference in concentration for a compound i in a certain medium, this compound will undergo a diffusion process according to Fick's first law:

$$M_i = D_i * A/L * (C_{i1} - C_{i2}) * t \qquad \text{EQ.1}$$

M_i	=	mass of compound i	(ng)
D_i	=	diffusion coefficient of compound i	(cm^2/sec)
A	=	cross section of the diffusion path	(cm^2)
L	=	length of the diffusion path	(cm)
C_{i1}	=	concentration of compound i at location 1	(ng/cm^3)
C_{i2}	=	concentration of compound i at location 2	(ng/cm^3)
t	=	time	(sec)

If a material is placed at location 2 which is a perfect trap for compound i due to an adsorption or chemisorption process, the concentration C_{i2} will drop to zero. Then the concentration C_{i1} represents the concentration C_i outside of the prescribed diffusion path. This model may be regarded as a prototype of a diffusive sampler. A porous material serving as a permeable entrance for all compounds ensures a well-defined outer boundary of the diffusion process, which is thus independent of the direction of any air flow.

The term $D_i * A/L$ can be interpreted as a theoretical sampling rate S_i. S_i depends on the diffusion coefficient - which is tabulated or can be calculated - and the dimensions of the diffusive sampler (2, 3). Thus, S_i is constant for a particular compound and sampler.

There are two types of diffusive samplers: the badge-type and the tube-type. The dimensions and the shape of the diffusion path are the only distinct differences of both types. Because of the small cross section A of commercially available tube-types, the sampling rate of this sampler is at least a factor of 5 less than the sampling rate of the badge-type. Some consequences of this will be discussed later.

In the beginning of the 1980s, we chose badge-type diffusive samplers with charcoal as sorbing material and a mean sampling rate of about $30 cm^3/min$ (4). Originally a sampling time of about eight hours was anticipated to control TWA-thresholds in the mg/m^3 range. When analyzing compounds at lower concentrations, the sampling time has to be prolonged. For a sampling time of two weeks, the total volume sampled is in the range of $0.6 m^3$. After sampling, the pad of charcoal is eluted with a suitable solvent in the laboratory. The mass of eluted compounds is high enough to be analyzed by a large variety of analytical methods (e.g. mass spectrometry in the linear scan mode) even in the low $\mu g/m^3$-concentration range with an aliquot of the solution.

Although most of this paper deals with the badge-type diffusive sampler to monitor VOC in the indoor environment, a special, very convenient tube-type diffusive sampler to monitor formaldehyde is used in our laboratories as well. Originally this type of sampler was introduced by Palmes to monitor NO_2 (5). At the bottom of the upright standing tube a stainless steel screen impregnated with triethanolamine traps formaldehyde. After a sampling time of two days the tube is sent to the laboratory. The screen is washed with distilled water and the solution is treated according to a standard method to analyze formaldehyde (e.g. pararosaniline method) (6).

3. TESTING DIFFUSIVE SAMPLER

When introducing a new type of sampler in a laboratory, many test procedures have to be undertaken to discover deviations from the theoretical diffusion process according to Eq.1 and for verification purposes. Using a test gas generator and glass chambers of different size the complete method is investigated. These experiments exhibit a correcting factor r_i' to correct the total deviation from theoretical assumptions. This leads to a corrected sampling rate S_i', which contains the recovery rate r_i of the individual compound i due to incomplete elution as well as the recovery rate r' of the sampling process.

$S_i' = S_i * r_i'$ $\quad (cm^3/sec)$

$r_i' = r_i * r'$ $\quad$ total recovery of compound i

r_i' = recovery of compound i due to elution deficiencies

r = recovery of the type of sampler due to deviations from the theoretical diffusion process

$C_i = M_i / (S_i' * t)$ $\quad (ng/cm^3)$ $\quad$ EQ.2

The final equation, EQ.2, is used to calculate the concentration of compound i in the air sampled.

The 1m³ glass chamber used to test different samplers with various test conditions is shown in Figure 1. A zero gas supply with thermal mass flow controller and a permeation device serve as test gas generator. The figure shows the arrangement to test the influence of the location of the sampler on the measured concentration.

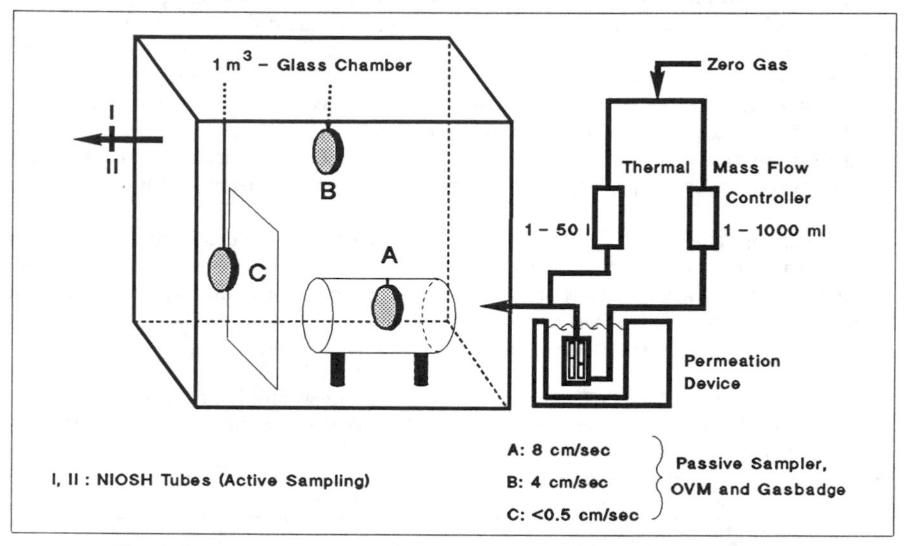

Fig. 1. Glass chamber with test gas generator and different air samplers located at sites with different air velocities

Table 1. Theoretical Concentrations and Results (deviations from theoretical concentrations in %) Received by Active Sampling and Diffusive Sampling of Different Badge-types (OVM, Gasbadge) at Different Air Velocities

	Glass chamber Theoretical concentrations	Long time* active sampling NIOSH-tubes		Position A 8cm/sec		Position B 4cm/sec		Position C <0.5cm/sec	
Compound	($\mu g/m^3$)	I	II	OVM	Gasb	OVM	Gasb	OVM	Gasb
Benzene	6.5 ± 12%	-11%	-9%	3%	14%	8%	11%	-2%	3%
Toluene	21.7 ± 6%	-12%	-19%	-9%	-3%	-9%	-7%	-15%	-11%
Xylene	3.2 ± 7%	9%	-19%	0%	-6%	0%	-9%	-16%	-22%
n-Hexane	58.8 ± 10%	-7%	-9%	-2%	-2%	-3%	-1%	-12%	-8%
n-Heptane	4.1 ± 8%	-17%	-12%	-10%	-7%	-7%	-5%	-15%	-15%
Mean value	±9%	-11%		-2%		-2%		-11%	

* 2 weeks with a sampling rate of 10cm³/min

When using diffusive samplers as stationary indoor air monitors, one uncertainty remains. Since the value of the air velocity is rather low and normally unknown, the mass flow rate will tend to decrease due to a decrease of the concentration C_i at the entrance of the diffusive sampler. This effect will increase with increasing sampling rates. On the other hand, tube-type diffusive samplers with their rather small sampling rates should show only marginal dependency on air velocities under practical conditions. Special investigations on indoor air velocities had shown values from < 1cm/sec to 57cm/sec with a median of 5.3cm/sec (7). In the test chamber, as shown in Figure 1, diffusive samplers were arranged at positions A, B and C with different air velocities. The weighing differences of the permeation tubes of the test gas generator are used to calculate the theoretical concentrations and their standard deviations. The values are tabulated in Table 1 with deviations of the analyses of different samples from theoretical concentrations. These samples are derived from active sampling with very slow sampling rate and diffusive sampling with different badge-types at positions with various air velocities.

These results are supplemented by special indoor air analyses. Duplicate diffusive sampling in almost stagnant air (low ventilation) was done in parallel with duplicate diffusive sampling in an open tube equipped with a very small ventilator (high ventilation) at nearly the same location.

The data in Tables 1 and 2 show a decrease of no more than about 15% at sampling sites with no measurable air velocity (detection limit: 0.5cm/sec) relative to sites with an air velocity of about 10cm/sec. Generally, this minor decrease is not relevant to the hygienic assessment of the VOC content in indoor air. On the contrary, this decrease should be kept in mind, if threshold levels, e.g. for workplace monitoring, are to be controlled.

Table 2 Comparison of Mean Concentrations (n=2) and Concentration Differences (μg/m³) of Parallel Indoor Air Analyses with Badge-type Diffusive Samplers at Air Velocities of <0.5cm/sec and 4cm/sec

	4cm/sec (high vent.)		<0.5cm/sec (low vent.)			
Compound	$Mean_{high}$	$Diff._{high}$	$Mean_{low}$	$Diff._{low}$	$Mean_{low}$	$-Mean_{high}$
Benzene	4.49	0.04	4.21	0.06	-0.28	-6%
Toluene	17.0	0.20	15.9	0.65	-1.1	-7%
n-Decane	28.0	2.0	25.5	1.3	-2.6	-9%
Limonene	60.8	2.6	53.6	2.8	-7.2	-12%

When prolonging the sampling time, it is necessary to prove that the sampling rate remains constant. This was tested using the above mentioned test gas generator generating a mixture of VOC with known concentrations in the range of 10–100μg/m³ over a period of several weeks (see Figure 2). A sampling time of two weeks is a convenient compromise between detection limit achieved (1μg/m³) and sampling duration. At normal indoor concentrations, no breakthrough or back diffusion is likely to occur for the compounds tested and similar ones.

In the analysis of diffusive samplers, all analytical steps, such as sample pretreatment, desorption with carbon disulfide, separation, detection, data analysis and quality assurance, are the same as for standard charcoal tubes according to NIOSH or VDI methods. Using these nearly identical procedures, the reproducibility within one laboratory is at least as good for diffusive samplers as for active samplers. For example, measurements of tetrachlorethane concentrations using tube-type diffusive samplers (n=9) exhibited a standard deviation of ±3% compared to a standard deviation of ±5% for NIOSH tubes (n=7) at a concentration of 3.6mg/m³.

To supplement the laboratory results the reproducibility of diffusive samplers was also evaluated under field conditions. The results of duplicate field samples taken with OVM samplers and of parallel samples of OVM and Gasbadge samplers are shown in Table 3. It can be seen that non-polar compounds exhibit small standard deviations. Although the standard deviations are higher for polar compounds, they do not exceed about 20%.

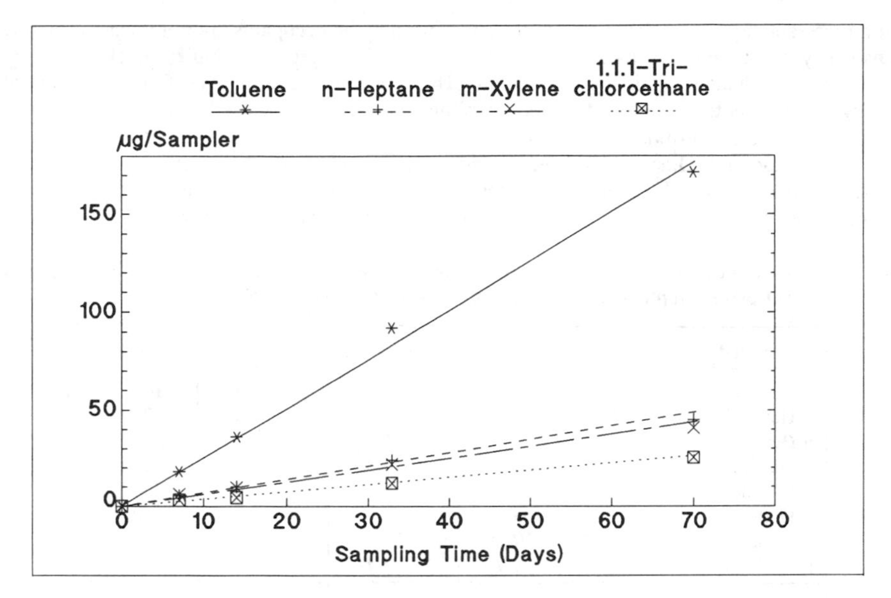

Fig. 2. Linearity of the sampling rate
Badge-type passive samplers (n=2), chamber with constant concentrations

Table 3. Standard Deviations of Selected VOC from Duplicate Field Samples ($\mu g/m^3$)

	OVM - OVM (n=27)		OVM - Gasb. (n=17)	
Compound	S_D	Concentration range	S_D	Concentration range
n-Hexane	± 0.9	(5 - 18)	± 0.5	(3 - 9)
Benzene	± 1.2	(13 - 31)	±1.4	(<1 - 12)
111-Trichloroethane	± 0.3	(1.4 - 6.5)	± 1.7	(2 - 93)
Ethylacetate	+ 2.2	(4 - 60)	± 9.2	(<1 - 52)
n-Butanol	± 2.5	(<1 - 25)	± 9.0	(<1 - 45)

These findings indicate the reliability of the diffusive samplers used. Some limitations for polar compounds are described in detail elsewhere (8). It should be mentioned that our intention normally was not to get the utmost accuracy for single compounds but to get as much information as possible about a complex and variable VOC composition at the sampling site with an effort that can be done on a routine basis. In general, this sampling method can be used as standard method for single measurements as well as for extensive epidemiological studies.

4. USE OF DIFFUSIVE SAMPLER TO MONITOR INDOOR AIR

In most cases, we use diffusive samplers in conjunction with other sampling methods for short special monitoring campaigns in Berlin. For use on a large scale or outside of Berlin, we mail the samplers together with brief instructions to a well-trained sampling team or to the local authorities who are involved in the case to be evaluated. They prepare the sampler and fix it at a representative site in the room of interest. Normally the occupants are not impaired by this absolutely quiet sampling method. Therefore they will forget the sampler quickly and the probability of changing their lifestyles is low. This is important, especially if the mean normal living situation is to be assessed. After sampling, the sampler is closed and sent back to our laboratory. The gas chromatographic analysis of the solution with capillary columns of different polarity, including FID and ECD detection and retention index calculation gives qualitative and quantitative data for all calibrated compounds which represent about 60–80% of all VOC sampled (with some limitations for polar compounds) in the range of n-hexane to n-tridecane.

Table 4. Concentrations of VOC in the Air of Two Rooms in an Office Building ($\mu g/m^3$). Diffusive Sampling, Two Weeks (formaldehyde: two days), Normal Working Conditions

Compound	Room No.1	Room No.2	Median*	Max*
n-Hexane	6.0	4.6	7.5	140
n-Heptane	1.7	1.3	5.2	170
n-Octane	1.4	2.1	3.1	92
n-Nonane	4.5	5.8	5.1	140
n-Decane	31	35	8.3	240
n-Undecane	69	79	6.0	120
n-Dodecane	29	51	4.2	72
n-Tridecane	5.7	7.4	5.0	79
Cycloalkanes	3.6	3.3	14	670
Benzene	13	11	7.7	90
Toluene	23	22	65	1700
Ethylbenzene	5.7	4.9	7.5	160
m+p-Xylene	7.7	7.7	17	300
o-Xylene	3.8	3.3	5.0	45
Styrene	5.7	5.4	1.1	41
Methylethylbenzenes	6.0	5.7	8.4	330
Trimethylbenzenes	7.7	7.0	11	500
Naphthalene	1.6	2.5	2.3	14
111-Trichloroethane	19	11	4.9	260
Tetrachlorethane	1.7	1.5	4.8	810
α-Pinene	36	39	6.8	120
Carene	15	13	no data	no data
Limonene	35	38	15	320
Ethylacetate	11	4.1	6.3	200
Hexanal	15	18	1	11
Butanols	350	410	1	52
4-Phenylcyclohexene	7.7	12	no data	no data
Formaldehyde	74	70	53	310

* Results of an epidemiological study (488 households, diffusive sampling, two weeks, more than 50 calibrated VOC) (9)

In the following, the results of an indoor air analysis (Table 4) will be used as an example to demonstrate the feasibility of the complete method. This analysis was requested, because occupants of an office were complaining about unspecific symptoms like headache and irritations of eyes and throat. To evaluate the results, the concentrations found in the air of two rooms are compared with results of an epidemiological study of German households chosen at random. Quite obviously, several compounds, some with irritating or annoying properties, exhibit relatively high concentrations, thus a relation between complaints and concentration levels cannot be excluded.

Because of the simple and inconspicuous use, diffusive sampling is an ideal sampling method for following indoor air concentrations over a long time with a sampling frequency of, for example, two weeks. In conjunction with the above mentioned study we investigated the seasonal variation of concentrations of VOC in households carefully chosen according to building age and smoking habits of the inhabitants (10).

There are many very interesting concentration profiles due to changes of ventilation rate, emission rate and activity of the inhabitants (e.g. renovations). As an example, the concentration profile of methylbenzoate, an odorous compound perceived at a very low concentration level, is shown in Figure 3 (11).

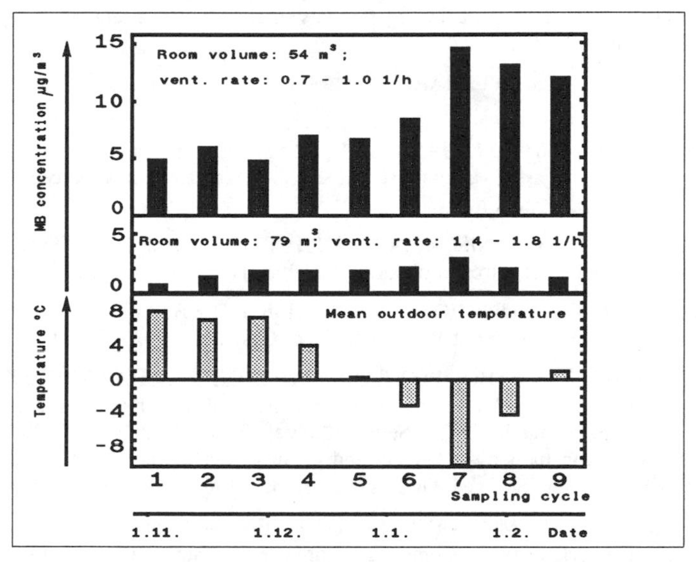

Fig. 3. Variations of the bi-weekly mean concentration of methylbenzoate and the outdoor temperature

Methylbenzoate is a compound of medium volatility emitted from special calorimetric measuring devices, which are fixed at the surface of radiators of many German flats. The expected dependency of the concentration on various parameters such as emission rate, room volume and ventilation rate can be proven qualitatively. The spacious room with high ventilation rates exhibits a lower concentration level of methylbenzoate than the small room with low ventilation rates. In addition, if a decrease of the mean outdoor temperature causes a rise of the radiator temperature to maintain a constant warm indoor climate, the emission rate of methylbenzoate is increased exhibiting higher indoor concentrations and vice versa.

5. CONCLUSIONS

Diffusive sampling with badge-type samplers equipped with charcoal pads proved to be a convenient and reliable method, in conjunction with standard gas chromatographic procedures, to monitor the mean concentrations of VOC in indoor air. Such averages give valuable information needed to assess the general VOC burden of occupants.

All results of diffusive sampling are of special interest, if the health impacts of VOC with long-term effects rather then with short-term effects are being considered.

Results of a large number of randomized samples taken at sites of no complaints can be used as a reference in evaluating the results of samples taken at sites with manifested complaints.

REFERENCES

(1) BERLIN, A., BROWN, R.H. and SAUNDERS, K.J., (Eds.), DIFFUSIVE SAMPLING - An alternative approach to workplace air monitoring, Proc. of an Int. Symp. held in Luxembourg, 22-26 September 1986: Royal Soc. Chem., Burlington House, London, 1987.

(2) LUGG, G.A. (1968). Diffusion coefficients of some organic and other vapors in air. Anal. Chem. 40, 1072-1077.

(3) GUENIER, J.P. and FERRARI, P. (1981). Cah. de notes doc. (INRS) no. 105, 493–507.

(4) SEIFERT, B. and ABRAHAM, H.-J. (1983). Use of passive samplers for the determination of gaseous organic substances in indoor air at low concentration levels. Int. J. Environ. Anal. Chem. 13, 237-253.

(5) PALMES, E.D., GUNNISON, A.F., DIMATTIO, J. and TOMCZYK, C. (1976). Personal sampler for nitrogen dioxide. Am. Ind. Hyg. Assoc. J. 37, 570-577.

(6) PRESCHER, K.E. and SCHÖNDUBE, M. (1983). Die Bestimmung von Formaldehyd in Innen- und Außenluft mit Passivsammlern. Ges.-Ing. 104, 198-200.

(7) MATTHEWS, T.W., THOMPSON, C.V., WILSON, D.L., HAWTHORNE, A.R. and MAGE, D.T. (1987). Air velocities inside domestic environments: an important parameter for passive monitoring. In Seifert, B. et al. (Eds.), INDOOR AIR '87 - Proc. 4th Intern. Conf. on Indoor Air Quality and Climate, Berlin, 17-21 August 1987, Vol.1: Institute for Water, Soil and Air Hygiene, Berlin, 1987, 174-178.

(8) DE BORTOLI, M., MØLHAVE, L., THORSEN, M.A. and ULLRICH, D. (1986). European Interlaboratory Comparison of Passive Samplers for Organic Vapour Monitoring in Indoor Air. (Report nr. EUR 10487 EN), Luxembourg

(9) KRAUSE, C., MAILAHN, W., SCHULZ, C., SEIFERT, B. and ULLRICH, D. (1987). Occurrence of volatile organic compounds in the air of 500 homes in the Federal Republic of Germany. In Seifert, B. et al. (Eds.), INDOOR AIR '87 - Proc. 4th Intern. Conf. on Indoor Air Quality and Climate, Berlin, 17-21 August 1987, Vol.1: Institute for Water, Soil and Air Hygiene, Berlin, 1987, 102-106.

(10) SEIFERT, B., MAILAHN, W., SCHULZ, C. and ULLRICH, D. (1989). Seasonal variation of concentrations of volatile organic compounds in selected German homes. Environ. Int. 15, 397-408.

(11) ULLRICH, D., LIU, Z. and SEIFERT, B. (1987). Assessment of the indoor air concentration of an organic substance emitted from a continuous source using methylbenzoate as an example. In Seifert, B. et al. (Eds.), INDOOR AIR '87 - Proc. 4th Intern. Conf. on Indoor Air Quality and Climate, Berlin, 17-21 August 1987, Vol.1: Institute for Water, Soil and Air Hygiene, Berlin, 1987, 174-178.

1424

ALDEHYDES IN THE NON-INDUSTRIAL INDOOR ENVIRONMENT

V.M. Brown, D.R. Crump, M.A. Gavin and D. Gardiner

Materials Division, Building Research Establishment
Watford, Herts, UK

SUMMARY

Aldehydes and formaldehyde in particular occur in indoor air at concentrations higher than outdoors. Their sources include building materials, furnishings, combustion processes and consumer products. Studies of indoor concentrations of formaldehyde have used a number of different sampling methods and strategies. This makes comparison and interpretation of data difficult. There is very little information about concentrations of other aldehydes.

Passive samplers are most suited to the study of population exposure. Experience with one type of passive sampler is reported and the results of measurements in 100 homes are discussed.

INTRODUCTION

Aldehydes are organic compounds of the general formula RCHO. They are introduced into the atmosphere through a variety of pathways. Natural sources include volcanic gases, but much more important are the emissions due to animal excretions. The main anthropogenic sources are rubbish incineration and exploitation of fossil fuels. In addition, photo-oxidation of organic compounds produces aldehydes and this source is dominant over primary sources(1).

In ambient air in areas of low pollution only very low concentrations of formaldehyde (< 0.4-13 μgm^{-3}) and acetaldehyde (0.54-2.3 μgm^{-3}) have been reported. In polluted areas, the Los Angeles region in particular, daily maxima of 36-60 μgm^{-3} of formaldehyde and 36-63 μgm^{-3} of acetaldehyde have been measured together with lesser amounts of propanal, butanal, 2-butanone and benzaldehyde(2).

The presence of aldehydes in the atmosphere of non-industrial buildings is of interest because nearly all are irritants of the skin, eyes and respiratory mucosa and particularly those with low molecular weights and those with unsaturation. This paper reviews information about indoor sources of aldehydes and the concentrations present in indoor air. It considers the problems associated with obtaining representative measurements of formaldehyde in particular, and BRE's experience with one type of passive sampling device.

SOURCES OF ALDEHYDES IN INDOOR AIR

Formaldehyde in the non-industrial indoor environment originates predominantly from a wide variety of indoor sources and not outdoor air(3). For example, it is a constituent of tobacco smoke and of combustion gases from heating stoves and gas appliances,it occurs in cosmetics and disinfectants, as an additive to water based paints and as a treatment for textiles. The bulk of industrially produced formaldehyde is used for resins contained in various products and in particular wood products such as particleboard. These sources are either intermittent in nature and closely related to occupant behaviour, such as use of particular products, cigarette smoking and use of heating fuel, or are more long term, such as emission from particleboard or urea formaldehyde cavity wall insulation. The emissions from these longer term sources will vary with time both due to ageing of the product and changes in environmental conditions. This has been extensively studied for wood products as reviewed by Myers[4]. The resulting concentration of formaldehyde in indoor air will depend both upon source emission rate and losses by adsorption to surfaces and ventilation to outdoors.

Sources of other aldehydes have been less extensively studied, except for tobacco smoke[5, 6]. Acetaldehyde is the most prominent aldehyde in both main stream and side stream cigarette smoke. Other significant compounds include acrolein, crotonaldehyde and formaldehyde. Consumer products such as cosmetics and air fresheners, fuel combustion, paints, lacquers and carpets can all emit aldehydes[6, 7].

CONCENTRATION OF ALDEHYDES IN INDOOR AIR

Results of studies of formaldehyde in indoor air in European countries have been summarised in a recent publication[8]. Most of the data relates to small localised studies or investigations of particular problem buildings. A variety of analytical and sampling methods have been applied. A survey in the Federal Republic of Germany during 1984-1986 is the study most closely approaching a random national survey [8]. This used a passive sampler and a 48 hour sampling period. Published results in the UK describe 3 sampling methods applied to a total of about 280 buildings since 1980. This is as comprehensive as most other countries, but small compared with the size of the building stock. Typical concentrations are 20-60 μgm^{-3}, but can exceed 100 μgm^{-3}[6,9].

Assessment of the significance of this data requires guidance on health effects. The World Health Organisation reviewed evidence in 1987 and concluded that "in order to avoid complaints of sensitive people about indoor air in non-industrial buildings, the formaldehyde concentration should be below 0.1 mgm^{-3} (100 μgm^{-3}) as a 30 minute average"[10]. Available data does not allow a good estimate of the proportion of the European community's population exposed to concentrations exceeding this value, either occasionally or frequently. The German study reported that 2% of readings exceeded 120 μgm^{-3} in a randomised survey and 8% in cases investigated because of complaint about air quality[8].

Only two studies have investigated concentrations of aldehydes in indoor air using a compound specific DNPH based method. De Bortoli et al[11] undertook measurements in 15 homes in Italy and found mean levels of acetaldehyde of 16 μgm^{-3} with a mean indoor/outdoor ratio of 4.9. Hexanal and nonanal were observed in the majority of homes. In a study of 10 homes in the UK, acetaldehyde was detectable in most at levels not exceeding 20 μgm^{-3} and acrolein was measured occasionally at lower concentrations[6]. There are no WHO air quality guidelines against which the significance of these other aldehydes in indoor air can be judged.

SAMPLING STRATEGY TO DETERMINE ALDEHYDES IN INDOOR AIR

The appropriate strategy depends upon the objective of the study. To determine the exposure of a population measurements are required under normal living conditions, preferably continuously. However to investigate reports of poor air quality it may be more informative to test under worst case conditions of minimum ventilation, high temperature and humidity and possibly during activities such as cigarette smoking. Against any ideal strategy must be balanced the requirement for available resources to carry out monitoring. The European Commission have published a report outlining factors to be taken into consideration when designing a strategy for sampling chemicals in indoor air[12]. As a minimum requirement for formaldehyde, they refer to a sampling protocol suggested by Godish[13] which is based on a one-off measurement using the impinger/chromotropic acid method to investigate building-related health complaints. The WHO do not advise on appropiate strategies for testing for compliance with their air quality guideline.

ANALYTICAL AND SAMPLING METHODS

Two types of sampling technique are available; a) active (pumped) and b) passive (diffusive) methods. Table 1 lists the main types of method applied to studies of formaldehyde and other aldehydes in indoor air.

Method	Sampling/Analytical Procedure	Sensitivity to formaldehyde µg/L	Ref
Chromotropic acid	impinger	0.05 (1 hour) 0.002 (12 hour)	14 15
Pararosaniline	impinger molecular sieve cartridge	0.03 (1 hour) 0.025 (15 min)	16 17
MBTH	impinger	0.006 (1 hour)	6
2,4-DNPH a)aqueous b)organic (acetonitrile)	impinger	a)0.002 (1 hour) b)0.002 (1 hour)	6 11
3M Formaldehyde Passive Monitor	impregnated (sodium bisulphite) adsorbent/chromotropic acid	0.012 (24 hour)	13
Air Quality Research Inc Passive Badge ('Palmes tube')	impregnated (sodium bisulphite) adsorbent/chromotropic acid	0.012 (7 day)	13
Pro Tek Passive Badge	liquid adsorbent (sodium bisulphite/ chromotropic acid)	0.01 (24 hour)	13
570 Series Passive Formaldehyde Monitor	impregnated (2,4-DNPH) adsorbent	0.006 (8 hour)	18
ORNL Passive Membrane Dosimeter	liquid adsorbent(water)/ pararosaniline	0.03 (24 hour)	17
'Palmes' tube	impregnated (triethanolamine) adsorbent/pararosaniline	0.044 (48 hour)	19

Table 1. Sampling and Analytical Methods for Aldehydes in Indoor Air

For active methods a sampling time of 30 minutes to 1 hour is commonly used whereas for passive samplers at least 8 hours is required to collect sufficient sample. Passive samplers provide mean concentrations over periods that would require considerable resources to achieve by active methods, but do not inform about short term fluctuations in concentration. All involve analysis by spectrophotometry except

2,4-dinitrophenylhydrazine based methods which require High Performance Liquid Chromatography (HPLC).

Active Methods;

The most widely used active method particularly in the USA has been the chromotropic acid technique using an impinger for sample collection. This procedure is a modification of recommended methods to determine formaldehyde in the work place in the USA and UK. Within each category of analytical method different studies have used different designs of sampler, air flow rates and sampling times, all factors which might influence the accuracy of the measurement. A few studies have compared the performance of different analytical methods and assessed the consequences of possible interference by other chemicals in indoor air. These have shown the chromotropic acid method to be most and 2,4-DNPH least susceptible to interference by other chemicals(20-23).

The MBTH method determines total aldehydes and while formaldehyde may be the dominant aldehyde in most situations simultaneous measurement by a specific method would be required for confirmation. The most efficient method of determining individual aldehydes is to use a DNPH-based method. This type of method has been quite widely used for outdoor air, particularly for the study of photochemical smog, but applied to indoor air in only two studies(6,11). There are many variants of the method with different collecting liquids and solid media. Higher molecular weight aldehydes (>C_4) can be analysed by collection on solid adsorbents and either solvent or thermally desorbed for gas chromatrographic analysis as used for measurements of "VOC's" (volatile organic compounds) in indoor air.

Passive Methods;

Accurate measurement by passive methods requires that the uptake rate of the sampler is constant and not unduly influenced by temperature, humidity, air velocity, analyte concentration and presence of other chemicals. The UK Health and Safety Executive (HSE) and the National Institute for Occupational Safety and Health (NIOSH) in the USA have published protocols for the validation of diffusive sampler performance(24). NIOSH has published results of evaluations of the Pro Tek and 3M Formaldehyde monitors for suitability for work place monitoring(25). The 3M monitor failed the evaluation on 3 accounts; sampling rate and capacity, poor accuracy and significant face velocity and humidity effects. The Pro Tek badge performed much better, but did not fully meet the performance criteria because of significant face velocity and chemical interference effects. Evaluation experiments were conducted in formaldehyde concentrations of 0.3-8.4 ppm and with exposure periods of 7.5 minutes to 12 hours.

The "Palmes" type tubes and the ORNL samplers have not been subject to this detailed testing. An accuracy of 25% is reported for the Air Quality Research Inc badge(13). Levin et al(26) describe validation tests of a passive sampler consisting of a DNPH impregnated glass fibre filter contained in an adapted standard 37 mm two section air sampling cassette. The formaldehyde hydrazone formed is solvent desorbed and analysed by HPLC. This sampler was applied to a survey of formaldehyde concentrations in 12 homes in Sweden.

Levin et al(18) report a version of their sampler which is commercially available and described as the 570 Series Formaldehyde Monitor. A silica gel tape is coated with DNPH and contained within a polypropylene housing with a 2.9 mm thick screen. Samplers were validated in the laboratory using formaldehyde concentrations of 0.07-3.7 µg/L, sampling times of 15 mins-8 hours, relative humidities of 10-80% and air speeds of 0.05-1 m/s according to HSE and NIOSH protocols. Uptake rate was not influenced by sampling time, wind velocity or relative humidity. A small (<5%) but significant decrease in uptake rate was found at higher concentrations (about 3 μg/L). At high wind velocity perpendicular orientation of the sampler resulted in a slight increase in uptake rate. Field validation showed good agreement between the diffusive and a pumped method including a test at low air velocity (<0.02 m/s) and low formaldehyde concentration (0.027 μg/L) for a 20 hour sampling period. The sampling uptake rate is 25 mL/min and the sensitivity for an 8 hour sample is 0.006 μg/L. Any possible effect of temperature on sampling uptake rate is not covered by this paper although the commercial manufacturer reports no effect of temperature between 10 and 30°C[27].

In conclusion, of the several formaldehyde passive samplers used in non-occupational environments only the Pro-Tek and 570 Series monitors have performed well in evaluation tests according to the HSE and NIOSH protocols. This full evaluation has only been carried out at formaldehyde concentrations and air speeds appropriate to residential environments for the 570 Series. Arguably satisfactory correlations in the field for passive and pumped samples indicate that the Pro-Tek badge is also suitable though the 570 Series dosimeter does have advantages of greater precision, sensitivity and specificity for formaldehyde.

EXPERIENCE WITH THE 570 SERIES FORMALDEHYDE MONITOR

We wished to undertake a study of population exposure to formaldehyde in non-industrial buildings. Our requirement was therefore for a sampler applicable to the study of a large number of homes with minimum resources. A passive sampler that could be delivered and returned by post and used by householders after simple training and had proven technical performance was required. For pilot studies and a preliminary survey of 100 homes we selected to use the 570 Series Formaldehyde Monitor.

Analytical Method:

The analysis of the exposed badge described by Levin[18] involves adding acetonitrile to the sample and blank portions of the filter in vials and shaking for approximately 1 minute and analysis by HPLC. We found improved chromatography peak shape using methanol to desorb the filter and no loss of efficiency of sample recovery. HPLC analytical conditions were as follows; column : Zorbax ODS, C_{18}, 250 x 45 mm, mobile phase : gradient methanol/water mixture - 0 min-8 min at 60% methanol, rising to 90% at 13 min and declining after 25 min to 60% at 30 min, flow rate : 1 mL/min, detection : UV absorption at 345 nm. Samples were analysed within 8 hours of the extraction procedure.

Sampling Period:

For non-industrial indoor air a sampling time of at least 8 hours is required for sensitivity reasons. Longer time periods provide a mean concentration less influenced by short term changes in use of products, occupant behaviour and environmental conditions. We investigated the influence of exposure period in a temporary office building by comparing uptake of formaldehyde by badges exposed for 48, 72 and 96 hours with the mean amount collected by badges exposed for 24 hours. The results in Table 2 show the uptake is not influenced by exposure time over 3 days but may have declined after four.

Sample	Period (day)	Hours exposed	HCHO concentration (µg/L)	Temperature (°C)		Relative Humidity Range (%)
				Mean	Range	
A	1-2	24	0.069	22.7	20-24	35-40
B	2-3	24	0.073	23.2	21-25	35
C	3-4	24	0.068	22.9	20-25	35-36
D	4-5	24	0.095	23.8	22-26.5	35-37
Mean A+B	1-3		0.071			
E	1-3	48	0.072	23	20-25	35-40
Mean A+B+C	1-4		0.070			
F	1-4	72	0.069	23	20-25	35-40
Mean A+B+C+D	1-5		0.076			
G	1-5	96	0.052	23.2	20-26.5	35-40

Table 2. Formaldehyde Concentrations Measured by Samplers with Various Exposure Times

Repeatability:

Five replicate samples were placed in bedrooms of each of 3 houses and exposed for 3 days. The standard error of the measurements was 6.2-8.7%. This compares with a standard error for 5 replicates using active sampling by a DNPH based method of 11.2%.

Ease of use:

The badges are readily transported by mail being lightweight and strong. During dispatch within polyethylene bags inside envelopes they do not acquire significant formaldehyde. Householders are able to expose and close badges reliably after simple instruction. Following exposure and prior to analysis the badges are stable for at least two weeks.

Passive Versus Active Sampling:

We have not attempted to validate the passive method by simultaneous continuous active sampling over the 3 day exposure period. However to compare results given by one-off active samples (1 hour) we have taken readings in 5 locations in 3 houses under normal living conditions. Table 3 shows that differences between the results of the two methods and sampling strategies were quite small. Further tests of this kind would provide guidance as to the likelihood of a short term (eg 30 min) air quality guideline being exceeded within a longer term passive sampling period.

Measurements in 100 homes:

Samplers were exposed for 3 days in the main bedroom and living room of 100 homes during a period of normal living conditions. Figure 1 shows the distribution of concentrations in the bedrooms which has a mean of 0.023 µg/L and standard deviation of 0.016. There was no significant difference between concentrations found in the living rooms (0.021 µg/L, SD 0.013) and bedrooms. Of the 100 houses two had a formaldehyde concentration of more than 0.09 µg/L and must have exceeded the 30 minute WHO air quality guideline during the 3 day period. A further unknown number may have exceeded 0.1 µg/L for at least part of the time. A greater understanding of temporal variation is required in order to estimate this number. Further work is also required to assess longer term

	Formaldehyde Concentration (μg/L)					
	House 1		House 2		House 3	
Location	Active	Passive	Active	Passive	Active	Passive
Main Bedroom	0.035	0.026	0.029	0.020	0.044	0.033
Lounge	0.018	0.029	0.020	0.016	0.027	0.022
Kitchen	0.021	0.023	0.041	0.032	0.020	0.039
2nd bedroom	0.022	0.019	0.023	0.018	0.044	0.020
Bathroom	0.011	0.015	0.015	0.011	0.022	0.018
mean	.021	.022	.026	.019	.031	.026
standand deviation	.007	.005	.009	.007	.01	.008

Table 3. Formaldehyde Concentrations Detected by Active and Passive Samplers in 3 Houses.

Figure 1. Formaldehyde Concentration in the Main Bedroom of 100 Homes

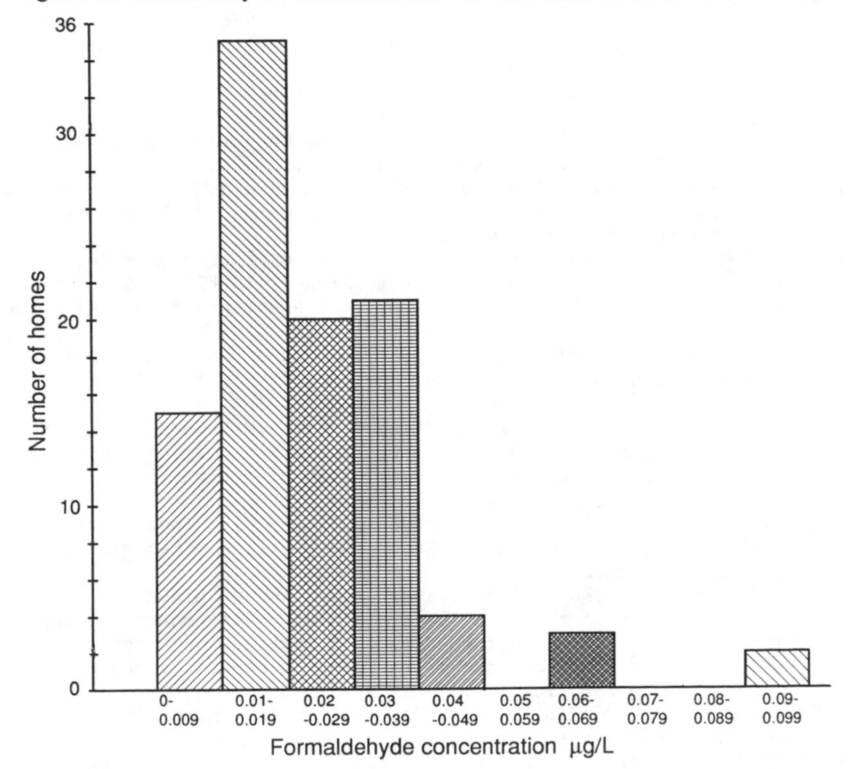

variations in formaldehyde concentration such as seasonal effects.

CONCLUSIONS

Formaldehyde is the predominant aldehyde in non-industrial indoor atmospheres and has many potential sources. There is little appropiate data for assessing the proportion of the population likely to be exposed to concentrations above the WHO recommended air quality guideline either continuously or intermittently. The different sampling methods and strategies applied in available studies makes interpretation and inter-comparison difficult.

Monitoring of population exposure to formaldehyde requires a relatively low cost technique. The BRE have applied the 570 Series Formaldehyde Monitor to a pilot study of 100 homes. More work is required on temporal fluctuation to determine relationships between short term fluctuation in formaldehyde with longer term mean concentrations given by passive samplers. Also, longer term fluctuations, such as seasonal trends, need to be assessed to determine the appropiate frequency of sampling to obtain a representative measure of concentrations.

Relatively little data is available concerning higher aldehydes though tobacco smoke is a significant source and compound specific methods are available which have been more commonly applied to outdoor air.

REFERENCES

(1) CARLIER, P., HANNACHI, H. and MOUVIER, G. (1986). The chemistry of carbonyl compounds in the atmosphere - a review. Atmospheric Environment, 20, 11, 2079-2099.
(2) GROSJEAN, D. (1982). Formaldehyde and other carbonyls in Los Angeles ambient air. Environ. Sci. Technol. 16, 254-262.
(3) NATIONAL RESEARCH COUNCIL (1981). Formaldehyde and other aldehydes. National Academy Press, Washington, D.C., U.S.A.
(4) MYERS, G. (1984). Effect of ventilation rate and board loading on formaldehyde concentration : a critical review of the literature. Forest Products J. 34, 10, 59-68.
(5) SCHLITT, H. and KNOPPEL, H. (1989). Carbonyl compounds in mainstream and sidestream cigarette smoke. In BIEVA, C.J., COURTOIS, Y. AND GOVAERTS, M. (ED.). Present and future of indoor air quality. Elsevier Science Publishers.
(6) CRUMP, D.R. and GARDINER, D. (1989). Sources and concentrations of aldehydes and ketones in indoor environments in the UK. Environment International, 15, 455-462.
(7) COLOMBO, A., DE BORTOLI, M., PECCHIO, E., SCHAUENBURGH, H., SHLITT, H. and VISSERS, H. (1990). Chamber testing of organic emission from building and furnishing materials. The Science of the Total Environment, 91, 237-249.
(8) COMMISSION OF THE EUROPEAN COMMUNITIES. (1990). Indoor air pollution by formaldehyde in European countries. Report EUR 13216 EN, EEC, Luxembourg.
(9) BROWN, V.M., COCKRAM, A.H., CRUMP, D.R., and GARDINER, D. (1990). Investigations of the volatile organic compound content of indoor air in homes with an odorous damp roof membrane. Proceedings of the Fifth International conference on Indoor air quality and Climate, volume 3, 575-580, Toronto, 29 July -3 August, 1990.
(10) WORLD HEALTH ORGANISATION (1987). Air quality guidelines for Europe. WHO European series No. 23, Copenhagen.
(11) DE BORTOLI, M., KNOPPEL, H., PECCHIO, E., PEIL, A., ROGORA, L, SCHAUENBURG, H, SCHLITT, H. and VISSERS,H. (1985). Measurements of

indoor air quality and comparison with ambient air - a study on 15 homes in Northern Italy. Report EUR 9695 EN, CEC, Luxembourg.

(12) COMMISSION OF THE EUROPEAN COMMUNITIES. (1989). Strategy for sampling chemical substances in indoor air. Report EUR 12617 EN, CEC, Luxembourg.

(13) GODISH, T. (1985). Residential formaldehyde sampling - current and recommended practices. Am. Ind. Hyg. Assoc. J. 46 (3), 105-110.

(14) CONSENSUS WORKSHOP ON FORMALDEHYDE (1984). Report on the Consensus Workshop on formaldehyde. Environmental Health Perspectives 58, 323-381.

(15) ATKINS, D., HEALY, C., and BISHOP, K. (1984). Measurements of formaldehyde concentrations in homes. Report AERE-R 11341, UKAE, Harwell, UK.

(16) MIKSCH, R., ANTHON, D., FANNING, L., HOLLOWELL, C., REVZAN, K. and GLANVILLE, J. (1981). Modified pararosaniline method for the determination for formaldehyde in air. Anal. Chem. 53, 2118-2123.

(17) HAWTHORNE, A., GAMMAGE, R. and DUDNEY, C. (1986). An indoor air quality study of 40 East Tennessee homes. Environment International, 12, 21-229.

(18) LEVIN, J., LINDAHL, R., and ANDERSSON, K. (1988). High-Performance Liquid-Chromatographic determination of formaldehyde in air in the ppb to ppm range using diffusive sampling and hydrazone formation. Environmental Technology Letters, 9, 1423-1430.

(19) PRESCHER, K. and SCHONDUBE, M. (1983). Die bestimmung von formaldehyd in innen - und Aussenluft mit passivsammlern. Gesundh. - Ing. 104, 198-200.

(20) ECKMANN, A., DALLY, K., HANRAHAN, L. and ANDERSON, H. (1982). Comparison of the chromotropic and modified pararosaniline methods for the determination of formaldehyde in air. Environment International 8, 159-166.

(21) PETREAS, M., TWISS, S., PON, D., and IMADA, M. (1984). Evaluation of two methods for measuring formaldehyde at ppb levels. Proceedings of the Third International Conference on Indoor Air Quality and Climate, volume, 3, 49-54. Stockholm.

(22) AHONEN, I., PRIHA, E. and MARJA-LIISA, M. (1984). Specificity of analytical methods to determine the concentration of formaldehyde in workroom air. Chemosphere 13, 4, 521-525.

(23) VAN der WAL, J., KORF, C., KUYPERS, A. and NEELE, J. (1989). Interference by chemicals in the determination of formaldehyde. Environment International 15, 517-524.

(24) HEALTH and SAFETY EXECUTIVE (1983). Protocol for assessing the performance of a diffusive sampler. MDHS 27, HSE, London, UK.

(25) KENNEDY, E. and HULL, R. (1986). Evaluation of the Du Pont Pro-Tek formaldehyde badge and the 3M formaldehyde monitor. Am. Ind Hyg. Assoc. J. 47 (2), 94-105

(26) LEVIN, J., LINDAHL, R. and ANDERSSON, K. (1989). Monitoring of ppb levels of formaldehyde using a diffusive sampler. J.A.P.C.A. 39, 44-47.

(27) GMD Systems Inc, (1989). 570 Series Formaldehyde Dosimeter Badge. GMD Systems Inc., Old Route 519, Hendersonville, PA 15339, USA.

Crown copyright 1991 - Building Research Establishment

1425

MICRO-ORGANISMS

BRIAN FLANNIGAN

Department of Biological Sciences,
Heriot-Watt University
Riccarton, Edinburgh EH14 4AS,
Scotland.

SUMMARY

Micro-organisms are important biopollutants of air in non-industrial workplaces. Viruses are spread mainly by personal contact, but rates of illness can be higher in buildings with HVAC systems than in naturally ventilated buildings. A range of moulds, yeasts and bacteria can grow on or in wall coverings, soft furnishings, soil, dust and HVAC systems, releasing spores and cells into the air. Cladosporium and Penicillium are usually the predominant moulds in the air and micrococci the predominant bacteria. Some micro-organisms can cause allergic diseases, including rhinitis, asthma, humidifier fever and extrinsic allergic alveolitis, but pathogens are seldom abundant, although Legionella can grow in and spread from water supplies and humidifiers. Inhalation of bacterial endotoxins, mycotoxins in mould spores and volatiles from moulds may possibly affect health, but the scale of problems which are directly attributable to micro-organisms and their products cannot be gauged. More attention must be given to the location, timing and methodology of sampling airborne microbes in investigations of apparently building-related health problems.

1. INTRODUCTION

The indoor air of private houses, non-industrial workplaces and public buildings contains three major types of particulate biopollutant which may be generated within the buildings and affect the health of their occupants. These particulates are mites and their faeces, dander from pets and other furred animals, and micro-organisms. Pollen is another type of particulate which can affect health, but most pollen in indoor air originates largely from plants outdoors and concentrations are generally much lower than outdoors. Consequently, pollen-related respiratory problems are generally associated with exposure to the higher concentrations outdoors. Insects are additional sources of airborne particulates. Particulates from cat fleas or clothes moths may cause allergy, and cockroach particulates may significantly affect socioeconomically disadvantaged groups. However, the numbers of cases involving these insect particulates is small in relation to those involving either the other animal categories or micro-organisms.

The main particulate biopollutants of possible medical significance which are likely to be released in non-industrial workplaces are micro-organisms. The airborne microbial burden comprises viruses; bacteria, including Actinomycetes; fungi, including moulds and yeasts; and perhaps also Protozoa such as amoebae which may grow in dirty humidifier reservoirs and drainage pans of heating, ventilation and air-conditioning (HVAC) systems, but about which little is known.

2. ECOLOGY OF MICRO-ORGANISMS

Viruses, of course, cannot lead an independent existence - they can only grow and multiply within a host organism - and direct person-to-person transmission is considered to be the most important means of spread of most viral diseases, e.g. mumps. However, the common cold, influenza and some other viral illnesses may be transmitted via indoor air, e.g. measles in schools (9). Aerosols of viruses are frequently generated by coughing and sneezing and the spread of infection is therefore likely to be greater in crowded areas. The rates of viral infection in buildings which have HVAC systems with recirculating air may also be higher than in naturally ventilated buildings (2).

The range of bacteria and fungi encountered in the indoor environment is wide. In addition to moulds and yeasts, the fungi include wood-rotting and plaster fungi in the Ascomycetes and Basidiomycetes. Most bacteria and fungi in buildings live as saprotrophs (4), being able to lead an independent existence by utilizing for growth and multiplication the organic matter in a wide range of materials and situations, including surface coatings on walls; wood; upholstered furniture, carpets and other soft furnishings; soil in plant pots; dust; air ducts, cooling units, humidifiers and drainage pans of HVAC systems; and foodstuffs. However, some micro-organisms are also opportunistic pathogens of man. For example, the mould Aspergillus fumigatus can grow saprotrophically on an extensive range of substrates, including wood and soil, but in exceptional circumstances may invade the lung tissue of weakened individuals and from there spread to other organs of the body, especially in the case of immunocompromised patients. Likewise, the bacterium Legionella pneumophila can grow and multiply saprotrophically in water supplies and HVAC equipment, but it can also invade the respiratory system, causing the outbreaks of Legionnaires' Disease which have been associated with hotels, hospitals and other large public buildings in particular.

Growth of these saprotrophic micro-organisms is affected not only by nutrient availability of the substrates, but also by moisture availability and temperature. The water activity (a_w) of the substrate is a key factor in determining which micro-organisms are able to grow. A_w is defined as the ratio of the vapour pressure of water in the substrate to that of pure water under the same conditions of temperature and pressure. In a closed system, there is a direct relationship between the a_w of a substrate and the relative humidity (R.H.) of the ambient atmosphere, e.g. in an atmosphere of 80% R.H.a substrate will come to equilibrium at a_w 0.80. It is often said that little or no microbial growth will occur in a building if the R.H. is kept below 70%. It is true that a substrate in equilibrium with a R.H. of 70%, i.e. with a_w 0.70, will only allow the (slow) growth of a few highly specialised micro-organisms, but even when the R.H. is <70% in the centre of a room the a_w at the surface of a poorly insulated wall or other cold surface may be well above 0.90 as a result of condensation on that surface, or within the matrix of the wall.

The minimum and optimum a_w for microbial growth differs from species to species and is affected by the temperature and the nutrient status of the substrate. For example, the minimum a_w at which a number of moulds associated with damp walls in houses would grow was found to be higher at low temperatures (Table 1). The minimum a_w for growth of the same organisms was lower when papered surfaces were coated with carboxymethyl cellulose to simulate surface soiling or painted with mould-susceptible emulsion paint (5).

From a study of moulds commonly found growing on surfaces in buildings (5) we concluded that could be divided into three groups according to their moisture requirements:

(i) primary colonisers, able to grow at a_w < 0.80, e.g. Aspergillus repens, A.

versicolor and Penicillum brevicompactum;

(ii) secondary colonisers, able to grow at a_w 0.80-0.90, e.g. Cladosporium spp.;

(iii) tertiary colonisers, able only to grow at a_w >0.90, e.g. Ulocladium spp. and Stachybotrys atra.

By infra-red imaging and distribution mapping of moulds, we can also observe that a succession of these moulds (from primary to tertiary colonisers) can develop on areas of the inside surface of outer walls which became progressively more moist with condensation as the temperature declines during the course of a winter. As well as the possibility of a temporal succession of moulds developing, there is also the possibility of the same moulds growing around localised sites of water penetration according to their particular moisture requirements, with the tertiary colonisers at or near the site of penetration and the primary colonisers towards the less wet margins of the affected area.

For bacteria and yeasts, the minimum a_w for growth varies but is generally >0.90, so that these organisms normally only grow on extremely wet surfaces or in humidifiers and drainage pans. Some bacteria, such as Pseudomonas spp. are able to survive in such environments with few nutrients, but other types are more fastidious, having specific requirements for particular nutrients. Although a very high a_w is required for bacterial growth, bacteria such as micrococci and Actinomycetes may survive at much lower a_w levels.

Table 1: Minimum water activity allowing growth of moulds isolated from houses on emulsion-painted woodchip paper [after Grant et al. (5)].

Species	Minimum a_w at 12°C	Minimum a_w at 18°C
Aspergillus versicolor	0.87	0.79
Penicillium brevicompactum	0.87	0.83
P. chrysogenum	0.87	0.85
Cladosporium sphaerospermum	0.93	0.92
Ulocladium consortiale	0.94	0.92
Stachybotrys atra	0.98	0.97

3. THE INDOOR AIR SPORA

The indoor air spora is related to the outdoor spora in both its magnitude and its composition, with nearby plants or organic debris being major contributors (4). When windows are open in summer, outdoor and indoor levels of airborne micro-organisms roughly correspond, but air-conditioning reduces the indoor level (Table 2). With natural ventilation, the total burden of viable spores over a year may be <20% of the outdoor level. In summer, when the outdoor level is at its highest, the level indoors may be reduced to <2% by closing doors and windows. Some types of mould which contribute spores to indoor air are primarily associated with leaf surfaces and other parts of plants, e.g. Alternaria and Epicoccum (Table 3). These "phylloplane" fungi are usually only found in appreciable numbers in indoor air during the summer growing season. Clearly, however, the growth of moulds (and other micro-organisms) indoors must add to the air spora within buildings, and Penicillium, Aspergillus and some other moulds which are among the commonest types growing on damp walls are regarded as "indoor" fungi. The differential between outdoor and indoor counts of these genera is much lower than for the phylloplane types. However, there may be marked differences in the indoor:outdoor ratios for individual species within particular genera, e.g. Penicillium . One of the other genera which occurs most frequently in indoor growths is Cladosporium. It is a phylloplane fungus, but its growth on damp surfaces indoors ensures that it is a prominent component of the air spora in winter, when there is little growth outside and the outdoor spore counts are therefore low.

Table 2: Numbers of viable airborne bacteria and fungi at sites within two buildings [after Austwick et al. (1)].

Site	n	Mean cfu*/m3 air			
		Bacteria		Fungi	
		37°	Maximum†	25°	37°
Building 1 (June)					
Air-conditioned	2	200	200	97	22
Naturally ventilated	1	234	234	447	8
Building 2 (November)					
Air-conditioned	2	130	173	138	1
Naturally ventilated	1	190	613	352	23

* colony forming units; † samples for bacteria incubated at both 15 and 37°

Cladosporium and Penicillium are usually the predominant moulds in the air of both naturally ventilated and air-conditioned workplaces (Tables 3 and 4). The commonest of the aspergilli in houses, Aspergillus versicolor, can also be prominent, and Basidiomycetes (possibly including Sistotrema brinkmannii) may also be noteworthy (Table 4). The pathogenic mould, A. fumigatus, may very occasionally be prominent in the air (Table 4).

Table 3: Fungal air spora in naturally ventilated and air-conditioned offices in summer.

	Percentage composition of air spora				
	Naturally ventilated building	Air-conditioned building			
		Office 106		Office 329	
	Monday	Monday	Friday	Monday	Friday
Alternaria alternata	5	-	3	9	2
Aspergillus versicolor	-	1	-	-	-
Cladosporium spp.	59	71	77	73	42
Epicoccum purpurascens	<1	-	-	-	-
Mucor spp.	1	1	-	-	-
Penicillium spp.	29	18	5	2	54
Unidentified non-sporing isolates	<1	9	6	16	2
Yeasts	3	-	9	-	-

The bacterial flora of indoor air is normally dominated by micrococci, including

Staphylococcus epidermidis shed from the body with skin scales, whilst the dominant types outdoors are those associated with plant surfaces, e.g. *Pseudomonas*. In both environments, however, a wide range of bacteria will be encountered. Some idea of the range can be seen in the list of types identified in one investigation (Table 5), in which *Micrococcus*, *Flavobacterium* and *Staphylococcus* were most abundant (1). Very few Actinomycetes were found in this investigation, but appreciable numbers of various thermophilic Actinomycetes have been isolated elsewhere in indoor air when they have colonised humidifiers.

Table 4: Principal fungi (expressed as cfu/m^3 air) in air-conditioned and naturally ventilated sites in two buildings [after Austwick *et al*. (1)].

	Building 1 (June)			Building 2 (November)		
Species	a.-c.*		n.v.†	a.-c.*		n.v.†
Alternaria sp.	-	-	4	-		6
Aspergillus fumigatus	45	-	10	-	-	8
A. versicolor	2	-	-	81	122	39
Aureobasidium pullulans	-	-	-	6	-	4
Botrytis cinerea	-	-	4	-	-	2
Cladosporium spp.	30	45	209	2	2	39
Phoma fimeti	-	8	12	-	-	-
Penicillium spp.	2	4	12	31	64	2
Yeasts	4	-	12	-	-	6
Unknown small	-	-	-	-	-	24
Unknown white	12	20	8	1	-	22
Basidiomycetes	2	4	2	-	-	-

* air-conditioned, two sites; † naturally ventilated, one site

4. HEALTH EFFECTS

The effect of micro-organisms in indoor air on human health has been the subject of recent reviews (4,7). Some can cause allergic diseases, including rhinitis, asthma, humidifier fever and extrinsic allergic alveolitis (EAA). European studies indicate that relatively few patients with suspected respiratory allergy react strongly to fungi. However, those who are particularly susceptible, i.e. atopic individuals, show Type I symptoms (rhinitis and/or bronchial asthma) to moulds encountered in buildings. *Alternaria* appears to be the most important of these fungi in U.S., but in Europe species of *Penicillium* and *Aspergillus* are apparently of greater significance. Other moulds which can grow in buildings and cause rhinitis or asthma include *Cladosporium*, *Mucor*, *Rhizopus* and *Ulocladium*.

EAA, also known as hypersensitivity pneumonitis, is an extremely serious Type III allergy affecting the lower respiratory tract, and can be suffered by both atopic and non-atopic people who encounter large numbers of allergenic micro-organisms. EAA is normally associated with the massive concentrations of fungal or actinomycete spores inhaled during handling of mouldy agricultural materials; farmer's lung and malt worker's lung are classic examples. However, concentrations of spores of the dry rot fungus, *Serpula lacrymans*, as high as 36 x 10^4/m^3 air have been recorded, and caused both asthma-like symptoms and

EAA in an atopic individual. Growth of allergenic moulds such as Aureobasidium, Cladosporium, Mucor and Penicillium on damp walls and ceilings, on flooring and floor covering wetted by leakage, in humidifiers and in heating systems have all been linked to individual cases of EAA. It has also been suggested that "humidifier lung" may be caused by Alternaria and Aureobasidium in humidifiers. Bacteria in humidifiers which have been implicated in such cases of EAA include Pseudomonas aeruginosa and the thermophilic Actinomycetes associated with farmer's lung, i.e. Thermoactinomyces vulgaris and Faenia rectivirgula. Although EAA due to allergenic organisms in the air of the non-industrial workplace or home is apparently rather rare, it is extremely distressing for those affected and shows clearly the potential hazard of unchecked growth of bacteria and fungi.

Table 5: Types of viable bacteria isolated isolated from indoor air in an investigation of sick buildings [after Austwick et al. (1)].

Actinobacter calco var. lwoffi	Ps. cepacia
Aeromonas hydrophila	Ps. fluorescens
Flavobacterium sp.	Ps. paucimobilis
Micrococcus spp.	Ps. vesicularis
Moraxella sp.	Staphylococcus aureus
Pasteurella haemolytica	Staph. epidermidis
P. pneumotropica	Streptococcus spp.
Pseudomonas aeruginosa	C.D.C. Group VE

Although the bacterial flora of indoor air is usually dominated by Gram-positive micrococci, it has been suggested that endotoxins of Gram-negative bacteria could present a health hazard, since symptoms similar to acute stages of "humidifier fever" can be provoked on inhalation challenge with extracted endotoxin or whole cells of some Gram-negative bacteria. The mould Stachybotrys atra (Table 1), is both allergenic (causing asthma attacks) and toxigenic, and it and other moulds growing in buildings release spores containing very high concentrations of mycotoxins. The toxins of S.atra were thought to have caused chronic health problems in a Chicago home (10) and, together with Trichoderma viride, in a Montreal hospital. Other toxigenic species which may be prevalent include Aspergillus versicolor and Penicillium spp. such as P. brevicompactum and P. viridicatum. The mycotoxins produced by some of the penicillia, and trichothecenes like those in S.atra, are acutely toxic to alveolar macrophages (10). It is therefore possible that inhalation of high concentrations of spores could affect health by inhibiting functions such as phagocytosis of living and non-living foreign particles in the lung.

Since it has recently been concluded that a non-allergic mechanism may account for the increased respiratory symptoms among adults in damp houses (3), the possiblity of a wider role for mycotoxins in respiratory health must be looked at further. Another factor which might also be involved is the production of odorous volatiles by moulds. Some, including 1-octen-3-ol, show a low-order toxicity to experimental animals (10), and in humans acute respiratory responses to fungal volatiles may range from a feeling of stuffiness to wheeze. However, much more information will be needed before objective assessment of the impact of microbial toxins and volatiles on health can be made.

Micro-organisms have been invoked as a cause of sick building syndrome (SBS), but it was concluded from one investigation of 15 buildings that there were very few qualitative or quantitative differences between different buildings, irrespective of the levels of symptoms among the workforce (1). Although several types of bacteria isolated (Table 5) were pathogenic, the numbers were low and whether they had any clinical significance in SBS was doubted. Similar doubts were expressed about the fungi (Table 4), as the numbers were too low even to be associated with allergic symptoms in atopic or other sensitized persons. The overall conclusion of the authors was that the airborne microbial content of the air of office buildings, whether air-conditioned or not, was unlikely to play even a contributory role in inducing symptoms of SBS, except possibly when there were >10^4 cfu /m^3 air.

5. AIR SAMPLING FOR MICRO-ORGANISMS

The collection of the objective microbiological data necessary to establish relationships between airborne microbial aerosols and health is problematic, and traditional approaches to assessment of microbial pollutants must be questioned. In many investigations, the need for qualitative and quantitative comparison with the air spora of outdoor air which can indicate the presence of indoor sources of moulds, yeasts and bacteria is ignored. More attention must be given to the spatial and temporal fluctuations in numbers of airborne micro-organisms (Tables 3, 4 and 6), which are clearly recognised in houses as being dependent on human activities (6,11). A single spot air sample reveals very little about exposure of individuals to micro-organisms in the course of a day or a working week.

Table 6: Counts of micro-organisms in a restaurant assessed by different methods.

Number of micro-organisms/m^3 air

	Spot samples (Andersen)*			Continuous samples (filter method)			
				Plate count*		DEFT method†	
	Before open	Mid-day	Early evening	a.m.	p.m.	a.m.	p.m.
Pantry							
Moulds	212	141	129	3160	920	10200 (82)	7260 (43)
Bacteria	224	152	208	138	625	34910 (18)	5100 (82)
Kitchen							
Moulds	812	334	506	n.s.	n.s.	n.s.	n.s.
Bacteria	212	127	527	n.s.	n.s.	n.s.	n.s.
Main restaurant							
Moulds	35	565	24	521	348	2010 (0)	12360 (100)
Bacteria	106	244	88	417	1180	43250 (9)	13580 (60)

* Method based on culture on agar plates; † Direct epifluorescence (microscopical) technique
Values in parentheses are percentage indicated viable by acridine orange stain
n.s., not sampled

Methods based on culture are undoubtedly extremely valuable in enabling living micro-organisms in indoor air to be identified, but it is not just those spores and cells which can readily be determined by culture methods that are important. The total numbers of allergenic or toxigenic microbial particles are important; non-viable spores can cause the same allergies or toxicoses as viable spores. There may be considerable variation in the percentage of particles of different fungi and bacteria in air which are culturable; in some cases, such as

Stachybotrys atra, this may be only 1-2% of the total. The disparity which there can be between counts given by methods based on culture and a method giving total counts (8) is highlighted in Table 6. Andersen or other "viable" samplers should be used in conjunction with "non-viable" methods such as filter techniques (8) or samplers such as the Burkhard which reveal the total bioburden in air. Attention must also be paid to the media used in methods based on culture. For example, it is now being recognised that moulds which are able to grow at around a_w 0.70, i.e. xerophilic moulds such as Wallemia sebi, are not readily detected on the high a_w agar media traditionally used. Such moulds are common in dust, may be thrown into the air when dust is disturbed and are of potential allergenic importance.

Despite the limitations of "viable" air-sampling methods, their value for identification of airborne organisms must be stressed. Accurate identification, coupled with knowledge of the ecology of the organisms, can reveal potental problems by indicating the existence of hidden growth within buildings. Although threshold limits for airborne micro-organisms in non-industrial workplaces have not been set, controllable growth of moulds and bacteria is unacceptable. Buildings must be regularly inspected and managed so that such growth on and in building materials, furnishings and HVAC equipment is prevented.

REFERENCES

(1) Austwick, P.K.C., Little, S.A., Lawton, L., Pickering, C.A.C. and Harrison, J. (1989). Microbiology of sick buildings. In: Airborne Deteriogens and Pathogens (ed. B. Flannigan), Kew, Surrey, The Biodeterioration Society, pp. 122-128.

(2) Brundage, J.F., Scott, R.N., Smith, D.W., Miller, R.N. and Lendnar, W.M. (1988). Building-associated risk of febrile acute respiratory diseases in army trainees. Journal of the American Medical Association **259**, 2108-2112.

(3) Dales, R.E., Burnett, R. and Zwanenburg, H. (1991). Adverse health effects in adults exposed to home dampness and molds. American Review of Respiratory Disease **143**, 505-509.

(4) Flannigan, B., McCabe, E.M. and McGarry, F. (1991). Allergenic and toxigenic micro-organisms in houses. Journal of Applied Bacteriology **70** (supplement), 61S-73S.

(5) Grant, C., Hunter, C.A., Flannigan, B. and Bravery, A.F. (1989). The moisture requirements of moulds isolated from domestic dwellings. International Biodeterioration **25**, 259-284.

(6) Hunter, C.A., Grant, C., Flannigan, B. and Bravery, A.F. (1988). Mould in buildings: the air spora of domestic dwellings. International Biodeterioration **24**, 81-101.

(7) Miller, J.D. (1990). Fungi as contaminants in indoor air. In: Indoor Air '90, Volume 5 (ed. D.S. Walkinshaw), Ottawa, CMHC, pp. 51-64.

(8) Palmgren, U., Ström, G., Blomquist, G. and Malmberg, P. (1986). Collection of airborne micro-organisms on Nuclepore filters, estimation and analysis - CAMNEA method. Journal of Applied Bacteriology **61**, 401-406.

(9) Riley, E.C., Murphy, G. and Riley, R.L. (1978). Airborne spread of measles in a suburban elementary school. American Journal of Epidemiology **107**, 421-432.

(10) Sorenson, W.G. (1989). Health impact of mycotoxins in the home and workplace: an overview. In: Biodeterioration Research 2 (ed. C.E. O'Rear and G.C. Llewellyn), New York, Plenum, pp. 201-215.

(11) Verhoeff, A. P., van Wijnen, J. H., Boleij, J.S.M., Brunekreef, B., van Reenen-Hoekstra, E.S. and Samson, R.A. (1990). Enumeration and identification of airborne viable mould propagules in houses: a field comparison of selected techniques. Allergy **45**, 275-284.

INDOOR AIR QUALITY ASSESSMENT: ANALYTICAL METHODOLOOGY

M.J. BERENGUER, X. GUARDINO and Collaboration Group on Sick Building Syndrome*
Instituto Nacional de Seguridad e Higiene en el Trabajo
Centro Nacional de Condiciones de Trabajo
Dulcet, 2-10. E-08034 Barcelona
Spain

SUMMARY

This paper intends to provide an overview of the analytical work that the INSHT is carrying out as a part of its program to prepare a protocol for the investigation of indoor air quality. A brief description of the revision that is being undertaken on the analytical methodology available in our laboratory for industrial hygiene application and a discussion of its state and analytical possibilities for the most significant indoor air pollutants are given. Measured concentrations of environmental samples, collected during the previous studies, are also presented as preliminary results.

1. INTRODUCTION

The increasing awareness of health concerns related to exposure to low levels of pollutants has become a pressing issue in labour relations in non industrial indoor environments, such as office, commercial and public buildings. It is recognized that apparently clean interiors can contain complex mixtures of pollutants that can affect people´s health in different degrees. Since 1988 the Instituto Nacional de Seguridad e Higiene en el Trabajo has been carrying out preliminary studies about indoor air quality (IAQ). Now, a program to evaluate the problems related to the sick building syndrome is being developed. The purpose of the program is the preparation of a protocol for investigating IAQ complaints. This involves both the design and validation of appropriate questionnaires, that will allow the interviewers to detect the problems, and the preparation of the analytical methodology to carry out, when necessary, airborne measurements.

To face the analytical necessities for indoor air (IA) monitoring the Institute has undertaken the revision of analytical methods used in the laboratory for industrial hygiene applications, validated in the range of occupational safety and health standards, and is considering the feasibility of methods for IA developed by other specialized institutions.

2. METHODOLOGY

Pollutants we are first considering include both those that have appeared as indoor problems to be solved during routine work and those known to be more significant in IAQ.

A wide variety of measurement devices and analytical methods with potential application to IAQ monitoring are described in the bibliography. However, their selection requires careful consideration of many factors. Mobility of sampling devices (personal, portable or stationary) as well as analytical performances must be considered. Taking our experience, as a basis, we start with those methods that include sampling acquisition of pollutant(s) in a suitable collector or on a support and its subsequent analysis and quantitation in the laboratory. In some cases, colorimetric tubes and monitoring instruments are also considered.

Threshold Limit Values proposed by ACGIH were used in the development of our methods for measuring airborne concentrations in occupational environments. Our challenge is to benefit from the work done and try to adapt them to indoor recommended values suggested by WHO to protect public health.

Some of the methods are being field tested and preliminary results have been obtained.

3. RESULTS AND DISCUSSION

Measurement methods for air quality monitoring

A brief description of the major pollutants and our standard measurement methods were given, as well as modifications and new possibilities that we are going to study.

The compounds reviewed have been: asbestos, carbon dioxide, carbon monoxide formaldehyde, metals, nitrogen dioxide, ozone, sulfur dioxide and volatile organic compounds. General data, including estimated limits, are refered to the work conditions established in our methods.

Monitoring results

Measured concentrations of environmental samples collected during the studies carried out on request and/or to prepare the protocol for investigating IA complaints are shown in the Table 1 .

Table 1: Ranges of significative concentrations for selected airborne pollutants collected indoors

Compound	Building Type	Indoors	Remarks
Asbestos	School	16×10^3 - 210×10^3 F/m^3	(1)
Benzene	Office	1 - 15 µg/m^3	(2)
Carbon Monoxide	Office	1 - 20 mg/m^3	(2)
Carbon Dioxide	Office	900 - 2200 mg/m^3	(3)
Formaldehide	Public	100 - 150 µg/m^3	(4)
Mercury	Public	2 - 100 µg/m^3	(5)
Nitrogen Dioxide	Residence	10 - 400 µg/m^3	(6)
Ozone	Office	1 - 3 µg/m^3	(2)
Sulfur Dioxide	Residence	40 - 100 µg/m^3	(6)
Tetrachloroethylene	Office	2 - 20 µg/m^3	(2)
Toluene	Office	40 - 100 µg/m^3	(2)
Trichloroethylene	Office	3 - 100 µg/m^3	(2)

(1) School situated in an industrial area where a fibrocement factory made use of asbestos fibers. Outdoor values were in the range 20×10^3-50×10^3 F/m^3. School moved to another area.
(2) Sampled with active and/or passive devices in office buildings. More VOCs were also identified and quantitated in significative amounts: sum of aliphatic hydrocarbons (80-600 µg/m^3), sum of Ethylbenzene and Xylenes (20-400 µg/m^3), sum of other alkylbenzenes (100-500 µg/m^3), sum of α- and β-Pinene (10-200 µg/m^3) and 1,1,1-Trichloroethane (30-800 µg/m^3).
(3) Mean indoor values reported are in the range 1000-1600 mg/m^3 and usually the outdoor value is about 500 mg/m^3.
(4) Redecorated office
(5) Public building where Hg was manipulated in a closed circuit with important leaks near the outdoor air intake. Arrangements were made to lower the level.
(6) Measures done in an IAQ study of pollution in home residences.

4. CONCLUDING REMARKS

Since our methods are validated in the range of occupational industrial standards they are not therefore, in general, applicable to the great majority of IAQ measurements. Among the compounds considered in this paper, only laboratory methods validated for asbestos, carbon dioxide, carbon monoxide and ozone can be used without further modifications.

A number of methods could be *a priori* used with only very few changes such as a greater sampling volume. This can be achieved with higher sampling times and/or higher flow rates as for formaldehyde, As, Hg, V, nitrogen dioxide and sulfur dioxide but sometimes more sorbent for air sampling would be necessary as for instance for VOCs if active sampling devices are used.

For metals such as Cd, Cr, Pb, Mn and Ni, instead of AAS-Flame, a more sensitive technique, for instance AAS-Graphite furnace, would be necessary.

For other compounds our methods are unsuitable for IA measurements and new methods should be developed.

Using the methodology available we could analyse about 80% of the compound listed in the WHO Air Quality Guidelines. Furthermore other methods allowing the determination of short term peaks to identify sources of pollutants and long term averages to control exposures with acceptable reliability should be developed and validated.

The search for better performance would be based on several monitoring systems and/or techniques, such as HPLC, IC, polarography, GC with specific detectors (FPD and electron capture detector), and GC-MS, to obtain more sensitivity and specificity.

The data obtained with passive methods for sample collection shows that their utilisation is attractive in IA monitoring because of their ease of use and its applicability for large surveys.

Field tests of colorimetric tubes have shown that they are not useful in IA measurements. Their applicability in this area is limited to preliminary surveys when a new IAQ investigation is initiated.

Although considerable progress has been made in recent years in the adaptation of existing measurement methods and development of new methods, further research and development are needed in this area. A major effort needs to be undertaken to standardize IAQ measurement methods. Standardized IAQ monitoring protocols, sample collection and analysis methods, and methods for data summary and interpretation can greatly improve the value of results. The work initiated has allowed measuring indoor air in school, offices, residences and public buildings and will be an important item of our protocol for investigating IAQ problems.

REFERENCES

- ACGIH Threshold Limit Values and Biological Exposure Indices for 1990-1991. American Conference of Governmental Hygienists, Cincinnati, Ohio, 1990.
- ASHRAE Standard 62-1989, Ventilation for Acceptable Indoor Air Quality, American Society of Heating, Refrigeratingand Air-Conditioning Engineers, Inc., Atlanta,GA.,1989.
- FORTMANN, R. C., NAGDA, N. L. and KOONTZ, M. D. "Indoor Air Quality Measurements" Sampling and Calibration for Atmospheric Measurements. ASTM STP 957. J.K. Taylor, Ed., American Society for Testing and Materials, Philadelphia, 1987, pp 35-45.
- I.N.S.H.T. Métodos de Toma de Muestras y Análisis. Propuestos, aceptados y recomendados.1980-1991.
- NAGDA, N. L., RECTOR, H. E., and KOONTZ, M. D., Guidelines for Monitoring Indoor Air Quality, Hemisphere Publishing Corporation, New York, 1987.
- LEWIS, R. G. and WALLACE, L., Workshop:Instrumentation and Methods for Measurement of Indoor Air Quality and Related Factors, Design and Protocol for Monitoring Indoor Air Quality, ASTM STP 1002, N. L. Nagda and J. P. Harper, Eds., American Society for Testing and Materials, Philadelphia, 1989, pp. 219-233.
- WHO Air Quality Guidelines for Europe, European Series No. 23, 1987.

---

* Analytical and Field Studies C. G.: A. Freixa, E. Gadea, A. Hernández, A. Martí, M. C. Martí, A. Pascual, M. G. Rosell and C. Santolaya.

1427

AEROBIOCOLLECTORS: EFFICIENCY OF MICROORGANISM COLLECTION IN MONOMICROBIAL AEROSOL AND POLYCONTAMINATED AIR

H. FRICKER*, S. PARAT**, P. GAILLARD***, A. PERDRIX**
J. MICHEL***, R. GRILLOT*

*Département de Parasitologie et Mycologie Médicale et Moléculaire
Université J. Fourier, Domaine de la Merci
38043 Grenoble Cedex 9, France

**Institut Universitaire de Médecine du Travail et d'Ergonomie
Hôpital A. Michallon BP 217, 38043 Grenoble Cedex 9, France

***Centre Technique des Industries Aérauliques et Thermiques
Plateau du Moulon, 91400 Orsay, France

SUMMARY

Four biocollectors (air sampling instruments) are compared in a controlled test installation and in a natural environment. The collection efficiency results for RCS, Ochlovar, SAS biocollectors are higher in the natural environment than in the test installation, in part, because of particle size and air speed differences. The Andersen is the most efficient biocollector for both situations.

1. INTRODUCTION

Infectious and allergic diseases caused by air conditioning systems are increasing (1). Therefore, it is very important to check airborne microorganism numbers and species in indoor ambient air, especially in a hermetic environment. Complete standardization of microorganism evaluation has never been proposed (2). The collection of airborne microorganisms is carried out with biocollectors, particle samplers through impaction or filtration (3). Several publications (3, 4, 5) comparing evaluations of commercial samplers, have highlighted a great dispersal concentration of viable particles measured in the same atmosphere.

The aim of this study is to evaluate four biocollectors: Andersen, RCS, SAS, Ochlovar, first in a test installation with controlled parameters, then in a natural environment (real conditions of use) and to compare the results in both installations.

2. METHODOLOGY

The test installation was a channelled flow of sterile air (size: 0,8 x 0,8 x 6m). The Collison nebulizer aerosolized a quantitative suspension of bacteria (*Staphylococcus epidermidis*) fungus spores (*Penicillium*), or inert particles (sodium chloride). Different parameters were controlled: temperature, humidity, pressure. Air speed was 0.5 or 2m/s.

Then, the biocollectors were tested in a natural environment (laboratory room) where the air was polycontaminated. The air speed was <0.1m/s except the RCS for which the air speed was <0.1m/s and 0.5m/s.

The biocollectors were the Andersen (6 stage impaction), the RCS (Reuter Centrifugal Air Sampler), the Ochlovar (1 stage impaction), the SAS (1 stage impaction). They were compared with a sampler considered as an inert capture particle reference: filtration on membrane.

Bacteria were cultured on trypticase soja agar (48h at 35°C). Fungi were cultured on Sabouraud gentamicine (72h at 27°C). Colonies were counted and translated in CFU/m^3 (Colony Forming Unit per cubic metre). Inert particles were collected on a filter placed in Petri boxes or directly in the sampler. The filter was weighed before and after sampling.

3. RESULTS

Table. 1. Comparison of Collection Efficiency (percentage of biocollector and capture reference results) in the Test Installation and Natural Environment

BIOCOLLECTORS	TEST INSTALLATION			NATURAL ENVIRONMENT	
	Bacteria	Fungi	Inert Particles	Bacteria	Fungi
Andersen	-	66,1 ± 9 (n=11) 72,7 ± 8 (n=12)	78,7 ± 6,3 (n=5)	63,5 ± 31 (n=10)	87,9 ± 28 (n=12)
RCS	<7 ± 3,7 (n=8) <4 ± 2 (n=9)	21,6 ± 10 (n=15) 15,9 ± 7 (n=15)	-	95,7 ± 58 (n=11) 135,4 ± 47 (n=5) 111,8 ± 47 (n=5)	82,2 ± 32 (n=12) 78,6 ± 25 (n=12)
SAS	-	-	6,3 ± 0,7 (n=5)	54,8 ± 12 (n=12)	63,1 ± 16 (n=10)
Ochlovar	-	-	1,8 ± 0,4 (n=4)	18,1 ± 5 (n=10)	3,1 ± 3 (n=12)

$$\text{Collection efficiency} : \frac{Cb}{Cr} \times 100$$

Cb : number of Colony Forming Units (CFU) with test biocollector

Cr : number of CFU with reference sampler

The Andersen impactor gave the same results in the test installation and in the natural environment. It was less efficient than the reference sampler. In the test installation, the RCS collected very low quantities of bacteria and fungi, but in the natural environment it collected the same quantities as the reference sampler (efficiency 100%). SAS and Ochlovar showed poor efficiency with, however, higher results in the natural environment than in the test installation. We observed a higher standard deviation in the natural environment than in the test installation, where the aerosol was more homogenous (mono-dispersed).

4. DISCUSSION

Different hypotheses arise concerning efficiency results in the test installation and the natural environment: difference of particle size and air speed (4).

In publication (3), the RCS efficiency depended on particle size: low efficiency in the collection of small particles (diameter $<5\mu$). In our study, particle distribution on each of the six stages of the Andersen shows that bacteria particles are smaller in the test installation (poor agglomerators), and fungi particle sizes are less homogenous in the natural environment. Moreover only one species is aerosolized in the test installation and several species are isolated from the ambient air (fungi: *Cladosporium, Alternaria, Aspergillus fumigatus, Aspergillus flavus, Aspergillus niger, Penicillium, Verticillium, Bothrytis*). In addition, in the natural environment, the microorganisms adhere to inert particles (3) but the monomicrobial aerosol is without inert particles.

Air speed was <0.1m/s in the laboratory ambient air, it was fixed at 0.5m/s or 2m/s in the test installation. We tested the RCS in the natural environment with the air speed at 0.5m/s. The air speed explains in part the difference of RCS efficiency for the fungi only.

In conclusion, we found that in both the test and natural environment installations the Andersen biocollector is the most efficient. In the natural environment with air speed <0.1m/s, RCS and Andersen are the best. The difference of results in the test installation and natural environment could be explained in part by differences in particle size and air speed.

REFERENCES

(1) PARAT, S.,PERDRIX, A., GRILLOT, R., CROIZE, J. (1990). Prévention des risques dus à la climatisation. Stratégie d'intervention. Arch. Mal. Prof. **51**, 27-35.

(2) COMTOIS, P. (1990). Indoor mold aerosols. Aerobiologia. **6**, 165-176.

(3) CHANTEFORT, A. (1988). Moyens de contrôles de l'état microbiologique de l'air et des surfaces. Sci. Tech. Anim. Lab. **13**, 1-8.

(4) MARCHISIO, V.F., CARAMIELLO, R., MARIUZZA, L. (1989). Outdoor airborne fungi: sampling strategies. Aerobiologia. **5**, 145-153.

(5) PINEAU, S. (1990). Factors to consider during aerofungal sampling. 4th international conference on aerobiology. Stockholm-Sweden.

SAMPLING EFFICIENCY OF AMBERLITE XAD-2 WITH GLASSFIBRE PREFILTER AND ACTIVATED CHARCOAL AS BACK-UP SECTION FOR COLLECTION OF MIXTURES OF ORGANIC AIR POLLUTANTS AT THE $\mu g/m^3$ LEVEL

MAJ-LEN HENRIKS-ECKERMAN

Turku Regional Institute of Occupational Health
Hämeenkatu 10, 20500 Turku, Finland

SUMMARY

Sampling efficiency of the collection device was studied by determining the recovery of representative organic air pollutants (5–100μg of each, 150–200 L of air). The desorption efficiency of purified XAD-2 resins was studied by the phase equilibrium technique. The total recovery was over 80% for most tested compounds. Polar, semi-volatile compounds had an unacceptably low total recovery. The commercial XAD-2 resin had the highest desorption efficiency, but the recoveries were only 56–76%. The sampling efficiency was not affected by humidity or sampling rate for most of the tested compounds. Results from field measurements are mentioned.

1. INTRODUCTION

Complaints about irritation and malodours are common in both work and indoor environments. The compounds responsible for the discomfort are usually at the $\mu g/m^3$ level. They are often collected on Tenax or XAD-2 adsorbent tubes and thermally desorbed. When no equipment for thermal desorption is available, the $\mu g/m^3$ level sensitivity can still be reached for solvent desorbed samples provided that splitless or on column injection is used and that large air volumes are sampled.

2. EXPERIMENTAL

In the first recovery experiment the influence of air humidity (20 and 85% RH) and flow-rate (1 and 3.6 L/min) on sampling efficiency was studied. The sampling device was an XAD-2 tube (300mg) with prefilter and a SKC charcoal tube as back-up section (Table 1).

Table 1. Adsorbents and Filters Used in the Recovery Experiments and in the Field Measurements

Adsorbent/filter	Amount (mg)	Origin
Amberlite XAD-2	300	Serva pract. (purified)
Amberlite XAD-2	150 + 75	SKC-226-30-05
Charcoal	400 + 200	SKC-226-09
Charcoal	100 + 50	MSA
Glassfibre filter	AP4001300	Millipore

The tested compounds were methylmethacrylate, hexanal, i-propyl-benzene, phenol, butylmethacrylate, decane, 2-ethyl-hexanol, acetophenone, 2,4-dimethylphenol, 2-amino-acetophenone, octanoic acid, bisphenol-A (5 and 100μg of each), i-butanol (200 and 4000μg), toluene (100 and 2000μg) and m-xylene (150 and 3000μg). The air volume was 200 L. The equipment for the recovery experiments, the desorption and the gas chromatographic analysis are described elsewhere (1).

The tested compounds in the second recovery experiment were furfurylalcohol, i-propylbenzene, α-pinene, decane, 2-hydroxyacetophenone, resorcinol, 4-hydroxy-acetophenone, bisphenol-A, di-2-ethylhexylphthalate (5μg of each), N,N-dimethylcyclohexylamine (20μg), octanoic acid (10μg), 2-phenyl-2-imidazoline (11μg) and pentachlorophenol (20μg). The sampling device consisted of an XAD-2 tube (purified XAD-2 resin or SKC-tube) with a glassfibre prefilter. The sampling rate was 1 L/min and 200 L of air were sampled (150 L for SKC-tubes). The relative humidity was 85%. The same compounds were used to study the influence of XAD-2 resin clean-up procedures on the desorption efficiency.

3. RESULTS

Recovery Experiments

The results from the first experiment have been published (1). The total recovery was over 80% for most tested compounds. Exceptions were hexanal and small amounts (5-10μg) of phenol, octanoic acid and bisphenol-A. At 85% RH the recoveries were slightly lower than at 20% RH, especially for compounds trapped by filter and charcoal. The flow-rate had a negligible influence on the sampling efficiency of the studied collection device.

In the second recovery experiment the tertbutylmethylether extracted resin had the best sampling efficiency (recoveries of 79–103%). The commercial SKC resin had the lowest sampling efficiency (recoveries of 56–76%). The compounds that were only trapped by the XAD-2 tubes were i-propylbenzene, α-pinene, N,N-dimethylcyclohexylamine, decane and 2-hydroxyacetophenone.

The total recovery (8–67%) of compounds that were not completely evaporated from the filter to the XAD-2 was unacceptably low for quantitative sampling. These compounds were octanoic acid, resorcinol and 4-hydroxyacetophenone. Only the isophthalate and phenylimidazoline were quantitatively recovered from the filter (80–90%). The recoveries of pentachlorophenol and bisphenol-A from the filter were 46 and 71%. The detection limits in the splitless mode were 0.2μg/sample for non-polar and 2–5μg/sample for polar compounds corresponding to 1–25μg/m^3 (200 L).

Desorption Experiments

The desorption efficiency of different XAD-2 resins was determined by the phase equilibrium technique. The resins were Soxhlet-extracted with methanol followed by tert.-butylmethylether, diethylether or dichloromethane. The desorption efficiency of all resins was good for the isophthalate and for the compounds trapped by XAD-2. The commercial resin had the highest desorption efficiency for all tested compounds. For the compounds partly or fully trapped by the filter, the diethylether purified resin had the lowest desorption efficiency. Pentachlorophenol was efficiently desorbed only from the commercial resin. A slightly better desorption efficiency was noticed for the oven dried (60°C, overnight) resins, but the vacuum dried (50°C, 2h) resins were purer.

Field Measurements

The field measurements with the studied collection device were performed in both work and indoor environments. 148–1222 L of air were collected. The trapped compounds were mainly aliphatic and aromatic hydrocarbons and alcohols. For example, with a sampled air volume of 432 L and 25mg/m^3 of trapped compounds, only ethanol, acetone and i-propanol could be detected from the second charcoal layer. The sampling rate was 3.6 L/min.

4. CONCLUSIONS AND RECOMMENDATIONS

Glassfibre filter - XAD-2 - charcoal can be used as a universal collection device for small amounts of organic air pollutants, provided that the desorption efficiency of the XAD-2 lot is tested before use with the compounds of interest. High flow-rates (3–4 L/min) can be used to shorten the sampling time. Small amounts of polar compounds of low volatility are not easily detected and not accurately quantified. Glassfibre filter is not needed in indoor air sampling. The problems with compounds (phenol, aldehydes, ketones) that are not fully trapped by XAD-2 and that cannot be recovered from charcoal due to irreversible adsorption or decomposition, should be considered when choosing the total air volume and the type of charcoal tube. For large air volumes the breakthrough should be checked by analyzing the second charcoal layer.

REFERENCES

(1) HENRIKS-ECKERMAN, M-L.(1990). Sampling efficiency of Amberlite XAD-2 with glassfiber prefilter and activated charcoal as back-up section for collection of mixtures of organic air pollutants at the μg/m^3 level.Chemosphere vol. 21, No.7 889-904.

1429

AIR PROFILE - A METHOD TO CHARACTERIZE INDOOR AIR

J. KRISTENSSON

Analytical Chemistry Department, University of Stockholm
S-106 91 Stockholm, Sweden

A system for analysis of volatile organic compounds in indoor air has been developed. The system is based on diffusive sampling on solid adsorbents followed by gas chromatography - mass spectroscopy analysis. The obtained results are treated in three different steps.

In the first step the 'air profile' is calculated. Air profile is a screening technique in which the total concentration of volatile organic compounds (TVOC) is calculated as toluene equivalents. The 'air profile' is a simplified presentation of the chromatogram as a bar graph, with five bars or sections in the graph. The calculation is similar to simulated distillation technique. The contribution from all volatile organic compounds that can be analyzed by the used technique, is included in the 'air profile'.

In the second step the concentration of each compound can be calculated from the stored data. The second step is not performed if the 'air profile' indicates a low total concentration of volatile organic compounds in the indoor air.

In the third step the sampling at specific places or samples from different materials such as vinyl carpets, concrete or wood can be analyzed for the emission of volatile organic compounds.

In a 'healthy building' the TVOC concentration is typically 100–200ng/l, in toluene equivalents. In a 'sick building' the TVOC concentration is 200–1000ng/l. In outdoor air the TVOC concentration is 20–100ng/l. The limit of TVOC concentration of 200ng/l as limit for 'sick building' has been estabished with correlation to questionnaires.

The air profile gives a simple method for screening measurements to find 'sick buildings'. The sampling technique is simple but effective and often performed by the occupants in the building. The analysis is fully automated.

LABORATORY AND FIELD VALIDATION OF A THERMALLY DESORBABLE SAMPLER FOR VOLATILE ORGANIC COMPOUNDS IN INDOOR AIR

A.M. LAURENT, A. PERSON, B. FESTY
Laboratoire d'Hygiène de la Ville de Paris
11, rue George Eastman 75013 PARIS (France)

SUMMARY
This paper reports on the main results related to the behaviour of a thermally desorbable tube used to collect a wide range of VOCs in a gaseous synthetic mixture (laboratory tests) and in a multipollutant atmosphere (field experiments) in various experimental conditions. The exposure time is the most critical parameter.

1. INTRODUCTION

In order to evaluate the compliance of volatile organic compounds (VOCs) concentrations with air quality guidelines in industrial atmospheres, several types of passive samplers have been developed and are now commercially available. It is not the case for the assessment of much lower VOC concentrations (usually between a few ppb and a few tens of ppb) measured inside non industrial spaces like homes, offices, schools,...

The purpose of this study is to describe laboratory and field experiments in order to validate a thermally desorbable passive tube for the estimation of short and mean exposure to low levels of VOCs (< 1 ppm).

2. METHODOLOGY

The sampler tested functions on the basis of molecular diffusion. The diffusive sampler consists of a glass tube (A=12,5 mm^2, l=160 mm) which contains a 2 cm layer of Carbotrap (100 m^2/g, 20-40 mesh) or Tenax TA (35 m^2/g, 60-80 mesh). The adsorbent is held between a pad of silanized quartz wool and a 250 mesh stainless screen whose position, 2 cm from the edge, defines the diffusion length (L). The active sampling (5 $cm^3.min^{-1}$) is carried out with tubes filled with 300 mg of Tenax. The tubes are stored hermetically by teflon ferrules and swagelock type connectors. Analysis is performed by capillary gas chromatography - FID detection after thermal desorption (350 °C - 15 min).

The laboratory experiments are carried out in a small test chamber (2,5 L capacity). For each test, 5 samplers are exposed to a mixture of gaseous pollutants in a wide range of volatility (50 - 200°C), mainly apolar and moderately polar compounds: aliphatic, aromatic, halogenated and oxygenated compounds. The synthetic mixture, released by permeation and diffusion tubes, is diluted by purified air. The behaviour of the sampler is tested in various experimental conditions. The main parameters studied are : the geometry (diffusion length and amount of adsorbent), the type of adsorbent, the exposure concentration (50-500 $\mu g.m^{-3}$ per compound), the exposure time (2 hours-7 days), the air humidity (10 to 90 %), the air velocity (0.1 to 5 $m.s^{-1}$) and the position of the sampler, the reverse diffusion, the analytical recovery, the storage stability and the sensitivity.

For the field experiments, 3 passive and 3 active samplers (Carbotrap) are exposed (2 to 24 hours) in a large stainless test chamber (V=30 m^3, T=20°C, RH= 20-50-80%, air exchange rate=0 ach, vertical air velocity=0.1 to 0.3 $m.s^{-1}$) to a complex atmosphere generated by an environmental tobacco smoke. The source strength is comprised between 8 and 40 cigarettes burning together in the unventilated chamber.

3. RESULTS

The laboratory experiments lead to the main following conclusions:
- The adsorbent layer, the air velocity and the direction of airflow have no influence on the sampling rates in the conditions tested;
- With a 2 cm diffusion length, VOCs are collected in the range 15-25 $cm^3.h^{-1}$ according to the compound. Sensitivity (GC-FID analysis) is at least 1 $\mu g.m^{-3}$ if exposure time >24 hours;
- Ideally, the sampling rate of the passive tube is constant during the whole exposure time (So=A.D/L, D=diffusion coefficient). Actually, for a given couple adsorbent/adsorbate, the experimental sampling rate (S) decreases with time (Table 1).
Referring to Tenax, Carbotrap is slightly better (Fig.1) for some compounds (e.g. benzene).
- In order to assess a long term exposure, tests were performed with Tenax by reducing the sampling rate (diffusion length equal to 12 cm). Results are improved (Fig.1) only for compounds with a breakthrough volume at least equal to a few hundreds of liters per gram (i.e. toluene, tetrachloroethylene).

		R (*CV*)					
		2 HOURS		24 HOURS		7 DAYS	
	So	T	C	T	C	T	C
Dichloromethane	24.9	0.92 (*8*)	0.94 (*6*)	0.52 (*7*)	0.55 (*6*)	0.09 (*14*)	0.07 (*15*)
Isopropanol	22.7	0.81 (*10*)	0.92 (*9*)	0.40 (*5*)	0.45 (*14*)	0.11 (*12*)	0.11 (*20*)
1,1,1-Trichloroethane	19.3	0.90 (*17*)	0.93 (*11*)	0.62 (*11*)	0.67 (*7*)	0.15 (*17*)	0.18 (*10*)
Trichloroethylene	19.7	0.96 (*7*)	0.99 (*9*)	1.03 (*3*)	0.78 (*6*)	0.45 (*6*)	0.61 (*6*)
Benzene	21.4	0.90 (*10*)	1.03 (*8*)	0.68 (*4*)	0.91 (*2*)	0.26 (*7*)	0.50 (*6*)
Toluene	18.7	0.99 (*7*)	1.01 (*9*)	0.90 (*5*)	1.04 (*4*)	0.59 (*5*)	0.75 (*7*)
Heptane	16.2	1.03 (*2*)	1.08 (*6*)	0.80 (*5*)	0.93 (*3*)	0.56 (*5*)	0.81 (*3*)
Tetrachloroethylene	17.7	0.95 (*9*)	0.99 (*10*)	0.76 (*9*)	0.93 (*5*)	0.54 (*4*)	0.95 (*16*)
1,2-Dichlorobenzene	16.5	0.97 (*6*)	0.95 (*4*)	0.86 (*5*)	0.97 (*5*)	0.76 (*6*)	0.89 (*7*)
m-Xylene	16.9	1.04 (*2*)	1.02 (*6*)	0.98 (*6*)	1.01 (*4*)	0.77 (*8*)	0.95 (*6*)
1,3,5-Trimethylbenzene	14.9	1.04 (*5*)	1.07 (*6*)	1.03 (*7*)	1.07 (*3*)	0.91 (*8*)	1.04 (*5*)
Dodecane	11.3	1.02 (*3*)	1.06 (*10*)	0.97 (*2*)	1.15 (*9*)	0.86 (*12*)	0.99 (*7*)

T=Tenax C=Carbotrap
RH = 40%, T=22°C,
concentration =200 to 500$\mu g.m^{-3}$ (per compound)

R = Sampling efficiency = Sexp / So
Sexp, So : Experimental, ideal sampling rate ($cm^3.h^{-1}$)
(*CV*) = Coefficient of variation (%)

Table 1 : Influence of exposure time on the sampling efficiency

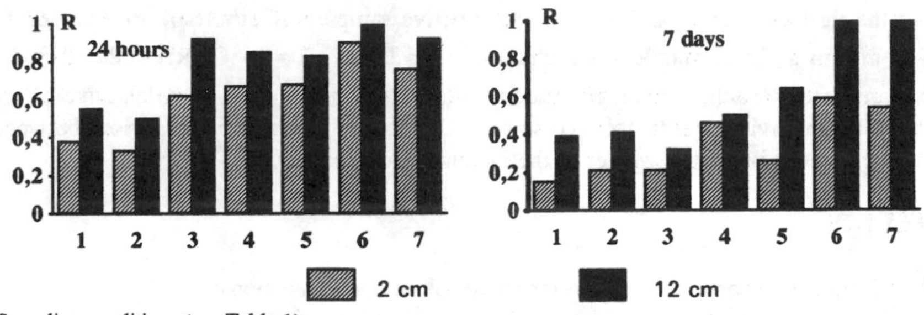

Sampling conditions (see Table 1)

1:butanone 2:hexane 3:trichloroethane 4:trichloroethylene

5:benzene 6:toluene 7:tetrachloroethylene

Fig.1 :Sampling efficiency (R) related to diffusion length

- Concerning the field experiments (Table 2) carried out in the complex smoky atmosphere, results agree only for non polar compounds (aromatics, alkanes >C7, limonene,..). Biases are finding out for some polar or basic compounds and searches have to be lead to explain this discrepancy.

		Duration = 3Hours		**Duration = 24 Hours**			
		RH=50%		**RH = 20%**		**RH = 80%**	
		P	**A**	**P**	**A**	**P**	**A**
Benzene	**a**	-	-	163 *(8)*	196 *(5)*	169 *(8)*	210 *(3)*
	b	1022 *(10)*	870 *(7)*	693 *(6)*	725 *(5)*	619 *(9)*	696 *(6)*
Toluene	**a**	-	-	305 *(7)*	300 *(10)*	313 *(3)*	320 *(3)*
	b	1407 *(5)*	1227 *(8)*	902 *(4)*	1137 *(14)*	826 *(5)*	1023 *(24)*
O-Xylene	**a**	-	-	44 *(3)*	43 *(10)*	56 *(2)*	51 *(4)*
	b	142 *(10)*	136 *(11)*	125 *(7)*	163 *(8)*	120 *(6)*	159 *(7)*
Pyridine	**a**	-	-	10 *(11)*	15 *(3)*	6 *(7)*	13 *(25)*
	b	59 *(12)*	69 *(15)*	55 *(28)*	93 *(24)*	45 *(7)*	107 *(20)*
Pyrrole	**a**	-	-	140 *(2)*	208 *(13)*	64 *(21)*	112 *(27)*
	b	728 *(10)*	904 *(8)*	457 *(6)*	962 *(1)*	187 *(11)*	415 *(7)*
Furfural	**a**	-	-	47 *(11)*	56 *(9)*	78 *(7)*	82 *(20)*
	b	825 *(9)*	671 *(8)*	257 *(7)*	900 *(13)*	223 *(4)*	824 *(10)*
Limonene	**a**	-	-	108 *(5)*	140 *(5)*	128 *(15)*	125 *(4)*
	b	670 *(8)*	645 *(7)*	426 *(7)*	500 *(7)*	391 *(10)*	480 *(6)*
3-Ethenyl	**a**	-	-	56 *(5)*	155 *(7)*	69 *(17)*	121 *(17)*
pyridine	**b**	367 *(8)*	467 *(6)*	206 *(7)*	479 *(8)*	148 *(6)*	218 *(6)*

( )= Coefficient of variation (%) RH=Relative Humidity Temperature=20°C

Triplicate sampling **a** = 20-50 $\mu g.m^{-3}$ **b** = 200-500 $\mu g.m^{-3}$ per compound

Table 2 : Comparison of VOCs mean concentrations ($\mu g.m^{-3}$) measured by passive (P) and active (A) sampling in relation to exposure time and humidity

1431

DETERMINATION OF 2-METHYLISOTHIAZOLONES IN AIR-CONDITIONED INDOOR AIR

Nagorka, Regine, Ullrich, Detlef, Matissek, Reinhard*
Institut für Wasser-, Boden-, Lufthygiene des Bundesgesundheitsamtes, Corrensplatz 1, D-1000 Berlin 33
* Institut für Lebensmittelchemie der Technischen Universität Berlin, Gustav-Meyer-Allee 25, D-1000 Berlin 65

Abstract. Humidifier water in air-conditioning systems is often treated 2-3 times weekly with biocides. In order to record biocide introduction into the air-conditioned rooms, we developed 2 methods, which allow determination of some of the most common used biocides (2-methylisothiazolones) in indoor air. For routine procedure the methylisothiazolones are determined by ion pair high performance liquid chromatography after pre-column derivatisation. The alternative method is performed by using gas chromatography/mass spectrometry after thermal desorption. Detection of the biocides is carried out in the selected ion monitoring mode. During the winter season, we observed concentration levels in 11 air-conditioned buildings up to 3.4 $\mu g/m^3$.

Introduction

In order to minimize microbial growth in humidifier units of air-conditioning systems, the humidifier water is often treated with biocides. Some of the most common biocide formulations used for this purpose are aqueous mixtures based on 2-methylisothiazolones, with active ingredients between 1.5-13.9 %. The major compounds 2-methyl-3(2H)-isothiazolone (Ia) and 5-chloro-2-methyl-3(2H)-isothiazolone (Ib), always occuring in the concentration ratio 1:3 (Ia:Ib), are shown in Figure 1.

During the last years there has been increasing concern over the contact sensitizing potential for adverse health effects from 2-methylisosthiazolones. Nevertheless up to now no study exists on biocide levels in indoor air caused by spray or contact humidification. In this paper we report 2 different methods to determine 2-methylisothazolones in indoor air. Measurements of both compounds in several buildings were conducted using these procedures.

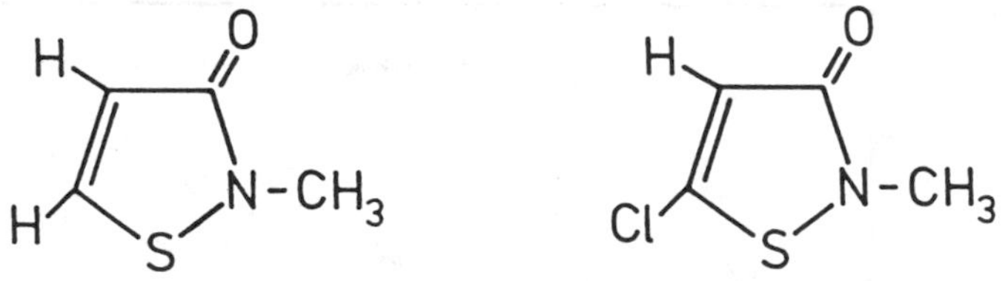

Fig. 1: Structures of 2-methyl-3(2H)-isothiazolone (Ia) and 5-chloro-2-methyl-3(2H)-isothiazolone (Ib)

Procedure

Determination by high performance liquid chromatography: For sampling, air is passed through 2 sorption tubes, filled with (200-220 mg) silica gel bonded with cyanopropyl (flow rate 1.5 ml/min, sampling volume about 300 l). Treatment of the samples is performed by pre-column derivatisation. The isothiazolones are converted spontaneously to ß-thio-substituted acrylamide derivatives by addition of sodium hydrogensulphite solution (1.5 ml in 0.5 ml steps, 0.08 m) to the sorbent packing. The derivatives are eluted with the reactant solution. Separation of the ionic derivatives is carried out by isocratic ion pair-HPLC in the aqueous reversed phase mode. Both products are detectable in the low UV-range.

Determination by gas chromatography/mass spectrometry: For sampling, air is passed through a sorption tube, filled with Tenax GC (60/80 mesh, 150 mg, sorption distance 10 cm); a flow rate of 0.2 ml/min is recommended. Treatment of the sorption tube is carried out by thermal desorption. The 2-methylisothiazolones are separated on a capillary column with a basic support. Detection is performed in the selected ion monitoring mode by scanning 2-3 m/z values for each compound: the parent ion and the fragments after loss of CO (M^+-28) resp. NCH_3 (M^+-29).

Results of investigation in air-conditioned buildings

During the winter season, indoor air has been investigated in 12 buildings with air-conditioning systems (spray humidifier). Water treatment with methylisothiazolone formulations was performed periodically as shock dosage several times weekly. Therefore, air samples has been collected directly after the biocide application.

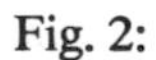
Fig. 2:

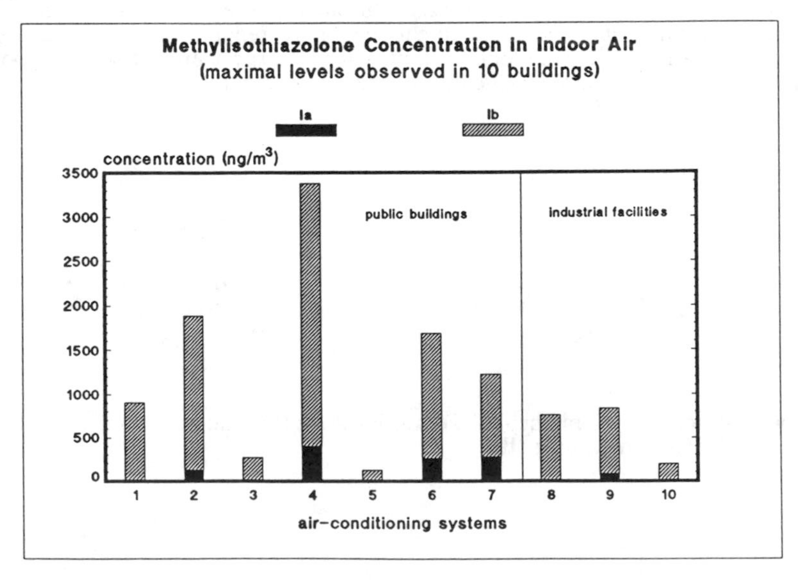

The chlorinated compound was detectable in every measurements. The biocide levels reached up to 3.4 μg active ingredients/m^3 (illustrated in Figure 2). Contrary to expectation a shift occured in the concentration ratio from 1:3 (Ia:Ib) in the humidifier water to 1:15 (Ia:Ib) in the humidified air; this is probably caused by higher polarity of the unchlorinated compound.

The biocide level mainly depends on:
- the methylisothiazolone concentration in the humidifier water
- the temperature of the outdoor air (influencing the humidifier operation)

INDOOR POLLUTION BY POLYCHLORINATED BIPHENYLS

F. NOLTING, E.-R. KRÖNKE, A. REIMER, M. WENZEL AND G.-J. WENTRUP
Technischer Überwachungs-Verein Hannover e.V.
Am TÜV 1, D-3000 Hannover 81

SUMMARY

During the last years around 60 measurements of indoor air were performed to determine the concentration level of airborne PCBs. The detected levels vary over more than two orders of magnitude from 15 to 2147ng/m^3 with an average of 237ng/m^3 as calculated from the standard six reference PCBs. The number distribution of the total PCB concentration shows that 20% of the measurements exceed the proposed target value of 300ng/m^3. From this result for precautionary reasons the need for experimental monitoring is established whenever positive hints for the use of PCBs exist.

1. INTRODUCTION

Polychlorinated biphenyls had a widespread use because of their numerous advantages for industrial purposes. Their high chemical stability and their high electrical resistance were of special importance. These features made them very useful as additives for the dielectric of phase shift capacitors of fluorescent lamps and they are often found in larger buildings, e.g. offices and schools. Although the manufacturing and installation of such capacitors have been stopped for more than 15 years because of their risks to human health, they are still in use in numerous cases.

Although the PCBs are encapsulated in 'closed' metallic cabinets very often these compounds can be detected in indoor air. This might be due to aged capacitors leading to heating of the capacitors. This in turn gives rise to small leakages for the liquid PCBs. This leads to contamination of the air with PCBs. In some cases the capacitors even explode and large amounts of PCBs are released with the risk of contamination of furniture, carpets etc.

During the last years the TÜV Hannover have performed more than 50 measurements of indoor air. The major part of them were thought of having previously mounted capacitors that contained PCBs. On the other hand there were also measurements for checking the rooms to be 'free' of PCBs requested by the owners as a precaution. The experimental results are presented in the following.

2. EXPERIMENTAL

The sampling unit is a cylinder made of glass. The enrichment of the PCBs is performed by a fibre filter for the deposition of aerosol particles and a four stage polyurethane (PU) plug. Typically a sample of 10m^3 of indoor air is taken.

The analytical characterization follows DIN 51 527, Part 1 (1). The filter and the PU plugs are extracted for 24h with n-hexane. The extract is cleaned up by a two-stage column chromatography on a benzenesulphonic acid and a silica column. The eluate is analyzed by gas chromatography and the amount of six reference PCBs (PCB 28, 52, 101, 153, 138 and 180) is quantitatively determined by ECD detection. Samples that are heavily polluted by chloroorganic compounds need further clean-up.

With a DB 5 column (60m x 0.32mm, d_f = 0.25μm) with hydrogen as carrier gas detection limits of 0.3ng/sample (PCB 28, 52, 101, 153 and 138) or 0.2ng/sample (PCB 180) are achieved.

3. RESULTS

In order to determine the 'total' amount of airborne PCBs the sum of the mentioned six reference PCBs is formed and is multiplied by a factor. Depending on the composition of the technical product this factor varies between 3 and 8.5. Table 1 summarizes the relative concentrations of the six reference PCBs in relation to PCB 138.

Table 1. Summary of 60 Measurements of Indoor Air.
The average relative concentration of the six reference PCBs (relative to PCB 138) is shown together with maximum and minimum values.

	Relative Concentration of Congeners (relative to PCB 138)					
	PCB 28	52	101	153	138	180
Average	36.4	24.0	7.0	1.3	1.0	<0.5
Standard deviation	54.0	51.1	16.1	1.5	0.0	0.7
Maximum	297.6	291.6	91.6	9.3	1.0	<5.0
Minimum	0.3	0.3	0.1	0.5	1.0	0.0

The data in Table 1 show that the lighter PCBs are more frequently observed with higher concentrations compared to the heavier congeners. This is not very surprising because of the higher vapour pressure of the lighter compounds. On the other hand from the large values for the standard deviation and from the large span between maximum and minimum values one can expect a great variation in the pattern of the congeners.

From these results and because the observed composition of the airborne PCBs very often does not agree with the composition of common technical mixtures, a factor of 8.5 is used for precautionary reasons to calculate the total PCB concentration in indoor air.

Under these conditions an average value of 237ng/m^3 is obtained for 60 measurements with a standard variation of 408ng/m^3. The maximum value is 2.147ng/m^3, whereas the minimum value amounts to 15ng/m^3. Thus a variation of more than two orders of magnitude is observed.

Figure 1 presents the number distribution of the measurements at intervals of 50ng/m^3. The proposed target value of 300ng/m^3 is denoted by an arrow. Most frequently concentrations beneath 50ng/m^3 were observed. 25% of all measurements can be assigned to this interval, whereas for higher concentration levels the number distribution levels off. Nevertheless, it has to be considered that 15 measurements (i.e. 25%) were performed for precautionary reasons only without any indication for a use of PCBs.

Compared to the approximate value of 300ng/m^3, around 20% of the concentrations exceed this value, two of them by more than a factor of 6. This percentage further increases (to about 25%) when taking into account the mentioned precautionary investigations.

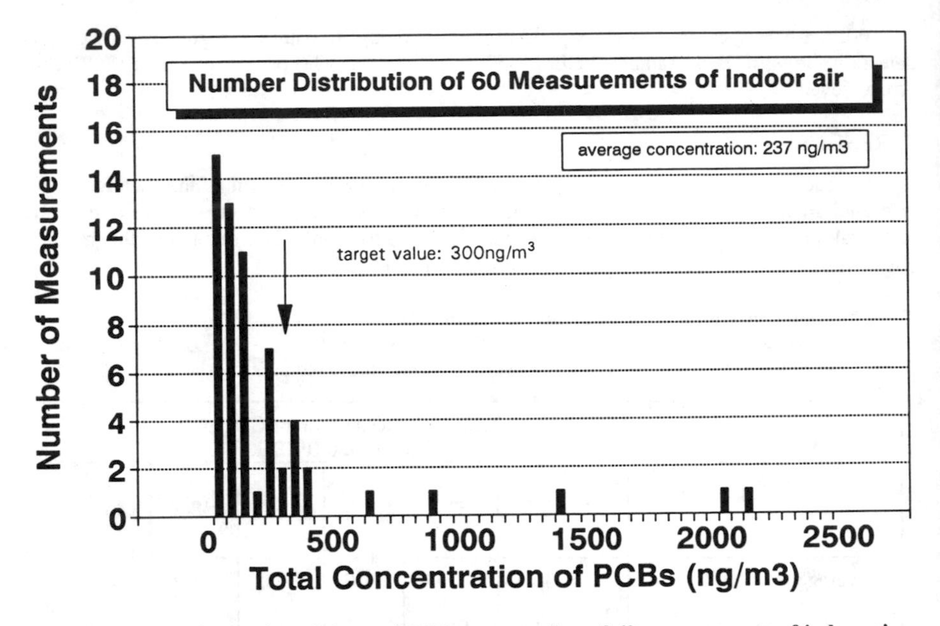

Fig. 1. Number distribution of the total PCB concentration of 60 measurements of indoor air. The proposed target level of 300ng/m^3 is marked with an arrow. Around 20% of the investigations show higher concentrations.

4. CONCLUSIONS

These data show that one takes a high risk in a judgement of the PCB concentration without any measured data. From these results the conclusion is drawn that measurements of indoor air are highly recommended even on a small suspicion of the incidence of PCBs. In addition, the measurements would also indicate the use of sealing compounds containing PCBs. In numerous examples it has been shown that the amount of PCBs as softeners reached up to 40% in these materials.

This need for experimental monitoring especially holds because in his opinion the expert has to answer very often the question for the possibility of **any** risk of indoor pollution affecting health of the inhabitants. In most cases this question arises at school buildings used by children. Because of their special biological system increased care should be taken for precautionary reasons in spite of high costs for measurements. In addition, this especially holds in view of the continuous debate on target or limit values for indoor pollution with a clear tendency to lower values.

REFERENCES

(1) DIN Deutsches Institut für Normung, DIN 51 527, Teil 1 (1987), reprints: Beuth Verlag GmbH, D-3000 Berlin 30.

PORAPAK S ® AS COMPLEMENTARY ADSORBENT TO TENAX TA ® FOR THE DETERMINATION OF VOC IN NON-INDUSTRIAL BUILDINGS

P. A. WÄGER, H. ROTHWEILER and C. SCHLATTER

Institute of Toxicology of the Swiss Federal Institute of Technology and University of Zürich, Schorenstrasse 16, 8603 Schwerzenbach

SUMMARY

Porapak S ® was evaluated as complementary adsorbent to Tenax TA ® in laboratory experiments and field measurements. A better adsorption behaviour towards several compounds (e.g. α-pinene, cyclohexane, propyleneglycoldiacetate) was confirmed, effects influencing sampling and analysis of VOC (displacements, chemical reactions) could be visualized. With respect to a toxicological interpretation, however, the use of Tenax TA ® alone would not have led to other results.

INTRODUCTION

The evaluation of possible health risks with respect to volatile organic compounds (VOC) in non-industrial buildings is of growing importance. In practice, enrichment on adsorbents (mainly Tenax TA ®) with subsequent thermal desorption is being increasingly used for this purpose. However, especially when using only one adsorbent (which is common), one has to be aware of several limitations: At fixed conditions, compounds will be underestimated due to non - ideal ad - and desorption behaviour (breakthrough, incomplete desorption). Moreover, depending on the composition of the sample, chemical reactions or displacement effects as a result of co-adsorption may lead to misestimations. In our field studies we therefore introduced Porapak S ® as additional adsorbent to Tenax TA ® and studied its ad- and desorption properties in laboratory experiments and field measurements.

EXPERIMENTAL

Thermal desorption for conditioning of the adsorbents and for analysis was performed on a Carlo Erba TDAS 5000 , cryotrapping of the desorbed compounds with a Carlo Erba MFA 515.

Laboratory Experiments

As already had been done for Tenax TA ® and Carbotrap ® [1], laboratory experiments were made in order to determine the recoveries of selected compounds on Porapak S ®.: The compounds (500 ng in 0.5 µl acetone) were applied onto each of three adsorbent tubes at room temperature. Measurements were performed at two different air sampling volumes (1 l and 5 l).

Cleaning and desorption conditions for Porapak S ® were previously optimized with respect to its blank value, which showed to be mainly temperature-dependent: After Soxhlet-extraction with hexane and dichloromethane (each 24 h) and drying under vacuum at 60 °C (4 h), the adsorbent (143 mg) was packed into glass tubes (10 cm x 0.4 cm i.d.) and conditioned at 175°C (He-flow: 14 ml/min) in two steps (110 min + 10 min). A desorption temperature of 150 °C (desorption time: 10 min; He-flow: 7 ml/min) during analysis led to sufficiently clean blank values (see Fig. 1) for our measurements on a high resolution GC / MS - system. Tenax TA ® (155 mg) packed into glass tubes (10 cm x 0.4 cm i.d.) was conditioned at 280 °C (110 min + 10 min; He-flow: 7 ml/min) and desorbed at 260°C (10 min; He-flow: 7 ml/min).

Field Measurements

Results obtained in laboratory experiments were reconsidered in field measurements performed in several new and recently renovated buildings. The adsorption tubes were handled the same way as in laboratory experiments. For Porapak S ®, however, both conditioning and desorption temperatures were higher in measurements a-h (250 °C and 230 °C, respectively). Here, due to the higher blank values, compounds with boiling points above about 210 °C were not considered. During sampling, identical air sampling volumes (a-h: 5 l; i-l: 4 l) were simultaneously drawn through Tenax TA ® - and Porapak S ® - adsorption tubes for each sampling set. After storage at -80 °C (usually several days), the samples were thermally desorbed (10 min), cryofocussed at -150 °C, separated by gas chromatography on an OV 1701-OH glass capillary column (60 m x 0.303 mm) and detected with a Finnigan MAT 4510 quadrupole mass-spectrometer.

RESULTS AND DISCUSSION

Laboratory Experiments

Table 1 shows recoveries measured for selected compounds on Tenax TA ® and Porapak S ®. Compared to Tenax TA ®, recoveries of measured terpenes and chlorinated hydrocarbons were good or better on Porapak S ®. On the latter, several oxygen containing compounds (propyleneglycoldiacetate, 2-butoxyethanol, propyleneglycolmethyletheracetate) could be recovered in the range of 80 - 100%. In contrast to Carbotrap ® [1], we had no indications for reactions having occurred.

Table 1 Recoveries of selected compounds measured on Tenax TA ® and Porapak S ®

compound	V_S	Recovery [%] ± S.D.			
		Tenax TA		Porapak S	
n-hexanal	1	97.3 ± 6.3	(3)	97.0 ± 9.4	(3)
	5	101.6 ± 3.7	(3)	87.0 ± 9.1	(2)
1,1,1- trichloroethane	1	28.2 ± 4.3	(3)	91.6 ± 4.5	(3)
	5	0.0 ± 0.0	(3)	86.0 ± 7.0	(3)
α-pinene	1	71.2 ± 18.8	(3)	96.9 ± 2.3	(3)
	5	11.8 ± 1.8	(3)	97.4 ± 1.0	(3)
cyclohexane	1	-		83.2 ± 6.7	(3)
	5	-		78.5 ± 2.1	(2)
propyleneglycoldiacetate	1	-		99.2 ± 15.0	(3)
	5	-		89.6 ± 7.4	(2)

V_S: (purified) air sampling volume — in brackets: number of measurements

Field Measurements

Table 2 lists concentration - ratios and - ranges of selected compounds found in new and recently renovated buildings in Switzerland. Concentration - ratios of toluene, n-decane, ethylbenzene and ethylacetate do not indicate a significant difference between both adsorbents over a wide concentration range. Compounds like cyclohexane or α-pinene, instead, were better retained on Porapak S ®, which is consistent with our recovery - data and their known better retention behaviour on the latter adsorbent [2]. Repeated measurements made in a renovated building showed that propyleneglycoldiacetate was better retained on Porapak S ® at higher concentrations (3 mg/m3; see table 2 a,c), whereas at lower concentrations (< 550 µg/m3; see table 2 b,d) there was no observable difference between both adsorbents. It remains unclear, if this effect is solely due to the higher amount of propyleneglycoldiacetate, or if displacement effects as a result of a co - adsorption of other compounds are involved.

Table 2. Concentration - ratios and - ranges for selected compounds sampled on Porapak S ® and Tenax TA ® in new and recently renovated buildings (objects a-k)

Pollutant	Concentration - ratios Porapak S / Tenax TA												Concentration range[1)] [μg/m³]
	a	b	c	d	e	f	g	h	i	j	k	Mean ± SD	
n-decane	1.11	1.36	1.16	1.05	0.97	1.10	0.97	0.96	1.15	0.84	0.85	1.05 ± 0.15	26 - 933
cyclohexane	1.90	2.70	2.48	2.44	n.d.	n.d.	n.d.	n.d.	2.64	n.d.	1.69	2.31 ± 0.41	88 - 7401
benzene	n.d.	0.87	1.00	1.00	1.10	1.11	>1.80	0.90	1.4	1.17	1.22	1.16 ± 0.27	6 - 27
toluene	1.19	1.02	1.07	0.96	1.09	1.09	1.00	0.95	n.d.	0.82	0.90	1.01 ± 0.11	26 - 7691
ethylbenzene	n.d.	n.d.	n.d.	n.d.	0.99	0.68	n.d.	n.d.	1.18	0.95	0.90	0.94 ± 0.18	23 - 719
α-pinene	n.d.	n.d.	n.d.	n.d.	n.d.	1.23	1.30	1.61	n.d.	n.d.	n.d.	1.38 ± 0.20	74 - 867
ethyl acetate	n.d.	n.d.	n.d.	n.d.	1.17	>1.20	n.d.	n.d.	0.83	1.08	1.01	1.06 ± 0.15	6 - 660
n-hexanal	1.93	0.84	1.19	1.40	1.08	1.05	0.71	0.66	1.56	n.d.	0.92	1.13 ± 0.40	8 - 1750
propyleneglycoldiacetate	1.85	0.94	1.93	0.74	n.d.	n.d.	n.d.	n.d.	n.d.	n.d.	n.d.	1.35 ± 0.60	1633 - 3240
TVOC	1.28	1.13	1.60	1.20	0.96	0.82	0.60	1.00	1.13	0.92	1.40	1.09 ± 0.28	1549 - 34457

1) : referred to the internal standard toluene - d8 (500 ng) n.d.: not detected at concentrations above 5 μg/m³
SD: standard deviation

The variability observed for n-hexanal - and benzene - concentration - ratios could be explained by displacement effects or chemical reactions taking place on the adsorbent. At least for benzene, displacement effects in the presence of other compounds were already observed [3,4]. Total volatile organic compound (TVOC) - ratios depend on the composition of the air samples: In those with significant amounts of higher boiling compounds, Tenax TA ® is favourized due to the cut-off at 210 °C (measurements a-h) and lower desorption efficiencies (measurements i-k) on Porapak S ®.. Porapak S ®, again, leads to higher TVOC - values in the presence of significant amounts of e.g. terpenes, cyclohexane or propyleneglycoldiacetate, compounds frequently measured in practice.

CONCLUSIONS

The better adsorptive behaviour of Porapak S ® towards several compounds (e.g. α-pinene, cyclohexane, propyleneglycoldiacetate) was confirmed. The use of two adsorbents (Porapak S ®, Tenax TA ®) in field measurements visualized difficulties occurring during sampling and analysis of VOC, which can result from displacement mechanisms or chemical reactions. However, due to its lower thermal stability limiting the temperature range of compounds desorbed, Porapak S ® cannot fully replace Tenax TA ®. Although the use of Porapak S ® was adequate in several objects, the use of Tenax TA ® alone would not have led to other results with respect to a toxicological interpretation.

REFERENCES

[1] ROTHWEILER, H., WÄGER, P.A. and SCHLATTER, C. (1991). Comparison of Tenax TA and Carbotrap for sampling and analyisis of volatile organic compounds in air. Atmos. Environ. 25B (2), 231-235.

[2] FIGGE, K., RABEL, W. and WIECK, A. (1987) Adsorptionsmittel zur Anreicherung von organischen Luftinhaltsstoffen. Fresenius Z. Anal. Chem. 327, 261-278.

[3] WALLING, J.F., BUMGARNER, J.E., DRISCOLL, D.J., MORRIS, C.M., RILEY, A.E. and WRIGHT, L.H. (1986) Apparent Reaction Products desorbed from Tenax used to sample ambient air. Atmos. Environ. 20 (1), 51-57.

[4] VEJROSTA, J., ROTH, M. and NOVÀK, J. (1983) Interference effects in trapping trace components from gases on chromatographic sorbents. J. Chromat. 265, 215-221.

1434

MONITORING OF ORGANOHALOGEN AND ORGANOPHOSPHORUS COMPOUNDS IN INDUSTRIAL/INDOOR ATMOSPHERES

S. HASTENTEUFEL*, S.A. HAZARD** and W.R. BETZ**

* Supelco D GmbH, 6380 Bad Homburg, Germany
**Supelco Inc., Supelco Park, Bellefonte, PA 16823-0048, USA

SUMMARY

The necessity to monitor volatile compounds in industrial processes is regulated by international and national safety and health organisations. Research efforts focus on the development of trapping techniques suitable for routine analysis of volatile and semi-volatile compounds.

1. INTRODUCTION

Important factors dominating the choice of an appropriate adsorbent are the capacity of the adsorbent(s) for the adsorbates, the adsorption mechanism, sampling rate and time, and the design of the sampling tube. The analytical method and the detectors used to analyze the compounds of interest can be limiting factors in the choice of the adsorbent material.

Solvent desorption with CS_2 and activated charcoal as the adsorbent medium is still the method commonly used. Alternative adsorbent materials in combination with appropriate desorption solvents and detector systems provide better analytical results compared to this non-specific technique.

Adsorbents used in this study have been characterized, using gas-solid chromatography (GSC). This technique had led to a classification scheme, developed by Kiselev, which provided insight into the adsorbate/adsorbent interactions. Table 1 summarizes this classification scheme. Several of the adsorbents chosen for this study are presented in Table 2.

Table 1. Classification of Adsorbents and Adsorbates

Molecules	Adsorbents		
	Type I without ions or active groups	Type II localized positive charges	Type III localized negative charges
Group A: • Spherically symmetrical shells • σ-bonds	Nonspecific interactions/ dispersion forces		
Group B: • Electron density concentrated on bonds/links • π-bonds Group C: • (+) Charge on peripheral links Group D: • Concentrated electron densities • (+) Charges on adjacent links	Nonspecific interactions	Nonspecific + specific interactions	

Table 2. Classification of Adsorbents

Adsorbent	Surface	Classification (Kiselev)
Graphitized Carbon Blacks	Graphitic Carbon	Class I
Carbon Molecular Sieves	Amorphous Carbon	Weak Class III (can approach Class I)
Activated Silica Gel	Oxides of Silica Gel	Class II
Activated Charcoal	Oxides of Amorphous Carbon	Class III
Porous Polymers	Organic "Plastics"	Weak → Strong Class III

2. EXPERIMENTAL

This work involved a two-fold approach. First, adsorbent tubes were developed (in this laboratory or by Federal agencies) and in-house clean-up procedures were optimized and tested for these tubes. Second, desorption efficiency evaluations were for the six chosen tubes.

3. RESULTS AND DISCUSSION

The results of the clean-up procedures are best exemplified by the chromatograms shown herein. The detection levels at which these adsorbents/adsorbent tubes have been evaluated typically were focused on an on-tube concentration of the analyte and represented 1/20 of the threshold limit value (TLV) for the representative analyte.

The results of the desorption efficiency studies for several of the adsorbents chosen for this study are presented in Table 3.

Table. 3. Desorption Efficiency Data for the Adsorbent Tubes

Adsorbent	Analyte	Desorbing solvent	% Recovery
Amberlite® XAD-2	chlordane	toluene	101
Amberlite XAD-2 urethane plug	dichlorvos	toluene	94
	diazinon	"	100
	malathion	"	100
	dursban	"	107
	parathion	"	99
Florisil®	PCB (Aroclor 1248)	hexane	96
Carboxen-564	methyl ethyl ketone	carbon disulfide	98
Carbotrap B	2,3-dibromopropanol	carbon disulfide	96
Activated charcoal	chloroform	carbon disulfide	98

The recovery values indicate that the adsorbents utilized in this study effectively released the adsorbates of interest. Furthermore, the clean-up procedures employed provided adsorbents without background interference. For example, Figure 1 depicts various common industrial solvents desorbed from a high purity activated charcoal, using CS_2. Recent work has focused on the use of carbon molecular sieves as replacement adsorbents for the more volatile organic solvents (Figure 2). The tube represented contains Carboxen-564, a carbon molecular sieve with adsorbing properties superior to those of activated charcoal.

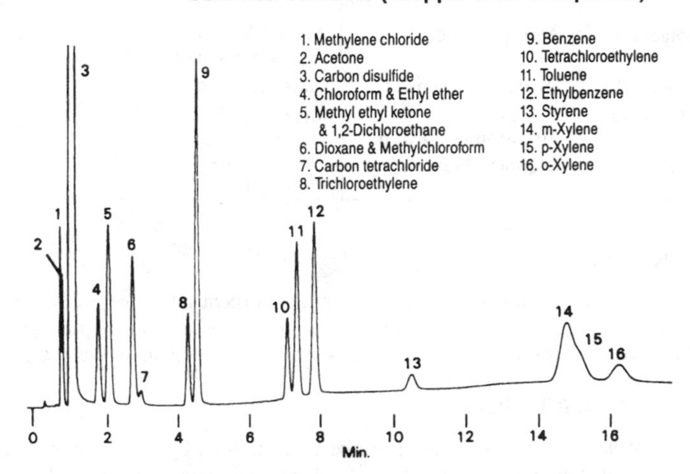

Fig. 1. Common industrial solvents/activated charcoal

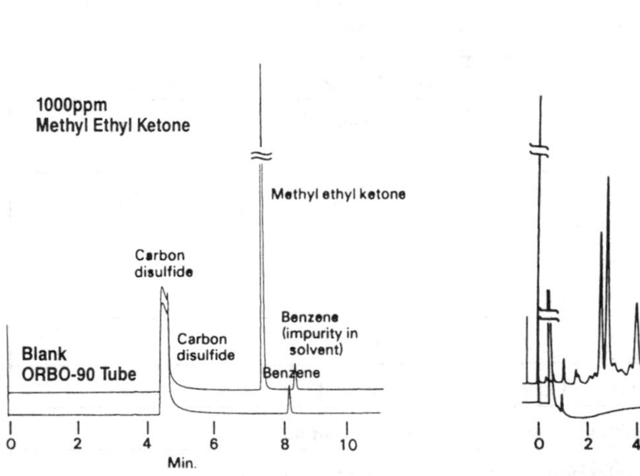

Fig. 2. Methyl ethyl ketone/ carbon molecular sieve

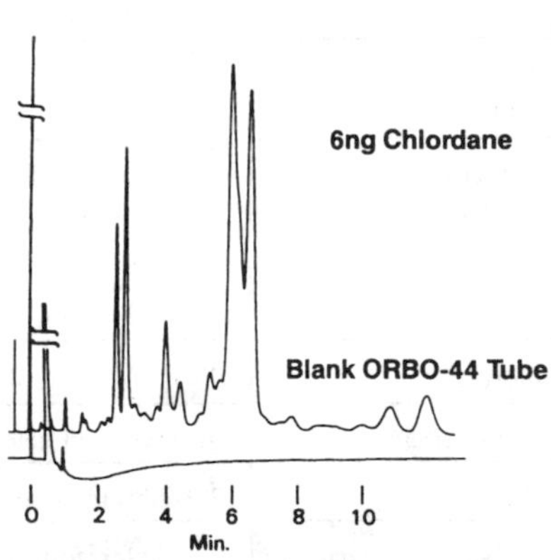

Fig. 3. Chlordane/Supelpak 20 (high purity Amberlite XAD-2)

The quest for adsorbents for organohalogen and organophosphorus compounds has led to the use of high purity Amberlite XAD-2 resin. This adsorbent is currently used in many sampling methods; an example (a chlorinated pesticide) is shown in Figure 3.

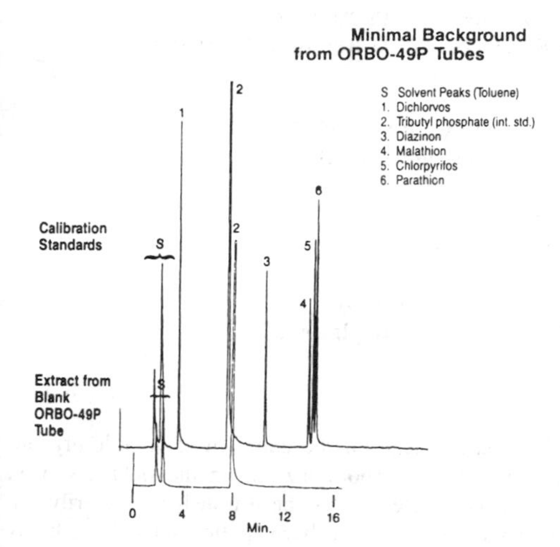

Fig. 4. Organophosphorus pesticides/ urethane plug, Amberlite XAD-2

A recent alternative to Amberlite XAD-2 is polyurethane foam adsorbent plugs in conjunction with XAD-2. Recent methods focusing on these are exemplified in Figure 4.

Inorganic adsorbents such as magnesium silicate have proven effective for monitoring PCBs. Figure 5 depicts the chromatographic profiles of several PCBs desorbed from Florisil using hexane.

Another recent advance has been the development of adsorbent tubes containing several non-specific graphite carbons. Figure 6 depicts a chromatographic profile of 2,3-dibromopropanol extracted from a graphitized carbon black.

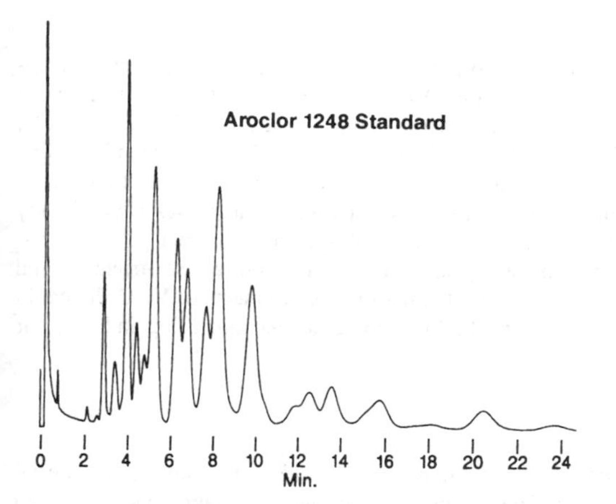

Fig. 5. PCBs/Florisil

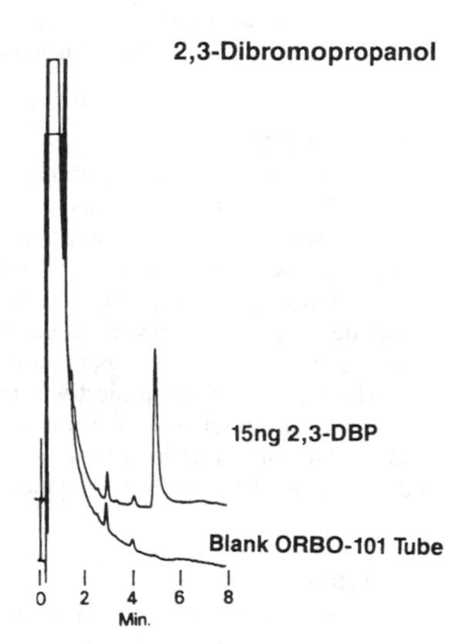

Fig. 6. 2,3-dibromopropanol/ graphitized carbon black

4. CONCLUSION

The use of adsorbent tubes has evolved over the past two decades as a viable technique for monitoring indoor air atmospheres. An effective understanding of the thermodynamic and kinetic properties of this dynamic technique has greatly assisted in understanding the mechanisms of adsorption/desorption.

A Micro Test Cell for the Measurement of Volatile Organic Compound Emissions from Building Materials

WOLKOFF, P., CLAUSEN, P.A., RASMUSEN, E.
Danish National Institute of Occupational Health

NIELSEN, P.A.
Danish Building Research Institute

GUSTAVSSON, H., JONSSON, B.
Swedish National Testing and Research Institute

SUMMARY
A micro test cell of stainless steel for measurements of the emission of volatile organic compounds from planar building materials in the laboratory and in the field has been designed, constructed, and a test protocol elaborated. It has been tested satisfactorily for repeatability and reproducibility in the laboratory and it has been applied in the field. It has also shown satisfactorily correlation with a small emission chamber. It may be applied to materials emitting by diffusion and unaffected by ventilation, e.g. most building materials.

1. INTRODUCTION

Building materials and consumer products are often major sources of volatile organic compounds (VOC) present in the indoor environment.[1,2] Since VOC may be one of the causative agents of the sick building syndrome there is a general demand in the public for healthy building materials, i.e. materials with low emissions or emission of VOC of low toxicity.[3] Measurement of building material emissions is therefore an important method to be able to reduce human exposure to VOC by product development of low emission building materials.[4] Several research activities have already elaborated standardized methods to measure the emission from building and consumer products.[5-8]

The purpose of this project was to design, test, and document a versatile, user-friendly, and relatively cheap emission cell for both laboratory and field emission measurements of VOC. It should be applicable for research as well as for use by the building material manufacturing industry for product control and product development.

2. METHOD

A micro emission cell is made of stainless steel. It is circular with a maximum test material surface area of 17.7 cm^2 and a volume of 35 ml. The test material becomes an integral part of the cell itself by placing it onto the test material, see Figure 1. The air is introduced into two diagonally positioned inlets at the perimeter of the cell wherefrom the air is evenly distributed over the test material surface. The air leaves the cell at the top of its center. The loading factor is maximum 506 m^2/m^3. The sealing material is made of emission free silicon rubber foam. All tubes and couplings are made of high quality stainless steel. A subunit has also been constructed to accomodate for testing of liquid and rugged materials like carpets and paints, see Figure 1. The cell is supplied with clean and humidified air from an air supply unit. The air velocity and humidity are controlled by two air inlet flows in a separate air supply unit (not shown here). The outlet is a union cross with a 9 cm tube to avoid false air intake during sampling.

Duplicate air samples were taken on Tenax TA using pumps with flow control. The tubes were desorbed thermally and analyzed by gas chromatography.

Four vinyl floor carpet samples from the same batch have been tested simultaneously in Denmark and Sweden with two micro emission cells (air exchange rate = 171/hour, temperature = 23°C and rel. humidity = 50%) and two climate chambers (234 L, air exchange rate = 0.25/hour, temperature = 23.0°C and rel. humidity = 45%) over a ten days period.

3. RESULTS

The analytical repeatability was less than 0.8% relative standard deviation for cyclohexanone and phenol but about 6% for TXiB (2,4,4-trimethyl-1,3-pentanediol diisobutyrate). The cyclohexanone and phenol concentrations in the emission cells correlated with $r^2 = 0.992$-0.996, respectively. The estimated initial masses and the first order decay rate constants of the emission in the emission cells showed not to be greatly different for cyclohexanone and phenol. Comparison of the emission cell with the 234 L chambers showed a significant discrepancy with regard to the initial mass indicating sample inhomogeneity or insufficient air mixing. The rate constants, however, were comparable.

The micro cell has been applied in the field. Simultaneous field sampling of VOC of the room air and measurements of the emission from floor and table surfaces provided source identification of VOC from two potential emitting materials, and the apportionment thereof. In another experiment the equilibrium concentration of 2-ethylhexanol from a vinyl wall to wall carpet was reached after 20 hours. The was a fair agreement between the measured room air concentration of 2-ethylhexanol and that calculated based upon the measured emission assuming an one compartment model.

A full report of this field and laboratory emission cell (FLEC) will be published elsewhere.[9]

REFERENCES

(1). WALLACE, L.A.; PELLIZARI, E.; LEADERER, B.; ZELON, H.; SHELDON, L. (1987). Emissions of volatile organic compounds from building materials and consumer products. *Atmospheric Environment*, **21** pp 385-393.

(2). TUCKER, W.G. (1991). Emission of organic substances from indoor surface materials. *Environmental International* **17** 357-363.

(3). WORLD HEALTH ORGANIZATION. (1989). Indoor Air Quality: organic pollutants. EURO Reports and Studies 111, WHO Regional Office for Europe, Copenhagen.

(4). WOLKOFF, P. (1990). Proposal of methods for developing healthy building materials: Laboratory and field experiments. *Environmental Technology*, **11** pp. 327-338.

(5). American Society for Testing and Materials. (1990). Standard Guide for Small-Scale Environmental Chamber Determinations of Organic Emissions From Indoor Materials/Products. Designation: D 5116-90. Philadelphia, Pa.

(6). European Concerted Action - COST 613. (1989). Report No. 2. Guideline for the Determination of Steady State Concentrations in Test Chambers. Brussels-Luxembourg: Commision of the European Communities. EUR 12196 EN.

(7). European Concerted Action - COST 613. (1991). Report No. 8. Guideline for Characterization of Volatile Organic Compounds Emission from Indoor Materials and Products Using Small Test Chambers. Brussels-Luxembourg: Commision of the European Communities. EUR 13593 EN.

(8). NORDTEST. (1990). Building materials. Emission of volatile compounds, chamber method. NT Build 358.

(9). WOLKOFF, P.; CLAUSEN, P.A.; NIELSEN, P.A.; GUSTAFSSON, H.; JONSSON, B.; RASMUSEN, E. (1991). Field and Laboratory Emission Cell: FLEC. Proceedings of IAQ 91. *American Society of Heating, Refrigerating and Air-Conditioning Engineers Transactions*. In press.

Test	FLEC-DK	FLEC-S	Chamber 1	Chamber 2
Homogenity of outlet	+[a]			
Repeatability RSD[b] % cyclohexanone phenol TXiB[c]	 0.6 (N=4) 0.8 (N=4) 7.2 (N=4)	 1.4 (N=5) 3.1 (N=5)		
Cyclohexanone N;r^2 M_0 mg/m² (d) k_1 h^{-1} (d)	12;0.990 250(7) 0.0044(3)	10;0.997 230(5) 0.0042(2)	12;0.97 136(8) 0.0047(5)	12;0.95 416(35) 0.0036(5)
Phenol N;r^2 M_0 mg/m² (d) k_1 h^{-1} (d)	10;0.992 14.0(5) 0.0041(2)	10;0.997 17.0(5) 0.0032(2)	12;0.95 15(2) 0.0024(4)	12;0.93 42(7) 0.0017(3)

a) T-test for accordance between VOC concentrations at sampling and outlet positions gave p = 0.95, p = 0.76 and p = 0.17 for cyclohexanone, phenol, and TXiB, respectively. **b)** Relative standard deviation. **c)** 2,4,4-trimethyl-1,3-pentanediol diisobutyrate. **d)** Values in parenthesis are the standard error on the last digit.

TABLE 1: Results of homogenity and repeatability with two FLECs. Initial mass M_0 and first order rate constant of emission for FLECs and 234 L climate chambers using a vinyl floor carpet in a parallel and simultaneous test period.

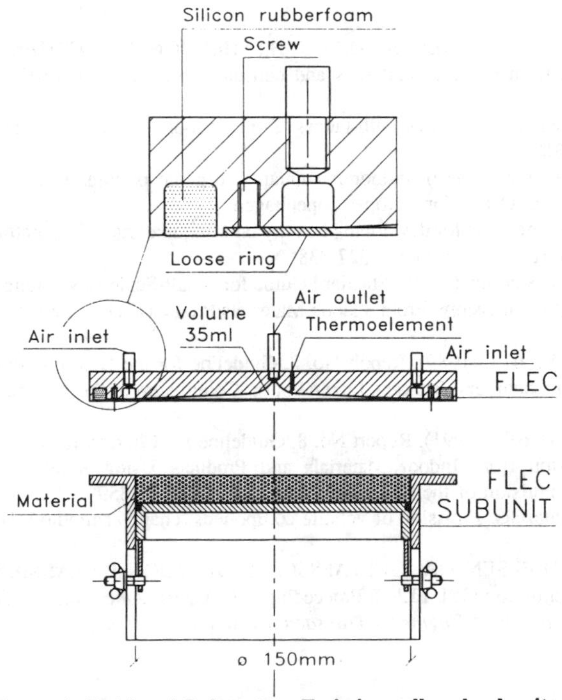

Figure 1. Field and Laboratory Emission cell and subunit.

Session VII

MEASUREMENT QUALITY

KEYNOTE PAPERS

POSTERS

1436

IS EVERYTHING UNDER CONTROL ?

B. GRIEPINK and S. VANDENDRIESSCHE

Commission of the European Communities,
Community Bureau of Reference (BCR), Brussels, Belgium

SUMMARY

A survey of the various methods to assure good accuracy in the chemical laboratory is presented as well as the means to monitor the control over the measurement procedure. Efforts have to be made in order to increase the reliability of results of chemical analysis.

1. INTRODUCTION

Measurements are a basis for decisions, whether this is on the acceptance of products, the need for medical treatment, or the compliance with legislation to protect health, safety and environment. Correct decision making therefore requires accurate results.

The standardisation of methods and the automation of equipment have in the past decades often led to the impression that nothing could go wrong and that the results delivered by the computer were correct above any doubt. In many fields of testing, measurement, analysis, this euphoristic feeling has been shattered by the outcome of intercomparisons which showed large discrepancies (up to orders of magnitude).

In many fields of analysis, insufficient accuracy of results has been demonstrated (e.g. refs. 1,4,5,6). One example may be given here. There is a strong public concern about the exposure of population to dioxins; consequently authorities have set low limits e.g. in food which in some cases led to a close-down of waste incinerators supposed to emit too high amounts of dioxins that finally entered in the food chain (e.g. milk). Typical concentrations considered in milk fat are 0.5-10 pg/g. In a BCR-intercomparison, however, in which experienced laboratories participated and received all the same sets of calibrants for the analysis of a fly ash extract (levels 1000 times higher than a milk fat), the results of table 1 were observed.

It is now increasingly realised that efforts are required to avoid errors and quality assurance must be essential part of the measurement practice.

Table 1 : Dioxins in incinerator fly ash extract

CONGENER	RANGE* OF RESULTS
2,3,7,8-PCDD	6
2,3,7,8-PCDF	6
1,2,3,7,8-PCDD	6
1,2,3,7,8-PCDF	7

* Highest value divided by lowest value.

2. HOW TO ACHIEVE ANALYTICAL ACCURACY ?

Accuracy can not be bought from a supplier; it is the result of working with due consideration and care, both at the stage of organising the laboratory and at the stage of performing the measurements.

Skillful design of the laboratory and of its functioning, skillful selection of methods and instruments, adequate training of staff, adequate maintainance of equipment and working environment are prerequisites for quality of an analytical laboratory, but are not sufficient. The laboratory should continuously monitor its own performance and take appropriate action.

The first thing which a laboratory must require from itself is that it is in control of its results, i.e. that its repeatability (consistency of measurements under the same conditions) and reproducibility (consistency of measurements under changed conditions, e.g. other technician, new column, new calibrant) are known and expliccable.

Control of results means also that the a priori expected random variance obtained upon quadratic addition of the expected errors of all single steps of the whole procedure should be in agreement with the observed variance in daily practice. This means that in typical HPLC or gas chromatographic determinations coefficients of variation (CV) in the order of 2-5 % are expected.

There are only few applications where reproducibility is sufficient. In trade as well as in environmental monitoring or in measurements to check for compliance with limit values, systematic error is as unacceptable as random error. Whilst, theoretically speaking, trade partners could agree as soon as they have the same bias, systematic error must in most cases - including compliance measurements - be kept as low as is feasible.

Control of results is easily demonstrated on a statistical basis in cases where the laboratory can prepare a homogeneous and stable batch of samples which are reasonably similar to those normally analysed. Samples of such an internal reference material are analysed at regular intervals and after every change in measurement conditions; any significant deviation from the previously accepted value, or any significant increase in standard deviation, should lead to corrective action.

Absence of systematic error is difficult to prove by a single laboratory although attempts have been made (e.g. 7). The true value of a parameter being measured is seldom known so that direct comparison of the true value and the result of measurement is seldom possible. The next best method to demonstrate the absence of significant bias is to compare in a round robin various methods based on widely different principles and various approaches to calibration starting from compounds of verified purity and stoichiometry.

Such intercomparisons are expensive and very limited in number, and the results are not available in time for corrective action. Again in cases where a homogeneous and stable batch of samples which are reasonably similar to those being analysed can be prepared, an intercomparison of the above type can be used - once agreement between the various methods of measurement and of calibration has been achieved - to certify a reference material. Each laboratory performing the same type of analyses can then, at any time and at low cost, use samples of this CRM and compare his results with the certified value.

3. INTERNAL AND EXTERNAL QUALITY CONTROL

The quality-minded laboratory implements a system of regular checks using certified and non-certified (internal or laboratory) reference materials in a quality assurance (QA) programme which is laid down in a QA manual. Different types of "control charts" can be used to check long-term consistency and accuracy and to define alarm levels and intervention levels. Alertness is required as soon as the overall uncertainty becomes significantly larger than expected on the basis of the cumulated random errors which are inherent to the method; indeed this indicates that something is not fully under control. In routine measurements, intervention is necessary when the total uncertainty might exceed the levels imposed by either economic or legal considerations; in fact it would be sensible in all cases to intervene as soon as the performance is not under statistical control.

A control chart provides a graphical way of interpreting the method's output in time, so that the repeatibility of the results and the method's precision over a period of time can be evaluated at a glance. To do so, one ore several standards of good homogeneity and stability should be analysed with each batch of unknown materials. Some 5-10 % of all measurements should be dedicated to this purpose[8].

In a Shewhart control chart[9] the laboratory plots for each standard analysed at a time the obtained value (X) or the difference between duplicate values (R). The X-chart additionally presents the lines corresponding to a risk of 5 resp. 1 % that the results are not comprised in the whole population of results. These lines are of "warning" and "action" resp. (fig. 1 and 2).

Figure 1 : Example of an X chart

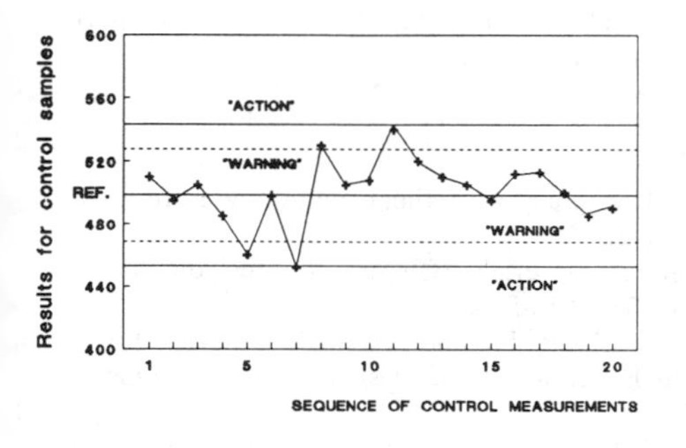

Figure 2 : Example of an R chart

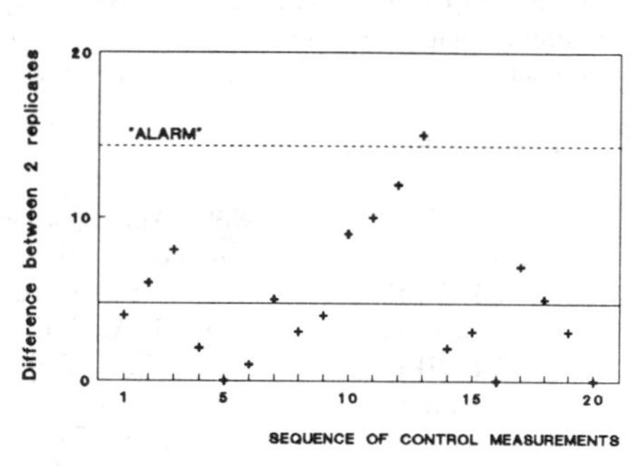

The results of a method are considered to be out of control:

- if the upper or lower control limit is exceeded ("action"),
- if the same "alarm-line" is exceeded twice in succession,
- if eleven successive measurements are on the same side of the "mean X" line.

X- and R-chart systems are also being tested in which the weighing factor for each point of the curve decreases according to a geometric progression (e.g. let the value at moment i be X_i, then the value plotted in a chart for a moment i is $X_i + (X_{i-1}) \cdot 0.2 + (X_{i-2})\, 0.16 + \ldots$). This gives an earlier detection of an upward or downward trend.

Cumulative sum (cusum) charts are frequently used, particularly as an early detection of a drift in the results. In these charts the cumulative sum of the differences between found value and probable value are plotted against the result obtained at moment i:

$$S_i = \sum_{i=1}^{i} (X_i - X_{ref})$$

in which S_i is the cumulative sum, X_i is the measured value and X_{ref} is the reference value. Dewey and Hunt[10] presented a practical evaluation of the use of cusum-charts. It should be emphasised that good results can only be obtained if the value of X_{ref} is accurately known. A certified reference material of good quality should therefore be used. The evaluation of a cusum-chart is done using the V-mask method[11-13].

The use of a Shewhart-chart enables detection as to whether or not a method is still in control; it is not able, however, to detect a systematic error which is present from the moment of introduction of the method in a laboratory. If possible results should be verified by other methods.

An independent method therefore can be used to verify the results of routine analysis. If results of both are in good agreement, one may conclude that the results of routine analysis do not contain a contribution of systematic nature (e.g. insufficient digestion). This conclusion is strongest when the two methods differ widely (e.g. are based on different principles of measurement).

In addition to maintaining quality by within-laboratory measures, the quality-minded laboratory also participates in external quality initiatives (such as proficiency testing schemes or occasional intercomparisons) so that it can demonstrate to its clients or partners that it performs adequately. In an increasing number of cases, participation in such initiatives is even a requirement for accreditation of the laboratory.

4. THE BCR PROGRAMME

The Community Bureau of Reference (BCR) is a research programme of the CEC aiming at improved quality of measurements. This programme has been quite active in the whole field of environmental analysis (largely by organising intercomparisons and certifying reference materials) and will pay increasing attention to improvement of methods and support to quality assurance in the area of workplace air monitoring.

REFERENCES

(1) GRIEPINK, B. (1989). Aiming at good accuracy with the Community Bureau of Reference (BCR). Quimica Analitica 8, 1-21

(2) GRIEPINK, B.(1990). Certified reference materials (CRMs) for the quality of measurement. Fresenius J. Anal. Chem., 337, 812-816.

(3) GRIEPINK, B., MAIER, E., QUEVAUVILLER, Ph. and MUNTAU, H. (1991). Certified reference materials for the quality control of analysis in the environment. Fresenius J. Anal. Chem., 339, 599-603.

(4) BERMAN, S.S. and BOYKO, V.J. (1987). ICES 6th round Intercalibration for Trace metals in Sea Water. JMG 6/TM/SW - ICES, Copenhagen

(5) SUBRAMIAN, K.S., STOEPPLER, M. (1986), Fres. Z. Anal. Chem. 328, 875

(6) HERBER, R.F.M., STOEPPLER, M. and TONKS, D.B. (1984) Fres. Z. Anal. Chem. 317,246

(7) WOLTERS, R. and KATEMAN, G. (1988), Anal. Chim. Acta 207, 111

(8) TAYLOR, J.K. (1985), Handbook for SRM-Users, NBS Special Publication 260-100

(9) SHEWHART (1931), Economic control of quality of the manufactured product D. Van Nostrand, New York.

(10) DEWEY, D.J. and HUNT, D.T.E. (1972), The Use of Cumulative Sum Charts (Cusum Charts) in Analytical Quality Control. Water Research Centre, Medmenham (UK) Technical report TR 174

(11) KATEMAN, G. and PIJPERS, F.W. (1981), Quality Control in Analytical Chemistry. John Wiley & Sons (New York)

(12) EDWARDS, R.W.H. (1980), Ann. Clin. Biochem. 17,205

(13) WESTGARD, J.O., GROTH, T., ARONSSON, T. DE VERDIER, C.D. (1977), Clin. Chem. 23,1981

1437

BCR PROJECTS IN THE FIELD OF WORKPLACE AIR MONITORING

S. VANDENDRIESSCHE

Commission of the European Communities, Community
Bureau of Reference (BCR), Brussels, Belgium

SUMMARY

The BCR programme comprises projects to improve the accuracy of the determination of volatile organic compounds in workplace air by personal monitoring techniques based on reversible sorption. Sources of error are detected in a sequence of intercomparisons and technical discussion meetings; when excellent analytical performance is achieved, a reference material is certified. One project has already reached this stage (CRM 112 - aromatic hydrocarbons on Tenax). Similar work is going on for aromatic hydrocarbons on charcoal, and for chlorinated hydrocarbons (thermal and solvent desorption). Sampling is studied in specially designed intercomparisons. Future projects will include reactive compounds (e.g. aldehydes, isocyanates), sampling of polar compounds, mixed phase sampling, and sampling and characterisation of aerosols.

1. THE BCR PROGRAMME

The Community Bureau of Reference (BCR) is a scientific programme of the CEC (with participation of Finland, Sweden and Switzerland) and has the task to improve the quality of measurements where this is of interest to the Community: implementation of Directives or European norms; elimination of technical trade barriers; areas where the unability to measure accurately causes risk or damage to health and safety, environment or economy.

The field of action of the BCR programme covers literally all types of testing, measurement and analysis:

- metrology (intercomparisons between national metrology institutes);
- industrial metrology (e.g. measurement of gas flows; micrometrology; mechanical testing; sound attenuation by ear protectors; etc.);
- analysis of food and agricultural products;
- biomedical analyses;
- analysis of industrial products;
- microbiology;
- environmental analyses.

Projects supported by the programme can consist of:

- studying sources of errors in methods
- comparing the performance of different methods
- development of measurement methods
- technical research which appears necessary for establishing European standards for measurement methods
- support to cooperation between national initiatives for external quality control
- certification of reference materials as tools for quality assurance or as support to the implementation of directives or standards

As an R&D programme, the BCR covers technical problems of measurements; it is not involved in the preparation of Directives or standards, nor in routine work of proficiency testing.

In many projects, the starting situation is one where methods exist but where the results from various methods and from various laboratories do not agree; in these cases, most errors can often be detected and eliminated by successive intercomparisons, each one followed by a meeting where the measurements are discussed in great detail. This procedure will usually lead to such an improvement that the group of laboratories will reach agreement on results, using different measurement methods; when possible, this achievement will be disseminated to other laboratories through a reference material certified by that group of laboratories.

Industries where workers might be exposed to potentially harmful vapours or aerosols spend considerable efforts in air monitoring, but the results are often unreliable. Draft European standard prEN 482 sets a minimum requirement of 30 % overall accuracy, but it is obvious from intercomparisons that this objective is often not met. The BCR has therefore set up several projects to enable laboratories to test and improve their performance.

Limit values exist for several hundreds of compounds, and many more are monitored even though no legally imposed limit values exist. It is neither feasible nor desirable to consider all of them for BCR projects; it was therefore decided to select some (e.g. 10) classes of compounds which would be representative from two points of view:

- measured in a wide range of industries;
- covering the various types of technical difficulties encountered in workplace air monitoring.

It seems reasonable to assume that good performance for such a range of cases would indicate good general performance.

2. COMPLETED AND ONGOING PROJECTS

The subjects considered first were vapours which could easily be trapped on sorbing agents and subsequently desorbed for GC analysis, either thermally or by solvent extraction, i.e. aromatic hydrocarbons and chlorinated hydrocarbons. In each of the projects described below, all participants were leading laboratories from government or major chemical companies. It is therefore found useful to list some typical sources of error; other laboratories might have similar difficulties.

Aromatic hydrocarbons - thermal desorption

The earliest BCR work in this field dealt with benzene, toluene and m-xylene trapped on Tenax TA. Three homogeneous and stable batches of charged tubes (approx. 1 μg of each compound per tube) were prepared and used in intercomparisons with 11 to 13 participants (see ref. 1). In the first intercomparison, two laboratories had some 40 % error and several had 10-15 % error for what was felt to be a very easy task. Some errors were identified (including the use of malfunctioning equipment) and avoided at the next occasion, but in the second round several laboratories still made errors of some 10 %. A new technical discussion revealed that these errors were mostly made in the calibration:

- evaporation of liquids during the preparation and manipulation of the calibrants;
- use of non calibrated syringes;
- use of a non calibrated and poorly maintained analytical balance.

In the third round, all laboratories used properly maintained and calibrated equipment and prepared their calibrants in such a way as to avoid evaporation losses. This led to excellent agreement between the laboratories (coefficients of variation between laboratories of approx. 0.02); their results also agreed with the value calculated from the charging of the samples.

The samples used for the third round were taken from a homogeneous batch of 1000 samples. As a consequence of the analytical work performed, the content of each compound in each tube was very accurately known (1.5 % uncertainty) and the batch was certified as a reference material (CRM 112) of which each interested laboratory (worldwide) can buy samples to check its own performance (ref. 2). These samples are stainless steel tubes (Perkin-Elmer type) of which the commercial caps have been replaced by Swagelok compression fittings with PTFE ferrules.

Aromatic hydrocarbons - solvent desorption

Similar work is going on for aromatic hydrocarbons sorbed on activated charcoal (levels of 10-20 μg for benzene and 100-200 μg for toluene and m-xylene). Two intercomparisons were held with nearly 30 participants. Sources of error which have been identified through these intercomparisons and the subsequent technical discussion meetings include:

- use of a gas chromatograph of which the septum leaked intermittently (this problem was not detected through daily quality control !)
- use of malfunctioning flame ionisation detector (drifting response factors had been ignored)
- discarding part of the desorption solution (to get rid of the charcoal) but calculating the results as if the original volume of solvent was used
- correcting in the opposite sense for the benzene blank in the CS_2 used for extraction
- correcting in the opposite sense for the desorption efficiency
- ignoring sorption of the internal standard, added to the desorbing solvent, to the charcoal
- the use of volumetric glassware, which is calibrated for water at 20 °C, for carbon disulphide taken from the refrigerator
- the use of simplified methods (most often the phase equilibrium method) to determine the desorption efficiency.

The latter two sources of error are difficult to avoid in many laboratories and will be studied in a third intercomparison. At the same time a reference material is being prepared (see paper by Flammenkamp, Ludwig and Kettrup in this volume); it is expected to be certified and available in 1993.

Aromatic hydrocarbons - sampling

Sampling techniques were studied in an intercomparison where the participants sampled all at the same time from the same air which was charged with known levels of pollutants and which was distributed by a system built for this purpose (see paper by Goelen and Rymen in this volume). Seven controlled test atmospheres were offered to the participants with benzene, toluene and m-xylene in concentrations ranging from < 0.1 to > 2 times the limit values. Figures 1 to 3 present the results for three selected cases:

- benzene in an atmosphere with 1.76 ppm benzene (ppm = 10^{-6} mole/mole), 21.4 ppm toluene, 16.9 ppm m-xylene and 17.8 ppm ethylbenzene in air with 30 % relative humidity (fig. 1).
- toluene in an atmosphere with 107 ppm toluene in air with 46 % humidity (fig. 2)
- m-xylene in an atmosphere with 4.9 ppm benzene and 17.4 ppm m-xylene in air with 46 % humidity (fig.3)

Each diagram presents, for each participating laboratory and for each sampling technique used, the mean and standard deviation of a variable number of replicate measurements; the vertical line presents the known value; the left-hand column contains, for each set of data, the laboratory number and a method code. The first part of this code represents the type of sampling: A = active; D1 = diffusive sampler for thermal desorption (stainless steel tube); D2 = cylindrical diffusive sampler; D3 = round, flat (badge type) diffusive sampler. The additional indication given in some cases after the A is intended to discriminate between slightly different sampling techniques applied by the same laboratory. The second part of the code represents the sorbent (C = charcoal; G = graphitised carbon used for thermal desorption; 6 = Chromosorb-106; T = Tenax TA). It should be noted that the D2 type samplers were, in the exposure chamber, supported by a wire which may have disturbed the diffusion pattern; the data presented in figures 1 to 3 should therefore not be invoked as evidence that the D2 type samplers give systematically low results.

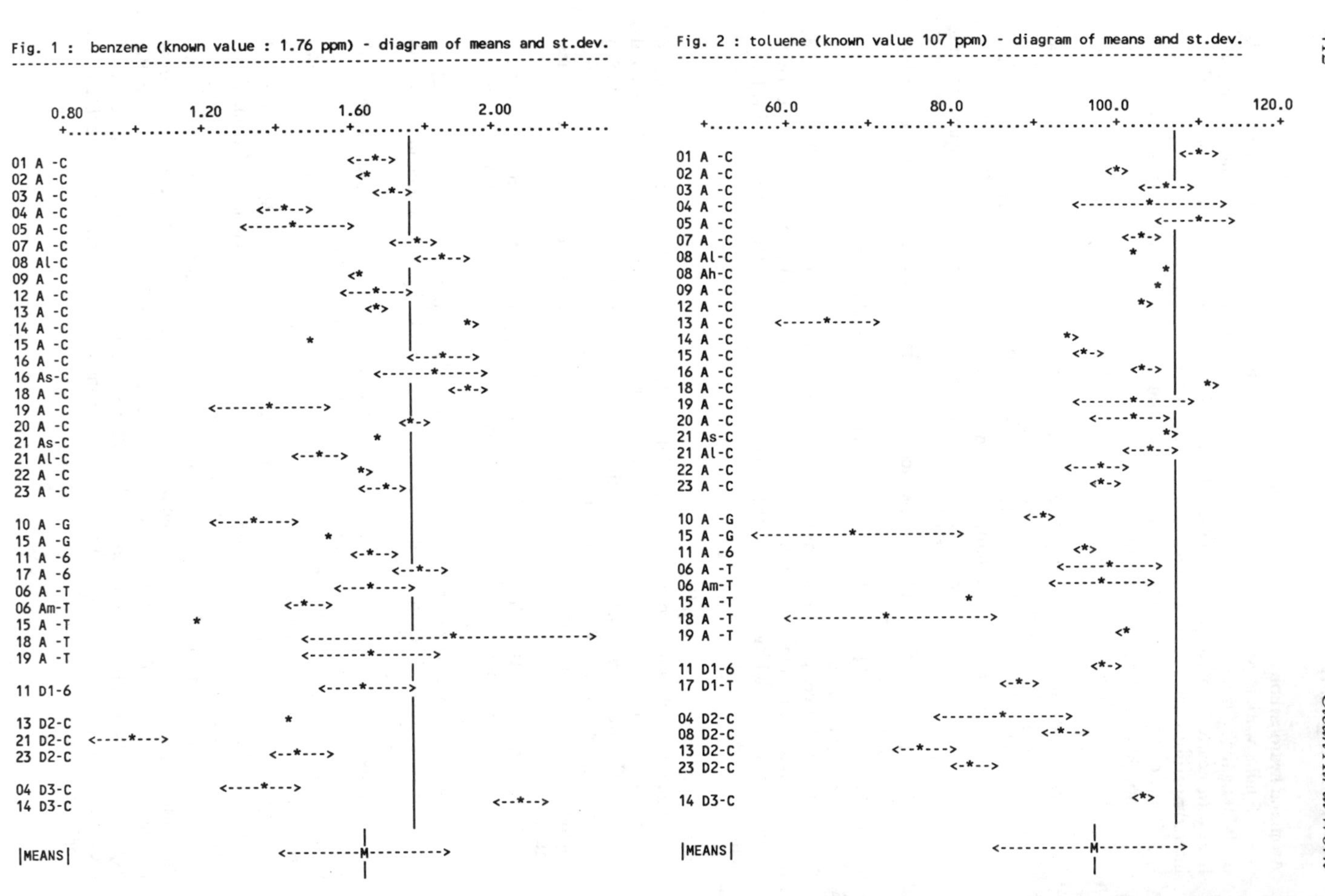

Fig. 1 : benzene (known value : 1.76 ppm) - diagram of means and st.dev.

Fig. 2 : toluene (known value 107 ppm) - diagram of means and st.dev.

Fig. 3 : m-xylene (known value 17.4 ppm) - diagram of means and st.dev.

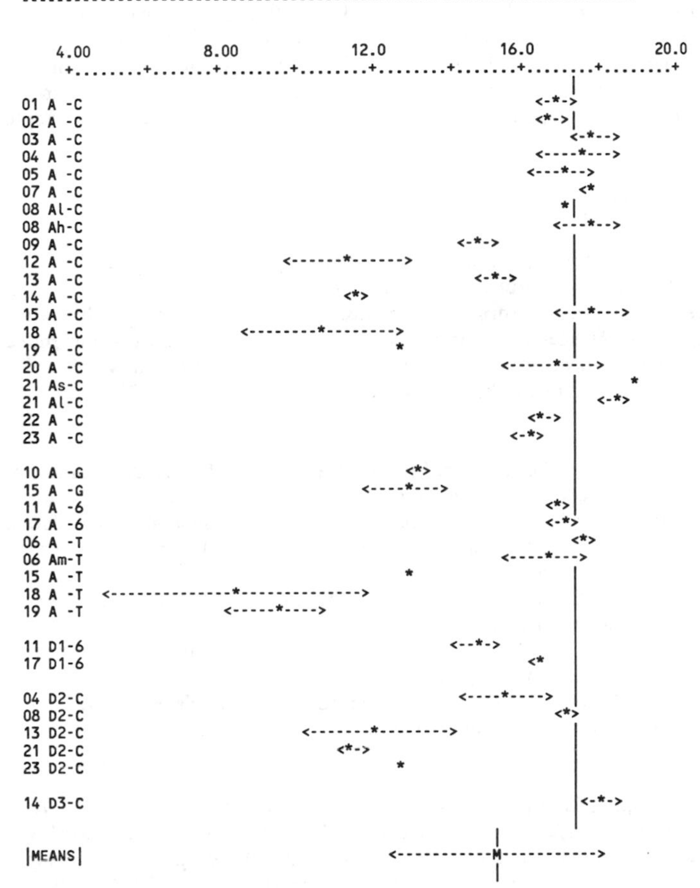

A technical discussion meeting after the exercise revealed that:

- several laboratories had used sampling pumps of which the flow rate had varied by more than 5 % (which will be unacceptable following a CEN standard under preparation);
- one lab (nr 15) had used this occasion to try out thermal desorption; the poor results obtained with this technique may be due to a lack of experience;
- one lab (nr 13) had injected, in some cases (high levels), solutions which were more concentrated than their highest standard solution and was not sure to remain in the linear range of their detector;
- one lab had to correct (by up to 20 %) for a leak in their on-column injection system;
- one lab (nr 21) used for the D2 samplers uptake rates and diffusion coefficients which were different from those recommended by the manufacturer;
- many laboratories had used a simplified method (the phase equilibrium method) to determine the desorption efficiency of the charcoal. This problem is still being studied.

Average errors due to sampling were 9 % for benzene, 24 % for toluene and 18 % for m-xylene. No cases were identified of methods which failed as such to meet CEN criteria. A second exercise is scheduled for November 1991.

Chlorinated hydrocarbons

The same types of work are in progress for chlorinated hydrocarbons. Intercomparisons for thermal desorption have shown larger errors than with aromatic hydrocarbons, apparently because of unexpected difficulties in the calibration. Several sorbents are being tested as it is hoped to include dichloromethane, 1,1,1-trichloroethane, trichloroethene and tetrachloroethene. Intercomparisons for sampling and for solvent desorption are scheduled for September 1991 and mid 1992, respectively.

3. FUTURE WORK

In March 1991, a workshop was held in which 60 experts from governments, independent research institutes and control organisms, consultancies and major chemical companies were invited to inform the Commission on areas where the quality of measurements in the field of workplace air monitoring had to be improved. They recommended in total 24 technical projects; those which received considerable support and which are likely to be accepted within two years are briefly presented.

Aldehydes

The compounds of interest are formaldehyde, acetaldehyde, acrolein and glutaraldehyde, and also acetone. The work will consist of:

a) preparation of batches of charged glass fibre filters simulating samplers used to trap formaldehyde with 2,4-dinitrophenylhydrazine (DNPH); stability tests, intercomparisons, certification of reference materials;
b) preparation of batches of hydrazones with excess DNPH in acetonitrile; stability tests, intercomparisons, certification of reference materials;
c) preparation of pure hydrazones in crystalline form; certification of reference materials;
d) intercomparisons for sampling.

A study of the long term stability of possible reference materials is already in progress; a first intercomparison is planned for the spring of 1992.

Isocyanates

The compounds of interest are 4,4'-diphenylmethane diisocyanate (MDI), 2,4- and 2,6-toluene diisocyanate (TDI) and hexamethylene diisocyanate (HDI). The work will consist of:

a) preparation of batches of dry solid simulating the evaporation residue of the 1-(2-methoxyphenyl)-piperazine (2-MP) solution (ready for HPLC) obtained after the working up of samples; stability tests, intercomparisons, certification of reference materials;
b) preparation of pure 2-MP derivatives in crystalline form; certification of reference materials.

A study of the long term stability of possible reference materials is already in progress; a first intercomparison is planned for mid 1992.

Welding fumes

Intercomparisons for the analysis of welding fumes on filters were proposed. Such intercomparisons will be organised in the framework of an existing BCR project on the determination of Cr species in environmental matrices; the first exercise with filters is planned for the spring of 1992.

Representative sampling of welding fumes is also problematic; therefore research for the development of in-helmet welding fume samplers was requested.

Sampling of polar compounds and of mixed phase aerosols

In situations where polar and non polar compounds are present together (e.g. painting, printing), or where the compounds of interest are present both as vapour and as prepolymer aerosol (e.g. isocyanates), multiple sampling is often applied, which is hardly acceptable because of the uncomfort to the worker. Research was requested to avoid the need for multiple sampling by:

- identifying sorbents which would trap all compounds of interest, and processes to desorb them all from a single sorbent;
- developping samplers which would contain a filter and a sorbent in the same flow of sampled air.

In a first stage, information is being collected and exchanged to evaluate the need for technical research. The next stage may involve studies of sorbents; studies of desorption techniques; development of sampling systems; intercomparisons and reference materials for selected mixed phase compounds and/or for selected polar compounds.

Nitrosoamines

At present, only one practicable measurement technique (based on trapping on and desorption from a patented device, followed by "thermal energy analysis") is available, without any possibility to check its reliability by independent means.
The following actions were therefore requested:

- development of (an) alternative measurement technique(s);
- validation of the current method (which requires either an alternative method or controlled test atmospheres).

Aerodynamic diameter measurements

The development of a reference material consisting of solid, spherical, polydisperse particles was requested. This need may be covered by an existing BCR project in which a range of polydisperse aerosols with widely varying particle sizes and optical properties will be certified. Each powder consists of high density spheres with smooth surface finish and very low tendency to agglomerate; the materials used for their production are glasses, metal oxides, carbon or resins according to the needed optical properties; they are easily dispersed in liquids and in air. Several materials with size range 0.05 or 0.1 μm to 1.0 μm and 0.3 μm to 3 μm (colourless transparent, coloured transparent, white opaque, coloured opaque) are expected to be certified by 1993; materials with size range 0.01 μm to 0.5 μm by 1994.

Sampling of aerosols

The experts consulted are convinced that the currently used techniques for the sampling of inhalable dusts perform poorly and are uncritically applied. Research to improve the sampling methods was therefore strongly requested.

In view of the current development of European standards, it was proposed to start with a pilot study on the application of the new CEN performance protocol to inhalable samplers. The next phase could aim at the development of a simplified system for wind tunnel testing of inhalable dust samplers as mounted on a worker's body.

It was also requested to study techniques to separate coarse dust from the air stream sampled for the determination of fibres.

Various

Other projects which may be considered for the near future include:

- intercomparisons for the infrared characterisation of quartz on filters;
- intercomparisons and reference materials for aromatic amines;
- study of the performance of ozone monitors in harsh environmental conditions; comparison of O_3 standards.

REFERENCES

(1) VANDENDRIESSCHE, S., GRIEPINK, B., HOLLANDER, J.C.Th., GIELEN, J.W.J., LANGELAAN, F.G.G.M., SAUNDERS, K.J. and BROWN, R.H. (1991). Certification of a reference material for aromatic hydrocarbons in Tenax samplers. The Analyst 116, 437-441.

(2) VANDENDRIESSCHE, S. and GRIEPINK, B. (1989). The certification of benzene, toluene and m-xylene sorbed on Tenax in tubes: CRM 112. Report EUR 12308 (Office for Official Publications of the European Communities, Luxembourg).

SURVEY OF QA INITIATIVES IN EUROPE

B. HERVE - BAZIN
Institut National de recherche et de Sécurité (INRS)
Avenue de Bourgogne, BP 27, 54501 - Vandoeuvre (France).

SUMMARY

Quality and quality assurance are receiving more and more attention, especially since the adoption of the international standards in the ISO 9000 and the EN 45000 series. Formal recognition of the competence of laboratories (i.e. accreditation) may not be sufficient to guarantee that results obtained by different laboratories would be the same on identical samples. This is why the laboratory should participate in a programme of measurement audits, proficiency testing or inter-laboratory comparisons specified by the accreditation body, and should make the results available for scrutiny (EN 45001).
This paper presents the main European experiences in proficiency testing, concentrating on the analysis of prepared samples that are generally supposed to represent air samples at the workplace in real conditions. Difficulties are mentioned and suggestions made to progress the harmonization of such schemes.

1. INTRODUCTION

Quality and quality assurance are receiving more and more attention, especially since the adoption of the international standards in the ISO 9000 and the EN 45000 series. These texts define key principles and requirements for accreditation, i.e. "formal recognition that a laboratory is competent to carry out specific calibration or tests or specific types of calibrations or tests" (EN45001). Accreditation relies primarily on a detailed review of laboratory facilities, equipment, staff, methods and organization, which are related to what is called "quality control". However important, it may not be sufficient to guarantee that results obtained by different accredited laboratories on identical samples would be the same or comparable. So "the laboratory should participate in any programme of measurement audits, proficiency testing or inter-laboratory comparison specified by the accreditation body, and should make the results available for scrutiny" (EN 45001).

Proficiency testing is the matter dealt with in this presentation, excluding other possibly related topics such as the use of reference materials or the analysis of biological samples. This paper will concentrate on the analysis of prepared samples that are generally suppposed to represent air samples at workplace in real conditions, and will ignore the questions of sampling techniques and strategies, storage and transportation. The contribution, role and experience gained from the WASP system (UK) will be the subject of another presentation, so will not be dealt with in this paper.

The following general conditions are common to most quality assurance schemes :
- a preliminary study is completed before launching the first round (homogeneity, storage stability, reproducibility);
- voluntary participation, sometimes free of charge;
- the names of the participants are kept anonymous by the laboratory in charge of the organization;
- time schedules must be observed for the return of results;
- each participant receives his results and the overall picture of results including all participants.

2. PROFICIENCY TESTING IN SELECTED EUROPEAN COUNTRIES

SPAIN

Y (a)	Sample kit characteristics	N (b)	N. L. (c)
1985	**PICC-VO** : 4 active charcoal tubes (100 + 50 mg) spiked with 1 - 4 organic solvents in varying quantities and proportions (esters, ketones and chlorinated or aromatic hydrocarbons).	5	20 (1986) 34 (1988)
1987	**PICC-MET** : 4 cellulose esters filters loaded with solutions of Pb (10 - 80 µg/filtre) and Cr (10 - 50 µg/filtre) compounds, dried at room temperature.	4	8 (1987) 27 (1990)
1988 (d)	**PICC-FA** : 2 - 3 slides from sampling asbestos fibers in a representative site.	8 - 10	27 (1990)

Table 1. Programa interlaboratorios de control de calidad (PICC), organized by : Instituto Nacional de Seguridad e Higiene en el Trabajo (INSHT).
(a) Year of first round.
(b) Number of rounds per year.
(c) Number of laboratories participating (year).
(d) Accredited laboratories should be "acceptable" (> 75% of results with no deviation greater than 35% from the reference value).

Details on PICCs (Spain)

- Organic solvents

Organic solvents are spiked onto charcoal tubes with an automatic dispenser. Mixtures are considered as acceptable if storage stability tests, made at room temperature as well as in a refrigerator, provide coefficients of variation not greater than 2% for each pollutant on a 2 months storage duration.

After being organized within the institute in 1985, rounds were opened to external participations from 1986. They were interrupted from 1989 on, to change from in-house prepared tubes to tubes produced commercially.

Processing of results : the arithmetic mean and coefficient of variation are calculated by taking all laboratories' results after excluding results using Winsor's test (coefficient of variation greater than 7%). Results are then considered as acceptable if they are within 3 standard deviations from the mean.

Identical samples are sometimes provided and results presented using Youden's diagrams.

Quality criteria were progressively modified since the first round, taking into account the growing number of participants and the experience gained. The criteria used in 1988 were:
- furnish at least 60% of results (former round included);
- have at least 50% of results less than 1 SD from the mean;
- have at least 80% of results less than 2 SDs from the mean;
- have no more than 5% outliers.

- Asbestos fibres

Circulated slides are first chosen after a test on homogeneity of fibre density on the filters.

The method does not take into account laboratory practice (sample preparation, internal controls, etc). This is considered as secondary in view of counting variability. Results of automated fibre counting using an image analyzer for the samples in each kit (or lot) are sent to the laboratory with the slides. Results must be sent back within 10 to 15 days.

Processing of results : raw data are standardized; the resulting distribution is considered a normal one. Standardized results are considered as acceptable if they are within 35% of the reference value (0,65 - 1,35). Experience shows that this criterion allows the greatest number of acceptable laboratories while keeping an acceptable variability. Beyond this, the number of acceptable laboratories does not increase linearly with accepted variability.

Quality criteria : A laboratory is considered as acceptable if at least 75% of its results are acceptable on 32 consecutive samples (i. e. with at least 24 acceptable results). This is a permanent requirement for accredited laboratories.

Results are presented as raw and standardized data, with the reference value and acceptability limits, inter- and intra- laboratory performance indices.

- Metals on filters

Loadings are chosen to be representative of what could be sampled with exposures nearing the limit value. Metal aqueous solutions are spiked on filters with an automatic despenser (deposit variability : coefficient of variation = 2,54%).

Results must be sent back within 25 to 30 days.

Processing of results : a theoretical coefficient of variation is set at 12%, in view of the observed values of deposit reproducibility and the variability of the analytical method.

If the coefficient of variation is greater than 12%, the Winsor's test is used to replace 2,5% of the extreme (lowest and highest) results by the nearest ones. The arithmetic mean and standard deviation are calculated. Results are acceptable if they are within 3 SDs from the reference value, others being considered as outliers.

Results are presented the usual way. Identical samples are sent occasionally and results presented using Youden's diagrams.

FRANCE

Y (a)	Characteristics of a round	N (b)	N. L. (c)
1988(d)	Lead analysis. 4 filters loaded with a standard solution, 2 with dust from ironworks, 2 with ceramic dusts; 3 blanks.	1	42 (1988) 11 (1990)
1979 (e)	21 slides (asbestos fibres).	0,5	39 (1990)
1990	Counting filters by transmission electron microscopy.	1	3
1987 (d)	Sampling 3 active charcoal tubes in a controlled atmosphere (benzene, saturated hydrocarbons, water vapour)	1	35 (1987) 24 (1990)
1979	Active charcoal tubes with different mixtures each year.	1	8 (f)
1990	PAH analysis : - 6 or 12 in solutions (testing columns or dosing 7); - in charcoal (dosing 7).		8 (f)

Table 2. Proficiency tests organized by : Institut national de recherche et de sécurité (INRS), avenue de Bourgogne, BP 27, 54501 VANDOEUVRE CEDEX (FRANCE).

(a) Year of first round.

(b) Number of rounds per year.

(c) Number of participants (year).

(d) organized for the ministery of work, who homologates or not the laboratories depending on their results.

(e) Applying common counting rules. Organized for the Ministry of Labour. After each round, 3 groups are defined according to results. The first two groups are approved, the third is not.

(f) Quality assurance organized with the regional laboratories from the regional insurance bodies. "PAH" stands for polycyclic aromatic hydrocarbons.

Details on some quality assurance schemes (France)

- Lead

Randomly selected concentrations of lead are prepared on filters in ranges from 9 to 37 $\mu g.cm^{-2}$ for standard solutions, from 70 to 291 $\mu g.cm^{-2}$ for dusts from ironworks, and from 12 to 52 $\mu g.cm^{-2}$ for ceramic dusts (lead silicates). Each laboratory may use its own analytical method.

Processing of results

If x is the theoretical value (known from sample preparation) and y the analytical result (each in $\mu g.m^{-3}$ for a supposed 240 litre air-sample), a "maximum permissible deviation" D is calculated for each filter : $D = a.x + b$, in $\mu g.m^{-3}$, with :

$a = 0$ and $b = 25$ if $x < 100\ \mu g.m^{-3}$,

$a = 0{,}25$ and $b = 0$ if $x > 100\ \mu g.m^{-3}$.

If the absolute value of the difference (y - x) is greater than D, the laboratory is rejected. If not, a "result index" RI is calculated for each filter :

$RI = ABS\ (y - x).25 / D$, and a laboratory index, LI :

$LI = \Sigma\ RI.10 / n$, where n is the number of analyzed filters.

Results are considered as acceptable if LI < 100. A comparison was made with the WASP scheme and showed the latter is more severe (in a practical case, the WASP way of calculation eliminates 5 laboratories out of 16 acceptable according to the RI value; other laboratories were identically classified). A simulation study showed that lead concentration has little influence on a laboratory's performance.

- Asbestos fibre counting

Counting results must be sent back within 15 days. The n = 21 results L'_i are divided by the mean L_i of counts of the reference group, and the mean L' calculated ($L' = \Sigma\ L'_i/(n.L_i)$). Counters are classified in 3 groups depending on their mean L' and coefficient of variation of their standardized results, CV. Group III is defined by $L' < 0{,}5$ or $L' > 2$, or CV > 40%, and calculations remade if group I composition has to be modified; group II by $0{,}50 < L' < 0{,}75$ or $1{,}33 < L' < 2$. Group I, $0{,}75 < L' < 1{,}33$, becomes (or remains) the reference group.

Differences with the RICE proficiency scheme (U.K.) were studied, but showed little influence on laboratory classification, excepting when a high CV appears, since this parameter is not taken into account by RICE.

- Mixture containing benzene

This scheme is generally used by the French Ministry of Labour, to give, depending on the results, authorization for internal controls or agreement (allowing controls to be made in other sites).

Benzene < 15 ppm, saturated hydrocarbons ca. 200 ppm, relative humidity around 50%.

6 tubes are sampled (3 by INRS) in parallel on 8 hours, 2 are kept by the laboratory for a later analysis, 3 are analyzed by INRS.

30 participants are expected for 1991.

- Exchanges with B.I.A.

In 1990, INRS sent to B.I.A. charcoal tubes containing bezene for comparison, and received from B.I.A. 3 tubes spiked with 1,1,1-trichloroethane, trichloroethylene and tetrachloroethylene. INRS received also 500 mg of dust to analyze As, Ni and Pb. Methods of chemical digestion and analysis were indicated but not imposed.

GERMANY

Y (a)	Characteristics of a round	N (b)	N. L. (c)
1989	3 active charcoal Dräger type B tubes spiked with different amounts of tetra-chloroethylene (89), 1,1,1-trichloro-ethane and trichloroethylene (90).	1	50 (1989) 67 (1990)
1990 (d)	500 mg dust samples containing As, Ni, Pb.	1	61
1987 (e)	14 (87/88) or 20 (89/90) slides (asbestos fibres).	0,5	12

Table 3. Proficiency tests organized by : Berufsgenossenschaftliches Institut für Arbeitssicherheit (B I A), Alte Heerstrasse 111, Postfach 2043, 5205 Sankt Augustin 2.
(a) Year of first round.
(b) Number of rounds per year.
(c) Number of participating laboratories.
(d) Analytical guidelines are given (chemical digestion method, atomic absorption), but not imposed. Results are to be given in mg/filter. The first round started in the end of 1990.
(e) Applying standardized counting rules.

Details on some quality assurance schemes (Germany)

Active charcoal tubes

(1989) 3 charcoal tubes (2 sections) were loaded with tetrachloroethylene pumped at 0,2 l/min from a standard atmosphere (Wösthoff pump). Concentration and essential parameters were continuously recorded; tubes were discarded if concentration was more than 5% from its mean value. Samples included blanks (new tubes, opened and not yet opened) and a tube loaded with a known (and documented) quantity. Analytical guidelines were given: CS_2 desorption, gas chromatographic analysis with a flame ionisation detector. The NIOSH method was recommended but not imposed, and was followed by more than 50% of the laboratories. Future rounds will include esters and ketones.

Processing of results : for each of the 3 concentration levels, the arithmetic mean of all results was calculated (in a case, 2 tubes only have been analysed) and was taken as the reference value for this level. Outliers were then determined according to Grubb's test at the confidence level of 99%. 9 results were eliminated out of a total of 149. Arithmetic means, standard deviations and coefficients of variation were calculated without these outliers. For the analysis of a given desorption solution, repeatability was better than 7%; reproducibility between 2 laboratories was better than 40% in 95% of cases. This allowed repeatability between 2 separate identical samples to be estimated at around 30%, probably reflecting variations in handling and desorption methods.

Individual results, including outliers if any, were standardized (i.e. divided by the reference value) and performance indices calculated to assess laboratory performance. The latter are identical with those of the "WASP" proficiency scheme. Three categories were defined : 11 laboratories were in category I (PI < 16), 17 in category II (16 < PI < 65), and the others in category III (PI > 65). A detailed study indicated that splitless injection, desorption with CS_2 and the flame ionisation detector seem preferable to other techniques. Also, outliers were characterized as those presenting more than 18% deviation from the reference value, after standardization.

(1990) For the 1990 round, samples were a mixture of tri- and tetra-chloroethylenes and 1,1,1-trichloroethane. Applying the "not more than 18% deviation" rule, 64 results were judged as outliers on a total of 600. A detailed study of the results : i) does not contradict the preceding conclusions; ii) shows that reproducibility has become better than 23% in 95% of the cases; and iii) indicates that internal is preferable to external standardization. More rounds are needed to confirm these indications.

Fiber counting

Results must be given in fibers.mm^{-2}. Two blank filters are provided ($0 < n < 6$ f.mm^{-2}). Results are standardized; outliers are detected according to Nalimov's test. Inter and intra-laboratory indices are calculated. The first gives the difference between the lab mean value and the reference value (in percentage points), the second gives the coefficient of variation of standardized results. The percentage of results in the interval 0,70 - 1,70 is also indicated (according to the "RICE" proficiency tests, it should be at least 75%).

BELGIUM AND THE NETHERLANDS

The Belgian Ministry of Labour has organised a quality assurance scheme for counting asbestos fibres since 1988. The adopted rules are comparable to those of the RICE and AFRICA schemes. The arithmetic count of at least 15 counts for the same sample, after excluding outliers (based on Dixon's test) is taken as the reference value. Eight samples are exchanged within 3 groups of 5 – 7 laboratories; samples are then changed from one group to the other.

The laboratories taking part are so called "recognised laboratories" (by the Ministry of Labour) and those from industries using asbestos. This participation is not (yet) regulatory. Employers and employees wishing to cooperate with a laboratory are encouraged to ask for its results in the scheme.

A quality assurance scheme for asbestos fibre counting has also been set up by TNO on governmental subsidies . This organization tried to pick up the best part of each method (RICE/AFRICA, U K - INRS, France). Reproducibility is the only parameter used for classification of the laboratories. The mean of all counts from a reference sample is taken as the best estimation of the "true value". Image analysis systems (like Magiscan) were found to produce substantial systematical errors and are not used for calibration purposes. 8 "real life" reference samples (mostly low concentrations) are sent to the labs every 3 months.

3. Questions and possible perspectives

3.1. Some questions

A) The application of specific analytical method for proficiency testing.

In general, the analytical method is not imposed. A laboratory uses his usual method. When BCR organized an intercomparison, it obtained far better results by allowing each laboratory to use their best methodology than in ISO intercomparisons where the method was specified.

In some cases, result depends strongly on the method; a detailed reference method should then be applied by all laboratories (e.g. fibre counting).

Some labs take part in too many rounds; part of the staff is sometimes specialized to make only the tests for proficiency. This clearly does not give an exact image of the laboratory performance.

B) National laboratory approval schemes

Some laboratories are approved by ministeries or independent associations, according to criteria differing from those for accreditation or proficiency testing (France, Spain).

This is gradually changing in France, although there are regulatory difficulties in relation with a role officially attributed to "social partners".

Accreditation practices differ within member states (see ILAC, WELAC and others), although some arrangements for mutual recognition have been established (e.g. between RNE - F - and NAMAS - U.K.).

C) Roles and coordination of international organizations.

Perhaps too many organizations deal with similar subjects with insufficient coordination.

D) Participation fees and difficulties with intercomparisons with many potential participants.

This point was developed by Dr. Vandendriessche (BCR).

E) Towards the harmonization of sampling methods in workplace air.

Occupational hygiene comparisons between member states cannot be valid if sampling methods differ. Sufficient harmonization does not seem guaranteed by the work of CEN TC 137 or other bodies. Shall we wait until each country defines his own "good sampling practice" before trying to harmonize them?

3.2. Some suggestions in order to make progress

A) Cooperation between national institutes

Cooperation between national institutes already exists to a certain extent : HSE (U.K.), B I A (Germany), and INRS (F) have exchanges of information (research programmes, on-going studies). BIA and INRS participate to the WASP scheme and send each other samples (spiked charcoal tubes, dust).

Exchanges should be maintained at all levels at determined intervals and could be placed in the frame of mutually agreed upon programmes after determining priorities, level of work acceptable, and aims to be pursued.

B) Reflexions on the themes "by pollutant" or "by kind of method".

3 main types may be distinguished as being most generally performed : analyzing organic compounds adsorbed on a solid sorbent, or metals on filters, or counting fibers. If, for example, we examine the one dealing with organic solvents, we see some common points; e.g. the chemicals used are aromatic hydrocarbons (benzene, toluene, sometimes xylenes) and/or chlorinated solvents (trichloroethylene, 1,1,1-trichloroethane, tetrachloroethylene); common esters and ketones seem a generally accepted posibility too.

Analyses are performed using gas chromatography, generally using a flame ionization detector.

Subject matter :	Organic compounds on a solid sorbent	Metals	Fiber counting
Possible choices	- active charcoal tubes; - spiked with a syringe; - rules for processing results .	- On filters (or not?); - chosing elements and compounds (Cr, Ni, Pb); - Atomic absorption spectrometry; - Rules for processing results.	- Detailed counting rules; - rules for processing results; - sampling techniques.
In all cases, pay specific attention to concrete conditions for homologation and their links with accreditation.			

Table 4. Examples of possible orientations for the technical harmonization of schemes.

Do we really have interest in trying to harmonize as closely as possible the technical aspects of such proficiency schemes? It seems difficult to impose analytical methods and equipments on every participating lab, excepting when they have such an influence on the results, that it cannot be avoided (e.g. fiber counting by optical microscopy). There is no interest or mandatory reason in harmonizing sample preparation by organizing labs.

So this "vertical harmonization" should not necessarily be a short-term goal. Perhaps a way to make progress would be to recommend the adoption and application of an international standard. The processing of results should then show if labs applying the standard perform generally better than others.

C) "Horizontal harmonization"

Perhaps a more promising way would be to propose a "horizontal harmonization", e.g. on processing, interpreting and presenting results. This is the way proposed by a UK working group in a discussion document presented at the IUPAC/ISO/AOAC harmonisation meeting in Geneva in May this year.

D) Different participation levels depending on laboratory and country?

An international level for the main institutes in the member states, a national level and possibly a regional level for large size countries?

REFERENCES

(1) D. BREUER et al. (1990). Ringversuch Tetrachlorethen. Qualitätssicherung im Rahmen des Erfahrungsaustausches ausserbetrieblicher Messtellen. Staub, 50, 203-209.
(2) Procedure for multicentre trials. EAC for offsite measuring stations. State of discussion after first multicentre trial. Information for participants. HSE translation n° 13883b, not for publication. (Durchführung von Ringversuchen EAK ausserbetrieblicher Messstellen. Stand des Diskussion nach dem 1. Ringversuch. Informationen für die Teilnehmer).
(3) Leaflet "Programa interlaboratorios de control de calidad", Instituto nacional de seguridad e higiene en el trabajo (INSHT).
(4) WASP - Workplace analysis scheme for proficiency. Information for participants. HSE committee on analytical requirements, May 1988,17 pages.
(5) J. N. TEJEDOR (1990). Programa interlaboratorios de control de calidad para vapores organicos. Internal document of INSHT (Madrid).
(6) C. A. BUEZO (1990). Programa interlaboratorios de control de calidad de fibras de amianto. Internal document of INSHT (Vizcaya).
(7) J. B. SANCHEZ (1990). Programa interlaboratorios de control de calidad para analisis de metales en aire. Internal document of INSHT (Barcelona).
(8) E. KAUFFER (1989). The French asbestos counting quality control scheme. Proceedings of the AIA 6th international colloquium on dust measurement technique and strategy, Jersey, Channel Islands, p. 156-163.

Acknowledgements
This work was made possible owing to the help of Drs. Breuer (BIA, Germany), Firth (HSE, UK), Quintana (INSHT, Spain), Saunders (HSE, UK), and Vandendriessche (BCR, EEC).

Annex: The Danish external quality assurance scheme (DEQAS)

The Danish National Institute of Occupational Health is the Danish reference laboratory for workplace investigations on occupational exposure. The Institute is responsible for the external quality assessment of laboratories measuring concentrations of e.g. toxic metals in human biological samples or concentrations of toxic compounds in air samples.

For atmosphere controls in occupational settings, DEQAS offers samples of :
- metals in welding fume. A specially designed air sampler exposes 100 cellulose nitrate filters (25 mm dia., pore size 8 μm) simultaneously around the threshold limit values. The metals are iron, manganese, and titanium.
- organic solvents. Charcoal tubes are exposed to 5 different concentrations of 3 different organic solvents (amongst a choice of 15) with a gas generator.

5 samples are provided for analysis, with 5 concentation levels. Samples for measuring blind values are delivered. Samples for recovery are delivered with the solvents. 2 rounds are performed each year.
A tolerance interval TI is calculated on the basis of an appointed standard deviation for the analytical method e.g. $TI = x \pm (ax/100)k$, where $a = 15$ and k is a constant dependent on the number of participating laboratories. Measurements out of the TI are stated as outliers. The estimated true value is the arithmetic mean of all results after elimination of outliers.

The scheme is designed to provide the laboratories an opportunity to locate systematic errors and an evaluation of the analytical method.

There is a fee for participating.

1439

EXPERIENCE WITH THE WORKPLACE ANALYSIS SCHEME FOR PROFICIENCY (WASP)

H.M. JACKSON AND N.G. WEST

Occupational Medicine and Hygiene Laboratories
Health and Safety Executive
403, Edgware Road
London NW2 6LN, UK

SUMMARY

The Workplace Analysis Scheme for Proficiency (WASP) was set up in 1988 as a voluntary scheme designed to test the proficiency of laboratories undertaking occupational hygiene analyses. Performance is assessed in terms of a quantitative performance index which is a function of the total error of a participants results. Performance targets have been set for each analyte on the basis of what is reasonably achievable by a typical laboratory taking account of the accuracy needed by the end-user of the results. The work of CEN Technical Committee 137 Working Group 2 'General Requirements for Measurement Procedures' is of particular relevance here. Results to date have shown that a significant minority of laboratories (~30%) perform below the standard which is being sought. The main reason for this is a lack of consistency from round to round which reflects poor internal quality control procedures. There is an emerging consensus that proficiency testing is the single most effective means of establishing mutual confidence between analytical laboratories and to facilitate comparison of performance a draft protocol aimed at harmonising proficiency testing schemes has been prepared by ISO/IUPAC. In the European context there is now a need to establish a forum to discuss the various initiatives in proficiency testing with the aim of encouraging collaboration and harmonisation.

1. INTRODUCTION

In the field of occupational hygiene the control of workers' exposure to hazardous substances is of major and growing concern. Legislation has been introduced in many countries to address this problem, for example, in Great Britain the Control of Substances Hazardous to Health (1988) Regulations (COSHH) came into force in October 1989. These regulations lay a duty on employers to prevent or control exposure to hazardous substances. Most toxic substances present a hazard through inhalation and the assessment of the risk or the demonstration of adequate control may require the measurement of airborne concentrations. The results of such measurements are used to make decisions and, if the measurements are in error, wrong decisions can be made. Workers' health may be put at risk or the employer may be burdened with unnecessary expenditure on control measures.

© British Crown Copyright 1992

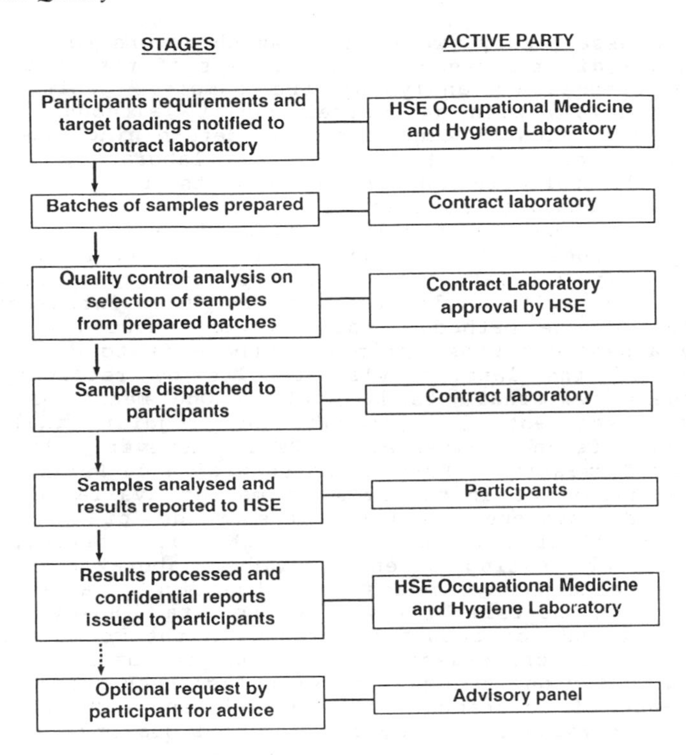

Figure 1: Outline quarterly cycle of scheme in operation.

The variability of replicate samples must be well below the typical analytical variability for the assessment of analytical performance to have any meaning. The types of sample presently available in the scheme are listed in Table 1 and were chosen on the advice of occupational hygienists on the Steering Committee. For the aromatic hydrocarbons, samples prepared on both charcoal and Tenax media are available to mimic pumped and diffusive sampling respectively. The analyte loading ranges are intended to reflect the levels that would be collected during sampling of workplace air at 0.2 to 2.0 times the relevant occupational exposure limit (6). The processing of results and the issuing of reports is done by HSE rather than the contract laboratory, to avoid any suggestion of conflict of interest or abuse of confidential information.

There are essentially two factors which govern the quality of a workplace air measurement, the process of taking the sample and the process of analysing the sample. This paper is concerned solely with analytical quality although errors in sampling are equally important in determining the overall measurement accuracy (1) (2). For a laboratory to produce consistently reliable analytical results it must implement appropriate quality assurance procedures (3). Analytical methods must be thoroughly validated and staff trained properly in their use. Internal quality control is essential to ensure that the laboratory produces consistent results from day to day, within limits set by the intra-laboratory precision of the method. This is complemented by external quality assessment whose primary purpose is to determine the accuracy of the laboratory's, results by reference to an independent, external standard (4). One means of external quality assessment is through the regular analysis of Certified Reference Materials (CRM). However, although the Europeans Commission's Bureau Communautaire de Reference (BCR) has made important initiatives in making available CRMs for occupational hygiene analyses, the range of materials is limited and their cost relatively high (5). Participation in a proficiency testing scheme provides laboratories with an effective alternative means of objectively assessing and demonstrating the reliability of their data. However, because of the multitude of substances found in the workplace only a representative cross-section of analytes/matrices can be included in any one scheme. Nevertheless such schemes cover a wider range of substances than do existing CRMs and overall offer better value for money for external quality assessment.

Concern about the quality of occupational hygiene analyses led the Health and Safety Executive (HSE) to set up the Workplace Analysis Scheme for Proficiency (WASP) in 1988. It is a voluntary scheme open to any laboratory and financed through modest fees charged to participants. A small Steering Committee having representatives from various professional bodies including the Royal Society of Chemistry and the British Occupational Hygiene Society oversees the running of the scheme and ensures that participants views are taken into account. The primary purpose of the scheme is to provide participants with an objective measure of their analytical performance and to encourage poor performers to improve. A secondary purpose is to inform policy decisions made by HSE in relation to the COSHH Regulations.

OPERATION

The scheme operates to a quarterly timetable and the various stages in each round are set out in Figure 1. For each analyte four samples are provided having a range of concentration typical of that for 'real' samples. Preparation of batches of samples is undertaken by a contract laboratory and subject to strict quality control by HSE to ensure that, for any sample type, participants receive samples having as near as possible identical compositions.

Analyte	Medium	Loading Mass Range (μg)	RPI Reference Value	Performance Category Limits
Lead	glass-fibre	15 - 150	36	16, 65
Cadmium	or membrane	5 - 50	36	16, 65
Chromium	filter	50 - 500	36	16, 65
Benzene	charcoal	40 - 400	81	35,146
Toluene	charcoal	500 - 5000	36	16, 65
m-Xylene	charcoal	500 - 5000	36	16, 65
1,1,1-Trichloroethane	charcoal	2500 - 12000	36	16, 65
Trichloroethene	charcoal	700 - 7000	36	16, 65
Tetrachloroethene	charcoal	400 - 4000	36	16, 65
Benzene	Tenax	2 - 15	144	62,259
Toluene	Tenax	20 - 200	36	16, 65
m-Xylene	Tenax	20 - 200	36	16, 65

Table 1: Range of analytes and samples, with performance criteria. See text for an explanation of the terms RPI Reference Value and Performance Category Limits.

The statistical protocol used to calculate a quantitative performance index (PI) for each analyte is presented schematically in Figure 2. A critical part of the protocol is the estimation of the 'true' result for each sample. There are a number of ways by which this can be done but for relatively straightforward samples of the type in the WASP scheme the preferred option is to base it on the mean of the participants' results (3), (7). However, it is necessary to exclude outlier results which might unduly bias this consensus mean. To facilitate the processing of results from four samples having different loadings, each of the participant's four results is divided by the corresponding 'true' result. This ratio is termed the standardised result and should lie close to one if the analytical performance is good. The four standarised results are then combined as shown in Figure 2 to to give a PI for that round. The PI is a measure of the total variance of the standarised results about unity; the factor 10000 is introduced to avoid small numbers. In practise, the PI is the square of the coefficient of variation of the results, for example, a PI of 36 represents a coefficient of variation of 6%. Typical performance indices lie in the range 0 to 200, the lower the figure the better the performance.

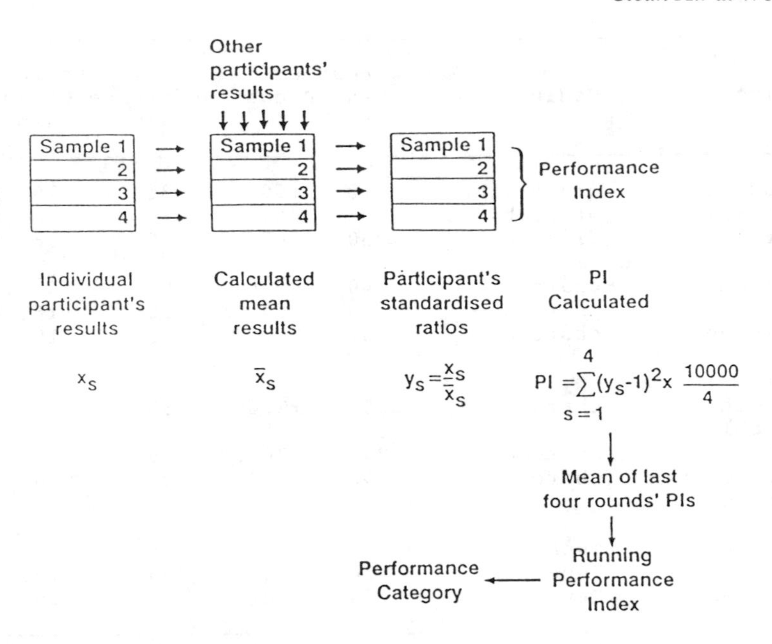

Figure 2: Results processing scheme for each analyte.

The calculation of a PI leads naturally to consideration of the quality of that performance. Categorisation of performance is based on a running performance index (RPI) which is the mean of the last four rounds' PIs. This provides a more reliable indicator of performance since it is based on sixteen rather than just four results. The performance categories are defined with respect to a <u>reference</u> or <u>target</u> performance level and three categories are recognised; category 1 which is significantly better (lower) than the reference RPI, category 2 around the reference RPI and category 3 significantly worse (higher) than the reference RPI (7). The reference RPIs and the performance category limits for the various analytes are listed in Table 1. The reference RPIs were set originally on the basis of performance data from existing schemes and the experience of BCR in their programme developing CRMs. Detailed reports are sent to participants after each round and poor performers have the option of seeking advise from a panel of analysts having extensive experience of occupational hygiene analysis.

3. EXPERIENCE TO DATE

Three years after its launch, clear patterns are now emerging in the way the scheme is developing. Table 2 shows the numbers and types of participants in round 1 (September 1988) and a recent round (11) in April 1991. Both the initial numbers in the scheme and the growth in those numbers from 110 to 161 have been most encouraging.

The scheme is by far the largest of its type outside the USA and an increasing proportion of laboratories are from outside the UK notably Germany with 11 participants. The distribution of types of laboratories has remained fairly constant over the years with nearly half associated with industry. This high proportion from industry is particularly welcome given the intention of the scheme to help employers make accurate measurements under the COSHH Regulations.

Numbers of participants	Round 1	Round 11
Total	110	161
UK	103 (94%)	130 (81%)
Other EC	4 (4%)	18 (11%)
Outside EC	3 (3%)	13 (8%)
Types of participants		
Industrial	51 (46%)	74 (46%)
Central Government	23 (21%)	33 (20%)
Local Government	15 (14%)	17 (11%)
Educational	5 (5%)	12 (7%)
Consultant, Contract & Research Organisations	16 (15%)	25 (16%)

Table 2: Numbers and types of participants

Regarding the performance of participants, the proportion of PIs and RPIs in each of the three categories are shown in Table 3. After a very poor start with half the PI scores in category 3, a very significant improvement is evident by round 11. However, performance categories are formally based on the RPIs and here the picture is not nearly so encouraging. There has been only modest improvement since RPIs were first calculated in round 4 and over 30% of scores are still in category 3 and less than 20% in category 1. In analysing these trends, allowance must be made for the fact that the population of laboratories is not constant and that as a voluntary scheme, there are no real penalties on poor performers. Even so, it would have been hoped that the proportion in category 3 would have fallen further by now. A closer examination of the performance data points to the fact that the main failing of many laboratories is a lack of consistency. They are able to achieve a good performance in a single round but are unable to sustain it over the four rounds which contribute to an RPI.

This is demonstrated clearly in the disparity between the percentage of PIs and RPIs in category 1 for round 11. The obvious conclusion to be drawn is that many laboratories do not have adequate internal quality control procedures in place.

Round	Category 1	Category 2	Category 3	Total
PIs - percentages in each category range				
1	22	29	49	487
2	32	30	38	569
3	32	29	39	684
4	32	28	40	714
5	32	33	35	740
6	37	27	36	738
7	40	28	32	731
8	31	30	38	898
9	35	30	35	910
10	37	31	32	931
11	42	28	30	908
RPIs - actual categories				
4	12	46	42	383
5	15	41	44	524
6	14	44	42	567
7	18	50	31	499
8	16	48	36	573
9	18	46	36	578
10	15	47	38	699
11	17	50	33	739

Table 3: Categorisation of performance for all analytes.

4. DISCUSSION

In discussing the performance of laboratories in the scheme, the main concern of both participants and organisers is the continuing high proportion of RPIs in category 3. It should be noted that HSE regards category 3 as an unacceptably low standard of performance for its own laboratories. The question to arise is whether the performance standard is reasonable or has it been set too high? There are two ways to approach the setting of criteria for acceptability. The first is to ask whether a particular standard is routinely achievable by laboratories as a whole, assuming some norm for quality of equipment, staff etc. The second approach is to ask what standard of analysis is actually needed by the end-user of the data to ensure that correct decisions are made.

On the question of practicability, it has already been stated that the reference RPIs have been based on performance in other schemes and on experience in the BCR. It is also the case that a significant proportion of RPIs are in category 1 indicating that some laboratories are able to perform at a level <u>consistently better</u> than the reference RPI. This strongly supports the view that the standard being sought is not unrealistic for a laboratory with proper quality assurance procedures.

On the second issue of whether the analytical standard is 'fit for the purpose' this should be viewed in the context of the work of CEN Technical Committee 137 Working Group 2. 'General Requirements for Measurement Procedures'. For compliance monitoring at 0.5 to 2.0 times an exposure limit, the Working Group recommends that the 'relative overall uncertainty' of the total measurement procedure should not be greater than 30%. The 'relative overall uncertainty' is defined as the bias (expressed as a percentage) plus twice the coefficient of variation (CV) of six or more measurements. In the optimum case of no bias this implies a maximum CV of 15% to cover both sampling and analysis. Now in general it is considered much easier to control analytical error than sampling error but if it is assumed that they are divided equally then the maximum allowable analytical CV would be 11%. For most analytes the WASP scheme requires a rather better performance (CV less than 8%) to avoid a category 3 rating. However, it is important to recognise that performance in the WASP scheme is likely to be better than that on samples passing routinely through a laboratory. Apart from the fact that they are known to be for proficiency testing purposes, WASP samples are also of the simplest possible type with no dissolution/desorption problems or unknown interfering substances.

Thus if the setting of the performance standard is considered from the viewpoint of either achievability or 'fitness for purpose', the present standards set in the WASP scheme are seen to be reasonable.

5. EUROPEAN COLLABORATION

Experience with the WASP scheme has shown the need for proficiency testing in hygiene analysis to enable poor performers to be identified and encouraged to improve. Similar initiatives have been taken in other European countries including Germany, Denmark, France and Spain. There is now a need to consider how to encourage harmonisation and collaboration between the various national schemes as part of the drive towards mutual recognition of results from laboratories across the European Community. Formally this will be done through agreement between national accreditation bodies but it is anticipated that accreditation assessments will increasingly utilise objective information on analytical performance produced from proficiency testing (8).

The way forward now is for a discussion forum to be set up preferably under the auspicies of either the BCR or Directive General V. This would enable the national experts to exchange information and make proposals to harmonise the existing schemes especially with respect to the development of a single statistical protocol.

6. REFERENCES

(1) HSE (1981-91) Methods for the determination of hazardous substances (MDHS) in series, Health and Safety Executive, Bootle, UK.

(2) HSE (1989) Guidance Note EH42, Monitoring strategies for toxic substances, HMSO, London.

(3) ISO/IUPAC (1991) Draft harmonised proficiency testing protocol, International Standards Organisation, Geneva.

(4) HSE (1991) Analytical quality in workplace and monitoring, MDHS 71, Health and Safety Executive, Bootle, UK.

(5) Vandendriessche, S. et al. (1991) Certification of a reference material for aromatic hydrocarbons in Tenax samplers. The Analyst 116, 437-441.

(6) HSE (1991) Occupational Exposure Limits 1991, HMSO, London.

(7) Jackson, H.M. and West, N.G. (Initial experience with the Workplace Analysis Scheme for Proficiency (WASP), Ann. Occ. Hyg. In press.

(8) NAMAS (1990) Information sheet 45, Accrediation for chemical laboratories. National Measurement Accrediation Service, National Physical Laboratory, Teddington, UK.

LABORATORY ACCREDITATION FOR WORKPLACE AIR MONITORING

J.G. FIRTH

Occupational Medicine and Hygiene Laboratory
Health and Safety Executive
403-405 Edgware Road, London NW2 6LN, UK

SUMMARY

Accreditation is a process in which the internal structures and procedures for measurement of a laboratory are assessed. A judgement is made as to whether or not they are of such a standard that the results produced by the laboratory can be fully relied upon. This paper describes the functions in the laboratory which are assessed and the directions in which accreditation is developing.

1. INTRODUCTION

The taking of measurements of concentrations of substances in workplace air is a process which like any other measurement process provides information. Information is all that a measurement provides. This information is then used to make decisions, e.g whether or not to install control systems, whether or not to use protective equipment or whether or not the workplace environment is satisfactory. It is the actions following these decisions which have serious implications for costs and for the health of employees. In some instances, and probably increasingly so in the future, measurements will be the main factor in making such decisions. It is therefore essential that the measurement results are reliable. It is the economic and social consequences of incorrect decisions made on the basis of incorrect measurements which are the main driving force for ensuring that the results of measurements are reliable and appropriate for their purpose.

A second factor in the drive for reliability in measurement is the professional credibility of the scientists carrying out the measurement. The days when the public or the courts accepted results from laboratory measurements without question are rapidly disappearing. The processes used in processing and recording samples and results will increasingly be questioned and when doubt can be cast on any part of the procedures used so that the reliability of the results can be doubted then, as in recent cases related to forensic work in the UK, the professional credibility of the scientists in question is severely damaged with 'knock-on' effects for other related areas of measurement. It is therefore increasingly essential that the procedures used in obtaining measurements can be shown to be reliable both to customers for the measurements and to a wider public.

Accreditation is intended to provide a laboratory and its clients with confidence in both the ability of the laboratory to carry out a range of testing and/or measurement applications and in the technical and commercial integrity of those operations. This is done by a review of the operations of the laboratory by assessors employed by a recognised Accrediting Authority. This review is carried out against a framework of requirements set by the Accrediting Authority. These requirements are based on those set by agreed international standards and guides – in particular the EN 45000 series of standards – 'General Criteria for the Operation of Testing Laboratories' and ISO Guide 25 – 'General Requirements for Technical Competence of Testing Laboratories'.

Accreditation shows that a laboratory produces results which can be relied upon to meet certain specified standards of performance. This gives confidence to customers that a quality result can be produced by the laboratory and also enables a laboratory to demonstrate, in relation to its competitors, that it is capable of producing reliable results. In the United Kingdom the main accreditation body is the National Measurement and Accreditation Service (NAMAS) and the aim is that accreditation of laboratories for all types of measurements in the UK will eventually rest with this one organisation. This will ensure a uniform standard of performance amongst the accredited laboratories and consistency in the degree of confidence that customers will be able to place on the results produced by these laboratories. The aim is then that the customers will be able to accept that, across a wide range of activities, results produced by a NAMAS Accredited laboratory are to the standard required by the customer. In addition, with mutual recognition of accreditation schemes with other countries, then test reports produced by accredited laboratories in one country will be acceptable in other countries and this development will reduce hidden barriers to trade.

The criteria for assessing laboratories for particular types of test, e.g. chemical testing, electrical testing, testing of personal protective equipment, etc, are developed by advisory committees composed of experts in those particular fields. The aim of these criteria is always to set a sensible and acceptable requirement for performance rather than looking for excessively and un-needed high standards. It is also essential that criteria are kept as simple as possible so that unnecessary burdens are not put on laboratories which are seeking accreditation. Nevertheless a range of activities in all laboratories seeking accreditation for whatever type of test has to be examined.

2. CRITERIA

Similar aspects of a laboratory structure and operation are assessed for accreditation for any type of test. These are:

Organisation and Management

It is essential that all staff in the laboratory are aware of the extent and limitations of their responsibility. In general a written job description and written specifications of the qualifications, training and experience necessary for a particular job are set down for key managerial and technical staff and for all other staff whose performance would affect the quality of the results produced. It is also essential that staff are given the authority and resources by the laboratory management needed to discharge their duties in a satisfactory manner.

One of the key requirements for a NAMAS Accreditation is that the laboratory shall identify a Technical Manager with overall responsibility for the technical operation of the laboratory and also a Quality Manager with a responsibility for ensuring that the requirements of the standards for accreditation are met on a day-to-day basis. If necessary these two functions can be discharged by the same person.

Quality System

The laboratory is required to operate a quality system appropriate to the type, range and precision of the test that it undertakes. It is essential that the system is such that the requirements of the standard for accreditation are fully met on a continuing basis. This system is formalised into a quality manual which is always maintained up-to-date. The quality manual contains statements by top management of the policy of the laboratory in regard to quality in all aspects of its work. It also includes amongst other things:

(a) the laboratory's scope of calibrations and tests,

(b) reference to the major equipment used,

(c) reference to the calibration and all test procedures used,

(d) charts defining the organisation of the laboratory,

(e) the laboratory's place in any parent organisation,

(f) the relationship between management, technical operations, support services and quality control,

(g) a list of the laboratory's approved signatories,

(h) the job descriptions of key managerial and technical staff.

Quality Audit and Review

The laboratory's activities and responsibilities have to be regularly audited by the laboratory management using suitably qualified staff. These audits have to be recorded and checked that the procedures being carried out on a day-to-day basis in the laboratory are those set out in the quality manual.

In addition to these audits, supplementary checks of the quality of results can be adopted by a laboratory. These checks should include:

(a) internal quality control schemes

(b) participation in proficiency testing schemes,

(c) regular use of standard reference materials,

(d) replicate testing,

(e) re-testing of retained samples.

Staff

Staff carrying out tests for which laboratories are accredited should be permanently employed by that laboratory. They should have the appropriate training for the test and where possible this would include academic qualifications which indicate the competence of the personnel to carry out the measurements. The changes in qualifications of the staff must also be recorded.

Equipment

The equipment used by the laboratory has to be demonstrated to the assessors as being suitable for the type of test or measurement being carried out. It has to be capable of achieving the accuracy required and records of the calibration and maintenance of the equipment have to be kept and produced when required. Equipment should also only be operated by authorised staff as named in the laboratory's procedures.

Measurement Traceability

NAMAS requires that all equipment used for measurement shall be calibrated with standards which, where possible, are traceable to national standards. In chemical testing, standard reference materials should be used in the calibration procedures. The effects of different matrices on the measurements should be taken account of in the choice of reference material. The calibration methods chosen have to be consistent with the accuracy required of the measurement.

Laboratory Environment

Accreditation requires samples and reference standards to be stored so that their integrity is ensured. In some laboratories special care will be needed where high level and low level measurement work is being carried out and in such cases special areas should be set aside for low level work.

Records

The laboratories are required to have a systematic documented record of all information relevant to calibration or the tests carried out. Records should be suitably cross-referenced and all observations and calculations should be clearly and permanently recorded at the time they are made. Where computer data processing is used, the original data has to be retained for a period of six years unless otherwise agreed.

Test Reports

The results of all tests or measurements have to be reported accurately, clearly, unambiguously and objectively. It is very important that reports make clear that the tests which are being reported on are those for which the laboratory has accreditation. There should be no implication in the report that accreditation has been given for other tests. It is also required that if subsequently faults are found in equipment used for measurements then clients should be informed promptly and in writing of the doubts about the results.

3. CURRENT DEVELOPMENTS

It can be seen from the above description that the basic requirements for accreditation are that well documented procedures are used in a laboratory on a day-to-day basis by identified staff. When a laboratory seeks accreditation the effort involved in producing the documentation required can be considerable even when generally the procedures are already being used in the day-to-day operations of the laboratory. This effort has been one of the main barriers to laboratories seeking accreditation. Nevertheless commercial and other pressures are increasingly requiring laboratories to seek such accreditation.

It will also be apparent that the procedures set out above are particularly relevant to physical measurement - particularly with regard to traceability. Chemical measurements of the type generally encountered in occupational hygiene operations are generally less precisely defined. Accreditation of chemical testing laboratories is being implemented and this has put increased emphasis on quality control procedures, calibrations and the use of standard reagents.

The level and type of quality control depends on the nature of the analysis, the frequency of the analysis, the batch size, the degree of automation, the difficulty of the measurement and its reliability. As a general guide, routine analysis requires use of quality control samples which represent not less than 5% of throughput or at least one per batch. A more complex procedure means that 20% or even 50% of the throughput may be quality control samples.

It is also recognised that one of the best ways for an analytical laboratory to monitor its performance is to participate regularly in proficiency testing or inter-laboratory schemes. For example, NAMAS strongly encourages laboratories to participate in proficiency testing as an integral part of their quality assurance protocols. Indeed for accreditation for asbestos measurement there is a requirement that a laboratory takes part in the RICE inter-laboratory proficiency testing scheme and achieves a minimum standard of performance in that scheme.

Chemical laboratory equipment can be divided into general service equipment, volumetric equipment, measuring instruments and computers and data processors. Particular attention should be paid to the servicing of general equipment, the contamination and calibration of volumetric equipment, the servicing, cleaning and calibration of measuring instruments with particular reference to stability and linearity of sources, sensors and detectors. It has to be emphasised that the correct functioning of a computer should not be taken for granted.

Up to now, accreditation has generally been confined to measurements within a laboratory. Obviously of key importance in any total measurement process is the taking of samples. Accreditation for sampling has been introduced in the UK by NAMAS and is currently limited to laboratories which also undertake measurements. The general requirements for accreditation for sampling are similar to those set out above for accreditation for measurement. They require a quality system to be in place, staff to be identified and a quality audit system to be operating. In addition, the requirements for maintenance and calibration of equipment for sampling are similar to those previously set out. Of crucial importance is the documentation of procedures for sampling. These procedures have to be agreed between the client and the laboratory and demonstrated to the assessors that they are suitable for the purpose concerned. In addition, particular attention has to be paid to the precautions necessary during the handling, transporting and storage of samples to prevent damage or contaminations.

4. FUTURE DEVELOPMENT

The procedures already outlined have been developed for accreditation for measurement and for sampling. They particularly apply to laboratories which are carrying out a limited range of measurements on a regular basis. Increasingly attention is being paid to how the laboratories can be accredited for samples which are done very infrequently. The main way in which this has been addressed is by the use of internal quality control systems with the use of standard reference materials in a well documented manner.

The possibility of accreditation of organisations where opinion is a major part of the result being reported is also being examined. The expertise of the person giving the opinion is of course of crucial significance although it is not unknown for two established experts to have opposite opinions of the same situation. It is therefore generally accepted that opinion can only be accredited where there are objective or well established procedures available for assessing that opinion, e.g. the interpretation of Spectra.

Accreditation of laboratories is increasingly being seen in many countries as a means of demonstrating the ability of laboratories to carry out particular tests. Many 'round-robin' comparisons of results in different countries and between laboratories in different countries have shown disturbing differences in the results reported. There is increasingly a recognition that these problems have to be dealt with and accreditation is one component of the solution. Independent assessment of a laboratory by a nationally recognised authority will probably be increasingly sought by clients for laboratory measurement. A number of countries have already established such bodies and other countries are in the process of doing so.

It is extremely important that the accrediting bodies in different countries - particularly within the European Community - set very similar criteria for accreditation. In principle, there is no reason why this should not happen but it is important that the same procedures and requirements are set by the different national bodies. This reduces the burden when organisations which operate internationally are seeking accreditation in different countries since procedures can be directly transferred between countries. It is also very important that national bodies recognise and accept accreditation by similar bodies in other countries. This means that tests carried out by a laboratory in one country within the EEC will be acceptable in any of the other countries. This will reduce barriers to trade between countries and simplify procedures and reduce costs for organisations which need to have measurements carried out.

QUALITY CONTROL OF AIRBORNE ASBESTOS FIBRE COUNTS IN SPAIN

M.C. ARROYO

Instituto Nacional de Seguridad e Higiene en el Trabajo
C° de la Dinamita s/n, 48903 Cruces-Baracaldo (Vizcaya) Spain

SUMMARY

Quality control of asbestos fibre counts in Spain has been performed since 1988 under the Programa Interlaboratorios de Control de Calidad de Fibras de Amianto (PICC-FA) established by the Instituto Nacional de Seguridad e Higiene en el Trabajo (INSHT). Circulation of permanent slides, reference values calculated as the average of results and limits of control at ±35% of reference values are the main features of PICC-FA. A description of preliminary interlaboratory tests concerning performance of Spanish laboratories and criteria applied to determine the operational characteristics and quality control limits in PICC-FA are presented.

1. INTRODUCTION

Quality assurance has always been a subject of particular interest for the Instituto Nacional de Seguridad e Higiene en el Trabajo (INSHT). With this aim INSHT began in the late 70s some interlaboratory tests to check the performance in its own laboratories. These tests provided the bases to establish a quality control program that is now available for any hygiene laboratory. The program called Programa Interlaboratorios de Control de Calidad (PICC) encloses a set of single programs for different types of contaminants: PICC-PbS (lead in blood), PICC-HgU (mercury in urine), PICC-VO (organic vapours), PICC-MET (metals in filters) and PICC-FA (asbestos fibres).

The establishment of PICC-FA was undertaken by INSHT being aware that asbestos fibre counts were much more variable compared to the analytical results for other substances and consequently a quality control program for asbestos fibre counts would be not only useful but also necessary. Another factor that supported the establishment of PICC-FA was that laboratory accreditation was becoming mandatory for evaluation of asbestos worker exposure in Spain. Therefore a national quality control program for asbestos fibre counts had to be developed and INSHT was engaged in this task.

2. PLANNING

Interlaboratory Test

The scheme of PICC-FA was implemented using the UK Regular Interlaboratory Counting Exchange (RICE) as a model (1). A set of 16 permanent slides was selected from routine samples and an interlaboratory test was carried out with participation of most Spanish laboratories. All of them used the same method according to the European Reference Method that had been issued by INSHT (2) and has also been published as national standard by the Spanish Standard Association (AENOR).

Results

An overall of 320 results (16 samples x 20 laboratories) were available. A Dixon test was applied to each sample count to reject outlier values. Following the same statistical treatment of RICE a mean count was calculated for each sample and each laboratory result was normalized to this mean.

The distribution of normalized values is shown in Figure 1. The hypothesis that the population is normal of mean 1.00 and standard deviation 0.24 cannot be rejected at the 95% confidence level. The resultant coefficient of variation is 24%. The mean interlaboratory index was ±12% and the mean intralaboratory index 22%. In spite of these quite good figures an analysis of variance gave evidence of differences between laboratories.

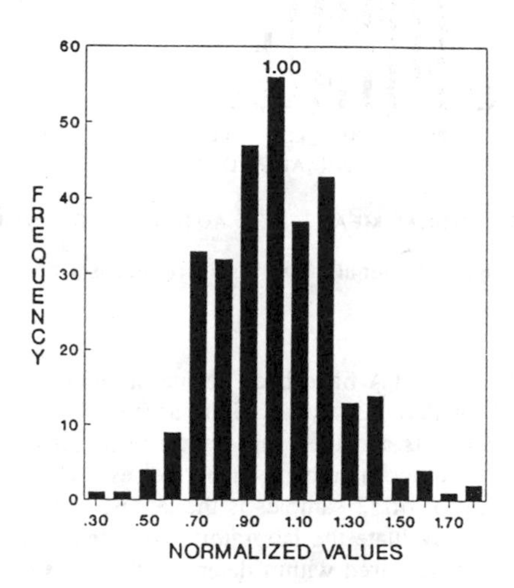

Fig. 1. Distribution of results

The test was considered satisfactory concerning operational characteristics. Differences between laboratories had to be reduced to improve national reproducibility and consequently a quality control program was established from these bases.

3. ESTABLISHMENT OF PICC-FA

Control Samples

Selected permanent slides were found useful with this purpose. Criteria for selection were based on quality, homogeneity and density of slides.

Reference Value

The national mean (excluding outliers) was considered the best estimation of the true value for each sample and defined as reference value.

Magiscan automatic counts were also studied for this purpose (3). The short experience in INSHT with this image analyzer did not advise its use in the first stage of PICC-FA. Further studies in progress with PICC-FA have shown a good automatic-visual count relationship (Figure 2). This subject will be taken into account in the future. Up to now the practical applications of Magiscan in PICC-FA have been the selection of control samples and the use of automatic counts to provide participants with a contrast value immediately. Laboratories find these contrast values very useful for a rough comparison of their results because experience has shown them to be very close to national mean.

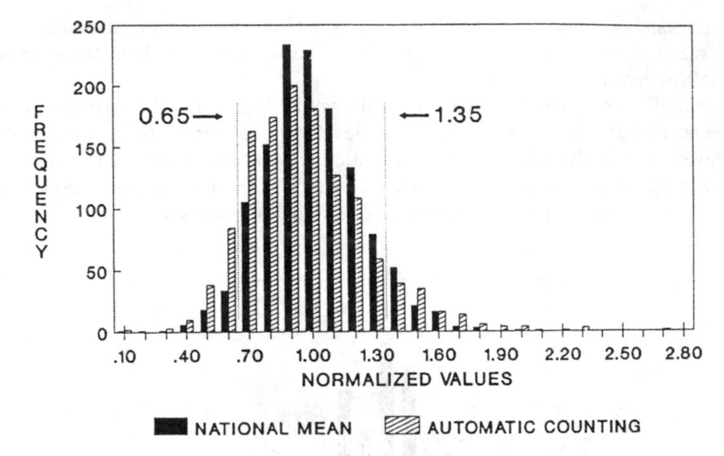

Fig. 2. Automatic and visual comparisons

Limits of Control

The limits of control in PICC-FA have been set up at ±35% of reference value. This means that normalized results within 1.35 and 0.65 are acceptable. These limits were inferred from the interlaboratory test results and they represent a practical compromise to match the minimum variation with the maximum number of laboratories.

According to RICE a number of 32 samples is the largest practical number that provides enough statistical confidence to evaluate the laboratory performance and ≥75% of results of these samples (≥24 results) are required within determined classes of limits for a laboratory to be classified in a certain category (4). In PICC-FA only two categorizations were used, so only one class of limits was set up. Results of the interlaboratory test were displayed on different classes of limits and the number of laboratories with ≥75% of their results within each class of limits noted. Table 1 and Figure 3 illustrate this point.

Table 1. Distribution of Overall Results in Different Classes of Limits

Classes of Limits		Laboratories with ≥75% of results within limits	
± % of reference value	Normalized values included	Number	%
5	1.05 - 0.95	0	0
10	1.10 - 0.90	0	0
15	1.15 - 0.85	1	5
20	1.20 - 0.80	3	15
25	1.25 - 0.75	10	50
30	1.30 - 0.70	13	65
35	1.35 - 0.65	17	85
40	1.40 - 0.60	17	85
45	1.45 - 0.55	19	95
50	1.50 - 0.50	19	95
55	1.55 - 0.45	19	95
60	1.60 - 0.40	19	95
65	1.65 - 0.35	19	95
70	1.70 - 0.30	19	95
75	1.75 - 0.25	20	100

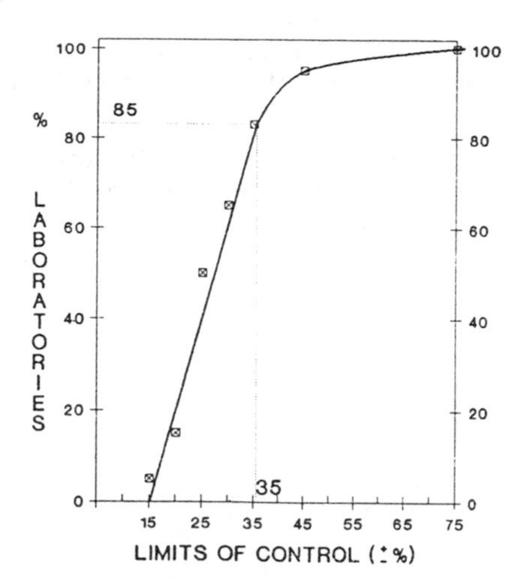

Fig. 3. Quality control limits in PICC-FA

The graphic representation begins with a straight line where short increments of limits width imply a large increment in the number of laboratories. The slope of the line breaks down dramatically in a point from which large increments in limits width are necessary to increase the number of laboratories. This point decided the selected ±35% limits.

4. PRESENT STATE

PICC-FA is now running satisfactorily and four circulations of 16 samples each have been concluded. Laboratories involved in asbestos fibre evaluation in Spain are obliged to participate in PICC-FA in order to obtain accreditation required by national asbestos regulations. Participation is at present 26 laboratories and more than 85% have achieved satisfactory performance. At the end of the current circulation results will be re-examined with a view to specifying tighter limits.

Acknowledgements.

The author is greatly indebted to all PICC-FA participants who contributed to this work. Thanks are due to P. Brown of the Institute of Occupational Medicine for valuable information about the RICE program. Also particular thanks are extended to L.C. Kenny for her cooperation with Magiscan and helpful suggestions during the development of PICC-FA.

REFERENCES

(1) CRAWFORD, N.P. and COWIE, A.J. (1984). Quality control of airborne asbestos fibre counts in the United Kingdom - The present position. Ann. occup. Hyg., vol 28, N° 4, 391-398.

(2) INSTITUTO NACIONAL DE SEGURIDAD E HIGIENE (1987). Métodos de toma de muestras y análisis. Determinación de fibras de amianto en aire- Método del filtro de membra na/Microscopia óptica. MTA/MA-010/A87 INSHT. Madrid.

(3) KENNY, L.C. (1984). Asbestos fibre counting by image analysis. The performance of the Manchester Asbestos Program on Magiscan. Ann. Occup. Hyg., vol 28, n° 4, 401-415.

(4) OGDEN, T.L. (1984). Statistical uncertainties in asbestos-laboratory classification through the RICE scheme. Ann. Occup. Hyg., vol 28, N° 4, 449-452.

1442

THE INSHT INTERLABORATORY QUALITY CONTROL SCHEME FOR LEAD AND CHROMIUM

J. BARTUAL

Instituto Nacional de Seguridad e Higiene en el Trabajo
Centro Nacional de Condiciones de Trabajo
Dulcet, 2-10. E-08034 Barcelona, Spain

SUMMARY

In 1986 the 'Instituto Nacional de Seguridad e Higiene en el Trabajo' established an interlaboratory quality control scheme for the analysis of lead and chromium III in occupational hygiene samples (PICC-MET). The operation of the PICC-MET scheme is explained and results from the first four years of the scheme's operation are reviewed. The performance obtained is assessed against the reliability established for the determination of lead in air by the Spanish legislation on lead in workplaces. The advisability of an accreditation program for laboratories which analyze lead in air is suggested.

1. INTRODUCTION

In 1981 the 'Instituto Nacional de Seguridad e Higiene en el Trabajo' (INSHT) established the PICC interlaboratory quality control scheme ('Programa Interlaboratorios de Control de Calidad') to provide an external quality assurance to laboratories which routinely analyze occupational hygiene samples.

Metals in airborne samples were included within the PICC in 1986 (PICC-MET), starting this quality control with the analysis of lead, according to the importance of this metal in occupational hygiene, and adding later the analysis of chromium (Cr III). The promulgation in 1986 of a national regulation on lead (1), adapted to the EEC Directive on the matter (2), was also a further reason to implement the PICC-MET.

2. OPERATION OF THE PICC-MET SCHEME

Preparation of Samples

Samples are prepared on 37mm diameter membrane filters Millipore-type AA. Filters are spotted with different volumes of lead and chromium solutions using a Hamilton programmable dispenser and a precision syringe.

The analyte loading ranges are 9-75μg for lead, and 6-60μg for chromium. The analyte loadings are designed to represent samples taken in a real workplace atmosphere at concentrations in the range 40-300$\mu g/m^3$ for lead, and the range 25-500$\mu g/m^3$ for chromium, with optional air sampling volumes between 120 and 240 litres.

Quality Control of Prepared Samples

Sixteen samples of the first round were analyzed for precision, and this control repeated occasionally. The coefficients of variation obtained were always below 5.0%. Therefore it is assumed that the variations in analyte loadings are small in relation to analytical errors.

Circulation of Samples

A set of four different samples and a blank filter are circulated to each participating laboratory on a three-monthly schedule, and results are requested within four to five weeks. Although metals may be analyzed by any appropriate method at the analyst's discretion, all the participating laboratories have been using the atomic absorption technique until round 17 when a laboratory which uses a polarographic (DPP) method joined the PICC-MET.

Evaluation of Results

The evaluation of results is based on the premise that individual results from the laboratories are normally distributed.

The evaluation involves the establishment of acceptable concentration ranges for each sample. Each reported value is then determined to be either acceptable, within the designated range, or outlier, outside the designated range.

The results are fed into a computer program, which calculates the mean and the coefficient of variation for each data set.

If the coefficient of variation of one set of data is higher than 12% (chosen value), then that set of data is Winsorized to eliminate the possible effect of some inconsistent results on mean and dispersion. Winsorization involves the ranking of the data. Those results which fall in the top $\alpha/2$ percent of results are replaced by a value equal to the highest result remaining in the set. Similarly those results which fall in the bottom $\alpha/2$ percent of results are replaced by a value equal to the lowest result remaining in the set. The α selected for this Winsorization is 10%.

The computer program recalculates then the mean and the coefficient of variation for the Winsorized data. The reference values equal the means calculated, and acceptable concentration ranges equal the means $\pm$ 3 standard deviations (3).

Data from participating laboratories are then compared against acceptable concentration ranges to determine acceptability. In addition outliers are further classified as 'high' ('AL') or 'low' ('BA').

Laboratory Outputs

Each participating laboratory is provided with a report which includes the results submitted by the laboratory, the reference values, the acceptable concentration ranges, the coefficients of variation, the Winsorized data and the number and distribution of outliers in the round.

Two Z-score plots are also provided which summarize for each metal a laboratory's performance over the last four rounds. The plotted value equals $(x-\bar{x})/s_x$, where x is the laboratory's reported value, $\bar{x}$ is the reference value and s_x is the standard deviation of a set of data. The 0 SD (0 D.E.) line represents the location of reference values and indicates the centre of the distribution. The +3 SD (3 D.E.) line represents the upper performance limit whereas the –3 SD (–3 D.E.) line represents the lower performance limit. Values are shown as odd multiples of 0.5 plotted into one of 12 intervals. Scores higher than 5.5 or lower than –5.5 are shown as 5.5 or –5.5 respectively.

Occasionally two similar samples are sent to participants in different rounds or in the same round. A Youden plot is then included in the laboratory output. Z-scores of each sample (X and Y) are plotted in a X-Y chart, each point representing a pair of results from a single laboratory. The corresponding point is identified in each laboratory output. An example of typical output provided to participating laboratories is displayed.

3. PARTICIPATING LABORATORIES

The PICC-MET scheme, as part of PICC, is both informal and completely confidential, and now is open to laboratories of all categories.

The number of participating laboratories has increased slowly from the outset in each succeeding round. A group of 36 laboratories from central and local government, industry and universities now takes part in the PICC-MET scheme.

4. RESPONSE OF LABORATORIES

In general, returns were made by 88% of the laboratories receiving samples.

Table 1 provides indication of the general level of performance over the first four years of the scheme's operation - 17 rounds - for lead and chromium. The figures show the maximum and the minimum coefficients of variation obtained from the four sets of samples analyzed in each round and the number of those results determined as outliers.

The values of the coefficients of variation fluctuate within the range 6.5–24.2% for lead, and the range 8.5–40.7% for chromium, with more variability for the maximum values and without a clear evidence for an improving trend. These facts can be explained in part by the continuous input of new participants and taking into account that, with only a small number of laboratories participating, the dispersion of a set of results could be significantly increased by a few poor results that have escaped being Winsorized. Nevertheless, it is apparent from the mean level of performance that there is reasonable room for improvement.

The number of outliers represents the 2.9% of results for lead, and the 4.4% of results for chromium. The distribution of outliers for laboratories shows that a minority of laboratories produce a majority of outliers.

Table 1. Performance for Lead and Chromium

Round	Lead			Chromium		
	Coeff. Var. (%)		Outliers	Coeff. Var. (%)		Outliers
	Min	Max		Min	Max	
1	8.8	10.4	0	-	-	-
2	9.3	20.9	5	-	-	-
3	6.9	19.1	6	-	-	-
4	7.5	17.3	4	-	-	-
5	9.7	18.7	2	16.0	22.0	5
6	11.4	22.8	0	8.5	19.4	4
7	11.8	14.8	0	16.3	19.7	4
8	10.7	13.7	3	14.8	15.7	3
9	9.5	24.0	2	16.5	20.7	4
10	7.3	11.2	1	13.3	30.1	4
11	14.0	24.2	5	9.7	25.5	4
12	13.3	23.1	4	11.8	40.7	4
13	9.2	11.5	2	11.1	19.0	5
14	8.7	18.6	1	10.4	11.3	1
15	10.0	15.4	1	12.0	23.6	2
16	6.5	11.6	0	10.5	12.5	2
17	11.9	14.7	2	10.7	16.4	5

The Spanish legislation on lead in workplaces (1) establishes a reliability equal or better than ±20%, with a level of confidence of 95%, for the determinations of lead in air, including sampling and analysis, at concentrations higher than 30μg Pb/m^3.

The performance shown by the results of PICC-MET provides enough reasons to question whether the analysis of lead in air, carried out routinely by some laboratories, fulfils that provision.

5. CONCLUSIONS

The PICC-MET scheme provides laboratories with a quality control which enable analysts to follow and improve their performance. Most of the participating laboratories are now producing correct results.

It is apparent from the results obtained over the first four years of the scheme's operation that the mean performance in analyzing lead is not good enough to assure that the quality requirements of the Spanish national legislation on lead in workplaces are always fulfilled.

Therefore the establishment of an accreditation program for laboratories which analyze lead in air could be advisable in order to control really the quality of these determinations. This initiative could be favoured if specific requirements on analytical quality assurance were included in the EEC Directives on chemical agents in workplaces.

REFERENCES

(1) Mº TRABAJO y SEGURIDAD SOCIAL (España). Reglamento para la prevención y protección de la salud de los trabajadores por la presencia de plomo metálico y sus componentes iónicos en el ambiente de trabajo. B.O. Estado Nº 98. 24 April 1986.

(2) EEC Directive 82/605/EEC. O.J. Num. L 247 of 23.8.1982, p.12.

(3) SCHLECHT, P. and SHULMAN, S. Proficiency Analytical Testing Program. Statistical Protocol. National Institute for Occupational Safety and Health. Cincinnati. Ohio. 1982.

1443

THE GENERATION OF VOC's IN AIR : A NEW APPROACH TO THE CAPILLARY DOSAGE TECHNIQUE

E. GOELEN and T. RYMEN

SCK/VITO, Department of Environmental Chemistry,
Boeretang 200,
2400 Mol, Belgium

SUMMARY

This paper describes how to use a capillary dosage device so that the concentrations generated by means of this technique can be considered as related directly to a primary standard. For this purpose, a home made capillary dosage unit is assembled which contains a built in weight sensor.

1. INTRODUCTION

The sampling and measurement of harmful pollutants in workplace air environments is of primary concern. There is still a need to perform research on analytical methods for occupational survey. It is important in this context to be able to generate gas phase calibration mixtures.

The capillary dosage technique is a continuous injection technique (1), developed to produce dynamic gas phase calibration mixtures of VOC's in a diluent gas at occupational hygiene levels. A new advance in this technique allows to calculate the amounts of pollutants emitted into the carrier gas, directly from gravimetric measurements. Compared to other currently applied methods e.g. the permeation method (2) and the diffusion tube method (3), the capillary dosage device offers some more possibilities. The other techniques have too low emission rates and cannot be used whenever a high gas flow rate (> 200 l/min) is required or whenever ppm levels of workplace air pollutants are desired in high gas flows.

2. DESCRIPTION OF THE CAPILLARY DOSAGE DEVICE WITH IN SITU WEIGHT SENSOR

Picture 1 shows an exploded view of the capillary dosage device together with read out equipment for the follow up of the weight sensor signal. The most important parts of the device are : a glass capillary, a pressure reducer, a miniature weight sensor and a stainless steel housing. The liquid phase of the pollutant to be generated is forced through the capillary by means of a pressure difference between both ends of the capillary. At the elevated pressure side, the capillary is immersed in the liquid. The reservoir containing the liquid is connected to a weight sensor. This approach allows a continuous measurement of liquid consumption at any time during its injection process. As a consequence, this technique is based on measurements related directly to a primary standard.

In order to obtain a broad concentration range of workplace air pollutants, a set of glass capillaries is available with internal diameters between 15 and 90 micrometer. The length is fully dependent upon the configuration of the capillary dosage unit, but is standard 22 cm. Pressure differences across the capillary are usually between 0.1 and 5 bar relative. Four capillary dosage units are each equipped with one weight sensor. The weight sensors have different operating ranges varying between 10 gr and 150 gr full scale.

Picture 1 : Exploded view of a capillary dosage unit together with read out equipment for the weight sensor

3. PERFORMANCE CHARACTERISTICS

The performance of home made capillary dosage units with in situ weight sensor is studied by distributing the spiked carrier gas through a glass manifold. The internal diameter of 40 mm is chosen to restrict the pressure difference across the entire lenght (46 meter) of the manifold. This design allows active sampling (30 sampling points available) along the manifold at approximately atmospheric conditions even with high gas flow rates through the distribution manifold : 180-220 l/min at standard conditions (max. 250 l/min). Two glass exposure chambers at the end of the manifold allow passive sampling.

A GC-FID with automatic sample loop measures continuously the concentration of the pollutants in the carrier gas and their stability. Stability tests (up to 24 hours) have been performed for some aromatic hydrocarbons and chlorinated hydrocarbons and showed a relative standard deviation of maximum 1 % in all cases and this at concentration levels from 50 ppb up to 200 ppm. Homogeneity tests showed no increase in relative standard deviations as compared to the stability test.

The performance of the in situ gravimetric determination of the liquid release of a capillary dosage unit has been demonstrated extensively. For example with pure toluene in the reservoir of the unit, a capillary of 88 micrometer and a pressure difference across the capillary of 1.000 bar relative, the mean value of four independent gravimetric determinations has a relative standard deviation of 0.8 %. Many such experiments with different compounds and settings indicated all a relative standard deviation between 0.4 % and 1.1 %.

The mean value (μg/min) of independent determinations and the flow rate of the carrier gas are finally used to calculate the exact vapour concentration in the distribution manifold. The carrier gas flow is mass flow controlled and measured by means of a home made mercury-sealed piston flowmeter.

The gravimetric experiments are performed for each capillary at different pressures using a pure liquid and thus result in calibration curves. These curves are essential for choosing the settings of a capillary dosage unit, settings required in order to obtain the desired concentration.

The overall uncertainty of a standard VOC concentration generated from a capillary dosage device is estimated at 2 to 3 %. This is confirmed when comparing the prepared concentration in the distribution manifold against the actual concentration as measured by a primary calibrated GC-FID. The on-line analytical gas chromatograph is primary calibrated by means of diffusion tubes according to ISO 6145.

4. CONCLUSION

It is clear from the experiments that an in situ weight sensor in a capillary dosage device allows to measure the amount of pollutant emitted into the carrier gas very accurately and this during the injection process without interfering with it. By using a set of capillary dosage devices one is able to prepare ppb and ppm levels of standard gaseous VOC concentrations in the same diluent gas flow. The nominal gas flow through the sampling manifold is 180 l/min at standard conditions. Frequent measurements of this gas flow are not required while the flow remains the same during experiments. A range of standard concentrations for calibration purposes is obtained by varying the pressure difference across the glass capillary and measuring accurately the weight loss of the pollutant during its injection.

The described experimental facility offers new possibilities regarding : the organisation of intercomparisons for workplace air sampling and analysis (up to 150 l/min gas available for active sampling); research and development of analytical methodology for sampling and analysis of workplace air; study of interference, precision and accuracy of personal samplers; active sampling versus passive sampling.

REFERENCES

(1) Preparation of calibration gas mixtures - dynamic volumetric methods, continuous injection method, ISO 6145 - Part 4

(2) KEEFFE O., ORTMAN, Anal. Chem. 38, 766 (1966)

(3) Preparation of calibration gas mixtures - dynamic volumetric methods, diffusion, ISO 6145 - Part 8

INTERLABORATORY COMPARISON OF DIFFUSIVE SAMPLERS FOR ORGANIC VAPOUR MONITORING IN WORKPLACE AIR

J.F. PERIAGO and B. URIBE

Instituto Nacional de Seguridad e Higiene en el Trabajo
c/ Lorca 70, E -30120 El Palmar (Murcia) Spain

SUMMARY

An interlaboratory experiment was made with passive samplers for volatile organic compounds in workplace air. 3M-3500 organic vapour monitors were exposed simultaneously in a test chamber to the vapours of n-hexane and toluene. Three levels of concentrations were studied and lots of two samplers of each level of concentration were mailed to participants. Ten laboratories were included in the experiment and the same analytical method - desorption with 1.5ml carbon disulphide and gas chromatography - was used. Repeatability and reproducibility were calculated according to ISO international standard 5725. We found repeatability values between 5.7% - 19.7% for n-hexane and 8.9% - 20.4% for toluene. The reproducibility was between 18.2% - 21.9% for n-hexane and 9.7% - 23.3% for toluene. Details of the test chamber, organization of the interlaboratory experiment and statistical treatment will be presented.

1. INTRODUCTION

Hexane and toluene are widely applied as solvents in shoe manufacturing. In order to increase the reliability of measurements of these solvents by means of diffusive samplers, a national interlaboratory test was organised by the INSHT (National Institute for Occupational Safety and Hygiene) between ten laboratories. The results and details of test chamber, organisation of the interlaboratory experiment and statistical treatment are described in this paper.

2. EXPERIMENTAL

Test Chamber

Test gas mixtures with known concentrations of n-hexane and toluene were generated by a dynamic volumetric method, using the continuous injection technique (syringe). Figure 1 illustrates the system used to generate the contaminated atmosphere.

Gravimetrically prepared mixtures of n-hexane and toluene were injected directly into the humidified air stream. The relative humidity of the test atmosphere was 50% and the relative standard deviation of concentrations was less than 4%.

General Procedure of the Experiment

Three concentration levels of n-hexane (29.6, 343.2 and 704.3mg/m^3) and toluene (30.7, 360.9 and 744.1mg/m^3) were established and 20 diffusive samplers 3M-3500 were introduced into the chamber during 120 minutes and two samplers for each level were mailed to the participants. Ten INSHT laboratories were included in the experiment and the same analytical method - desorption with 1.5ml carbon disulphide and gas chromatography - was used (1).

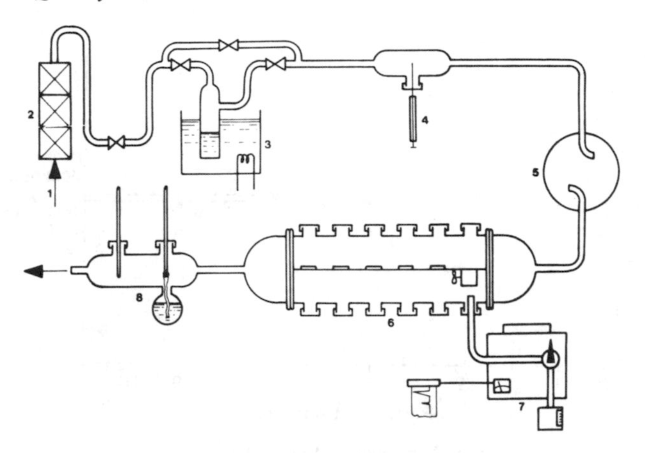

Fig. 1. Test chamber
1 Uptake air; 2 Filter; 3 Humidifier; 4 Automatic syringe injector
5 Mixture chamber; 6 Sampling chamber; 7 Gas chromatograph; 8 Dry and wet thermometer

3. RESULTS

Analytical Results

Graphic representation of the results of analysis of the samples are shown in Figures 2 to 7, including the true value of the test atmosphere and the mean value of the laboratories.

Statistical Evaluation of the Results

The results were evaluated according to ISO International Standard 5725 (2). Cochram's maximum variance test for determination of repeatability and Dixon's outliers test for determination of reproducibility were applied, according to the standard prescription. Two Dixon's outliers were rejected from lab 10 and no results were rejected according to Cochram's test.

Repeatability (r)

The repeatability is the value below which the absolute difference between two single test results obtained with the same method on identical test material, under the same conditions (same operator, same apparatus, same laboratory, and a short interval of time), may be expected to lie with a probability level of 95%.

Reproducibility (R)

The reproducibility is the value below which the absolute difference between two single test results obtained with the same method on identical test material, under different conditions (different operators, different apparatus, different laboratories, and different time), may be expected to lie with a probability level of 95%. The results of the statistical evaluation are shown in Table 1.

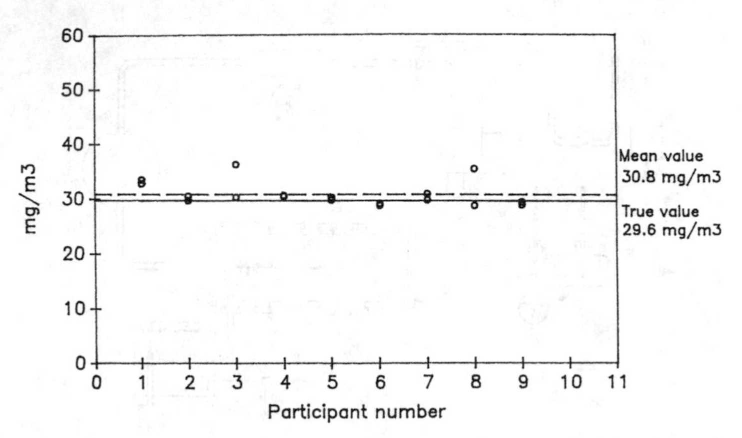

Fig. 2. n-hexane (Level 1)

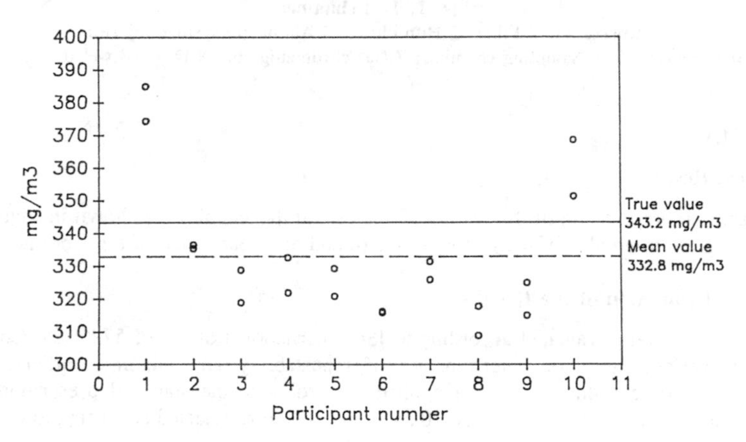

Fig. 3. n-hexane (Level 2)

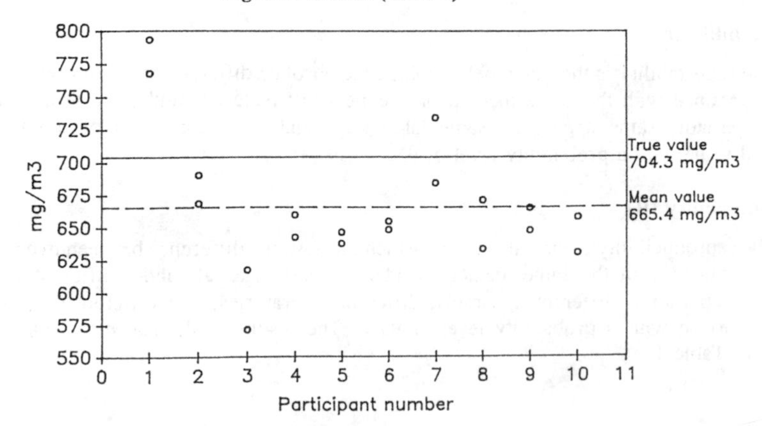

Fig. 4. n-hexane (Level 3)

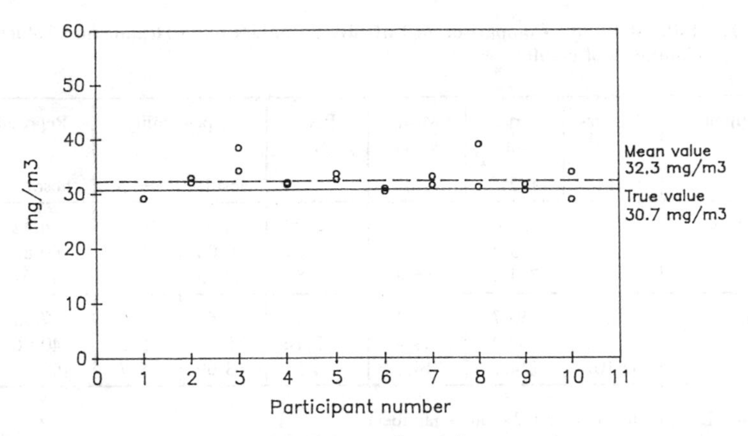

Fig. 5. Toluene (Level 1)

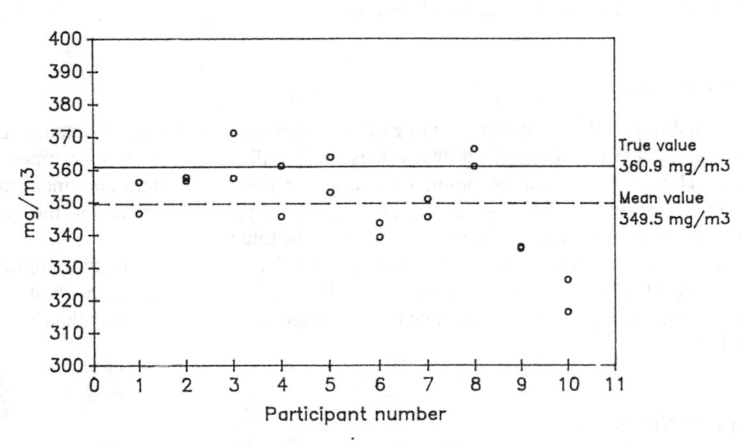

Fig. 6. Toluene (Level 2)

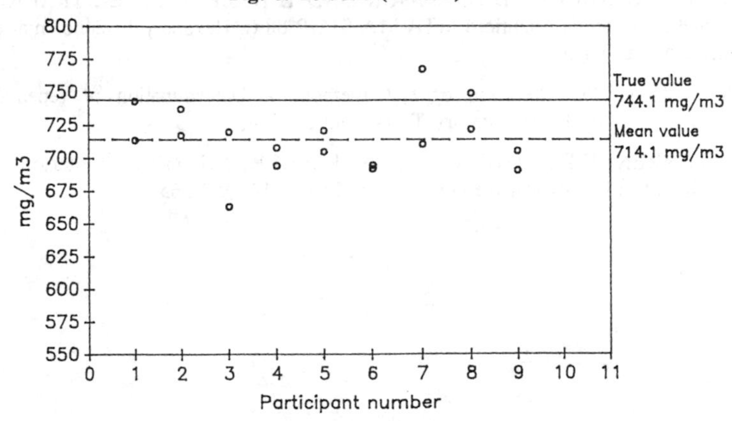

Fig. 7. Toluene (Level 3)

Table 1. Interlaboratory Comparison of Diffusive Samplers for n-Hexane and Toluene. Summary of results

Compound	Outliers (1)	True value (2)	Mean value (3)	Bias %	Repeatability		Reproducibility	
					absol.	rel %	absol.	rel %
n-Hexane	D(10)	29.6	30.8	+4.05	6.08	19.7	6.42	20.2
		343.2	332.8	-3.01	19.04	5.7	60.82	18.2
		704.3	665.4	-5.51	58.23	8.7	146.01	21.9
Toluene		30.7	32.3	+5.21	6.62	20.4	7.55	23.3
		360.9	349.4	-3.16	18.05	5.1	40.06	11.4
	D(10)	744.1	714.1	-4.02	63.69	8.9	69.63	9.7

1. D(): Dixon's test outlier (laboratory number)
2. Atmosphere concentration (mg/m^3)
3. Mean value of the results of all the laboratories

4. DISCUSSION

Repeatability and reproducibility are extreme values of the precision obtained by an interlaboratory test (2). Reproducibility values are in all cases greater than repeatability, as expected. However it should be pointed out that for lower concentrations most imprecision cames from the within laboratories variability. The opposite is true for the two greater concentrations, except in the greatest concentration of toluene.

In any case, the reproducibility is lower than 25% in the whole range of application. According to MacFarren *et al.* (3), it could be classified as an 'excellent method'.

Otherwise, the apparent bias obtained in the intercomparison was less than 5.5% for both solvents.

5. REFERENCES

(1) INSTITUTO NACIONAL DE SEGURIDAD E HIGIENE EN EL TRABAJO (1988): Manual de Metodos Analiticos MTA MA-015/R 88 (n-Hexano y Tolueno en aire mediante muestreadores pasivos).

(2) ISO 5725 (1986): Precision of test methods - Determination of repeatability and reproducibility by Interlaboratory Tests. 2nd Edition.

(3) MacFARREN, E.F., LISHKA, R.J. and PARKER, J.H. (1970): Criterion for judging acceptability of analytical methods. Anal. Chem. 42, 358-365.

1445

Quality Assurance of Workplace Air Monitoring and the Role of Laboratory Accreditation

B.KENT

NAMAS Executive, National Physical Laboratory
Teddington, UK

Introduction

The National Measurement Accreditation Service (NAMAS) was established by the UK Government in 1985 to assess and accredit the competence of laboratories to carry out defined types of tests, analyses or measurements. Accreditation is a formal recognition of that competence. NAMAS is the UK's third party quality assurance service for laboratories. It is administered by an Executive based at the National Physical Laboratory in Teddington, which is the largest of the Department of Trade and Industry's research establishments.

Accreditation is voluntary but laboratories are increasingly being required by their clients to demonstrate their competence through third party accreditation. Any laboratory carrying out objective tests or measurements is eligible for accreditation. Since April 1991 accredited laboratories have been eligible for accreditation by NAMAS for the taking of samples and defined sampling activities prior to the subsequent testing or analysis. In NAMAS terms testing and analysis are synonymous. Laboratories have already been accredited in a very wide range of fields of testing and calibration, including construction, electrical, chemical analysis, microbiology, asbestos, mechanical etc. Accredited laboratories include commercial third party test houses, manufacturing organisations, academic institutions and government laboratories. Chemical analysis for occupational hygiene, food testing, waters and environmental monitoring are among the growth areas of NAMAS accreditation.

Customers or clients of laboratories whether they are manufacturers, local or national government, enforcing authorities, employers or employees, need confidence in the validity of the results they are given. In many cases they are dependent on the quality of the laboratory concerned since it is often not possible to obtain valid samples for checking by another laboratory. Third party accreditation can provide the required assurance to the customer as well as giving the laboratory itself the confidence deriving from submission to independent assessment.

Accreditation Criteria

In order to achieve accreditation a laboratory must satisfy NAMAS and its assessors that it meets all the criteria contained in the NAMAS Accreditation Standard M10. This standard covers all aspects of the operation of a laboratory relevant to the quality of the testing, calibration or sampling which it undertakes. The main paragraph headings of M10 are:

Organisation & Management
Quality System
Quality Audit & Review
Staff
Equipment
Calibration & Traceability
Methods & Procedures
Accommodation & Environment
Handling of Test Items
Records
Test Reports & Certificates
Complaints & Anomalies
Sub-Contracting of Tests
Services & Supplies

Although the Accreditation Standard applies to all laboratories, in view of the diversity of the testing, the Standard requires to be interpreted for each field. Such interpretations are often published in the form of NAMAS Information Sheets, the NIS series. These NIS documents are usually prepared by NAMAS technical advisory committees or working groups which have expert representatives currently involved with the area concerned.

Although adequate documentation is essential for a good quality system and is one of the prime considerations of NAMAS assessment and accreditation, NAMAS is equally concerned with other aspects such as technical competence of analysts and samplers. The appropriate use of standards, reference materials and internal quality control and where appropriate external proficiency testing are other critical aspects. The test equipment used will often be of high investment value but it is only as good as the quality of the data it produces. Poor quality analytical results are very often not only useless but in many cases positively damaging. They need to be **adequate for the intended purpose**, or it's not worth switching the analyser on. Equally the analytical results are only as good as the quality of the sample analysed.

The prime document is the laboratory quality manual, which states how the laboratory operates with regard to aspects relevant to the work of the laboratory. In terms of NAMAS accreditation, the manual needs to be supported by other documentation including methods, standard operating procedures, equipment records, calibration programmes and other records such as training. Together these form the basis of the quality system. The laboratory says what it does and then needs to do what it says.

Accreditation Process

In summary the process involves: (1) Formal Application, (2) Preliminary Visit, (3) Initial Assessment, (4) Rectification of Non-Compliances, (5) Accreditation, (6) Continued Monitoring by Surveillance and Re-assessment Visits.

The process involves continuous consultation between the laboratory and the NAMAS Technical Officer with the assessors as appropriate.

Selection and Training of Assessors

Although NAMAS Technical Officers are all professionally qualified, it would not be possible, within the Executive, to cover the very wide range of testing and calibration activities encountered in laboratories. NAMAS therefore employs external experts in particular areas of testing to act as assessors. Assessors are drawn from a number of sources including public sector laboratories, universities and polytechnics, private and commercial laboratories and independent consultancies. When moving into new areas of accreditation, advice is sought from the relevant professional bodies on the most appropriate sources of assessors.

Potential assessors must satisfy the NAMAS requirements in terms of qualifications and experience, and must demonstrate acceptable personal qualities. All assessors must also attend a successful 4 day training course covering a number of different aspects of assessment. The use of technically expert assessors accepted as such by the laboratory is an important and distinctive feature of NAMAS accreditation.

European and International Aspects of Accreditation

In Europe, primarily to meet the needs of the Single Market, a European Standard for laboratory accreditation, EN 45001, has been published. The NAMAS Accreditation Standard M10 is consistent with this standard and also with ISO Guide 25, the international equivalent. As a result, any laboratory accredited by NAMAS can also claim to meet the requirements of EN 45001 and ISO Guide 25.

Third party laboratory accreditation is already being used, where testing or analysis is required, as an effective means of implementing EC Product Directives eg Toy Safety Directive. 17 out of the 18 EC and EFTA countries now have laboratory accreditation schemes in various stages of development. The Council of Ministers resolution on 'A Global Approach to Testing and Certification' has led to the setting up of WELAC, the Western European Laboratory Accreditation Cooperation, which involves accreditation bodies from the 17 European countries. Five WELAC Working Groups are producing proposals for (1) Guidance Documents for the interpretation of EN 45001 in specific fields of testing, eg Chemical Analysis; (2) Mutual Recognition agreements between National schemes; (3) Selection and training of assessors; (4) Development of Proficiency Testing in Europe and (5) Reference Materials.

Although the initial impetus for cooperative development arose from trade aspects of the Single Market, the 'Global Approach' can and will apply to 'non product' areas such as Occupational Hygiene and the Environment in its widest sense.

REFERENCE SCHEMES FOR AIRBORNE MAN-MADE MINERAL FIBRE COUNTING

PW Brown and NP Crawford

Institute of Occupational Medicine Ltd
8 Roxburgh Place, Edinburgh EH8 9SU, Scotland

SUMMARY

Two fibre counting quality assurance schemes for man-made mineral fibres are run by the Institute of Occupational Medicine. The UK National MMMF Reference Scheme has been introduced recently in response to legislative changes within the United Kingdom. The WHO/EURO MMMF Reference Scheme has for several years enabled laboratories in different countries to compare their fibre counting performances.

1. THE UK NATIONAL MAN-MADE MINERAL FIBRE REFERENCE SCHEME

Introduction

From 1 January 1991, any laboratory in the UK which evaluates airborne MMMF samples for compliance purposes using phase contrast optical microscopy is required to participate in the national MMMF reference scheme which has been established to enable inter-laboratory differences in results to be minimised.

Organisation

The UK national MMMF counting scheme is managed by the Institute of Occupational Medicine (IOM) on behalf of the Health and Safety Executive's Committee on Fibre Measurement. The membership consists of specialist MMMF counting laboratories. (Because of the short notice of the requirement for a national MMMF scheme, laboratories which evaluate MMMF samples and which also take part in the RICE scheme for asbestos are currently deemed to fulfil the requirement for participation in the MMMF scheme.) Workplace samples are used, incorporating various types of MMMF, spanning a wide range of fibre densities and permanently mounted on microscope slides. The scheme operates in a similar way to the RICE and AFRICA asbestos schemes (ie, batches of samples are circulated around groups of laboratories at regular intervals). Laboratory performance is assessed by comparison with reference counts. In the MMMF scheme each reference count is the consensus of a number of visual counts on that sample. In the first round of sample exchanges, counting performance of participants was satisfactory.

Further development

A questionnaire survey is being carried out to estimate the number of laboratories likely to participate in the MMMF sample circulations. In addition, a feasibility study is being conducted into sample production using a fluidised bed aerosol generator, aimed at producing sets of about 20 nominally 'identical' samples which could be used in radial exchanges.

2. THE WHO/EURO MAN-MADE MINERAL FIBRE (MMMF) REFERENCE SCHEME

Aim

To harmonise counting of airborne MMMF by European laboratories using phase contrast optical microscopy.

Organisation

Reference samples are sent to participants every six months for counting. The scheme is managed by the Institute of Occupational Medicine Ltd (IOM), under the auspices of the World Health Organization (Regional Office for Europe). Thirteen laboratories, located in Czechoslovakia, Denmark, France, Germany, Italy, Poland, Portugal, Spain, Sweden, the United Kingdom and Australia currently take part. Five of the most experienced laboratories constitute the Reference Group. Workplace samples from MMMF production plants in Europe (glasswool, rockwool, ceramic fibres) are used. Each sample is given a reference count (in the form of fibre density, fibres.mm^{-2}), which is the mean of previous counts made by the Reference Group. Batches of reference samples, usually 8 samples per batch, are circulated around groups of laboratories for evaluation at approximately six-monthly intervals. Once a laboratory evaluates the samples it sends its results to the IOM. Within a few days the laboratory is sent a report for each sample. The report provides the reference count, and the ratios of the laboratory/reference counts are listed. From the count ratios in the report the laboratory can see how close its counts are to the reference values. A system adapted from that used in RICE and AFRICA can also be applied. By this modified RICE system, a laboratory's performance can be regarded as satisfactory if at least 75% of the ratios lie within the range 0.5 to 2.0 times the reference counts. Alternatively, performance can be assessed using two indices: an inter-laboratory index and an intra-laboratory index.

Inter-laboratory index

This is a measure of the laboratory's agreement with the reference counts, and is calculated as

$$(M - 1) \times 100 \%$$

where M is the mean of the laboratory/reference ratios. A value close to zero indicates that the laboratory's counts are on average close to the reference counts. A large positive value shows that the laboratory counts high; a large negative value that it generally counts low.

Intra-laboratory index

This is a measure of variability of the ratios from the same laboratory. It is defined as

$$\frac{S.100}{M} \%$$

where S is the standard deviation of the laboratory/reference ratios. The intra-laboratory index should be as low as possible.

Results

In the early rounds, only the five members of the Reference Group participated. An improvement in the level of agreement between these laboratories was observed. The improvement is illustrated in Figure 1, which shows the inter-laboratory and intra-laboratory indices for one of the participants. Among the Reference laboratories, the maximum difference between inter-laboratory indices decreased from a factor of about 2.5 in the first round to 1.6 in the second round, and was maintained within the range 1.3 - 2.1 in the subsequent four rounds. In the course of the scheme individual laboratories' performances have deviated from the satisfactory range for one round. In most cases the problem has been rectified by the following round.

Figure 1. Performance Indices from one of the reference laboratories in rounds 1-6.

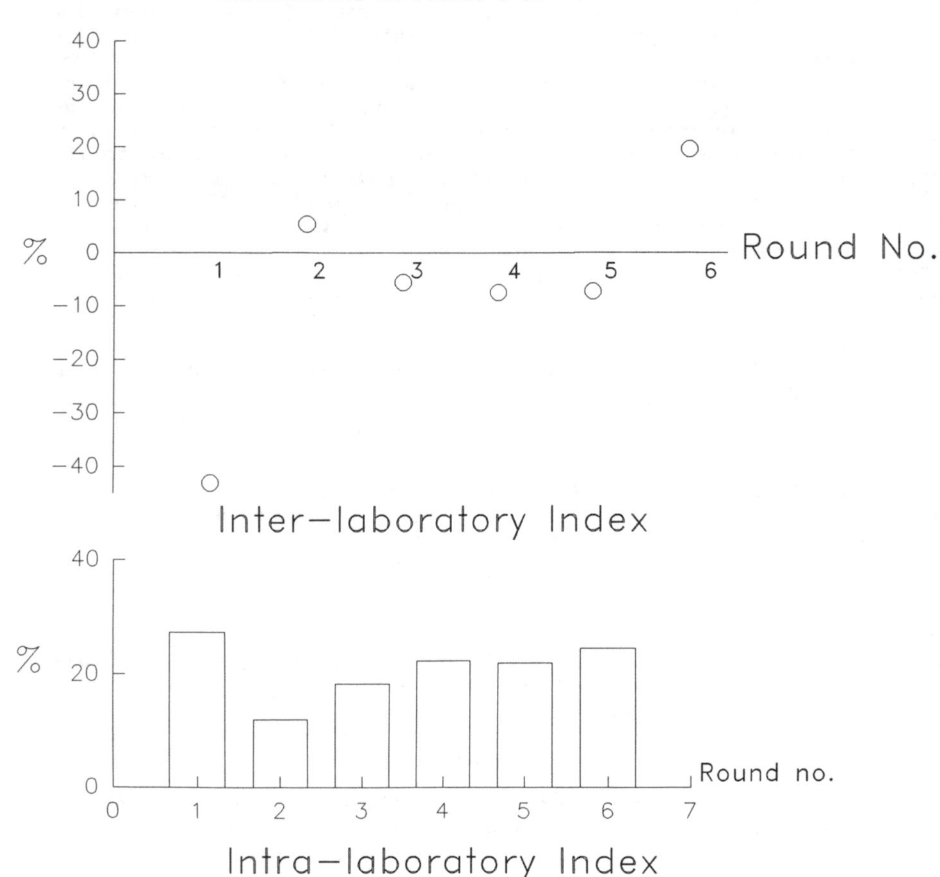

1447

TOWARDS DEVELOPMENT OF RICE TO INCLUDE LOW DENSITY SAMPLES FROM ASBESTOS CLEARANCE OPERATIONS

NP Crawford, PW Brown, AD Jones, WM Maclaren and BG Miller

Institute of Occupational Medicine Ltd
8 Roxburgh Place, Edinburgh EH8 9SU, Scotland

SUMMARY

An outline is presented of the operation of the RICE scheme which aims to improve asbestos fibre counting standards in the UK. Currently, the reference sample stock comprises mainly industrial samples and laboratory performance assessment criteria are based on counts made by an automated image analyser. Operational procedures are being improved, particularly to introduce asbestos clearance samples and visual reference counts. The counting variation obtained in supplementary exchanges between laboratories of clearance samples is described. Performance assessment criteria, based on a median of at least 15 visual counts, are proposed.

The Regular Inter-laboratory Counting Exchanges (RICE) Scheme was initiated in 1984 to minimise differences in asbestos fibre counts obtained by UK laboratories using phase contrast optical microscopy. Currently about 320 laboratories participate in the scheme. Reference samples are counted periodically by each laboratory. Counting performance is assessed as 'satisfactory' or 'unsatisfactory' by comparing laboratory counts with reference counts.

The reference samples, which currently comprise mainly industrial samples with fibre densities typically in the range 100-600 fibres mm^{-2}, are subdivided into batches (8 samples per batch) and the laboratories are formed into groups (5-6 laboratories per group). Every 3-4 months, a round of sample exchanges takes place, the IOM sending a batch of samples to each group for counting. Laboratories are recommended to use the counting methodology specified in the European Reference Method. All results are sent to the IOM for processing. Laboratory counts are compared with reference counts made by a Magiscan automated image analyser using software (the Manchester Asbestos Program, MAP) developed by the University of Manchester. Assessment of laboratory performance is formally based on the results averaged over four consecutive rounds (32 samples over a period of about a year). After each round, each laboratory is notified of its performance assessment.

The performance of a laboratory is judged by how consistently close its counts are to the reference counts (R). The present assessment criteria are as follows:-

Criterion	Rating (single round performance)	Category	Classification
		(Performance over 4 rounds)	
⩾ 75% of counts within 0.70-1.70 x R	1	1	Satisfactory
⩾ 75% of counts within 0.55-2.20 x R	2	2	Satisfactory
Otherwise	3	3	Unsatisfactory

Many laboratories have improved their counting performance since the scheme began. Alongside this, operational procedures have also been improved and further improvements are being made, the most notable being related to the nature of the reference samples and reference counts.

The Magiscan MAP can be replaced by a visual reference system, thereby permitting direct (rather than indirect) comparison of a laboratory's counts with those of its peers. In addition, removal of the Magiscan MAP component of the counting variation in RICE may permit even more stringent performance criteria to be defined and hence further improvements in counting standards. The median of visual counts is preferable to the geometric mean as a reference count since it is insensitive to the presence of outliers. The medians of small numbers of counts are imprecise estimates and only once the number of counts exceeds 12-15 does the variation begin to stabilise at an acceptably low level. A median calculated from at least 15 visual counts therefore provides a suitable reference count. The performance limits for a visual reference, corresponding to the limits based on Magiscan MAP, are

Inner limits: 0.65 - 1.55 x R

Outer limits: 0.50 - 2.00 x R

The present stock of industrial reference samples are atypical of the samples routinely encountered by most laboratories, which are mainly samples from asbestos clearance operations with fibre densities usually less than 100 fibres mm^{-2}. Four supplementary rounds of exchanges involving over 500 clearance samples were conducted in 1988-90 to

a) collect good quality clearance samples for incorporation into the reference stock;

b) collect sufficient visual counts on these samples to provide reference counts, and

c) define performance assessment criteria with such samples included.

The results in these exchanges indicate that the variation of visual counts is fairly constant on the scale of the square root of the fibre density. Also where fibre densities for clearance samples overlap with those of the industrial samples, there is continuity in the behaviour of the visual counts. This gives a basis for deriving a visual reference system and performance criteria for low density samples which link in with those derived for higher density samples.

For reference densities $\geqslant$ 63.7 fibres mm^{-2}, the performance limits are those which apply to the industrial reference samples (63.7 fibres mm^{-2} is the density corresponding to counting 100 fibres in 200 Walton-Beckett graticule areas). For reference densities $<$ 63.7 fibres mm^{-2}, the correspondidng performance limits are:

	Upper limit	Lower limit
Inner limits	$(\sqrt{R} + 1.96)^2$	$(\sqrt{R} - 1.57)^2$
Outer limits	$(\sqrt{R} + 3.30)^2$	$(\sqrt{R} - 2.24)^2$

It is planned to introduce clearance samples, visual reference counts and associated performance criteria into the scheme in early 1992. There is also scope for introducing more stringent performance criteria later.

Other established fibre counting reference schemes (e.g. AFRICA, WHO/EURO, MMMF) may be similarly modified.

REFERENCE MATERIAL: ACTIVE CHARCOAL TUBES CHARGED WITH DYNAMICALLY GENERATED MULTI COMPONENT TEST GAS OF THE THREE ANALYTES BENZENE, TOLUENE AND M-XYLENE

E. FLAMMENKAMP, E. LUDWIG and A. KETTRUP

GSF-Forschungszentrum für Umwelt und Gesundheit
Institut für Ökologische Chemie
Ingolstädter Landstr. 1, 8042 Neuherberg, Germany

SUMMARY

Quality assurance becomes more and more essential in analytical work. For quality assessment and for statistical quality control reference materials get a paramount role. The purpose of this work was to provide a reference material for benzene, toluene and m-xylene on charcoal. For this a multi component test gas was generated dynamically according to the saturation vapour pressure principle. A quantity of three hundred tubes was charged with these compounds. Every tenth tube was analyzed directly after charging to prove the homogeneity. A number of charged tubes was stored under different conditions and analyzed within one year. It appeared that the charging with this equipment was homogeneous and that no significant alteration was found within one year.

INTRODUCTION

Today quality assurance is an essential part of the measurement practice. The decision for the acceptance of products or for the need of medical treatment depends on accurate analytical results. Assessments of compliance to legislative limit values in the field of health, safety and environment presuppose reliable measurement values. Various means to achieve the accuracy are available. One of these is reference materials. They are applicable for calibration of equipment, achievement of traceability of calibration, improvement of measurement quality, verification of accuracy of results and daily quality control through use in statistical quality controls.

The use of solid sorbents is the most important method for workplace air monitoring. Active charcoal is successively used for the sampling of a large number of important compounds e.g. aromatic hydrocarbons. The collected substances were desorbed by solvent extraction mostly and analyzed gaschromatographically. Quantification results from direct injection of liquid standards on the sorbent or seldom by use of a multiple diluted gaseous standard. For the determination of the desorption efficiency the direct injection is practicable and easy to handle. The dynamic spiking method with a gaseous standard allows the determination of the collection efficiencies and the complete validation of a method. Most labs are not equipped for this type of calibration. So we provide a batch of active charcoal tubes charged dynamically with definite amounts of benzene, toluene and m-xylene (btx) as a reference material.

DESCRIPTION

The test gas was produced according to the saturation vapour pressure principle (1, 2). For the dynamically generated test gas synthetic air was used as carrier gas. Alternatively, inert gas can be used as carrier gas. The gas stream was split into two channels. Through the first channel the carrier gas was split to three receiver flask-condenser-containers, each filled with one of the btx-components. Control of the gas streams was achieved by three needle-/shut off-valve combinations. The enrichment of the carrier gas with the hazardous compound happened in the thermostatted receiver; enrichment of the carrier gas with the hazardous compound happened in the thermostatted receiver flasks. After this the gas reached the condensers, whose temperatures were about 20°C below the temperature of the receiver flasks. So the gas was supersaturated and the surplus of the compound was condensed or frozen out. From all three receiver flask-condenser-containers the flows were combined and with the help of the second channel the saturated gas stream could be diluted with pure carrier gas. The dilution ratio was variable by a further needle-/shut off-valve combination. The inlet pressure was accommodated at 1 bar regulated by a needle valve and controlled by a precision manometer (pressure regulator) to get a gas pressure high enough for the correct functioning of the equipment.

The homogeneity of the multi component test gas mixture was parallel controlled by an automatic working gaschromatograph with a gas sample loop of 0.150ml.

With this equipment a batch of 300 active charcoal tubes were loaded with 0.010mg benzene, 0.100mg toluene and 0.070mg m-xylene each. The time for charging was 10 minutes (3).

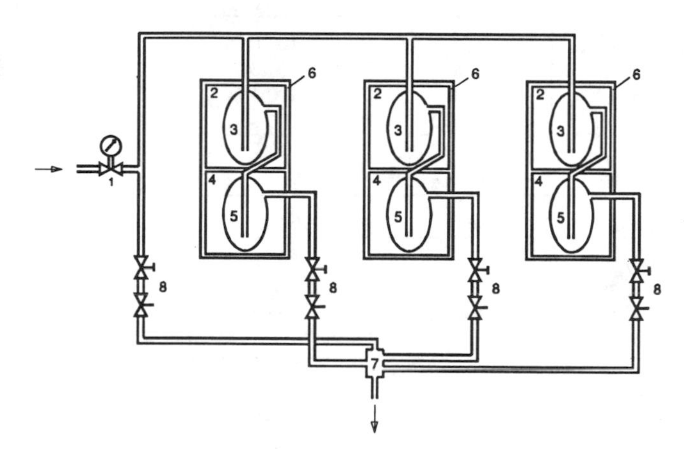

1 pressure regulator
2 thermostat for the receiver flask
3 receiver flask
4 cooling thermostat
5 condenser
6 vacuum jacket for the thermostats
7 mixing chamber
8 needle-/shut off valve combination

Fig. 1. Three component test gas apparatus

Advantages of this technique:

- Test gases are producible from liquid and solid substances with finite vapour pressure.
- The concentration of the working substance in the pure carrier gas is independently variable in the range of the MAK- resp. TRK-values.
- The test gas is produced dynamically with a fast equilibrium adjustment.
- Wall adsorption effects have no influence.
- The test gas volume is nearly unlimited.
- The test gas concentration can quickly be changed by varying the temperature of the condenser.
- The chosen concentration is stable for a long period.

RESULTS

To prove the homogeneity of the generation during the charging the multi component test gas was controlled by an automatic working gaschromatograph with a gas sample loop. The resulting serial chromatograms of the three component test gas showed a good homogeneity of the mixture. It appeared that a relative standard deviation better then 2% was obtained.

To test the homogeneity of the charging 30 tubes were selected randomly. The contents were determined immediately after the charging by headspace gaschromatography. The tubes were desorbed with 2ml benzyl alcohol and the substances detected gaschromatographically. Relative standard deviations of 2.2% for benzene, 2.3% for toluene and 2.7% for m-xylene show the homogeneity of the charging to be acceptable.

The stability of the charged tubes was investigated, too. Charged charcoal tubes were stored under various conditions: in the dark at 2°C and at 40°C and with occasional light at room temperature. The loaded tubes turned out to be stable over at least 12 months, independent of storage conditions. No negative slope versus time was found.

Table 1. Recovery Rates. Storage: 293 K, Light

Days	Benzene	Toluene	Xylene
30	99.4	99.7	102.0
90	103.1	100.0	100.3
180	99.7	100.1	100.1
360	99.6	102.3	101.3

Table 2. Recovery Rates. Storage: 313 K, Dark

Days	Benzene	Toluene	Xylene
30	100.5	98.7	101.0
90	99.7	100.3	100.7
180	100.3	98.4	99.0
360	96.0	100.4	100.1

Table 3. Recovery Rates. Storage: 275 K, Dark

Days	Benzene	Toluene	Xylene
30	99.1	100.6	101.8
90	99.0	99.3	99.1
180	99.1	98.9	98.4
360		103.5	102.5

So it was possible to show that active charcoal tubes can be charged homogeneously with this method and that the charged tubes do not show any loss of substance within one year.

REFERENCES

(1) A. KETTRUP. Equipment for preparing standard gases. GIT Fachz. Lab. 6, 556-558 (1982).

(2) H. WEBER, H. STENNER and A. KETTRUP. Development of an instrument for the dynamic generation of test gases. Fresenius Z. Anal. Chem. 325, 65-67 (1986).

(3) E. LUDWIG. Untersuchungen zur Kalibrierung von Probenahme- und Bestimmungsverfahren für Gefahrstoffe in Luft. Universität Paderborn, doctoral thesis (1989).

Session VIII

FUTURE NEEDS

Keynote Papers

Posters

Panel Discussion

1449

FUTURE TRENDS

J.G. FIRTH

Health & Safety Executive
Occupational Medicine and Hygiene Laboratory
403 Edgware Road, London NW2 6LN, UK

It is sometimes said that the future of any topic is a development from its present. This Conference reveals the present state of affairs across the field of measurement of workplace contamination by toxic substances. Many papers also show the way forward in particular areas of activity but it is useful to step back slightly and look at particular topics across the board with a view to identifying common features.

Each session of the Conference includes papers from many of the countries making up the European Community. One of the main impressions gained is that workers in many of the different countries of the Community are carrying out programmes of work to solve the same technical and operational problems. However many of these programmes have differences of approach and content even though the end result, in each case, is likely to be very similar. There is therefore a very strong case for cooperation between laboratories in the different EEC countries in order to make best use of the resources available, and to achieve answers to the problems more rapidly.

The time is now right to seek more seriously to initiate international collaborative programmes in the different areas related to air measurement and it might be useful to examine each of these areas in more detail. They can roughly be divided into:

(a) exposure limits

(b) sampling techniques

(c) laboratory measurement techniques

(d) quality control techniques for laboratory measurement

(e) field and instrumentation

(f) other concerns.

Exposure Limits

The setting of exposure limits is very important for workplace air measurement. The limits define the baseline against which estimates of satisfactory or unsatisfactory conditions are made. Most of the Community countries have their own sets of exposure limits, many of which can be traced back to limits originally set by the American Conference of Government Industrial Hygienists (ACGIH). More rigorous procedures are now being implemented in many countries to determine values which in the light of current knowledge represent true no-effect levels, that is levels below which there seem to be no adverse effects on workers. The European Community is also moving towards producing its own list of exposure limits values. It is important that on a European basis these activities are properly coordinated so that optimum use can be made of scarce toxicological and hygiene resources. With the growing internationalisation of industry across Europe it seems sensible that each country has the same limit for a particular substance as far as this is possible.

Legislative bodies have also to make clear quite how the exposure limits have to be interpreted when measurements are being made. In view of the uncertainty or errors in measurement, a requirement to demonstrate that measured concentration levels have exceeded a particular value is different from a requirement to demonstrate that a concentration has not exceeded the exposure limits. Thus, if a measurement technique has an error +30% then for an exposure limit of 5ppm the measurement result would have to be approximately 6.5ppm to be 95% confident that the limit had been exceeded or 3.5ppm to be 95% confident that the limit had not been exceeded.

Similarly it has to be borne in mind by limit setting bodies that where the legal requirement is that the exposure limit shall not be exceeded then the log-normal distribution of concentrations within a working environment means that the mean concentration has to be controlled to about one-third of the exposure limits to be 95% certain that the exposure is not exceeded in any particular time. It is not clear that such considerations are being taken into account by limit setting bodies.

Sampling Techniques

Most workplace measurement methods still involve sampling the atmosphere in the workplace followed by laboratory analysis of the sample. It is widely accepted that sampling is the greatest source of uncertainty in the overall measurement. This means that sampling techniques should be properly validated against an agreed protocol. Such protocols have been developed in a number of countries and these need to be harmonised through discussion in the appropriate technical committee of CEN. It seems sensible, in view of the inaccuracies in sampling, that, as far as possible, standard sampling techniques are used and that the range of techniques is reduced to a minimum. Thus for respirable dust the same sampling technique should be used throughout all Europe. The same argument applies to other types of samplers such as diffusive samplers for vapours. If this were achieved then an international programme of validation of techniques to the relevant protocols could be carried through quickly.

Laboratory Measurement Techniques

The work of CEN Technical Committee 137 is an excellent example of international collaboration. The development of performance standards for measurement methods rather than the development of standard methods, allows a flexible approach within different laboratories for the measurement of samples which have been collected in standard ways. In some laboratories it may be more economical to use highly sophisticated analytical measurement techniques whilst in others more labour intensive techniques would be appropriate. Nevertheless similar analytical methods are widely used for a number of substances and an international programme of validation of methods to the CEN performance criteria should be seriously considered.

It is suggested that standards methods for measurement should only be used where the substance being measured is defined by the analytical technique. An example is the measurement of rubber fume where the solvent used in the method defines the result obtained. In such situations standard methods should be used in all countries of the Community where the measurement is to be related to a Community exposure limit value.

Quality Control Techniques

The problems of reproducibility of analytical results between different laboratories are becoming increasingly recognised. Many countries have set up interlaboratory quality assurance or proficiency testing schemes to limit this variability eg WASP, RICE. It seems sensible that such schemes should be organised on a European-wide basis so that duplication between different countries is avoided and resources are efficiently used. Where this cannot be attained then similar national schemes should be harmonised so that assessments of the performance of laboratories in different countries have the same meaning. Obviously, interlaboratory quality assurance exercises involving large numbers of laboratories analyzing samples from a batch of one material is a way of producing a form of standard reference material. The sale of such materials can be used to offset the cost of the quality assurance scheme.

Accreditation schemes for laboratories are being set up in a number of countries and mutual acceptance of such accreditation between countries is taking place. These schemes examine qualifications of staff, management structures, recording of results etc, in the laboratories. It is important that in future the results of the performance of a laboratory in proficiency testing schemes are incorporated into the accreditation process.

Field Instrumentation

The vast majority of measurements taken in routine monitoring exercises are well below the exposure limit. The use of laboratories to obtain such results is an expensive waste of resources. Simple field methods have been developed for some substances which give qualitative results which are sufficient for 'screening' purposes. Such techniques need to be developed for a wide range of substances encountered in workplaces. They range from colorimetric techniques to portable instrumentation. The performance requirements for this type of instrument need to be established by the relevant CEN process.

Field instruments are also becoming available for the quantitative measurement-of atmospheres in the factory without recourse to laboratory measurement. This development should be encouraged since measurements obtained on the spot enable remedial action to be taken immediately. Again the performance requirements for this type of instrumentation needs to be established in CEN standards.

Other Concerns

Most of the papers in this Conference naturally deal with air monitoring. However it has to be recognised that toxic substances enter the body by routes other than the respiratory tract. The main alternative route is through the skin and very often material comes into contact with the skin because surfaces are contaminated. Biological monitoring, i.e. measurement of substances or their metabolites in blood, urine or breath, is a rapidly growing subject. Strategies need to be established to determine whether air monitoring, biological monitoring measurement or surface contamination or a combination of these techniques is the most effective procedure in particular circumstances.

New areas of concern are appearing in workplaces which have implications for measurements. Sensitisation of workers leading to allergies is an area which is receiving increasing attention. Sensitising agents include many reactive substances such as dyestuffs which combine readily with proteins. Sensitisers also include microorganisms as well as proteins from dead microorganisms and from animal or insect sources. The development of techniques for the measurement of such substances both separately and in the mixtures in which they more commonly occur will probably receive increasing attention in the next few years.

It has to be recognised that in many workplaces toxic substances are present in the atmosphere and on surfaces as mixtures. Procedures for identifying the total hazard represented by such mixtures will have to be developed in more detail and these procedures will have major implications for the way in which the measurement of such mixtures is carried out. The development of biologically based techniques which estimate the combined hazard of the components of mixtures, rather than the separate chemical measurement of each component, may well take place.

REQUIREMENTS FOR THE DESIGN AND PERFORMANCE OF FUTURE INSTRUMENTS

B. MILLER

Consultant, Runcorn, Cheshire, UK

SUMMARY

With the increase in industry in the need for accountability, traceability and accreditation there is an increasing need for intelligent, quality instrumentation. Also, as the numbers, experience and skills of people who operate and maintain laboratory and plant instrumentation decreases, the design and performance requirements will increase. The major need of a manufacturer will be the introduction of a Quality System. Instrument parameters such as quality and reliability, performance characteristics, diagnostics and communications will need to be reviewed and improved. Development departments will need to work closely with Quality Assurance departments and production facilities will need to be uprated to improve quality and keep costs down. Instruments in the future will be run using expert systems reducing the need for interaction with the operator.

1. INTRODUCTION

With the increase in industry in the need for accountability, traceability and accreditation there is also an increasing need for intelligent, quality instrumentation.

In order to meet the demands of the user in the future the manufacturer will have to improve the design and performance characteristics of instruments. This is a result of a reduction in the numbers of people available to operate and maintain the equipment combined with a reduction of skills and experience.

The manufacturer will need to install a Quality System which complies with the requirements of, for instance BS 5750 Part 1. Users seeking accreditation of their laboratory by a National body, such as NAMAS, will be helped by having quality equipment. Potential customers will buy equipment only from suppliers who have properly planned quality systems.

Instruments will need to include at least the following features: de-skilled operation; minimum down time; fewer parts making the equipment less mechanical; three level diagnostics; error recovery and logging; ease of maintenance; good communications; self validation and, particularly in the case of field instruments, self calibration.

Instrument performance will have to be considered in the light of improved detection; use of chemometrics; data handling; better accuracy and improved safety aspects.

Development departments will need to work closely with Quality Assurance departments and production facilities will need to be uprated to improve quality, reduce the number of rejects and make the workforce more flexible to reduce costs.

The instrument of the future will be a "black box" consisting of expert systems which will allow the instrument to think for itself, checking and correcting functions without reference to the operator.

2. QUALITY SYSTEMS

Any manufacturer wishing to produce and supply instruments in the future will need to install a quality system which complies with the requirements of BS 5750, Part 1 : ISO 9001 or EN 27001 (1). Quality has already become the most important issue facing a manufacturer who wishes to compete on an international basis. Legislation concerning fitness for purpose and the performance of products is increasing and major customers will buy equipment only from suppliers who have properly planned quality systems. Quality governs the process of manufacture and development, the more sophisticated the equipment the more difficult it is to maintain quality. In-house written procedures help employees and reduce rejects. Equipment produced to a Quality Standard will help the purchaser to comply with accreditation requirements, such as NAMAS in the UK, and so help to sell his product(2).

3. INSTRUMENT FEATURES

Instruments should incorporate at least the following features: de-skilled operation when the design takes into account the need for testing and the needs of the user. User interaction via the software facilities. This has greatly increased in recent years, for example, in the Perkin Elmer range of gas chromatographs, the Sigma 3 series used 3K of software whereas the current 8700 series uses 600K;

minimum down time;

fewer parts, making the equipment less mechanical;

diagnostics should be in three levels to cover initial switch on; user accessibility: remote access for the engineer enabling the user to use a telephone plug-in so that the engineer can investigate the problem prior to a visit and carry out a repair if required, storage of instrument history;

malfunction, description of the event, can the instrument continue to operate or is the malfunction fatal and the instrument stops;

error logging and error recovery;

ease of maintenance, preventative maintenance;

communications, for example, the Drager Polytron system when linked to DEDAS (Digital Environmental Data Acquisition System) enables the operator, using a fieldbus system, to interrogate each head separately, thus increasing flexibility;

self validation, instruments should be able to self validate using electronic means.

4. PERFORMANCE

Performance should, particularly in the case of the chromatograph, include improved detection and the use of chemometrics, deconvoluting peaks, better accuracy and more/better data handling to reduce dependence on chromatography.

One possible outcome would be that all instruments would be equipped with mass spectrometric detectors and that a detector will be developed to replace the electron capture detector.

In the case of the Polytron and similar instruments the development of an "intelligent" data storage card will enable transmission of all measured values and special events to a printer. For example, with a twelve channel system using different sensors on each channel it will be possible to "activate" the card which will look at information from channel cards and print out all the information available from the measuring heads including any alarm information. This information can be collected at any time and the system will replace the need for multi-channel recorders giving a hard copy report.

5. PRODUCTION

All production will need to be carried out to the requirements of BS 5750 Part 1 or equivalent. Manufacture will have to be controlled and use made of the "Just in Time" (JIT) system to reduce the need to carry excessive numbers of parts and thus reduce financial outlay. A streamlined, automated operation together with a flexible workforce will reduce costs, keep the price to the customer down and improve quality.

6. QUALITY AND RELIABILITY

As equipment becomes more sophisticated and thus more expensive, the purchaser expects fewer problems and it becomes more difficult to maintain quality.

The quality of information gathered using equipment is most important and requires the use of an Internal Quality Card (IQC), External Quality Assurance (EQA) and the use of certified standards for calibration (such as those available from BCR).

It is likely in the future that equipment, to be acceptable across Europe, will need to carry the "CE" mark as a guarantee of quality.

The use of a "drop test" on equipment packed for delivery will reduce the possibility of delivering equipment damaged in transit.

Suppliers can provide regular preventative maintenance checks via service contracts to prevent problems arising which could result in the instrument failing.

7. DEVELOPMENT

It will be necessary for development departments to work closely with QC departments to ensure the use of approved components and to consider materials used, screening and PC boards. Consideration will have to be given to regulatory requirements such as those for radio frequency (RF) and electronic discharge, also to European Standards dealing with, for instance, performance requirements. Safety regulations will need to be considered, such as CSA for the selection and use of approved components (3).

Instruments will have to be made more "idiot proof" such that the design takes into account the need for testing and the user. Self calibration, involving the use of input voltages and the construction of a curve electronically will replace the need for manual production and storage of a curve by the operator using, for instance, standard injections.

8. THE FUTURE

It is likely that the instrument of the future will become a "black box" containing expert systems enabling the instrument to think for itself, checking and correcting functions.

For example, in the case of chromatographs, if retention time starts to drift due to column ageing, the instrument will modify conditions to put peaks back into windows without reference to the operator. (Current practice is to flag up a warning to the operator who has to take the necessary action). Safety aspects will be improved. Mains effects will be covered using a "brown out" test to cover missing mains cycles and a "black out" test for complete loss of mains short term; a microprocessor will "tidy things up". Instruments will undergo a three day "soak test", that is exposure in a hot room (at about 40 ^{o}C) to accelerate any potential faults showing. When equipment is linked to automatic sampling equipment such as the Automatic Thermal Desorber (ATD) then a no peak alarm is needed to stop the sampling sequence and prevent the loss of results. It will be possible to feed methods from a computer into the instrument to check that conditions/configuration are correct.

For example, in the case of area monitors, using a field bus system to digitalise signals, flexibility can be improved by using programmable measuring ranges. One standard head with a selection of sensors, select sensor, plug in and programme system to detect that particular gas. Pre-calibrated, plug-in sensors can be developed (currently calibrated on installation) and signal prediction can be used to predict what may happen to levels being encountered using a built in rate of change from historical information. To test the complete system, instead of applying a gas manually, it would use automatic calibration (produce gas at the sensor), by telling the system electronically that a standard gas concentration was present.

In conclusion, I believe the instruments of the future will be like our modern day washing machine when the system can be set up to do a job simply by pressing a few buttons.

REFERENCES

(1) BS 5750: Part 1 (1987). British Standard. Quality systems. Part 1. Specification for design/development, production, installation and servicing.

(2) NAMAS Information Sheet NIS 45. Edition 1, October 1990. Accreditation for chemical laboratories.

(3) Canadian Standards Association, Regulation C22.2, No.151.

1451

STANDARDS AND INFORMATION MANAGEMENT

H. BLOME

Berufsgenossenschaftliches Institut für Arbeitssicherheit
Alte Heerstrasse 111, Sankt Augustin, Germany

ISO/IEC, as well as CEN/CENELEC, standards are very significant on the international level. While it is up to the member countries to implement ISO/IEC standards, European standards (EN) passed by a weighted majority of votes, have to be incorporated into national law. It is a basic rule of European standardization work that the national standards of the different member countries are harmonized wherever possible. On account of the European internal market in preparation and the largely interconnected world economy it appears to be useful to set up requirement- and test standards to which both manufacturers and users can refer; this is particularly interesting on account of the fact that a purely national implementation has become an exception. The aim is the free circulation of goods and services, regardless of their origin, within the European Communities, assuming that they meet European guidelines and harmonized European standards.

As to the field of air pollution measurement, there are two European draft standards:

- 'General requirements for the performance of procedures for workplace measurement' (prEN 482), and
- 'Specification for conventions for measurement of suspended matter in workplace atmospheres' (prEN 481; Definitions of inhalable, thoracic and respirable dust).

By defining dust fractions and determining generally applicable requirements in a primary standard, new ground was broken. Two urgent tasks result from the above-mentioned draft standards. Firstly, existing measuring techniques have to be investigated to check compliance with the defined requirements. Secondly, test conditions must be determined as in the case of dust-measuring devices.

The CEN central secretariat received draft standards for measuring strategies and requirements for diffusive samples.

In addition, the CEN/TC 137 plans to elaborate standards concerning:

- requirements for sampling pumps
- requirements for test tubes
- requirements for sorption tubes
- requirements for direct-reading instruments for gases and vapours
- test methods for size-selective sampling
- terminological definitions for the field 'air pollution measurement'
- list of equivalent terms.

The TC 137 has defined a basic principle to be considered when determining requirements for specific types of devices or parts of measuring systems. It says that the elaboration of methodological instructions for single substances shall be avoided; requirements applying to certain types of devices are going to be defined in separate standards ancillary to the general requirements.

In addition, the activities of TC 114 'Safety of machinery' are very important for the field of air pollution prevention. Working groups deal, for example, with the determination of conditions for the testing of pollutant-emitting machines. Technical modifications can be assessed with regard to their influence on the emission level and information on expectable exposure concentrations can be obtained by means of a test bench offering the advantage of reproducible framework conditions.

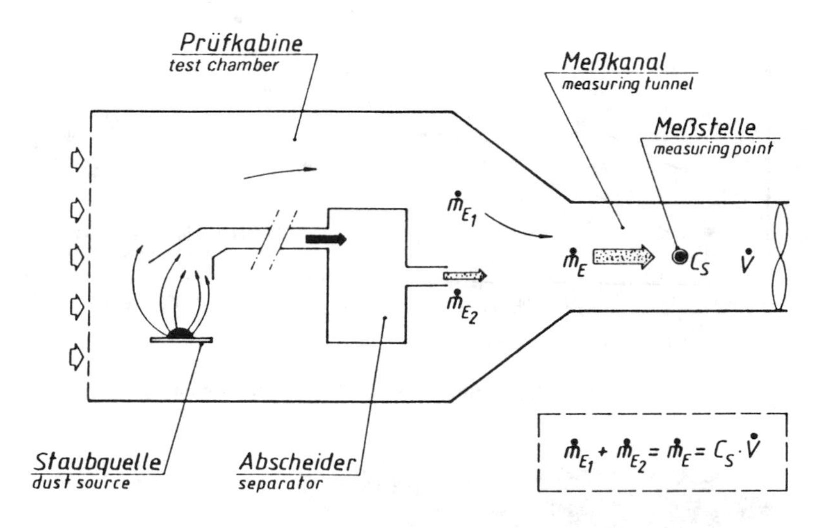

Fig. 1. Scheme of the test bench

Figure 1 is a schematic presentation of the test bench, consisting of a test chamber and a measuring tunnel through which a defined air volume flow is drawn. If, as illustrated in Figure 1, the machine under test is a dust-emitting device (the dust source) with a downstream separator, the total dust emission rate $\dot{m}_E$ consists of the dust quantity $\dot{m}_{E1}$ which is not captured by the suction device and of the dust portion $\dot{m}_{E2}$ which is not held back in the separator but reintroduced with the return air. The emission rate $\dot{m}_E$ results from the average concentration C_S at the measuring point in the tunnel multiplied by the volume flow $\dot{V}$ drawn through the test bench during the test time. The test chamber has to be dimensioned so that the device under consideration can be operated under field conditions. After determination of the emission rate, it is possible - based on a correlation analysis with respect to values measured in the field - to assess the devices in terms of compliance with concentration limits applicable at industrial workplaces.

Figure 2 illustrates the correlation between fibre emission in the test tunnel and fibre concentration in the test chamber determined within the framework of an investigation of 25 asbestos-cement-working machines in the BIA test bench. The fibre concentrations in the chamber correspond to worst-case conditions at the workplace.

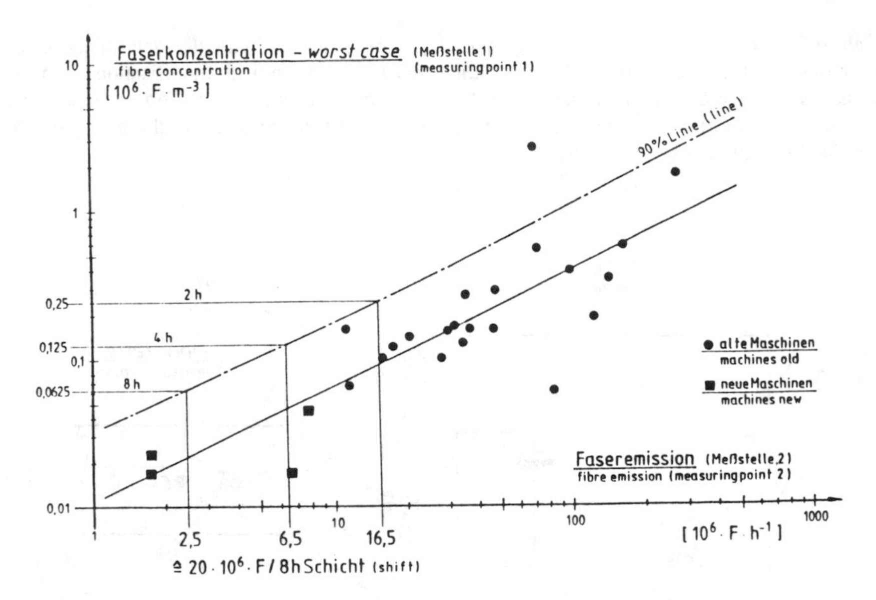

Fig. 2. Correlation fibre emission/fibre concentration

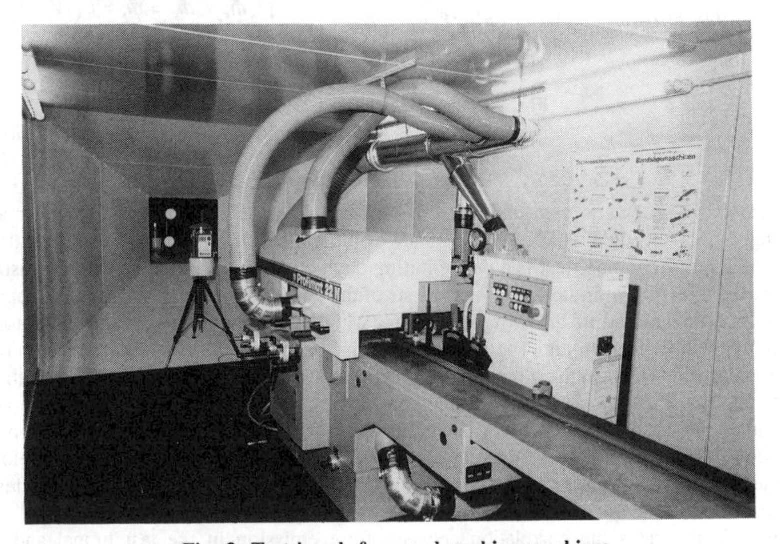

Fig. 3. Test bench for wood-working machines

The definition of other test methods for technical working means, machines and devices would be a recommendable approach:

- working-machines for brake linings (to estimate the workplace exposure to asbestos fibres)
- blasting-machines

- working-machines for dental protheses
- capture devices for welding fumes.

The international standardization work needs future-oriented developments which are trendsetting in the field of technical pollution prevention:

a) Model studies with a view to determining process-dependent pollutant emissions, as well as

b) Testing of solids with respect to dustiness.

ad a) There are many cases in which potential workplace risks are difficult to determine, workplace conditions being hardly reproducible. Among them count, for example, mobile workplaces or workplaces with manual activities which are often characterized by frequently changing working conditions. In such cases, model studies permit the determination of hazardous conditions. The results obtained in these studies can be used to develop test procedures covering the largest range of applications possible. In this way, it may be possible, in the case of certain technical procedures, to substitute or complement expensive exposure measurements or to develop strategies for simplified measurement. As they aim at elaborating field-oriented recommendations for workplace monitoring, model studies could be considered a valuable complementary aid for small and medium-sized enterprises.
ad b) Studies investigating the dustiness of solids by means of standardized instrumentation and procedures could also serve to provide valuable information for firms and supervisory bodies. It would be possible to analyze the degree of dustiness of certain solids and examine which portion of a specific dust fraction or of specific components could be transferred into the airborne state under defined framework conditions. Products of reduced dustiness could thus be recommended for use in the field. It would also be possible to determine whether certain components, such as asbestos in talc, are detectable in the airborne state under defined working conditions. In this context, numerous other examples can be given, *viz.* the use of large-grained materials without occluded fine dust instead of small-grained or plastic-laminated or impregnated solids.

Several of the above-mentioned arguments in favour of the development of new standards aim at elaborating suggestions for reducing exposure on the basis of the obtained results. One should perhaps think also of laying down in a standard some basic considerations with regard to exposure reduction.

Apart from the elaboration of an internationally harmonized terminology, it is also necessary to specify the hierarchy of the following protective measures: material substitution, obligations of the manufacturer, manufacturing technological measures, production parameters, maintenance, care, cleanliness, capture technology, air ventilation.

The knowledge gained in many different ways, by means of standardization activities, has to be implemented suitably, to make sure that firms, measuring services and supervisory authorities can make use of it.

Any expert working in the field of pollution prevention has to cope with the problem of information transfer:

- Simplification and implementation of new pertinent scientific knowledge in the field.
- Confrontation with a large variety of different substances nowadays used at workplaces (the EC register of chemical substances, EINECS, comprises more than 100,000 items).
- Questions concerning safe storage and disposal.
- Requirements in accordance with a large spectrum of laws, regulations and standards.
- Complexity and variety of possible technical approaches.

- Effects of the substances on man and reactivity with other products.

In this context, modern communication has to be used to provide the essential information. It is a support not only given to safety specialists in the factory, but also to technical inspectors, occupational hygienists and to those who deal with the administration of occupational medicine. In founding the common project GESTIS in 1989, the German statutory accident prevention and insurance institutions in industry decided to create an extensive information system on hazardous substances whose coordination lies with the BIA.

GESTIS is intended to reduce or avoid workplace exposure to hazardous substances. The system includes measuring activities, protective measures, handling of hazardous substances, information on potential risks and assistance in investigating occupational diseases.

Due to the system's structure comprising the categories

- data on substances and products
- exposure data
- data on professional diseases,

it is possible to obtain separate information on substances, products, professions, industrial branches and workplaces.

Apart from the elaboration of statistics on occupational accidents and diseases, the EC member countries are more and more concerned about the creation or further development of information systems or product registers. Such systems are meant to provide details on chemical substances, products and their formulations, quantities used, effects on human health, information on the identification of substances and on safety data sheets, material substitution and legal regulations.

Recommendations and guidelines containing quickly available, clear and practical instructions for measurement, assessment, implementation of regulations and technical measures could be part of such information systems, e.g.:

- Recommendations for workplace monitoring (which could embrace instructions concerning specific substances or workplaces, such as cooling lubricants, fibre mixtures, rubber industry).
- Analyses of model workplaces (which could serve as an example in similar cases).
- Information on specific measuring techniques (in addition to defining standards and manufacturer's specifications, a lot of measuring methods need supplementary explanation to be correctly and efficiently used in the field).
- Substance- and location-specific dossiers (containing precious information on exposure concentrations, technical and personal protective measures, legal bases as well as details on the elaboration of risk analyses etc).

In addition to the definition of requirements and test methods within the framework of standards it is absolutely necessary to make these regulations available to users and addressers of guidelines and standards so that the latter can be implemented in view of effective work safety strategies.

1452

NEEDS FOR LEGISLATION

R. HAIGH
Head of the Industrial Medicine and Hygiene Unit

Health and Safety Directorate
Commission of the European Communities
Jean Monnet Building
L-2920 LUXEMBOURG

SUMMARY

Action to safeguard the health and safety of workers was given a renewed emphasis by the introduction in 1987 of a reference to the establishment of minimum standards under Article 118A of the Treaty establishing the European Community. The impact of this new treaty basis for action is described in terms of the legislation now being developed and adopted. Other activities now taking place, or envisaged, which are likely to have the effect of increasing the extent of legal requirements concerning the measurement of exposure or its practise are discussed.

The range of technically very detailed papers presented at this conference shows the extent to which the problems of workplace air monitoring are now being addressed. It has been indicated quite clearly that there is not yet any room for complacency, in terms of ensuring that there are available all of the techniques needed to give reliable answers to all of the questions that are asked of the hygienist, the employer, employee, or even the enforcement officer.

The aim of this paper is to broaden the debate, and to look to the future in giving a view on the way that legislation, and the role of the Community, might develop in the next few years. It has to be the personal view of the author because whatever the Commission Services might propose has to be carried forward with the consensus and support of Member States. The Commission is sometimes described as the engine driving change. For legislation this might be apt but for the wider action discussed in this paper, the Commission and the Community might be described more aptly as catalysts helping to bring together the interest groups and to promote change and progress.

The first paper at this conference described the background of current legislation against which the Community activities on standards and measurement have developed. There are important changes, to which the paper referred, now taking place, which will be of increasing importance in the future.

Perhaps the key to these changes was the adoption in 1987 of Article 118A in the Treaty establishing the Commmunity. Article 118A gives extra emphasis to the role of the Community in matters concerned with the health and safety of workers (Fig. 1).

FIG.1

Article 118A of the EEC Treaty

1. Member States shall pay particular attention to encouraging improvements, especially in the working environment, as regards the health and safety of workers, and shall set as their objective the harmonization of conditions in this area, while maintaining the improvements made.

2. In order to help achieve the objective laid down in the first paragraph, the Council, acting by a qualified majority on a proposal from the Commission, in cooperation with the European Parliament and after consulting the Economic and Social Committee, shall adopt, by means of directives, minimum requirements for gradual implementation, having regard to the conditions and technical rules obtaining in each of the Member States.

Such directives shall avoid imposing administrative, financial and legal constraints in a way which would hold back the creation and development of small medium-sized undertakings.

3. The provisions adopted pursuant to this Article shall not prevent any Member State from maintaining or introducing more stringent measures for the protection of working conditions compatible with this Treaty.

But as can be seen from the second and third clauses, it also ensures that such action takes into account the needs of smaller undertakings, whose well-being is vital to the economy and social structure of the Community.

It also gives Member States the right to go further than the minimum standards set by the Community and is aimed at preventing those Member States who already have higher standards from back-sliding (e.g. by relaxing requirements where they are already more stringent than Community limit values).

Linked to the question of Minimum Standards and Community action is the notion of subsidiarity on which the President of the Commission, Mr Delors, has made a definitive policy statement. In essence the policy is that the Community should act only where it is necessary to do so to achieve equality of action, leaving points of technical detail or local needs to be taken into account by individual Member States, or putting it another way, only when there would be a net benefit, should Community measures be undertaken.

So too, social questions, such as worker protection, cannot be separated from the creation of the Internal Market. An important reference point of this Social Dimension is the Social Charter which, whilst not being binding, is nevertheless an indication of the willingness of the Community to address social rights.

This link between the Internal Market and the Social Dimension raises the question of the Commission policy on reference to standards and particularly CEN standards - in developing the policies based on harmonization using Article 100A of the Treaty.

It is indisputable that CEN and other standardisation bodies have a key supporting role - but it is a supporting role and should not become the leading role - a supporting role in which the Community legislator in cooperation with CEN decides that mutual benefit can be accrued by standards work.

Difficulties have arisen in the past over the role of CEN standards in the context of minimum standards in legislation (Article 118A) which can, or should, be built on. There have been differences of view on the context of standards and on the extent to which standards can be used in this way. It seems likely that their use in the future will be primarily in areas where Member States individually, in their laws, are unlikely to wish to be more restrictive, i.e. to impose higher legal standards, then the Directive on which the national law is based. Whether the Commission and CEN have yet learned how to cope with their joint responsibilities is not clear.

With these ideas in mind it is now relevant to comment on how this development in the Treaty base and in policy is now being put into effect in health and safety, and its impact on measurement of workplace conditions in the future.

A study of the Directives prepared up to 1988 shows that they were often quite detailed and prescriptive. The methodology and requirements for monitoring lead or VCM in air are good examples where even the fine detail of measuring strategies is given. For the future it seems logical to suggest that only the key principles should be incorporated into Directives. Technical details should be delegated to the Commission or a Standards organisation. In this way it should be more easy to keep the content of Directives up to date with developing techniques such as were described in the earlier papers on CEN and later in detail on measurement techniques, to the extent that there is a requirement for a Community methodology.

Based on Article 118A of the Treaty, there is now in place a Framework Directive 89/391/EEC (1) "on the introduction of measures to encourage improvements in the safety and health of workers at work". This Directive is written in very broad-ranging terms that make it the responsibility of the employer to protect, in every respect related to work, the health and safety of his/her employees.

There are a number of so-called "daughter" directives made under this Framework Directive that expand on this basic requirement - but still in terms of principle rather than detail.

For example there is a Council Directive adopted under the Framework Directive dealing with the "protection of workers from risks related to exposure to carcinogens at work" (90/394/EEC) (2) which makes provision for the establishing of Limit Values for carcinogens. As the Commission takes forward its work on these Limit Values it has to consider at the same time the availability of methods for measurement and the reliability of these methods, since there is no point in setting limits that can neither be applied by the employer, nor enforced by the competent authorities. In making the assessment of measurement methods, technical advice will be needed, some of which may come from the Scientific Expert Group on limit values and some of which will need help from the specialist sources available in Member States. If, as has been proposed, a Health and Safety Agency is established within the Community, it could be anticipated that it would be charged with gathering technical information of this sort. But given the numerous approximations inbuilt in the numerical limit it is necessary to guard against searching for perfection in methodology incompatible with the needs of the employer or legislator.

Resolving technical problems in ensuring measurement quality will also require the continuation and perhaps the extension of the excellent work being done in this respect by DG XII - the BCR - and which is covered by a separate presentation at the conference.

If what has been said about Article 118A and about carcinogens is now related more generally to chemicals and other pollutants in the workplace, there is a case for preparing a "consolidated" text on chemicals, under the framework of Council Directive 89/391/EEC. This would bring up to date the legislative base provided by

Council Directive 80/1107/EEC (3) "on the protection of workers from the risks related to chemical, physical and biological agents at work". This has already been done for biological agents and a Directive is under discussion for physical agents.

It is therefore possible to foresee in the fairly short term, so far as chemical agents at the workplace is concerned, a new broadly based text laying down the principles of the control of health and safety risks from almost all chemicals in almost all workplaces. There are some special cases that are likely to remain separate, e.g. where ionising radiations are involved, and in the case of some substances such as asbestos which it remains expedient to treat on a case-by-case basis.

Such a text on chemicals would provide a good framework on which to more rapidly develop Community wide exposure limit values, perhaps of a rather more binding nature than the 27 Indicative Limit Values already adopted. It could give a basis for describing and clarifying how the relevant CEN standards on measurement could be brought into the legal context of Article 118A and it could include a requirement for measurements to be made by reliable techniques wherever measurement is necessary for ensuring compliance with limit values. The work which a number of organisations are now doing to improve the quality of measurement would then be more firmly underpinned by Community legislative requirements.

But it is possible to go further. The earlier discussion in this conference shows that there is a need to think about how the whole Community can benefit from many of the more local initiatives now being taken in Member States. For example, to improve the availability of information on up-to-date methods, and to ensure that laboratories who undertake measurements and analysis are able to demonstrate the quality and reliability of their work through quality assurance. There is also a need to coordinate better data-gathering on exposure so that the policies and limit values that are derived from them, or to which the data contributes, are more soundly based. It is a regrettable fact that every time information is sought on exposure patterns in workplaces, it is difficult to get good data not because the measurement methods are not available, but because the pattern of measurement at workplaces is irregular - too much of the effort seems to go into too few workplaces - or because the data is not recorded in a readily accessible of consistent way across the Community. There have already been informal discussions between groups in some Member States but, perhaps with 1992 in mind - The European Year of Safety Hygiene and Health Protection at Work, the Commission can assist in some way to facilitate the development of these discussions leading to a greater uniformity of data management. It is interesting to note that the title of this conference "Clean Air at Work" has been adopted as one of the themes to be addressed in the year.

The Commission has also noted numerous attempts to prepare on diskette a composite database of exposure limits. It has been suggested that the Commission should consider whether a data base on exposure limits together with associated validated methods for measurement might usefully be developed with a European dimension, as more and more industry and technical support activities cross the national boundaries within the Community and beyond. Perhaps all European limit values and validated methods should indeed be accessible on one database.

The points made in this paper should have given an impression of Community action based on a legal framework but not constrained or limited to just legal matters. The aim, after all, is to promote the health and safety of workers as part of an important social dimension to Community policies, and this will only be successful if it achieves change at the place where the risks arise - at the workplace. The aim of this paper has been to stimulate thought and reflection on what more might be done at a Community level and to bring forward ideas in the panel discussion to follow - and to encourage to continue contribution in a positive way even after the conference becomes just a memory. The contribution of the people who are involved on a day to day basis is the seedcorn from which real benefits can be harvested in the future by cooperative action whether in the development of legislation, of standards, of new and better measurement techniques and equipment or in promoting workplace practice and awareness.

While the legal base remains as the key area for Community action, it seems sensible to use Community legislation not as the end but as the basis now to begin to develop initiatives to promote better understanding and application of the practice of health and safety - including initiatives related to "Clean Air at Work" - across the whole Community. This is the "catalyst" role described earlier in this paper. This is the subsidiarity of Mr. Delors; this should be the future trend.

REFERENCES

(1) Council Directive 89/391/EEC of 12 June 1989 on the introduction of measures to encourage improvements in the safety and health of workers at work (O. J. N° L. 183 of 26.06.1989, p.1).

(2) Council Directive 90/394/EEC of 28 June 1990 on the protection of workers against risks related to exposure to carcinogens at work (O. J. N° L. 196 of 26 July 1990, p.1).

(3) Council Directive 80/1107/EEC of 27 November 1980 on the protection of workers from the risks related to exposure to chemical, physical and biological agents at work (O. J. N° L. 327 of 3 December 1980, p.8).

PANEL DISCUSSION

Chairman:
R. HAIGH
Commission of the European Communities, Luxembourg

Rapporteur:
S.G. LUXON
Royal Society of Chemistry, London

Mr Haigh, in opening the Panel Discussion, invited the Chairman of each session to outline the main points which had emerged. These points were then assembled in a logical grouping and each of them discussed in some detail. The final conclusions are summarised below.

LEGISLATION

1. The interpretation of limit values needed clarification to take account of the errors in their determination which could be as much as $\pm 30\%$. There should be mean and maximum values or some other way of providing the necessary flexibility in their application to the control of workplace contaminants.
2. There was a need for additional training in occupational hygiene for all those who are involved in the evaluation of the working environment including those who carry out the monitoring procedures.
3. It was recognised that CEN were in the process of producing standards. The conference debated the value of such documents in providing technical specifications expanding the general provisions of Directives. On balance however it supported the general principle of adopting such consensus documents.
4. Smaller enterprises needed more simple direct guidance to enable them to understand the practical effect of legislation.

METHODOLOGY

1. It was agreed that sampling strategy was less developed than methodology and guidance was needed on the former.
2. Training and instruction was needed in simple easily operated methods.
3. More direct-reading instruments which are simple and robust were needed.
4. The validation of methods and sampling procedures together with guidance on good laboratory practice would ensure reliable results.
5. More use could be made of diffusive samplers, which have now been validated for the evaluation of long term exposures to many organic vapours.

AEROSOLS

1. There was a need to ensure that the values obtained by older sampling techniques can be compared with those obtained from more recently developed procedures, e.g. for asbestos.

QUALITY ASSURANCE

1. The view was expressed that the Commission should endorse the need for quality assurance and proficiency schemes. While it was agreed that the Commission could not be expected to initiate such activities there was a need for it to promote coordination within the Community.
2. A 'Euro-body' should be set up to accredit suitably qualified and trained 'occupational hygienists' but it was recognised that differing practices in member states might make such a universal approach difficult. It was agreed that the accreditation of the training programmes might be a better approach.

GUIDELINES AND STANDARDS

1. There was a need for more informative and readily understood guidelines on validation and testing protocols so as to achieve a uniformity of approach in member states.
2. A dictionary of equivalent terms in each Community language would avoid misunderstandings and promote the exchange of information. It was suggested that CEN might be asked to undertake such an exercise.
3. Line management should be involved in occupational hygiene if future demands requiring day-to-day supervision are to be implemented on the shop floor.

SMALL AND MEDIUM ENTERPRISES

1. Monitoring techniques are moving in the right direction in that they are becoming simpler to operate, more robust and direct-reading.
2. The need to consider the cost effectiveness of health and safety legislation and its implementation was emphasised.
3. There was a need for the removal of perceived barriers to action. The cost and complexity of the assessment and control of hazards was often perceived to be too complex for 'in-house' action.
4. A simpler approach was needed. This might take the form of a preliminary assessment of the hazard using smoke tubes and light beams followed by the implementation of simple cost effective control measures.
   Simple monitoring using detector tubes, diffusive samplers or direct-reading instruments could then be carried out 'in-house'. Only if there was still doubt would outside expert advice be necessary.
5. There was a need for a conference of representative bodies in each member state to promote an awareness that easily understood cost effective action was possible to control airborne contaminants.

ALLOCATION OF PRIORITIES AND COORDINATION

1. A way must be found of avoiding duplication of effort.
2. The protection of workers by the control of the hazard must be the first priority.
3. If other parts of the Commission are involved there must be coordination so that local action is simplified. This is particularly important in small and medium enterprises where one person will often be responsible for all health, safety and environmental matters.

INDOOR AIR

1. Offices and related workplaces employ over 50% of workforce.
2. There must be a multidisciplinary approach to health problems.
3. The reason why exposure limits for offices are set at a lower level than for other workplaces requires investigation.
4. There was a need to be selective in the choice of building materials.
5. Source control was the best approach.

SUMMARY REPORT

ACKNOWLEDGEMENTS

INDEX OF AUTHORS

INTERNATIONAL SYMPOSIUM "CLEAN AIR AT WORK"

LUXEMBOURG, 9-13 SEPTEMBER 1991

SUMMARY REPORT

A. Braithwaite (UK), R.H. Brown (UK), E. Buringh (NL), M. Curtis (CEC), J.C. Guichard (F), L.C. Kenny (I), H. Knöppel (CEC), S. Luxon (UK), R. Narayanaswamy (UK), K.J. Saunders (UK), B. Striefler (D) and S. Vandendriessche (CEC).

1. INTRODUCTION

1.1 ORGANISATION AND AIMS

This symposium was organised jointly by the Health and Safety Directorate (DGV) and the Community Bureau of Reference (DGXII) of the Commission of the European Communities, in conjunction with the UK Royal Society of Chemistry, Analytical Division, Automatic Methods Group and with the cooperation of the WHO and six national organisations involved in workplace health and safety.

The Commission of the European Communities in 1986 organised a symposium on diffusive sampling, which not only set out a new trend, but also catalysed cooperation in Europe in the field of workplace air monitoring. The importance of this cooperation has increased to such an extent that it was felt that there was a need for a new symposium with a much wider scope.

With this in mind, a team of speakers was brought together to outline the latest developments in regulations, in standards, in methods of assessment and measurement in both the industrial and non-industrial working environment, and in quality assurance for laboratory techniques.

It was the aim of the symposium in each of these areas to review the state of the art, predict future needs and, in particular, to promote European cooperation in the establishment of agreed performance targets, common methodology and consistent reporting and information management.

The symposium was attended by approximately 320 participants from all the Member States of the Community and a number of other countries such as Australia, Austria, Canada, Finland, Japan, Norway, Poland, South Africa, Sweden, Switzerland, and the United States of America.

This summary report is based on the reports of the Session Rapporteurs and gives an overview of the contributed papers and discussion, and contains the conclusions and recommendations for action agreed by the participants.

1.2 OPENING REMARKS

The Symposium was opened by W.J. Hunter, Director of the Health and Safety Directorate of the Commission of the European Communities. He stressed the renewed emphasis that the Community was placing on the importance of health and safety as an integral part of its social policy. In particular, he cited Mr Delors and Commissioner Mme Papandreou as saying that health and safety should not be allowed to fall behind the economic aspects in the development of the internal market. It was against this background that the Commission was pleased to support the Symposium and looked forward to a fruitful meeting.

1.3 LEGAL BACKGROUND

The development of Community legislation on workplace health and safety (apart from mining) dates from the mid-1970s with the setting up of the tripartite Advisory Committee on Safety, Hygiene and Health Protection at Work. This was followed in 1978 by the first Directive involving a chemical agent and the need to monitor exposure - Directive 78/610/EEC on vinyl chloride. Next came the 'framework' Directive 80/1107/EEC (amended by Directive 88/642/EEC) on the protection of workers from risks related to exposure to chemical, physical and biological agents at work. Implicit in the latter Directive is that measurements of exposure may be needed to assess whether exposure limits have been met - exposure limits that are themselves the subject of continuing Commission activity; Directive 91/322/EEC contains the first set of 27 Indicative Limit Values.

The Annex to Directive 80/1107/EEC contains details of the content and requirements for reference methods for measuring exposure covering

- (1) definitions of suspended matter;
- (2) definitions of Limit Values;
- (3) methodology for assessment and measuring strategy;
- (4) requirements concerning the reliability of measurements;
- (5) requirements regarding persons competent to take measurements.

There are two other 'specific' Directives, one on lead (82/605/EEC) and one on asbestos (83/477/EEC). These are more specific on what and how to measure (like the vinyl chloride Directive), and may need to be modified in the light of the more general provisions of the framework Directive 89/391/EEC.

1.4 CEN STANDARDS FOR WORKPLACE MONITORING

The Annex to Directive 80/1107/EEC refers to the use of CEN (or ISO) standards on measuring procedures, where they exist, and it is on the basis of this reference (rather than a formal mandate from the Commission), that CEN has undertaken the development of such standards, initially covering indents 1, 3 and 4 of section 1.3 above.

This CEN activity is undertaken by CEN/TC137, which has chairman H. Blome (D) and secretariat DIN. TC137 started in November 1989 and has four working groups.

Working Group 1, under the convenorship of R. Grosjean (B), has prepared a guidance document on the 'assessment of exposure to chemical agents in air at the workplace for comparison with Limit Values and measurement strategy'. Its purpose is to give practical guidance to those who have to carry out assessments. Annexes to the document give examples of measurement strategies that have been successfully applied in Member States.

Working Group 2, under the convenorship of S. Zloczysti (D), has prepared a draft standard (prEN 482) 'general requirements for measuring procedures'. This standard does not prescribe specific methodology for chemical agents, but sets minimum performance requirements that any method used must meet. The basic requirement is a measure of the confidence that can be placed in the measurement result - specifically the 'overall uncertainty' of the measurement; mathematically it is a combination of precision (or random error) and bias (or systematic error). The target overall uncertainty depends on the measurement task (as defined by WG1), but for comparison with Limit Values and for estimated concentrations between half and twice the Limit Value overall uncertainty should be less than $\pm$ 30%.

The WG2 draft standard does not define the methods of test for the performance requirements. These are in preparation in a supplementary series of draft standards covering, for example, diffusive samplers, pumped samplers and gas detector tubes.

Working Group 3, under the convenorship of T.L. Ogden (UK), has prepared a draft Standard (prEN 481) 'particle size fraction definitions for health related sampling'. This work has been done in close cooperation with ISO/TC146 and for the first time, world-wide agreement has been reached on the definitions of the various aerosol fractions that reach the lung compartments. The fractions are named (in English) inhalable, thoracic and respirable. 'Inhalable', which is now the most commonly accepted term, is equivalent to the term 'inspirable' used in the Directive (80/1107/EEC). Full definitions and the defining curves are given in the paper by Ogden in the main text of the Proceedings.

WG3 is also preparing a draft standard covering the methods of test for aerosol measuring instruments analogous to the WG2 supplementary series.

Working Group 4, under the convenorship of J.C. Guichard (F), has prepared sets of equivalent terms in workplace air monitoring/ industrial hygiene in the three official CEN languages.

1.5 World Health Organisation

In a separate, but coordinating role, the WHO has an as objective the promotion of the highest possible level of health for all; preventative action at the workplace represents an important element in meeting this objective. At the present time, there are known preventative techniques for the control of health risks at the workplace, by the application of good occupational hygiene practice, but these are not universally applied. The cause for this is partly a lack of political will, and partly the absence of trained personnel. The WHO is mounting an initiative to promote good occupational health practice and to stimulate the development of facilities for training. These activities are being focussed particularly among Member States covered by the European Regional Office of the WHO.

2. MEASUREMENT METHODOLOGY

2.1 PRACTICAL AND CHANGING NEEDS

In the development and validation of analytical methods, priority must be given to chemicals that have widespread use, and particularly those with assigned Limit Values. However, there are a significant number of chemicals in common usage for which properly validated analytical methods do not exist. In addition, Limit Values for many chemicals have changed or are changing to lower values as toxicity data becomes available and technology improves; for such chemicals, existing analytical methodology may need to be re-evaluated at the lower levels. Methods which have been validated may still not meet the full CEN protocol (1.4). For example, most NIOSH evaluations have been performed only between half and twice the limit value (i.e. TLV); CEN requires additional evaluation at one-tenth and/or ten times the Limit Value for certain measurement tasks.

Thus, much additional validation is needed, and it is expedient to share this work between interested parties in the Member States and possibly on a wider front.

The existence of a validated method does not of course solve all measurement problems. Different measurement tasks require different solutions; regarding measurement range, required overall uncertainty, specificity, and selectivity. Methods may be required for personal or background sampling. Methods may need to be continuous and/or direct-reading.

It is impossible to generalise, as different situations will require different methodology. In many instances, measurement will not be needed at all, for example if an initial assessment indicates that exposures will be much less than the limit value, or control or organisational measures are in place that will prevent significant exposure. Guidance in this area is offered in the WG1 document (1.4), but perhaps more needs to be done at Community level.

There are nevertheless tendencies for measurement methodology to develop in certain directions. The Directives (1.3) contain Limit Values based on personal exposure. Measurement methodology has tended to follow this trend. Although Directives at present refer only to Limit Values for a nominal 8-hour period (except those for vinyl chloride and lead), many national limit values also include short-term (10-15 minutes) limits or refer to 'ceiling' values. In both cases, fast-responding, near real-time monitors are of obvious value, but are not always compatible with personal monitoring. It goes without saying that both manufacturers and users would prefer simpler, less expensive and more reliable methodology/ instruments, and there are signs of real advances in this area, e.g. diffusive monitors and multi-sensor arrays.

2.2 GASES AND VAPOURS

This session was concerned primarily with the sampling and measurement equipment required for monitoring gases and vapours. Since the Luxembourg Conference on Diffusive Sampling in 1986, much experience has been gained in using this equipment and instrumentation, including sorbents for active and diffusive sampling, sampling media for reactive species, detector tubes, continuous plant monitoring systems and direct-reading instruments. Although specific techniques and monitoring procedures were discussed, there were a number of common themes that emerged.

As discussed in 2.1, there was a call for advice on sampling strategy; guidance on when sampling is required as part of a hygiene assessment, and if required, when, where, and how samples should be taken. There was also a call for more method validation; the development of agreed protocols, the sharing of the task and the sharing of information. Particular areas of concern in method development included some identified requirements for improved specificity and selectivity, improved reliability and improved response times of some instrumentation.

Adequate training and experience were seen as essential, both in sampling strategy and in measurement methodology, as indeed also in the interpretation of the results.

The session included in-depth studies of sorbents for active sampling, sampling of reactive species, diffusive sampling, direct-reading gas detector tubes, and direct-reading instruments. Each study included a review of the state-of-the-art, potential problem areas and future developments.

2.3 AEROSOLS

Inhalable particles play an important role in occupational hygiene. Dusts and aerosols occur in almost all industrial processes - traditionally in mining, metal-working and the ceramic and fibre-processing industries, but also in the food industry, wood-working industry and in the building trade. A large proportion of industrial disease can be attributed to dust, often from exposure some decades previously. Indeed, until about 1980, workplace monitoring concentrated very largely on dust problems - gases and vapours are a more recent concern! Because of the long historical background, there has been much diversification in measurement methodology, including inconsistency in defining particle size fractions, technical differences in the sampling devices used, and different methods used for collecting and storing exposure data. All these areas are now the subject of attempts to standardise at the European (and sometimes world-wide) level.

Foremost in these standardisation efforts is the (agreed) definitions of 'particle size fractions for health related sampling' developed in CEN/TC137/WG3 and noted above (1.4). The same group is also working on developing performance standards and test protocols for aerosol measuring instruments. This harmonisation is widely supported.

The existence of agreed standards, though, is only the first step. The various sampling methods in use and under development must be evaluated to determine their performance. Some devices are already well-characterised, but there is still much work to be done, and again it is hoped that European cooperation can spread the load. Measuring the performance of existing sampling methods will also give information which allows the existing exposure data to be interpreted in the light of new standards. This is vital for the continuity of exposure data and for the setting of new exposure limits. This continuity is made more difficult where standards necessitate a change from static to personal monitoring. However, the principle of personal monitoring is supported.

The session included in-depth studies of aspects of the health-related measurement of particulate fractions; respirable and thoracic dust, total and inhalable dust, and fibres. The last of these (fibres) included both air monitoring aspects and the measurement of bulk and surface contamination. Each study was a review of the state-of-the-art, potential problem areas and future developments.

Specific areas of concern included the sampling and analysis of fibrous dusts, which were seen to have many technical weaknesses requiring improvements in sampling instruments, analysis methodology and quality control.

2.4 INDOOR AIR

The session 'Indoor Air' was aimed at discussing air pollution in the non-industrial workplace environment as a special case of occupational hygiene. The session focussed mainly on the office environment, although 'indoor air' normally has a wider definition, including the residential environment. Already about 50% of the workforce is occupied in offices or similar non-industrial environments, and the proportion is increasing. Interest in industrial pollution tends to be characterised by relatively high concentrations of a few, process-related, chemicals, the effects of which are usually well-defined.

In contrast, in the indoor environment, there are a large number of pollutants, often at very low concentrations, from many different sources. The consequences of exposure - such as 'sick building syndrome' - are frequently ill-defined and indistinguishable, except in frequency of incidence, from general symptoms such as mild irritation of mucous membranes, headache, tiredness, lack of concentration, etc. Moreover, the causative agents are not always chemicals; they may be biological (particularly bacteria, viruses and fungal spores) or physical (air movement, temperature, humidity, lighting, radiation). There may be psychological causes underlying the complaints of poor air quality.

The very different nature of indoor pollution raises a number of questions. First, toxicological data for many of the pollutants are missing, or are insufficient to set Limit Values for the indoor environment. In addition, there is often no proven cause and effect relationship. Hence there are only a few guideline values. Second, methods for chemical measurement which may be appropriate for the industrial workplace may not be suitable for indoor air. The lower concentrations cause sensitivity problems and/or blank and contamination problems. Protocols for method validation are less standardised, although arguably PrEN 482 (1.4) can be applied. Methods for the assessment of microbiological pollution are limited.

A key part of the session was a description of the CEC Concerted Action "Indoor air quality and its impact on man", which has four main objectives:

- (1) identification and characterisation of indoor pollutants and their sources;
- (2) assessment of population exposure to indoor pollution;
- (3) assessment of health effects of indoor pollution;
- (4) investigations into complaints about indoor air quality in office buildings.

The session also included in-depth studies of some specific aspects; active and passive (diffusive) sampling of organic compounds, aldehydes (as an example of reactive species) and microorganisms. Diffusive sampling was noted several times as having a particular advantage in overcoming the sensitivity problem discussed above. Provided a strong sorbent is used, samplers can be exposed for long periods (several weeks) enabling a very large integrated sample to be taken. The length of the sampling period would have an influence on the sampling protocol.

Prevention is often better than cure, and considerable effort is put into identifying the source of emissions. In this connection methodology is required for example to determine emissions from wood-based materials or for the characterisation of volatile organic compounds emitted from indoor materials and products. These tasks are part of the remit of the Concerted Action Group.

3. DATA AND INFORMATION MANAGEMENT

This session was concerned with the implementation of databases and their use in the risk assessment and control of hazardous materials in the workplace. The session was introduced by a general presentation by the Chairman, followed by four presentations by speakers from France, United Kingdom, Germany and Norway - each describing database schemes that have been developed in the mid-1980s for national use. The schemes were respectively COLCHIC, NEDB, GESTIS and EXPO.

The procedure adopted by most schemes involved:

- (1) the gathering of information concerning materials, associated hazards, work practices, workplaces and types of worker;
- (2) the gathering of exposure data in the form of measurements of hazardous materials and health surveillance in relation to individuals or processes and in relation to the risk assessment and exposure limit;
- (3) storage of these data such that they can be retrieved and processed.

Functions of the databases included:

- (1) exposure assessments and records of individual workers;
- (2) exposure assessments for processes;
- (3) prediction of exposure patterns;
- (4) epidemiology of exposure;
- (5) determination of priorities for future assessments;
- (6) assisting in setting of Limit Values.

A common emphasis was the need for collaboration at the European level and for the evolvement of a common database on safety and hygiene that could be accessed (at an appropriate security level) by all Member States for their mutual benefit.

Moves in that direction had indeed already taken place, for a meeting had been held a few weeks previously to discuss the setting up of such a collaborative programme. This had been done under the umbrella of the European Foundation for the Improvement of Living and Working Conditions and a Working Group was formed to which the relevant occupational hygiene societies of the Member States would be invited. It was agreed that, in such an approach, it was essential that data input into a common database should be in a common format and be properly validated and documented. A potential problem area, however, was the question of confidentiality of technical information and of health surveillance data on individuals.

4. MEASUREMENT QUALITY

4.1 QUALITY ASSURANCE

The first part of the Measurement Quality session was concerned with quality assurance.

Measurements form the basis for decisions, which, in the context of occupational hygiene, may affect the health of workers or the financial burden on employers. Therefore, measurements must be reliable and consistent, no matter where they are made. This means that the error in the measurement result must not be greater than the level of confidence required to reach a correct decision. There must also not be a significant difference between results from different laboratories.

In practice, there is considerable evidence that in many areas of chemical analysis, measurement results do not meet these requirements. Indeed, some estimates of the proportion of results that are 'wrong' are quite alarming. A bad result may be worse than no result at all, because a wrong decision may result. A result which cannot be demonstrated to be within acceptable confidence limits creates problems both for the user and supplier of data, because of the need to repeat tests and to question decisions based on suspect data.

Common experience in the field of method verification and quality assurance is that different, apparently reputable, laboratories quite frequently get significantly different results when analysing identical materials. It is also common experience that 'performance' of individual laboratories (measured against the consensus values) improves as they take into account the likely source of errors. It is to be expected that other laboratories not yet involved in these collaborative exercises will produce results which are no better and possibly less reliable. As a result, it is now increasingly recognised that quality requires effort, in addition to having well-trained and experienced analysts. Any analytical laboratory must therefore continuously monitor its own performance and take appropriate action in the following key areas:

- (1) awareness of technical problems in chemical analysis;
- (2) development and use of well validated methods;
- (3) measurement traceability underpinned by Certified Reference Materials;
- (4) participation in proficiency testing schemes;
- (5) internal quality control schemes;

The session emphasised the unique role of the Community Bureau of Reference. The BCR is a scientific programme of the CEC(with participation from Finland, Sweden and Switzerland) and has the task of improving the quality of measurements where this is of interest to the Community; implementing Directives or European norms; eliminating trade barriers; and investigating areas where the inability to measure accurately causes risk or damage to health and safety, environment or economy. The field of action of the BCR programme covers all types of testing, measurement and analysis. As an R&D programme, the BCR covers technical aspects of measurement, including studying sources of measurement error, development of measurement methods, support for cooperation between national initiatives for external quality control and the certification of reference materials.

The session included a survey of quality assurance schemes in Europe. There was a call for the integration of these quality assurance schemes, leading possibly to a unified scheme and for greater attention to the reliability of measurements in the future.

4.2 LABORATORY ACCREDITATION

Accreditation is a formal recognition that a laboratory produces results that can be relied upon to meet specified standards of performance. The requirements are based on those given in ISO 25 and EN 29000 and EN 45000 series. The NAMAS (UK) accreditation scheme was explained as an example.

At the present time, the accreditation bodies in the different Member States differ in their organisation, requirements and procedures. However, there are already moves towards the mutual recognition of national schemes (at the moment on a one-to-one basis) and the advantages of extending inter-recognition were stressed by several speakers.

5. FUTURE NEEDS

The future needs session highlighted some of the points already made by previous speakers with the addition of the personal view of the contributors.

In the introduction, the Chairman noted a common theme throughout the Symposium - that what was carried out now at national level could in most cases be more efficiently and cost-effectively done at Community level. This applied to measuring strategy, measuring procedure development and validation, definitions of size fractions of particulates, reference methods for indoor air quality, databases of information, quality assurance, accreditation and training.

A great deal of cooperation already existed. Standards were being developed under CEN, and EC initiatives were already under way in the fields of indoor air quality and occupational exposure databases. The time was ripe for extending cooperation into other areas, particularly in method validation, quality assurance and accreditation.

What was less clear was the mechanism by which this should be done. Individual enthusiasm was there, but what was needed was an umbrella organisation, in some cases the European Commission perhaps, which could bring the parties together and initiate some action.

Other presentations concentrated on the specific requirements for future instrumentation, likely developments in standards and information management, and needs for legislation. These papers, together with ideas and recommendations that emerged from previous sessions were fully discussed in the final panel discussion, which resulted in the following conclusions and recommendations.

6. CONCLUSIONS AND RECOMMENDATIONS

With regard to legislation/ standards;

The importance and significance of Directive 88/842/EEC was recognised and the Commission encouraged in its task of preparing (initially Indicative) Limit Values.

There was a call for legislation to be more user-friendly to the small and medium sized enterprises, in line with the Commission's own committment to these enterprises.

The Commission was encouraged to take note of the work of CEN, particularly TC137, in developing technical standards in support of Directives.

In relation to these standards and/or the work of the Commission, more guidance was needed in defining when and where measurements were necessary in the context of occupational hygiene assessments.

In relation to the work of WG4, it was hoped that a list of equivalent terms (in occupational hygiene/ measurement) could be made available in all three CEN languages.

With regard to measurement methodology,

There was a call for more simple, easy and cost-effective methods, including diffusive methods.

There was a call for more direct-reading, real-time, portable monitoring instruments.

There was a need for more collaboration in the determination of the performance of (all types of) measuring instrumentation, but particularly of dust-measuring instruments. This should include the evaluation of existing methods against the CEN criteria and the development of new ones where needed.

With regard to measurement quality,

There was agreement that quality initiatives such as quality assurance schemes and accreditation schemes should be harmonised or mutually recognised.

With regard to databases,

There was agreement that the several national initiatives should be brought under an international umbrella as far as possible.

With regard to both measurement quality and databases,

There was a call for coordination at Community level.

With regard to indoor air,

It was recognised that the topic was a relatively new one, and much research still needed to be done. The problems needed to be more precisely defined and strategies evolved to deal with them.

Decisions should be made at Community level on whether indoor air quality should be controlled by (air) Limit Values or by controlling source emissions, or both.

The discrepancy between indoor air quality criteria and (industrial) workplace Limit Values needs to be explained. These are some orders of magnitude different, but are based on similar toxicity and epidemiology data.

With regard to priorities,

The main priority was to avoid duplication of effort, by sharing information, expertise and resources at European if not international level.

The ultimate objective is to protect the working population, through research, information and practical guidance and the promotion of better health and hygiene practices. In particular, we should address our efforts to the small and medium-sized firms, and to the training of the new generation of occupational hygienists and analysts.

ACKNOWLEDGEMENTS

The authors wish to thank all the participants who have contributed to the success of the Symposium.

Special thanks are due to the following invited speakers, rapporteurs, chairmen and organising committee:

Speakers:

L. Armbruster (D), P. Beaumont (UK), A. Berlin (CEC), D. Berger (D), H. Blome (D), R.H. Brown (UK), B. Carton (F), W. Coenen (D), D. Cottica (I), D.R. Crump (UK), J.F. Fabriès (F), J.G. Firth (UK), P.E. Fjeldstad (N), B. Flannigan (UK), B. Goelzer (WHO), B. Griepink (CEC), R. Grosjean (B), Th. L. Hafkenscheid (NL), R. Haigh (CEC), R. Hervé-Bazin (F), W.J. Hunter (CEC), E.R. Kennedy (US), H. Knöppel (CEC), K. Leichnitz (D), J.-O. Levin (S), B. Miller (UK), T.L. Ogden (UK), M.J. Quintana (E), G. Riediger (D), H. Rothweiler (CH), M. Sapir (B), K.J. Saunders (UK), T. Schneider (DK), H. Siekmann (D), D.C.M. Squirrell (UK), R. Stamm (D), D. Ullrich (D), S. Vandendriessche (CEC), N.G. West (UK), P. Wolkoff (DK), S. Zloczysti (D).

Rapporteurs:

A. Braithwaite (UK), R.H. Brown (UK), E. Buringh (NL), J.C. Guichard (F), L.C. Kenny (I), H. Knöppel (CEC), S. Luxon (UK), R. Narayanaswamy (UK), B. Striefler (D).

Chairmen:

A. Berlin (CEC), H. Blome (D), W. Coenen (D), J.G. Firth (UK), B. Griepink (CEC), R. Haigh (CEC), K.J. Saunders (UK), D.C.M. Squirrell (UK), P. Wolkoff (DK).

Organising Committee:

H. Blome (D), A. Braithwaite (UK), R.H. Brown (UK), B. Carton (F), D. Cottica (I), M. Curtis (CEC), J. Kristensson (S), K. Leichnitz (D), R. Lidgett (UK), M.J. Quintana (E), K.J. Saunders (UK), S. Vandendriessche (CEC), H. Walerius (CEC), P. Wolkoff (DK).

Administration (CEC):

V. Ammann, S. Blair, L.Eisen, D. Nicolay, I. Pichon and A. Poos.

INDEX OF AUTHORS

ASTON UNIVERSITY
LIBRARY SERVICES
WITHDRAWN
FROM STOCK